Sulfur Biogeochemistry—Past and Present

Edited by

Jan P. Amend
Department of Earth and Planetary Sciences
Washington University
St. Louis, Missouri 63130
USA

Katrina J. Edwards
Geomicrobiology Group
Department of Marine Chemistry & Geochemistry
Woods Hole Oceanographic Institution
Woods Hole, Massachusetts 02536
USA

Timothy W. Lyons
Department of Geological Sciences
University of Missouri
Columbia, Missouri 65211
USA

Special Paper 379

3300 Penrose Place, P.O. Box 9140 ▪ Boulder, Colorado 80301-9140 USA

2004

Published by The Geological Society of America, Inc.
3300 Penrose Place, P.O. Box 9140, Boulder, Colorado 80301-9140, USA
www.geosociety.org

Printed in U.S.A.

GSA Books Science Editor: Abhijit Basu

Library of Congress Cataloging-in-Publication Data

Sulfur biogeochemistry : past and present / [edited by] Jan P. Amend, Katrina J. Edwards, Timothy W. Lyons.
p. cm. — (Special paper ; 379)
Includes bibliographic references.
ISBN 0-8137-2379-5 (pbk.)
1. Sulphur--Research. 2. Sulphides--Oxidation. 3. Sulphur deposits. 4. Sulphur--Isotopes. 5. Marine sediments. 6. Sulphur cycle. 7. Environmental geochemistry. I. Amend, Jan P., 1964- II. Edwards, Katrina J., 1968- III. Lyons, Timothy W. IV. Special papers (Geological Society of America) ; 379.

QE516.S1S79 2004
546'.723--dc22

2004047224

Cover: Elemental sulfur precipitates at volcanically active La Fossa crater on Vulcano Island, Italy. (Photo by Everett L. Shock.)

10 9 8 7 6 5 4 3 2 1

Contents

Marine Sulfate over Geologic Time

Preface

This collection of papers has its roots in two thematic sessions organized for the 2001 and 2002 annual meetings of the Geological Society of America.[1] The presentations in both sessions were linked by a common theme—recent advances in the biogeochemistry of sulfur. The scales of observation ranged from micro to macro, covering everything from molecular-level analysis of bacterial metabolic pathways to ocean-scale sulfur burial over geologic time. The 2001 session was assembled to honor the career of William T. Holser and his many contributions to sulfur geochemistry.

Holser was among the first wave of sulfur geochemists who, more than four decades ago, began a systematic characterization and interpretation of long-term sulfur isotope variability in the ocean (Holser and Kaplan, 1966; Claypool et al., 1980; Schidlowski et al., 1983). These pioneers—building from the groundbreaking studies of the first sulfur geomicrobiologists (Harrison and Thode, 1958; Kaplan and Rittenberg, 1964; Kemp and Thode, 1968; Rees, 1973; Chambers et al., 1975)—understood that the isotopic composition of sulfate in the ocean tracks the cycling of sedimentary pyrite. Parallel studies were revealing how, when, and where sedimentary pyrite forms and its relationship to bacterial sulfate reduction and cycling of organic matter (Berner, 1970, 1984; Goldhaber and Kaplan, 1974; Rickard, 1975; Jørgensen, 1977; Froelich et al., 1979; Morse and Cornwell, 1987). Armed with an understanding of organic-carbon and pyrite-sulfur burial and weathering as the principal drivers of C and S isotope variation in the ocean, geoscientists were able to model the concentrations of oxygen in the Phanerozoic atmosphere (Garrels and Lerman, 1984, Berner, 1987; Kump and Garrels, 1986) and related paleoenvironmental variability (Berner and Raiswell, 1983).

In recent years, we have seen a renaissance in the number and diversity of studies devoted to sulfur biogeochemistry. Through new and refined approaches to experimental and theoretical sulfur microbiology (Detmers et al., 2001; Canfield, 2001) and sulfur geochemistry and mineralogy (Wilkin and Barnes, 1996; Rickard and Luther, 1997; Sinninghe Damsté et al., 1998; Benning et al., 2000), the cycling of sulfur in the environment is now more intimately known. This work in the lab and behind the computer is an ideal mate for the countless comprehensive studies of modern natural systems, including sites of water-column anoxia. Because of the interdisciplinary breadth and rigor of this work, essential but elusive microbial pathways are now revealing themselves in full mechanistic detail (Canfield and Thamdrup, 1994; Hoehler et al., 1994; Hinrichs et al., 1999; Boetius et al., 2000; Orphan et al., 2001). And this sharpened microbiological perspective, in combination with the emergence of new proxy methods (Burdett et al., 1989; Paytan et al., 1998; Farquhar et al., 2000), has spawned a new and highly effective generation of sulfur-based paleoenvironmental reconstructions, particularly for the Precambrian (Canfield and Teske, 1996; Canfield, 1998; Habicht et al., 2002; Hurtgen et al., 2002; Farquhar and Wing, 2003).

This volume provides broad coverage of much of what is new in sulfur research. At the risk of violating one of the central tenets of the emerging field of biogeoscience—that is, removing rather than building boundaries—we have grouped the papers into four thematic areas: (1) the microbial end-member; (2) sulfide oxidation

[1]2001—"Sulfur Cycling in Precambrian to Recent Ocean-Atmosphere Systems: A Session Honoring the Career of William T. Holser," Timothy W. Lyons and Alan J. Kaufman, presiding (sponsored by the Geochemical Society and GSA International Division). 2002—"Microbial Sulfur Transformations throughout Earth's History: Development, Changes, and Future of the Biogeochemical Sulfur Cycle," Jan P. Amend and Katrina J. Edwards, presiding (sponsored by the GSA Geobiology and Geomicrobiology Division).

in the environment; (3) sulfur intermediates in marine settings, and organic and inorganic sinks for reduced sulfur; and (4) isotope proxies for marine sulfate over geologic time. Obvious overlap exists among these four artificial divisions and each owes much of its strength and relevance to the work presented in the others.

The first set of papers primarily addresses microbial contributions to sulfur biogeochemistry. Brüchert relies largely on sulfur isotopes to interpret the interdependence of microbial sulfate reduction and organic matter conversion in marine sediments. He notes that mass-dependent sulfur isotope fractionations range between 2‰ and 47‰ in bacterial sulfate reduction; microbes that convert organic matter to acetate fractionate <18‰, and those that completely oxidize organic carbon to CO_2 fractionate >18‰. Key controls on these fractionations are the chemical composition of the electron donors (diverse simple and complex organic compounds as well as H_2) and the microbial community composition and abundance. In the next paper, Amend et al. review the energetics of chemolithoautotrophy—primarily sulfur oxidation and reduction—in several geologic systems, and also evaluate Gibbs free energies of 25 sulfur-redox reactions in the well-known shallow marine hydrothermal system off Vulcano Island (Italy). Sulfide and elemental sulfur oxidation are coupled to the reduction of O_2, NO_3^-, Fe(III), and CO_2, and sulfur and sulfate reduction are combined with the electron donors H_2, CH_4, carboxylic acids, NH_4^+, and Fe^{2+}. Many of the reactions considered are known to support the growth of thermophilic archaea and bacteria, but the authors also calculate the in situ energetics of S-redox reactions for which microbial catalysts are currently unknown. Bernhard and Buck expand our view of microbial sulfur biogeochemistry to include the eukaryotes. In this contribution, the abundance and diversity of eukaryotic communities are examined in three deep-sea sulfidic sites, the Santa Barbara Basin in southern California, the Soledad Basin off Baja California (Mexico), and the Cariaco Basin in Venezuela. At all three sites, eukaryotic biovolume and abundance were dominated by foraminifera, but nevertheless, significant differences at the species level and in symbiotic relationships were observed.

The three papers in the second section focus on sulfide oxidation in the environment. Schippers provides a broad review of sulfur oxidation, including sulfide minerals, elemental sulfur, and various intermediates that can be part of the complex oxidative sulfur pathways, with an emphasis on the mechanism of oxidation. It is pointed out that microbial oxidation of ferrous to ferric iron provides the oxidant for metal sulfides and sulfur intermediates, and further, that microbes completely oxidize various sulfur compounds to sulfate. This paper covers many of the environments where sulfur oxidation occurs, both terrestrial and marine. The next two papers focus entirely on the marine realm. Jørgensen and Nelson address sulfide oxidation in the coastal marine environment, with considerable attention paid to reconciling the relative importance of microbially mediated versus purely chemical (abiotic) reaction pathways. They also discuss the coupling between the sulfur and nitrogen cycles in marine sediments, via microbial nitrate reduction. This is a process that has only recently become widely recognized, and appears to be globally significant. Edwards' contribution considers yet another recently recognized process—microbiological weathering of massive sulfide deposits associated with deep-sea hydrothermal environments. Similar to the weathering of massive sulfide deposits in continental systems, as those discussed by Schippers, this process involves complex communities of both sulfur and iron oxidizing microorganisms, and in this study, Edwards places particular emphasis on the role of the iron oxidizers and their influence on weathering.

The next three papers explore the formation, distribution, recycling, and burial of various sulfur species in sedimentary systems. Through improved sampling and analytical techniques, Zopfi et al. examine the occurrence and fate of transient sulfur intermediates in modern sediments from the North and Black Seas. This work shows that the majority of sulfide produced in sediments is reoxidized to sulfate via a complex interplay of chemical and biological reactions involving sulfur of intermediate oxidation states, in particular, elemental sulfur. Schoonen revisits the mechanisms of sedimentary pyrite formation, with an emphasis on work over the past decade. In his thorough review of experimental, natural, and theoretical systems, the traditionally invoked requirement for reaction between iron monosulfide and sulfur intermediates is explored in light of controversial recent findings suggesting that pyrite can form by direct reaction between FeS and H_2S. Among many other timely topics, this paper addresses the multiple roles played by bacteria during pyrite formation—beyond the generation of hydrogen sulfide. Werne et al. give us an overview of organic sulfur research—past, present, and future—and an ideal complement to Schoonen's review of "inorganic" sulfur sinks. The complexities of organic sulfurization are tackled with a threefold agenda: (1) illuminating the available constraints on the timing and pathways of organic sulfur formation, including a connection to intermediate sulfur compounds and the relative importance of polysulfides; (2) understanding the relationships between organic sulfur content and petroleum generation; and (3) defining the impact of organic sulfurization on the preservation of organic compounds at bulk and molecular levels. All this work has benefited from new and improved analytical approaches, including the integration of sulfur isotope techniques, which are now being applied at the level of individual organic compounds.

Papers in the final section are linked by the theme that sulfur records biogeochemical conditions in the ancient ocean. Paytan et al. employ an exciting, recently developed paleoceanographic method—carefully extracted barite as a proxy for the $\delta^{34}S$ of seawater—to confirm that high barite levels in organic-rich sediments from Cretaceous Ocean Anoxic Events and Mediterranean sapropels are indeed the product of high biological productivity rather than secondary enrichment and remobilization. The authors explore the conditions that favor the preservation of biogenic barite in sulfate reducing environments, thus refining our understanding of why and where the barite proxy is a robust paleoenvironmental recorder. Using barite, Paytan et al. then summarize the $\delta^{34}S$ of seawater sulfate over the past 130 m.y. Lyons et al. investigate the fidelity of another proxy for the isotopic composition of seawater sulfate—carbonate-associated sulfate (CAS)—through a careful look at modern lime muds in South Florida. Most important, this modern calibration shows that bulk sediment at sites of extensive, early diagenetic carbonate dissolution and net precipitation can faithfully record the $\delta^{34}S$ of seawater sulfate. This retention of the seawater isotope value occurs despite the strong ^{34}S enrichments observed in the pore waters at a site of anomalously high rates of bacterial sulfate reduction. More generally, this is a study of the sulfur cycle in shallow platform carbonates and its relationship to calcium carbonate saturation. Hurtgen et al. use the CAS technique to reconstruct Proterozoic ocean chemistry as preserved in rocks from Death Valley, California. Concentrations of CAS suggest that the amount of sulfate in the late Mesoproterozoic to mid-Neoproterozoic ocean was only ~10% of that present today. Furthermore, the broad stratigraphic coverage provides an ideal, longer-term context for the anomalously large and rapid isotopic excursions observed in late Neoproterozoic sediments. These excursions are a possible consequence of "snowball earth" glacial events. Finally, Strauss has compiled a massive amount of data into a thorough review of the $\delta^{34}S$ of seawater sulfate over the past ~3.5 b.y. These data derive from a variety of sources, including CAS analysis. Strauss' interpretations tell us that the observed patterns of isotopic variability track the oxygenation history of Earth's surface and the corresponding balance between the burial and weathering of reduced sulfur.

Jan P. Amend
Department of Earth and Planetary Sciences
Washington University in St. Louis

Katrina J. Edwards
Department of Marine Chemistry and Geochemistry
Woods Hole Oceanographic Institution

Timothy W. Lyons
Department of Geological Sciences
University of Missouri

REFERENCES CITED

Benning, L.G., Wilkin, R.T., and Barnes, H.L., 2000, Reaction pathways in the Fe-S system below 100°C: Chemical Geology, v. 167, p. 25–51, doi: 10.1016/S0009-2541(99)00198-9.

Berner, R.A., 1970, Sedimentary pyrite formation: American Journal of Science, v. 268, p. 1–23.

Berner, R.A., 1984, Sedimentary pyrite formation: An update: Geochimica et Cosmochimica Acta, v. 48, p. 605–615, doi: 10.1016/0016-7037(84)90089-9.

Berner, R.A., 1987, Models for carbon and sulfur cycles and atmospheric oxygen: Application to Paleozoic geologic history: American Journal of Science, v. 287, p. 177–190.

Berner, R.A., and Raiswell, R., 1983, Burial of organic carbon and pyrite sulfur in sediments over Phanerozoic time: A new theory: Geochimica et Cosmochimica Acta, v. 47, p. 855–862, doi: 10.1016/0016-7037(83)90151-5.

Boetius, A., Ravenschlag, K., Schubert, C.J., Rickert, D., Widdel, F., Gieseke, A., Amann, R., Jørgensen, B.B., Witte, U., and Pfannkuche, O., 2000, A methane microbial consortium apparently mediating anaerobic oxidation of methane: Nature, v. 407, p. 623–626, doi: 10.1038/35036572.

Burdett, J.W., Arthur, M.A., and Richards, M., 1989, A Neogene seawater sulfur isotope curve from calcareous pelagic microfossils: Earth and Planetary Science Letters, v. 94, p. 189–198, doi: 10.1016/0012-821X(89)90138-6.

Canfield, D.E., 1998, A new model for Proterozoic ocean chemistry: Nature, v. 396, p. 450–453, doi: 10.1038/24839.

Canfield, D.E., 2001, Isotope fractionation by natural populations of sulfate-reducing bacteria: Geochimica et Cosmochimica Acta, v. 65, p. 1117–1124, doi: 10.1016/S0016-7037(00)00584-6.

Canfield, D.E., and Teske, A., 1996, Late Proterozoic rise in atmospheric oxygen concentration inferred from phylogenetic and sulphur-isotope studies: Nature, v. 382, p. 127–132, doi: 10.1038/382127A0.

Canfield, D.E., and Thamdrup, B., 1994, The production of ^{34}S-depleted sulfide during bacterial disproportionation of elemental sulfur: Science, v. 266, p. 1973–1975.

Chambers, L.A., Trudinger, P.A., Smith, J.W., and Burns, M.S., 1975, Fractionation of sulfur isotopes by continuous cultures of *Desulfovibrio desulfuricans*: Canadian Journal of Microbiology, v. 21, p. 1602–1607.

Claypool, G.E., Holser, W.T., Kaplan, I.R., Sakai, H., and Zak, I., 1980, The age curves of sulfur and oxygen isotopes in marine sulfate and their mutual interpretation: Chemical Geology, v. 28, p. 199–260, doi: 10.1016/0009-2541(80)90047-9.

Detmers, J., Brüchert, V., Habicht, K.S., and Kuever, J., 2001, Diversity of sulfur isotope fractionations by sulfate-reducing prokaryotes: Applied and Environmental Microbiology, v. 67, p. 888–894, doi: 10.1128/AEM.67.2.888-894.2001.

Farquhar, J., and Wing, B.A., 2003, Multiple sulfur isotopes and the evolution of the atmosphere: Earth and Planetary Science Letters, v. 213, p. 1–13, doi: 10.1016/S0012-821X(03)00296-6.

Farquhar, J., Savarino, J., Jackson, T.L., and Thiemens, M.H., 2000, Evidence of atmospheric sulphur in the martian regolith from sulphur isotopes in meteorites: Nature, v. 404, p. 50–52, doi: 10.1038/35003517.

Froelich, P.N., Klinkhammer, G.P., Bender, M.L., Luedtke, N.A., Heath, G.R., Cullen, D., Dauphin, P., Hammond, D., Hartman, B., and Maynard, V., 1979, Early oxidation of organic matter in pelagic sediments of the eastern equatorial Atlantic: Suboxic diagenesis: Geochimica et Cosmochimica Acta, v. 43, p. 1075–1090, doi: 10.1016/0016-7037(79)90095-4.

Garrels, R.M., and Lerman, A., 1984, Coupling of the sedimentary sulfur and carbon cycles—An improved model: American Journal of Science, v. 284, p. 989–1007.

Goldhaber, M.B., and Kaplan, I.R., 1974, The sulfur cycle, *in* Goldberg, E.D., ed., The sea (v. 5): New York, Wiley, p. 569–655.

Habicht, K.S., Gade, M., Thamdrup, B., Berg, P., and Canfield, D.E., 2002, Calibration of sulfate levels in the Archean Ocean: Science, v. 298, p. 2372–2374, doi: 10.1126/SCIENCE.1078265.

Harrison, A.G., and Thode, H.G., 1958, Mechanism of the bacterial reduction of sulphate from isotope fractionation studies: Transactions of the Faraday Society, v. 54, p. 84–92.

Hinrichs, K.-U., Hayes, J.M., Sylva, S.P., Brewer, P.G., and DeLong, E.F., 1999, Methane-consuming archaebacteria in marine sediments: Nature, v. 398, p. 802–805, doi: 10.1038/19751.

Hoehler, T.M., Alperin, M.J., Albert, D.B., and Martens, C.S., 1994, Field and laboratory studies of methane oxidation in an anoxic marine sediment—Evidence for methanogen-sulfate reducer consortium: Global Biogeochemical Cycles, v. 8, p. 451–463, doi: 10.1029/94GB01800.

Holser, W.T., and Kaplan, I.R., 1966, Isotope geochemistry of sedimentary sulfate: Chemical Geology, v. 1, p. 93–135, doi: 10.1016/0009-2541(66)90011-8.

Hurtgen, M.T., Arthur, M.A., Suits, N.S., and Kaufman, A.J., 2002, The sulfur isotopic composition of Neoproterzoic seawater sulfate: Implications for a snowball Earth?: Earth and Planetary Science Letters, v. 203, p. 413–429, doi: 10.1016/S0012-821X(02)00804-X.

Jørgensen, B.B., 1977, The sulfur cycle of a coastal marine sediment (Limfjorden, Denmark): Limnology and Oceanography, v. 22, p. 814–832.

Kaplan, I.R., and Rittenberg, S.C., 1964, Microbiological fractionation of sulphur isotopes: Journal of General Microbiology, v. 34, p. 195–212.

Kemp, A.L.W., and Thode, H.G., 1968, The mechanism of the bacterial reduction of sulphate and sulphite from isotope fractionation studies: Geochimica et Cosmochimica Acta, v. 32, p. 71–91, doi: 10.1016/0016-7037(68)90088-4.

Kump, L.R., and Garrels, R.M., 1986, Modeling atmospheric O_2 in the global sedimentary redox cycle: American Journal of Science, v. 286, p. 336–360.

Morse, J.W., and Cornwell, J.C., 1987, Analysis and distribution of iron sulfide minerals in Recent anoxic marine sediments: Marine Chemistry, v. 22, p. 55–69, doi: 10.1016/0304-4203(87)90048-X.

Orphan, V.J., Hinrichs, K.-U., Ussler, W., Paull, C.K., Taylor, L.T., Sylva, S.P., Hayes, J.M., and DeLong, E.F., 2001, Comparative analysis of methane-oxidizing archaea and sulfate-reducing bacteria in anoxic marine sediments: Applied and Environmental Microbiology, v. 67, p. 1922–1934, doi: 10.1128/AEM.67.4.1922-1934.2001.

Paytan, A., Kastner, M., Campbell, D., and Thiemens, M.H., 1998, Sulfur isotopic composition of Cenozoic seawater sulfate: Science, v. 282, p. 1459–1462, doi: 10.1126/SCIENCE.282.5393.1459.

Rees, C.E., 1973, A steady-state model for sulphur isotope fractionation in bacterial reduction processes: Geochimica et Cosmochimica Acta, v. 37, p. 1141–1162, doi: 10.1016/0016-7037(73)90052-5.

Rickard, D.T., 1975, Kinetics and mechanism of pyrite formation at low temperatures: American Journal of Science, v. 275, p. 636–652.

Rickard, D.T., and Luther, G.W., III, 1997, Kinetics of pyrite formation by the H_2S oxidation of iron (II) monosulfide in aqueous solutions between 25 and 125°C: The mechanism: Geochimica et Cosmochimica Acta, v. 61, p. 135–147, doi: 10.1016/S0016-7037(96)00322-5.

Schidlowski, M., Hayes, J.M., and Kaplan, I.R., 1983, Isotopic inferences of ancient biochemistries: Carbon, sulfur, hydrogen, and nitrogen, *in* Schopf, J.W., ed., Earth's earliest biosphere—Its origin and evolution: Princeton, Princeton University Press, p. 149–186.

Sinninghe Damsté, J., Kok, M., Köster, J., and Schouten, S., 1998, Sulfurized carbohydrates: An important sedimentary sink for organic carbon?: Earth and Planetary Science Letters, v. 164, p. 7–13, doi: 10.1016/S0012-821X(98)00234-9.

Wilkin, R.T., and Barnes, H.L., 1996, Pyrite formation by reactions of iron monosulfides with dissolved inorganic and organ sulfur species: Geochimica et Cosmochimica Acta, v. 60, p. 4167–4179, doi: 10.1016/S0016-7037(97)81466-4.

Printed in the USA

Geological Society of America
Special Paper 379
2004

Physiological and ecological aspects of sulfur isotope fractionation during bacterial sulfate reduction

Volker Brüchert
Max-Planck Institute for Marine Microbiology, Celsiusstrasse 1, 28359 Bremen, Germany

ABSTRACT

Reported mass-dependent sulfur isotope fractionations during bacterial sulfate reduction by pure cultures of sulfate-reducing bacteria range from 2‰ to 47‰. All sulfate-reducing bacteria that release acetate as the final product of organic substrate oxidation fractionate less than 18‰, whereas sulfate reducers that are capable of complete organic substrate oxidation to CO_2 consistently fractionate more than 18‰. The regulation of isotope fractionation occurs by different membrane transport mechanisms and by species-specific fractionations of sulfur intermediates during reduction by the enzyme adenosyl phosphosulfate reductase (APSR) and dissimilatory sulfite reductase (DSR). An unexplored aspect is the isotope effect of the coupling between the rate of membrane transport of electron donor and acceptor. Of particular importance for understanding the bulk isotope fractionation of the sulfate-reducing community in marine sediments is the variable abundance of specific sulfate-reducing bacteria, which are optimally adapted to the respective fermentation products generated during the transformation of complex organic matter. The relative balance between a hydrogen-formate-lactate–based versus an alkane-based, aromate-based, long-chain fatty acid–based, or acetate-based culmination during fermentation may cause significant shifts of up to 25‰ in the overall isotope fractionation by the sulfate-reducing community. A functional understanding of the overall isotope fractionation of the sulfate-reducing microbial community in marine sediments in different environments, therefore, also requires an understanding of the carbon transformation steps before terminal carbon oxidation by bacterial sulfate reduction.

Keywords: Stable sulfur isotopes, physiology, phylogenetic diversity, organic substrate, sulfate-reducing bacteria.

INTRODUCTION

The stable isotope analysis of sedimentary sulfur compounds is an integral part in the paleoenvironmental reconstruction of ancient marine environments from the Archaean to Cenozoic (e.g., Joachimski et al., 2001; Passier et al., 1999; Shen et al., 2001; Strauss, 1999; Werne et al., 2002). The isotope composition of sedimentary sulfides has been widely applied to distinguish environments with high or low rates of bacterial sulfate reduction and to reconstruct sulfate concentrations in ancient environments (e.g., Habicht et al., 2002). Over geologic time scales, the stratigraphic variation in the isotope composition of mineral sulfates, principally gypsum and barite, has been used as indicator of the extent to which sedimentary sulfide has been buried or to assess the influence of hydrothermal activity on the global sulfur cycle (Holser et al., 1988; Paytan et al., 1998).

Brüchert, V., 2004, Physiological and ecological aspects of sulfur isotope fractionation during bacterial sulfate reduction, *in* Amend, J.P., Edwards, K.J., and Lyons, T.W., eds., Sulfur Biogeochemistry—Past and Present: Geological Society of America Special Paper 379, p. 1–16. For permission to copy, contact editing@geosociety.org.

The resulting sulfur isotope time curves yield insights into the geological sources of sulfur to the oceans, specifically the relative burial versus weathering of reduced sulfur and the linkages to atmospheric O_2 (Berner and Petsch, 1998). Isotope shifts may be coupled to variations in primary productivity and may be used in reconstructions of water column stratification, restricted basin circulation, and the extents of water column anoxia. An additional important aspect has been introduced by the work of Canfield, Habicht, Thamdrup, and co-workers, who interpreted the isotope fractionations preserved in sedimentary sulfides in terms of sulfur cycling through the oxidative pathways in the sulfur cycle. These data provided new information on the redox chemistry of the Precambrian Ocean (Canfield and Teske, 1996).

The long history of broad application of stable sulfur isotopes implies a thorough understanding of the processes regulating the isotope fractionation during bacterial sulfate reduction, the key initial process for the expression of isotope differences preserved in sediments. Specifically, knowledge about the biochemical regulation of bacterial sulfate reduction and the associated isotope effects is required. However, studies of bacterial sulfate reduction have not been performed from the perspective of isotope fractionations at the biochemical or enzymatic level. Therefore, it has been necessary to draw on independent physiological studies of sulfate reducing bacteria involving metabolic rate, sulfate transport across the membrane, and the biochemistry of the reduction process, and to indirectly relate this information to the regulation of isotope fractionation. Experimental investigations have focused largely on measuring isotope fractionations of selected organisms that are capable of the dissimilatory reduction of sulfate (summarized in Canfield, 2001a). The isotope fractionations measured experimentally from these selected experiments were then extrapolated or directly adopted to the natural environment assuming that the isotope fractionation by the specific organism investigated at least approximates the overall process in situ. The shortcoming of this approach has long been known, and the first studies were complemented by experimental investigations that addressed the environmental variability imposed on the process (e.g., by varying temperature and substrate availability [Kaplan and Rittenberg, 1964; Kemp and Thode, 1968]).

There was also the need for obtaining isotope fractionations experimentally with natural communities of bacteria. Using time series experiments with sealed anoxic bags containing unamended sediments, conditions for the bacterial community were maintained as close to the natural conditions as possible (Habicht and Canfield, 2001). These studies likely yield the best information on the size of the isotope fractionations in natural environments. However, only a bulk response is measured. Microbial communities in marine sediments are complex and often consist of many species of sulfate-reducing bacteria (Llobet-Brossa et al., 2002; Ravenschlag et al., 2001). These bacteria, when isolated, have shown very diverse phenotypic and physiological characteristics, which implies that they are adapted to different conditions, may not all be active at the same time, and have different cell-specific sulfate reduction rates. Quantification of their relative abundances and activities in a natural environment remains a big challenge. The importance of a differential response of the microbial community to an environmental trigger (e.g., a substrate pulse following a phytoplankton bloom) is potentially underestimated. Rather than responding as a whole to an environmental perturbation, it is more likely that specific sulfate reducers with optimum adaptation show a differential response specifically tuned to the environmental conditions at hand. These organisms may then dominate the biogeochemical process and the resulting bulk isotope fractionation.

On the basis of the preexisting information on isotope fractionation during bacterial sulfate reduction, further information on the following questions is required.

1. How variable is the isotope fractionation between different sulfate-reducing bacteria?

2. What are the systematic relationships in isotope fractionation among different organisms?

3. What are the biochemical regulators for isotope fractionations at the cellular level?

4. What are the biogeochemical implications of isotope variability at the cellular and interspecies level?

This paper expands upon a recent review on isotope fractionation during bacterial sulfate reduction (Canfield, 2001a) and focuses on the physiology and microbial ecology of sulfate-reducing bacteria. A new aspect that has become clear from recent studies on isotope fractionation is the close interdependency between isotope fractionation and the oxidation steps of the organic electron donor or hydrogen. To address this issue, I present new data to support the linkage between fractionation effects and the processing of carbon by the anaerobic microbial food chain.

PURE CULTURE INVESTIGATIONS OF SULFUR ISOTOPE FRACTIONATION

Sulfur isotope fractionation during bacterial sulfate reduction has been known for almost 50 years. At present, sulfur isotope fractionations have been determined for over 40 species of sulfate-reducing bacteria (Table 1). The *Desulfovibrio* and *Desulfotomaculum* genera, in particular the species *Desulfovibrio desulfuricans* and *Desulfovibrio vulgaris*, stand out among the investigated sulfate reducers, likely due to the relative ease of isolating and cultivating them under different experimental conditions in the laboratory.

Canfield (2001a) summarized perspectives on isotope fractionations during dissimilatory sulfate reduction. Of particular relevance for this paper were the following observations.

1. There is a large range in isotope fractionation for pure cultures of sulfate-reducing bacteria, extending from 2‰ to 47‰ (Bolliger et al., 2001; Detmers et al., 2001a), even when growth conditions were optimized for each species.

2. On the basis of 16S DNA sequences, there is no relationship between phylogenetic relatedness and isotope fractionation behavior (Detmers et al., 2001a) (Fig. 1).

TABLE 1. CELL-SPECIFIC SULFATE REDUCTION RATES AND FRACTIONATION FACTORS OF SULFATE-REDUCING PROKARYOTES GROWN IN BATCH CULTURE AT OPTIMUM GROWTH CONDITIONS

Microorganism	Isolated from	Electron donor (mM)	cell-specific SRR (fmol cell^{-1} d^{-1})	ε (‰)	Reference[#]
Complete oxidizing					
Desulfonema magnum	Marine mud	Benzoate (3)	5.9	42.0	9
Desulfobacula phenolica	Marine mud	Benzoate (3)	125.0	36.7	9
Desulfobacterium autotrophicum	Marine mud	Butyrate (20)	64.8	32.7	9
Desulfobacula toluolica	Marine mud	Benzoate (3)	40.3	28.5	9
Desulfobacula toluolica	Marine mud	Toluene (0.25 mM)	8.9–390	19.8–46.9	8
Desulfospira joergensenii	Marine mud	Pyruvate (20)	0.9	25.7	9
"Desulfarculus baarsii"	Lake mud	Butyrate (20)	11.5	23.2	9
Desulfotignum balticum	Marine mud	Butyrate (20)	4.2	23.1	9
Desulfotomaculum gibsoniae	Freshwater mud	Butyrate (20)	13.0	27.8	9
Desulfofrigus oceanense	Arctic sediment	Acetate (20)	7.6	22.0	10
Desulfobacter sp. ASv20	Arctic sediment	Acetate (20)	6.6	18.8	10
Desulfobacca acetoxidans	Anaerobic sludge	Acetate (20)	17.0	18.0	9
Desulfococcus sp.	Marine mud	Pyruvate (20)	41.8	16.1	9
Desulfosarcina variabilis	Marine mud	Benzoate (3)	11.1	15.0	9
Incomplete oxidizing					
*Desulfonatronum lacustre**	Alkaline lake mud	Ethanol (20)	16.2	18.7	9
*Archaeoglobus fulgidus strain Z**	Submarine hot spring	Lactate (20)	63.8	17.0	9
Thermodesulfovibrio yellowstonii	Thermal vent water	Lactate (20)	24.0	17.0	9
"Desulfobotulus sapovorans"	Freshwater mud	Lactate (20)	26.0	16.5	9
*Desulfotomaculum thermocisternum**	Oil reservoir	Lactate (20)	310.0	15.0	9
*Desulfotomaculum nigrificans** = *Desulfovibrio desulfuricans*	Soil	Lactate	n.a.	6.7 ± 0.9, 14.6	7,5
*Desulfotomaculum orientis**	Soil	Lactate	n.a.	5.5 ± 0.9	7
Desulfomicrobium baculatum	Manganese ore	Lactate (20)	4.8	12.7	9
*Desulfotomaculum geothermicum**	Aquifer	Lactate (20)	8.1	12.5	9
Desulfohalobium redbaense	Saline sediment	Lactate (20)	434.0	10.6	9
Desulfocella halophila	Great salt lake	Pyruvate (20)	10.2	8.1	9
Desulfobulbus "marinus"	Marine mud	Propionate (20)	28.9	6.8	9
Desulfotalea arctica	Arctic marine sediment	Lactate (20)	4.2	6.1	10
Desulfobulbus elongates	Digester	Propionate (20)	35.3	5.5	9
Desulfovibrio sp. strain X	Hydrothermal vent	Lactate (20)	36.0	5.4	9
Thermodesulfobacterium commune	Thermal spring	Lactate (20)	45.4	5.0	9
"Desulfovibrio oxyclinae"	Hypersaline mat	Lactate (20)	69.1	4.5	9
Desulfotalea psychrophila	Arctic sediment	Lactate (20)	6.3	4.3	10
Desulfovibrio profundus	Deep-sea sediment	Lactate (20)	17.0	4.1	9
Desulfovibrio halophilus	Hypersaline microbial mat	Lactate (20)	33.1	2.0	9
Desulfovibrio desulfuricans	Anoxid mud	Lactate, pyruvate	n.a.	5.1–19.2	2,4,5,6,7
Desulfovibrio vulgaris = D'vibrio desulfuricans Hildenborough strain	Marine mud	Lactate	0.56–10.5	6.5-14.6	1,2,3,5,6,7
Desulfovibrio africanus	Mud	Lactate	n.a.	14.6	5
Desulfivibrio gigas	Mud	Lactate	n.a.	14.6	5
Desulfovibrio salexigens	Mud	Lactate	n.a.	8.9–14.6	5,7
Hydrogen and Formate					
Desulfobacterium autotrophicum	Marine mud	H_2 (1 bar)	34.8	14.0	13
Desulfonatronovibrio hydrogenovorans	Lake mud	Formate (20)	12.7	5.5	13
Desulfovibrio desulfuricans	Anoxic mud	H_2		5.2–14.6	1,5,6

*Under experimental conditions, these strains are considered as incomplete oxidizers.

[#]1—Kaplan and Rittenberg (1964); 2—Harrison and Thode (1958); 3—Fry et al., (1988); 4—Jones and Starkey (1957); 5—Chambers et al. (1976); 6—Kemp and Thode (1968); 7—McCready (1975); 8—Bolliger et al., (2001); 9—Detmers et al. (2001a); 10—Brüchert et al. (2001).

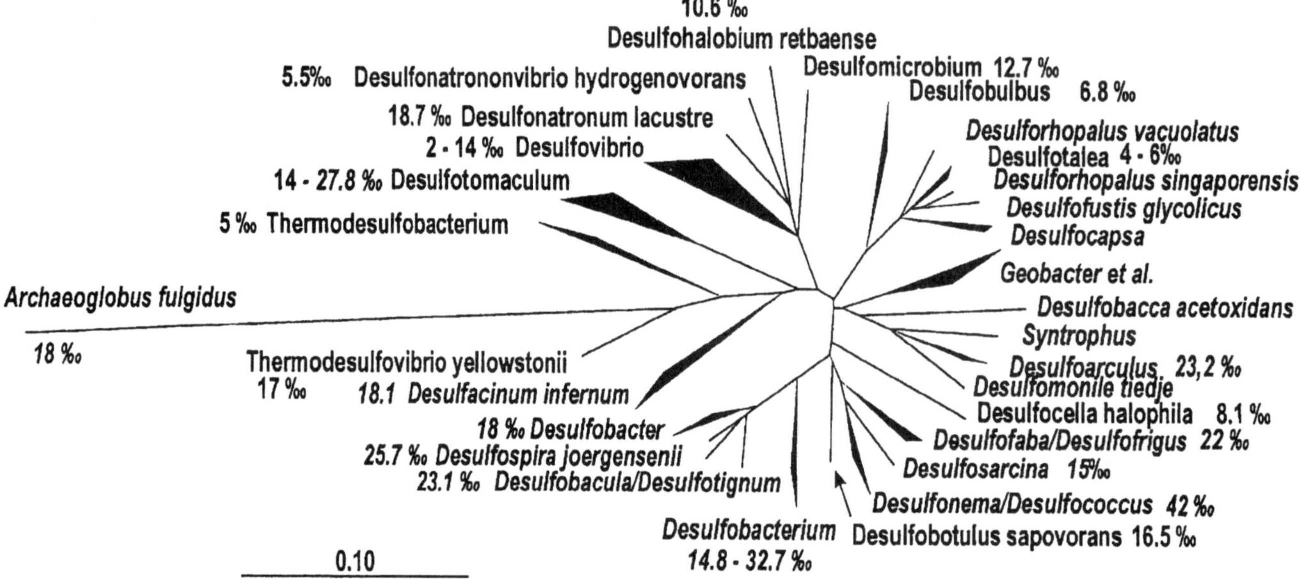

Figure 1. Phylogenetic distribution of sulfur isotope fractionations based on 16S rDNA sequence similarities (modified from Detmers et al., 2001a). Fractionations for all shown strains were determined on growing batch cultures with excess substrate under optimum growth conditions.

3. Isotope fractionations of specific sulfate-reducing bacteria vary in some but not all cases as a function of cell-specific sulfate-reduction rate (Kaplan and Rittenberg, 1964; Kemp and Thode, 1968; Detmers et al., 2001a; Böttcher et al., 1999). Likewise, temperature, a fundamental regulator of sulfate reduction rate, does not affect isotope fractionation in a unique way. Isotope fractionation increased with decreasing temperature in experiments with *Desulfovibrio desulfuricans* (Kaplan and Rittenberg, 1964); however, this strain does not grow at the lowest experimental temperatures used in this particular study and more likely exhibited a stress response. In more recent studies with strains capable of growth at all experimental temperatures, temperature had no effect on isotope fractionation (Brüchert et al., 2001).

4. There appears to be a close relationship between the pathway of organic electron donor and hydrogen oxidation and the isotope fractionation. Fractionations of <18‰ are observed for incomplete-oxidizing sulfate reducers, organisms that release acetate as the final product of carbon oxidation. Sulfate reducers that oxidize the organic substrate completely to CO_2 fractionate sulfate in excess of 18‰ up to 47‰ (Detmers et al., 2001a; Bolliger et al., 2001) (Table 1).

STEPWISE BACTERIAL SULFATE REDUCTION AND ASSOCIATED SULFUR ISOTOPE EFFECTS

Sulfate Transport and Enzymatic Reduction: Energetic Considerations

The first step is the transport of sulfate into the cell, which occurs together with two sodium ions for marine sulfate reducers and with protons in freshwater sulfate reducers (Cypionka, 1989). Sulfate transport is driven by the preexisting proton potential (the outside of the cell has lower pH) or a cross-membrane gradient of sodium ions (Fig. 2). In general, sulfate transport is proportional to the proton motive force or the electrochemical potential created by a sodium gradient. Freshwater species can concentrate sulfate within the cell up to 5000-fold relative to the surrounding environment (Cypionka, 1989). While there is clear evidence that sulfate transport is reversible, there is also an indication of membrane impermeability to preserve the proton motive force for adenosine triphosphate (ATP) synthesis. The details of membrane transport regulation are still under investigation, but it is clear that a fine-tuned regulation of transport at the genetic and activity level is required to prevent the loss of membrane potential due to the reversible transport of sulfate (Cypionka, 1994). This consideration is important since the transport of the negatively charged sulfate ion out of the cell would decrease the membrane potential. The lower the membrane potential, the less ATP can be synthesized, the less energy is available for biomass synthesis, and the cell would ultimately deactivate itself. Therefore, energetic considerations would predict that the cell would keep reverse transport of sulfate out of the cell low.

The main problem for sulfate-reducing bacteria is that despite their name, they cannot utilize sulfate directly for energy metabolism because of the high activation energy required to reduce sulfate to sulfide. Therefore, the imported sulfate must first be activated to form a high-energy compound that can be reduced. This activation is the phosphorylation of sulfate to form the high-energy compound adenosine phosphosulfate (APS) by the enzyme ATP sulfurylase. Pyrophosphate (PP_i) is the second

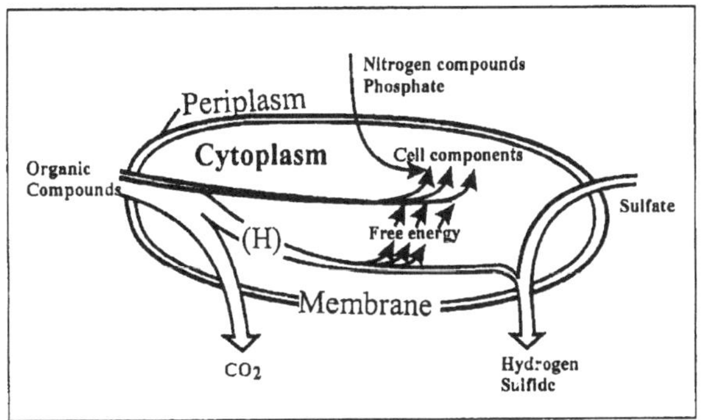

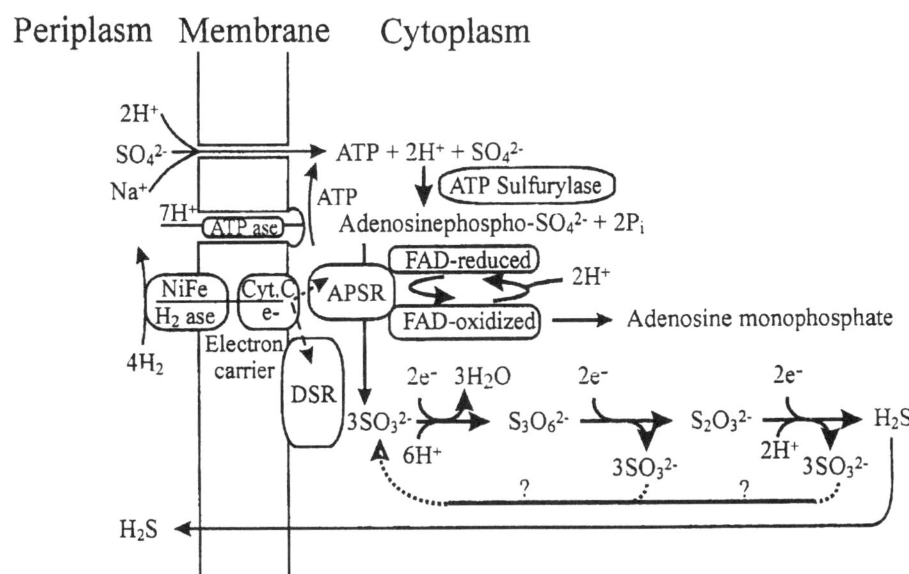

Figure 2. Schematic dissimilatory sulfate reduction with hydrogen as electron donor. ATPase—adenosine triphosphate synthase; NiFeH_2ase—periplasmic nickel iron hydrogenase; APSR—adenosine phosphosulfate reductase; DSR—dissimilatory sulfite reductase; Cyt.C—transmembrane electron carrier Cytochrome C; FAD—flavin adenine dinucleotide; FAD-oxidized, FAD-reduced—APSR residue flavin in oxidized and reduced form. Sulfate enters the cell together with Na^+ and H^+ along the cross-membrane ion concentration gradient. At the ATP sulfurylase, ATP is cleaved to release 2 P_i, and the formed adenosine monophosphate (AMP) reacts with sulfate to form adenosine phosphosulfate (APS). APS reduction occurs at a residue of the APSR, the reduced FAD. An adduct of $FADH_2$ sulfite is formed, which dissociates to oxidized FAD, protons, and releases sulfite. The enzymatic reduction of sulfite is described using the trithionate pathway, which is characterized by three 2-electron transfer steps. Trithionate dissociates to sulfite and thiosulfate, which also dissociates to sulfite and hydrogen sulfide. An alternative pathway is a single 6-electron transfer, in which no intermediates accumulate. Note the close spatial coupling between the electron carrier chain and the enzymes participating in sulfur compound reduction. The rate of the periplasmic hydrogenase controls the rate of electron donor and consequently, the rate of the electron acceptor flow.

product of this reaction (Fig. 2). This energy-consuming process consumes one mol ATP per mol sulfate (Cypionka, 1995). At high concentrations of APS and pyrophosphate, this step would be reversible. To pull the reaction toward APS, PP_i is hydrolyzed to phosphate (2 P_i), and concentrations of APS need to be less than 0.1 μM (Rabus et al., 2000), which can only be accomplished through further reduction. Although theoretically possible, the reverse reaction of APS back to sulfate is therefore also unlikely. Sulfate transport across the membrane and activation have been assumed, but not demonstrated, to occur without any significant fractionation.

In the third step, the APS complex is reduced by a cytoplasmic APS reductase to sulfite, which is further reduced to sulfide by the cytoplasmic enzyme dissimilatory sulfite reductase (DSR) (fourth step). The end product, hydrogen sulfide, is finally transported out of the cell. Since both reduction steps yield ATP, previously invested ATP for sulfate transport and the formation of APS can be regained (Rabus et al., 2000) (Fig. 2). The breaking of sulfur-oxygen bonds in the reduction of APS to sulfite and the subsequent reduction to hydrogen sulfide will require different activation energies for the molecules with the light and heavy isotope, respectively. The molecule with the lighter isotope requires less activation energy, is reduced faster, and is preferentially consumed.

Since a large intracellular pool of APS is considered unlikely, the isotope effect during APS reduction must be small. Larger isotope fractionations may occur at the dissimilatory sulfite reductase. Modulation of the isotope fractionation is possible if the reduction of sulfite proceeds via the trithionate pathway (Rabus et al., 2000). This pathway is outlined in Figure 2. In the trithionate pathway of sulfite reduction, reduction takes place in three 2-electron reductions. Sulfite is reduced to trithionate, which is reduced to sulfite and thiosulfate. Thiosulfate is subsequently reduced by a thiosulfate reductase to hydrogen sulfide. While the existence of recycled sulfite would provide an attractive agent for regulating isotope fractionation in a mechanism similar to disproportionation (Canfield and Thamdrup, 1994), this issue has not yet been resolved. Assimilatory sulfite reductases, although different in structure, appear to catalyze the reduction of sulfite in a single 6-electron transfer, without the formation of sulfur intermediates (Crane et al., 1995).

Isotope Fractionation Models

The early experimental results by Harrison and Thode (1958), Kaplan and Rittenberg (1964), and Kemp and Thode (1968) were used by Rees (1973) to construct a flow model for sulfate transport through the cell during bacterial sulfate reduction. Isotope fractionation can be expressed (*a*) during cellular uptake, (*b*) during reduction to sulfite, and (*c*) during reduction to sulfide (Fig. 2). Rees (1973) summarized the combined effects for isotope fractionations and introduced the concept of rate-limiting steps. According to this concept, isotope fractionations are not produced downstream from the rate-limiting step. If the rate-limiting step is the cellular uptake of sulfate across the cell membrane, the largest possible isotope fractionation is that associated with cellular uptake. Conversely, if the rate-limiting step occurs downstream in the sulfate reduction process (e.g., at the dissimilatory sulfite reductase), isotope fractionations can be produced at all steps upstream from and at the rate-limiting step. In general, it can be expected that the bacterial cell will tune the cellular uptake and the energy-consuming activation of sulfate to the rate of electron donor uptake. Consequently, the rate of sulfate and intermediate sulfur compound turnover depends on the rate at which the electron-transport chain operates. Furthermore, the rate of the electron-transport chain influences the electrochemical potential, which in turn regulates the rate of sulfate transport across the cell membrane.

These considerations are the basis for kinetic models of isotope fractionation during bacterial sulfate reduction. Generally, the overall isotope fractionation is regulated by two factors: (1) the rate of cellular uptake of sulfate relative to the sulfate demand by the enzymes catalyzing the reduction of APS and sulfite, and (2) the isotope fractionation of the reduction process. The overall isotope fractionation epsilon (ε_{BSR}) is the sum of these two effects and can be expressed by

$$\varepsilon_{BSR} = f \cdot \varepsilon_{transport} + (1 - f) \cdot \varepsilon_{reduction}, \quad (1)$$

where $0 < f < 1$. The implication of the Rees (1973) model is that isotope discrimination of sulfate occurs inside the cell and that the isotope composition of the sulfate transported into the cell is that of the ambient environment. Inside the cell, the specific enzymatic kinetics of the ^{34}S-APS and ^{32}S-APS molecule at the APSR and of the ^{34}S-sulfite and ^{32}S-sulfite at the DSR regulate the isotope composition of the sulfur compounds. The selective consumption of the light isotope would leave the residual pool of the intermediate sulfur compounds, i.e., internal sulfate, APS, or sulfite, enriched in ^{34}S. To preserve mass balance, at steady-state ^{34}S-enriched intermediate sulfur compounds or sulfate must continuously leave the cell.

There are energetic considerations that would support an alternative model for the cellular regulation of isotope fractionation. As indicated above in the section "Sulfate Transport and Enzymatic Reduction: Energetic Considerations," it is unlikely that a bacterial cell will only reduce part of the sulfate transported at the expense of ATP into the cell and then transport unused ^{34}S-enriched sulfate back out of the cell. Furthermore, it is energetically costly for the cell to lose sulfate from inside the cell back to the environment because reversed transport of a negatively charged species such as sulfate to the outside of the cell lowers the membrane potential. Reversed transport (i.e., leakage of ^{34}S-enriched sulfate from the cell) negatively affects the fundamental requirement for the maintenance of cell metabolism, because the cell de-energizes itself (Cypionka, 1995; White, 1995). Therefore, an alternative fractionation model would be to consider a sulfur isotope gradient across the cell membrane due to the fractionation created by the specific activity of the APSR and DSR. The specific activity of the two enzymes, in turn, is regulated by the rate of the electron-transport chain (Fig. 2). In this case, sulfate transported across the cell membrane is already fractionated (i.e., ^{32}S-enriched relative to the ambient environment) exactly to the extent that the two enzymes APSR and DSR fractionate APS and intermediate sulfur compounds. Equation 1 is also valid for this modification. If the rate of transport across the cell membrane controls the rate of sulfate reduction, then uptake is the rate-limiting step for sulfate reduction. In this case, sulfate availability in the micrometer-scale ambient environment around the cell is limited, with the consequence that the diffusing sulfate is not fractionated relative to sulfate in the ambient environment. Alternatively, if the rate of reduction is the rate-limiting step, then the enzymes can produce isotope fractionation. This is because sulfate availability in the ambient environment is unlimited, and the diffusing sulfate can be fractionated. Implicit to the model is that there are no intermediates and that that there is no reverse transport. This hypothesis should be tested with appropriate experiments.

ENZYME-SPECIFIC FRACTIONATION OF THE ADENOSINE PHOSPHOSULFATE REDUCTASE AND THE DISSIMILATORY SULFITE REDUCTASE

Although these data are central in our understanding of sulfur isotope fractionation, there is only one study that determined the isotope fractionation of the enzyme DSR, and in this study, only cell-free extracts were used (i.e., the enzyme was not purified). Kemp and Thode (1968) reported an isotope fractionation of 18‰ for the dissimilatory sulfite reductase of *Desulfovibrio desulfuricans*. At present, comparable information is not available for the adenosine phosphosulfate reductase (APSR). Rees (1973) derived an isotope fractionation of 25‰ for the APSR by difference after subtracting an isotope fractionation of 25‰ for the DSR from a total isotope fractionation of 47‰ measured experimentally (Kaplan and Rittenberg, 1964).

Differences in isotope fractionation may be deduced from the structural and compositional properties of the dissimilatory sulfite reductase enzymes. DSR enzymes are distinguished on the basis of their spectroscopic properties. There are four types of dissimilatory sulfite reductases known for sulfate-reducing bacteria: Desulfiviridin, Desulforubidin, Desulfofuscidin, and P-582 (Rabus et al., 2000). These enzymes share the presence of a siroheme complex responsible for the transfer of electrons to sulfite, which is exchange-coupled to a 4Fe-4S reactive center (Rabus et al., 2000; Steuber and Kroneck, 1998). Sulfite is ligated to the iron atom of the siroheme. The specific configuration of the siroheme complex and the exchange coupling of sulfite with the 4Fe-4S cluster are responsible for the size of the isotope fractionation. There is some indirect evidence for structural differences between DSR enzymes, which may be deduced from the phylogenetic distance of dissimilatory sulfite reductase genes of different sulfate reducers (Hipp et al., 1997). Gene sequence differences of DSR and APSR also affect the tertiary structure of the enzymes (Hipp et al., 1997). These sequence differences influence the three-dimensional structure of the docking sites for the APS and sulfite, and may modify the kinetic property of the enzymes (Steuber and Kroneck, 1998).

However, sequence analyses of the dissimilatory sulfite reductase genes of different sulfate reducers suggest that the sequences encoding the active center of the enzyme are highly conserved (M. Bauer, 2002, personal commun.), which points to a similar structure of the active center and suggests a general similarity of fractionation. A comparison of isotope fractionation and type of DSR enzyme present in 31 species of sulfate-reducing bacteria also suggested no relationship between isotope fractionation and enzyme type (Klein et al., 2001; Detmers et al., 2001a). Nevertheless, it is important to point out that sequence homology does not equal functional equality. Protein- or DNA-based amino acid sequences of different APSR or DSR enzymes may be statistically homologous, but the kinetic properties of homologous enzymes may not be identical. Multiple lateral gene transfers have occurred for the DSR and APSR (Klein et al., 2001; Friedrich, 2002). Lateral gene transfer results in sequence similarities of enzymes in distant lineages of sulfate-reducing bacteria. For this reason and from the arguments presented above, it will be difficult to link differences in isotope fractionation to gene sequence dissimilarity. Kinetic experiments with purified enzymes of different types and computational modeling of the binding configuration of the electron acceptor may provide a better understanding of the differences in isotope fractionation associated with a particular enzyme type.

EFFECT OF TEMPERATURE AND SULFATE CONCENTRATION ON ISOTOPE FRACTIONATION

Temperature has been considered one of the cardinal regulators for isotope fractionation. In general, temperature affects rates of sulfate reduction because the specific activity of the enzymes slows down (Rabus et al., 2002). Early experiments by Kaplan and Rittenberg (1964) suggested that the sulfur isotope fractionation increased as rates decreased with decreasing temperature. However, these experiments were performed with resting cells of cultures that were incubated at temperatures at which the bacteria could not grow and more likely exhibited stress responses rather than healthy physiological characteristics. Experiments with growing psychrophilic and mesophilic strains over temperatures at which the bacteria were viable have shown no effect of temperature on isotope fractionation (Brüchert et al., 2001).

These findings have implications for the regulation of isotope fractionation. The underlying assumption of the model by Rees (1973) was that a less stringent sulfate demand at low rates of sulfate reduction would enhance isotope fractionation. According to Equation 1, at low temperatures and high concentrations of sulfate and electron donor, the rate of reduction is the dominating regulator for isotope fractionation, since there would be no limitation in sulfate uptake. In fact, the constant isotope fractionation in the experiments cited above showed that over their viable temperature ranges these sulfate reducers never experienced uptake limitation, although the cell-specific sulfate reduction rates varied more than 10-fold (Brüchert et al., 2001). Even at the highest sulfate reduction rates, the isotope fractionations were the same as those at the lowest rates.

These experiments provide further evidence that the Rees (1973) model of reverse ^{34}S-enriched sulfate transport out of the cell requires modification. According to the model, the lower the rates of sulfate reduction, the higher the fractionation, with the consequence that more ^{34}S-enriched sulfate or other sulfur compounds have to be transported out of the cell. However, since the experiments by Brüchert et al. (2001) demonstrated constant fractionations independent of the rate of sulfate reduction, sulfur compounds that were transported out of the cell must have had the same isotope composition at all rates. These experiments clearly demonstrate that, at seawater sulfate concentration, the cardinal regulator for isotope fractionation is the rate of electron donor transport, which regulates the isotope composition of sulfate taken up. At all experimental rates, the cells were not sulfate-limited, with the consequence that isotope fractionations did not vary.

Similar considerations apply to the role of sulfate concentration on isotope fractionation. Recently, Habicht et al. (2002) demonstrated for *Archaeoglobus fulgidus* reduced isotope fractionation at sulfate concentrations below 200 µM. When sulfate concentrations exceeded 1 mM, there was no effect on isotope fractionation. To summarize, the effect of sulfate uptake on isotope fractionation for many environments is likely minor given the high concentrations of sulfate in the world ocean and in surface sediment pore waters. Fractionations will decrease only when sulfate concentrations of the ambient environment are extremely low. This may be the case in recent sedimentary environments with concentrations of sulfate less than 1%, the modern seawater concentrations of 28 mM, and in the early Precambrian ocean.

EFFECT OF ELECTRON DONORS ON ISOTOPE FRACTIONATION

The early studies on regulation of sulfur isotope fractionation by Kaplan and Rittenberg (1964) and Kemp and Thode (1968) suggested an effect of the electron donor on isotope fractionation. The smallest fractionations were observed for hydrogen, followed by lactate, acetate, and ethanol (Kaplan and Rittenberg, 1964). In the experimental design used by Kaplan and Rittenberg (1964), cell-specific sulfate reduction rates were slower with ethanol and acetate than with lactate, which led the authors to conclude that different substrates could result in different isotope fractionations insofar as cell-specific sulfate reduction rates varied. By contrast, recent studies with a large number of pure cultures capable of growth on a variety of electron donors have indicated no unique relationship between cell-specific sulfate reduction rate and electron donor (Detmers et al., 2001a). The new finding of this study was that the particular substrate oxidation pathways used by sulfate reducers are reflected in the sulfur isotope fractionation.

Substrate Oxidation Pathways as Regulators for Isotope Fractionation

Electron donor flow during bacterial sulfate reduction can be simplified into two categories: (1) the oxidation of hydrogen and the incomplete oxidation of lactate, propionate, or pyruvate to acetate involving a periplasmic hydrogenase, and (2) the complete oxidation of organic electron donors to CO_2 (Rabus et al., 2000).

Desulfovibrio species have a periplasmic hydrogenase, but they do not have a carbon monoxide dehydrogenase that can oxidize coenzyme A-bound acetate to CO_2. Acetate must therefore be released as the terminal oxidation product. The schematics of sulfate reduction coupled to hydrogen oxidation shown in Figure 2 illustrate the potential effect on isotope fractionation. The periplasmic location of the hydrogenase has the principal advantage that electrons are translocated through the cytoplasmic membrane by electron-carrying cytochromes, while the protons remain in the periplasm to generate a proton motive force that can be used for ATP synthesis (Fig. 2). Cytochrome C3 and an additional unknown electron carrier transport the electrons to the APSR and the DSR (Steuber and Kroneck, 1998). It is conceivable that this mechanism provides a direct electron shuttle to the APSR and DSR. The consequence is an efficient consumption of APS with little buildup of intermediate sulfite, which would minimize isotope fractionation. In agreement with this hypothesis are the higher growth rates of sulfate-reducing bacteria when growing on hydrogen, pyruvate, or lactate compared to acetate (Rabus et al., 2000). In contrast, there are two known pathways for complete electron donor oxidation. One, used by *Desulfonema* and *Desulfobacterium* species, involves the oxidation of acetate, bound as acetyl CoA, to CO_2 using a carbon monoxide dehydrogenase (Schauder et al., 1986). The other pathway is used by *Desulfobacter* species and employs a modified version of the citric acid cycle (Brysch et al., 1987). Both pathways involve significantly more intermediates and enzymes to catalyze the carbon transformation when compared to the oxidation of lactate-hydrogen-pyruvate. It is therefore conceivable, but difficult to demonstrate directly, that the type of electron donor pathway affects isotope fractionation because the rate of transfer of electrons along the electron transport chain to the APSR and DSR affects the residence times of the intermediate APS and sulfite pools. In essence, the hypothesis is that transport of electron donors and acceptors may be less tuned in complete-oxidizing than in incomplete-oxidizing sulfate-reducing bacteria. Some support for this hypothesis may be drawn from the free energy yield for the oxidation of different electron donors with sulfate (Detmers et al. 2001a) (Table 2). The incomplete oxidation of lactate and propionate and

TABLE 2. FREE ENERGY CHANGES AT STANDARD STATE WITH DIFFERENT ELECTRON DONORS FOR COMPLETE AND INCOMPLETE OXIDATION

Electron donor type of oxidation	Stoichiometry	$\Delta G^{0'}$ (kJ mol sulfate^{-1})
pyruvate (incomplete)	$2\ CH_3COCOO^- + SO_4^{2-} \rightarrow 2\ CH_3COO^- + 2HCO_3^- + HS^-$	–340.9
lactate (incomplete)	$2\ CH_3CHOHCOO^- + SO_4^{2-} \rightarrow 2\ CH_3COO^- + 2HCO_3^- + HS^- + H^+$	–160.1
hydrogen	$4\ H_2 + SO_4^{2-} + H^+ \rightarrow 4H_2O + HS^-$	–152.2
formate	$4\ HCOO^- + SO_4^{2-} + H^+ \rightarrow 4\ HCO_3^- + HS^-$	–146.9
ethanol (incomplete)	$2\ CH_3CH_2OH + SO_4^{2-} \rightarrow 2\ CH_3COO^- + HS^- + 2\ H_2O + H^+$	–146.6
pyruvate (complete)	$4\ CH_3COCOO^- + 4\ H_2O + 5\ SO_4^{2-} \rightarrow 12\ HCO_3^- + 5\ HS^- + 3\ H^+$	–106.3
lactate (complete)	$2\ CH_3CHOHCOO^- + 3SO_4^{2-} \rightarrow 6\ HCO_3 + 3\ HS^- + H^+$	–85.1
propionate (incomplete)	$4\ CH_3CH_2COO^- + 3\ SO_4^{2-} \rightarrow 4\ CH_3COO^- + 4\ HCO_3^- + 3\ HS^- + H^+$	–50.2
benzoate (complete)	$C_7H_5O_2^- + 3.75\ SO_4^{2-} + 4\ H_2O \rightarrow 7\ HCO_3^- + 3.75\ HS^- + 2.25\ H^+$	–49.7
butyrate (complete)	$CH_3CH_2CH_2COO^- + 2.5\ SO_4^{2-} \rightarrow 4\ HCO_3^- + 2.5\ HS^- + 0.5\ H^+$	–49.2
hexadecane (complete)	$C_{16}H_{34} + 12.25\ SO_4^{2-} \rightarrow 16\ HCO_3^- + 12.25\ HS^- + H_2O$	–48.7
acetate (complete)	$CH_3COO^- + SO_4^{2-} \rightarrow 2HCO_3^- + HS^-$	–47.6
decane (complete)	$C_{10}H_{22} + 7.75\ SO_4^{2-} \rightarrow 10\ HCO_3^- + 7.75\ HS^- + 2.25\ H^+ + H_2O$	–45.0

pyruvate yields more than three times as much energy per mole of sulfate oxidized compared to the complete oxidation of acetate or benzoate. Although not a direct relationship, ultimately the amount of free energy conserved by the oxidation of an electron donor will translate into the amount of ATP that can be formed for biomass synthesis and cellular maintenance.

Substrate Limitation as Regulator for Isotope Fractionation

In continuous culture experiments with *Desulfovibrio desulfuricans*, fractionations decreased from 35‰ to 18‰ as sulfate reduction rates increased from 0.7 femtomoles cell^{-1} day^{-1} to 8.1 femtomoles cell^{-1} day^{-1} (Chambers et al., 1975). Canfield (2001b) determined isotope fractionations on bulk sediment with natural communities of sulfate-reducing bacteria, which were supplied continuously with electron donors and acceptors. He observed lower fractionations at high sulfate reduction rates when lactate was not limiting, and high fractionations at lower rates of sulfate reduction when lactate was supposedly limiting. Interestingly, at the bulk community level, changes in electron donor affected whole-sediment sulfate reduction rates, but the overall isotope fractionation varied little. Both studies rationalized their results in terms of the rate of cell-specific sulfate reduction and variable exchange of sulfur intermediates across the cell membrane using the rationale of Rees (1973). At high rates, a greater fraction of the intracellular sulfate pool is consumed, with the consequence that the isotope composition of sulfide approaches that of the imported sulfate (Fig. 3). At low rates, only a fraction of the intracellular sulfate is consumed. Consequently, the potential enzymatic isotope fractionation during the reductive process is expressed. While the results of Chambers et al. (1975) indicate that cell-specific sulfate reduction rates influence isotope fractionation, the results of Canfield (2001b) allow multiple explanations. As indicated above, different sulfate-reducing bacteria likely consume lactate and acetate. Canfield (2001b) did not determine the end products of carbon oxidation or investigate shifts in microbial community composition despite the extended duration of his experiment. Pure culture investigations have demonstrated that the fractionations produced in the two metabolic pathways—lactate versus acetate oxidation—can differ by up to 40‰ (Detmers et al., 2001a), which raises the possibility that changes in isotope fractionation reflect changes in sulfate-reducing community composition.

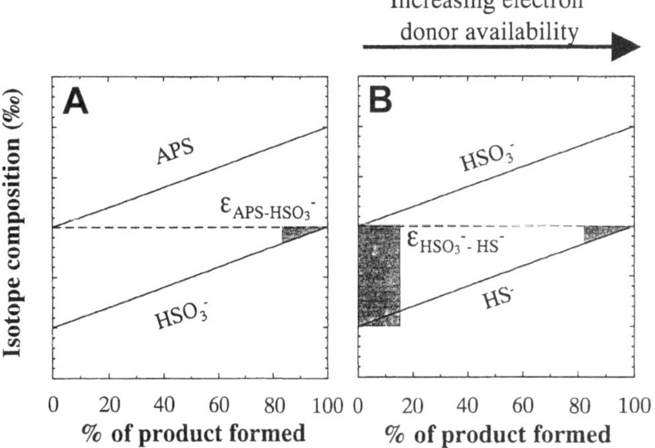

Figure 3. Regulation of kinetic isotope effects at the adenosine phosphosulfate reductase and the dissimilatory sulfite reductase in a steady-state open system. The relative percentages of reactant and product for the two enzymatically catalyzed reductions depend on the availability of electrons carried by the electron transport chain to the DSR (dissimilatory sulfite reductase) and APSR (adenosine phosphosulfate reductase). APS does not accumulate in the cell and concentrations in the cell are likely only nanomolar (Rabus et al., 2000). Therefore, the equilibrium for the reaction of adenosine phosphosulfate (APS)/sulfite is far on the side of sulfite (gray shaded area in Fig. 3A). Even if fractionation occurs during APS reduction, the near-complete consumption of APS prevents a strong isotope effect. Isotope effects can occur at the DSR during reduction of sulfite to hydrogen sulfide. The larger the intracellular sulfite pool, the larger the isotope effect, irrespective of whether sulfite reduction occurs via the trithionate pathway or the single 6-electron transfer. The size of the sulfite pool must be regulated by the availability of electron donors. At low electron donor availability, the isotope effect is large, at high electron donor availability the isotope effect decreases to zero (Fig. 3B). The dashed line indicates the isotope composition of external sulfate for reference.

New Experimental Results of Isotope Fractionations by Pure Cultures Grown on Multiple and Highly Refractory Substrates

To further test the relationship between substrate variability and isotope fractionation, I present results from experiments with pure cultures capable of growth on multiple substrates, some of which are extremely refractory organic compounds. Isotope fractionations were determined for four different strains capable of growth on multiple substrates, following the same procedures as described in Brüchert et al. (2001). The isotope fractionation was determined for *Desulfobacterium autotrophicum* for growth on hydrogen/CO_2 and butyrate, and for the cold-adapted strain *Desulfofrigus oceanensis* and the *Desulfobacter* strain ASv 20 for growth on lactate and acetate. In addition, the thermophilic strain TD-3 (Rueter et al., 1994) was used to investigate fractionation of refractory organic substrates such as n-alkanes, longer-chain fatty acids, aromatics, and crude oil. Briefly, experiments were performed with growing batch cultures with an excess of organic substrate. Time points were taken in the lag, exponential, and stationary phase of growth. At each time point, the abundance of live cells was counted. Sulfate and sulfide concentrations were determined by ion chromatography and spectrophotometry as described in Brüchert et al. (2001). The isotope composition of dissolved sulfate and sulfide was determined for each time point by EA IRMS using a ThermoFinnigan delta Plus Mass spectrometer coupled to a Euro 3000 elemental analyzer via a Conflo 2 interface. Accuracy and analytical precision were determined using the secondary standard $BaSO_4$ standard NBS 127 ($\delta^{34}S$ versus Vienna Canyon Diablo Troilite [VCDT] = 20.0‰ ± 0.3). Since the experiments were performed as closed systems, fractionation factor epsilon (ε) could be calculated using the Rayleigh relationship as described in Brüchert et al. (2001).

Results are summarized in Table 3. In all experiments, fractionations exceeded 18‰ whenever the terminal product of substrate oxidation was CO_2. Fractionations were <14.8‰ when acetate was released as the terminal product or when hydrogen was the electron donor. The experiment with *Desulfofrigus oceanensis* showed that this strain was capable of switching from complete to incomplete oxidation of organic substrate at high concentrations of substrate, a characteristic also known for *Desulfotomaculum* species (Kuever et al., 1999). In agreement with other experiments, the cell-specific sulfate reduction rates were of the same order as reported for other sulfate-reducing bacteria and showed no influence on isotope fractionation. Experimental results with strain TD-3 indicated very high fractionations, in particular when growing on complex substrate such as crude oil, in agreement with studies on refractory aromatic compounds (Bolliger et al., 2001).

BIOGEOCHEMICAL IMPLICATIONS

A mixed sulfate-reducing community with diverse species-specific isotope fractionations adds a new perspective to the regulation of sulfur isotope fractionation. Whereas sulfur isotope fractionation has been interpreted traditionally in terms of sulfate reduction rate and sulfate concentration, newer results emphasize the importance of microbial diversity for isotope fractionation (Detmers et al., 2001a). Sulfate-reducing bacteria, while sharing the exploitation of a common electron acceptor, must also compete for organic substrates or hydrogen. The ecological consequence is that different sulfate reducers exploit different ecological niches. An alternative strategy is to outcompete other sulfate reducers, at least temporarily, for a particular energy resource. Differences in the affinity of sulfate-reducing bacteria for a specific substrate can play a critical role in regulating community composition and whole-community isotope fractionation. Some sulfate reducers grow quickly, but only at high concentrations of organic substrate. Other species may be better adapted to low concentrations, but only grow slowly.

A temporal shift in substrate availability may be produced during the accumulation of a phytoplankton bloom at the sediment surface. In shallow-water organic-rich environments with anoxic bottom waters, such as coastal upwelling systems, bacteria with relatively low substrate affinity but high growth rates at high substrate concentrations will likely dominate the overall bacterial sulfate reduction process. The fast-growing *Desulfovibrio* species may successfully compete in such an environment. The dominance of *Desulfovibrio* species in the uppermost centimeters of intertidal sediments would support this hypothesis (Llobet-Brossa et al., 2002). The high abundance of particulate organic substrate in such an environment may also make a strategy of energy-inefficient complete substrate oxidation at least temporarily unnecessary. This is supported by the high concentrations of acetate found in many organic-rich, near-shore environments and the observation that the addition of high-molecular weight polysaccharides as carbon substrate often yields high concentrations of pore-water acetate, which is not readily consumed (Brüchert and Arnosti, 2003). Deeper in the sediment, where the readily consumable organic substrate is depleted, sulfate-reducing bacteria with different adaptations (i.e., high substrate affinity and the ability to oxidize a variety of refractory compounds, such as aromatic compounds, long-chain fatty acids, or even alkanes) may become predominant. Presently, little is known about the pathways that funnel complex organic substrates toward sulfate-reducing bacteria. Inhibition studies have repeatedly indicated the importance of acetate and hydrogen for terminal metabolism (Sørensen et al., 1981; Parkes et al., 1993). The other volatile fatty acids—propionate, butyrate, lactate, isobutyrate, glycolate, and pyruvate—and valerate appear to play a subordinate role (Albert and Martens, 1997; Hoehler et

TABLE 3. COMPARISON OF ISOTOPIC FRACTIONATION OF FOUR SULFATE-REDUCING BACTERIA GROWN ON DIFFERENT SUBSTRATES

Species	Environment	Substrates	Terminal C-product	Specific SRR fmol cell^{-1} day^{-1}	Isotope fractionation ε (‰)
Desulfo-bacterium autotrophicum	Coastal lagoon	H_2/CO_2	—	34.8	−14.8
		Butyrate	CO_2	64.8	−32.7
Desulfofrigus oceanense	Arctic sediment		Acetate	3.6	−10.6
			CO_2	7.6	−22.0
Strain TD-3	Hydrothermal vent	Hexadecane	CO_2	0.3	−30.4
		Decane	CO_2	0.6	−27.1
		Butyrate	CO_2	0.6	−26.8
		Crude oil	CO_2	2.7	−35.9
Desulfobacter strain ASv 20	Arctic sediment	Lactate	Acetate	3.2	−6.8
		Acetate	CO_2	6.6	−18.8

Note: SRR—sulfate reduction rate.

al., 1998; Parkes et al., 1989; Sørensen et al., 1981). In cold environments, thermodynamics would predict that hydrogen is the dominant terminal electron acceptor (Westermann et al., 1994; Conrad et al., 1986; N. Finke, 2002, personal commun.). On the other hand, anaerobic degradation of the aromatic compounds of lignin (i.e., terrestrial organic carbon sources) produces mainly acetate or monomeric aromates, for which only complete-oxidizing bacteria can compete.

In the following, these concepts are discussed in light of field and experimental data from various marine environments with different temperatures and organic carbon concentrations.

Substrate Amendment Experiments—Effects on Sulfate-Reducing Community Composition

The effect of substrate pulses was simulated by incubating surface sediment from intertidal flats of the Wadden Sea in sealed polyethylene bags, each amended with different substrates. In time course experiments, shifts in sulfate-reducing community composition were monitored with fluorescent in situ hybridization (FISH) using fluorescently-labeled oligonucleotide probes that target the 16S rRNA of sulfate-reducing bacteria (Amann and Ludwig, 2000). In this study, we selected two specific probes. One probe (DSS 658) specifically targeted the 16S RNA of *Desulfobacteriaceae*, which are exclusively complete-oxidizing sulfate-reducing bacteria. The other probe (DSV 658) targeted the group of *Desulfovibrioceae* (i.e., incomplete-oxidizing sulfate reducers). Makame algal material (1.5 g) was added to 1 kg of wet sediment. In one incubation, lactate was added to bring the final concentration to ~20 mM. In the other incubation, acetate was added to a final concentration of ~10 mM. The substrate amendments were expected to selectively stimulate incomplete- and complete-oxidizing sulfate reducers. A control incubation without substrate was monitored in parallel. Methods for the experimental design are described in Hansen et al. (2000) and Rosselló-Mora et al. (1999). Sulfate concentrations were monitored in time courses and determined by ion chromatography using methods described in Brüchert et al. (2001). As expected, incomplete-oxidizing *Desulfovibrio* species increased in the Makame- and lactate-amended incubations, whereas complete-oxidizing *Desulfobacter* species increased in the acetate-amended incubations (Table 4) and support the assessment that specific substrate levels favor selective sulfate-reducing species.

Isotope Fractionation in Sulfidic Sediments with High Rates of Bacterial Sulfate Reduction

Whole-community isotope fractionations can also be determined from sediment core profiles using the depth profiles of concentration and isotope composition of dissolved sulfate and sulfide when applying diagenetic diffusion-advection rate models to the measured isotope profiles (e.g., Berg et al., 1998). This was done using the depth profiles of the concentration and isotope composition of dissolved sulfide and sulfate in organic-rich sediments in the coastal upwelling zone off central Namibia. These sediments are in shallow water (<30 m), contain more than 18 wt% organic carbon, and are almost pure diatomaceous muds (Brüchert et al., 2003). In the profile shown in Figure 4, sulfate is consumed in the upper 8 cm of sediment, where sulfide concentrations have increased to 20 mM. These high pore-water sulfide concentrations and the presence of free hydrogen sulfide in the overlying water column inhibit the oxidative part of the sulfur cycle including disproportionation and ensure that the modification of the isotope signal is exclusively due to bacterial sulfate reduction (Thamdrup et al., 1993). Sediment sulfate reduction rates were measured with the ^{35}S incubation method and are highest near the sediment-water interface, where sulfate is still abundant (Brüchert et al., 2003). In this environment, the bacteria are not limited by sulfate.

Isotope profiles show extremely steep gradients in $\delta^{34}S_{sulfate}$ and $\delta^{34}S_{sulfide}$ (Fig. 4). The isotope profile of $\delta^{34}S_{sulfate}$ was recalculated to depth profiles for $^{34}S_{sulfate}$ and $^{32}S_{sulfate}$ using the concentration of sulfate, the $\delta^{34}S$ composition of sulfate, and the isotope ratio of 0.0450045 for the $^{34}S/^{32}S$ ratio in the Canyon Diablo Troilite Standard (Gonfiantini, 1984). Fluxes of $^{34}S_{sulfate}$ and $^{32}S_{sulfate}$ diffusing into the sediment were calculated with the modeling tool by Berg et al. (1998). The calculated isotope ratio of the diffusing $^{34}S_{sulfate}$ and $^{32}S_{sulfate}$ entering the sediment was +15‰ versus VCDT and depleted by only 5‰ relative to seawater sulfate. The resulting whole-sediment isotope fractionation factor ε is only 5.0‰. Such a low value strongly supports the existence of a sulfate-reducing community dominated by incomplete-oxidizing or hydrogen-consuming sulfate-reducing bacteria that produce small isotope fractionations. These data are important for two reasons. First, they demonstrate that low fractionations can be produced in environments with high concentrations of sulfate. Second, the low fractionation suggests a sulfate-reducing community dominated

TABLE 4. COMPOSITIONAL CHANGES OF SULFATE-REDUCING COMMUNITY IN INTERTIDAL SEDIMENT IN RESPONSE TO CHANGES IN ORGANIC SUBSTRATE

	% of all bacteria							
	Control		Acetate		Lactate		Makame	
Time point	Desulfo-vibrio sp.	Desulfo-bacter sp.	Desulfo-vibrio sp.	Desulfo-bacter sp.	Desulfo-vibrio sp.	Desulfo-bacter sp.	Desulfo-vibrio sp.	Desulfo-bacter sp.
t0	5.5	6.8	5.5	6.8	4.2	7.2	5.5	6.8
t1			4.8	6.5			5.2	6.5
t2	4.8	6.2	5.2	8.5	7.8	4.2	8.5	6.8

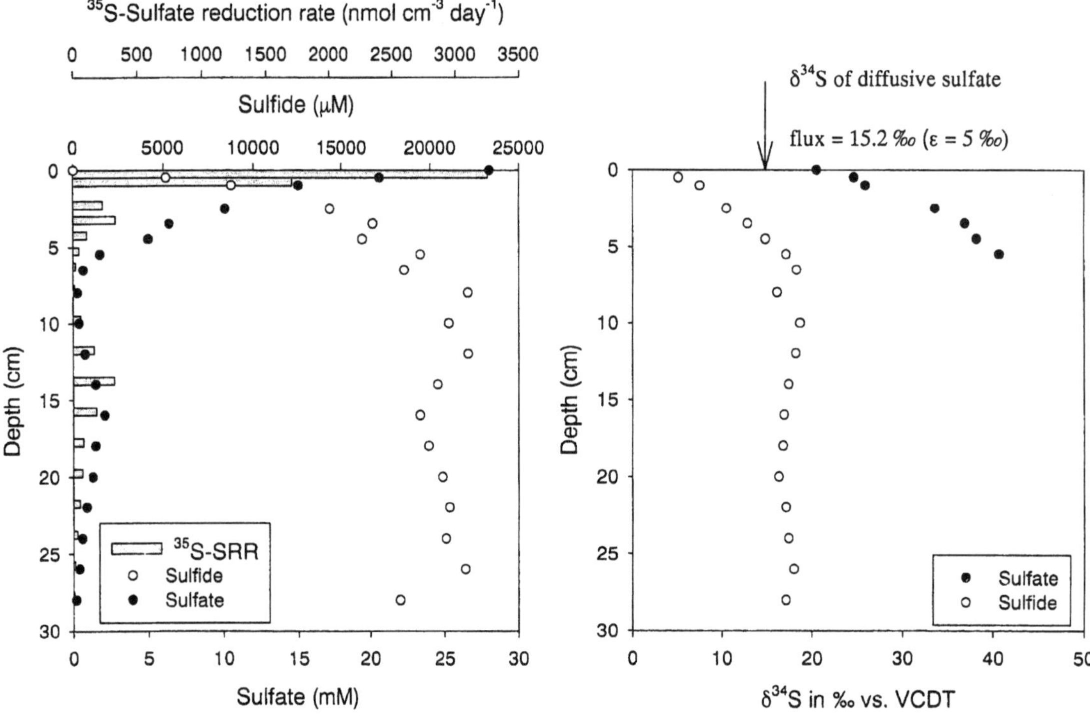

Figure 4. Depth profiles of concentration and sulfur isotope composition of dissolved sulfate and sulfide in organic-rich sediment from the anoxic Namibian shelf. Steep sulfate and sulfide concentration profiles are the consequence of high bacterial sulfate reduction rates and absence of sulfide oxidation. Isotope profiles of dissolved sulfate and sulfide are correspondingly steep. Note that this system is open to diffusion. The isotope data were fit to a diffusion-reaction model to calculate the diffusive flux of dissolved ^{32}S and ^{34}S-sulfate, from which an isotope composition of +15.2 ‰ versus VCDT for the diffusive flux of sulfate across the sediment-water interface was calculated. The resulting bulk sediment isotope fractionation is only 5‰, suggestive of a predominantly incomplete-oxidizing sulfate-reducing community.

by incomplete-oxidizing sulfate-reducing bacteria. Both findings are consistent with the measured sulfate reduction rates in this highly active near-shore coastal upwelling sediment.

Whole-Sediment Incubations in Arctic Environments

In a third study, isotope fractionations were determined in anoxic bag incubation experiments of whole sediment from fjord stations around Spitsbergen at 79°N (Station J, Smeerenburgfjorden), using methods described in Hansen et al. (2000). The sediment temperature was only 0.5 °C. Isotope fractionations were determined following the decrease in sulfate concentration with time and the associated shift in sulfur isotope composition of dissolved sulfate according to methods described in Habicht and Canfield (1997). The isotope fractionation factors (ε) were calculated using the Rayleigh equation (Brüchert et al., 2001). Isotope fractionations between 9‰ and 18‰ were determined for surface sediment (0–3 cm depth) (Fig. 5). Sediment from the 3–6 cm sediment depth interval, however, yielded a fractionation of 35‰. These results are consistent with the molecular ecological analysis of the community of sulfate-reducing bacteria in the upper 10 cm of sediment at this station, based on 16S rRNA-targeted FISH with oligonucleotide probes (Ravenschlag et al., 2000). Their results indicated the overall dominance of complete-oxidizing bacteria of the nutritionally versatile cluster *Desulfosarcina/Desulfococcus* spp. (>70%). However, incomplete-oxidizing bacteria, such as *Desulfovibrio* and *Desulfotalea* species, were relatively more abundant in the uppermost centimeter, consistent with the lower whole-sediment isotope fractionation. The whole-sediment sulfate reduction rate did not correspond to the change in isotope fractionation. These rates varied only between 5 and 15 nmol cm^{-3} day^{-1}, with the higher rates in sediment intervals with higher fractionations.

Sulfur Isotope Fractionation Associated with Coupled Sulfate Reduction and Anaerobic Oxidation of Methane

A unique marine environment is that associated with the anaerobic oxidation of methane (Valentine and Reeburgh, 2000). There is molecular genetic evidence that this process is catalyzed by a syntrophic association of Archaea and sulfate-reducing bacteria (Boetius et al., 2000). Hydrogen and formate have been proposed as potential interspecies transfer agents. In this environment, the substrate spectra for the sulfate-reducing bacteria are defined

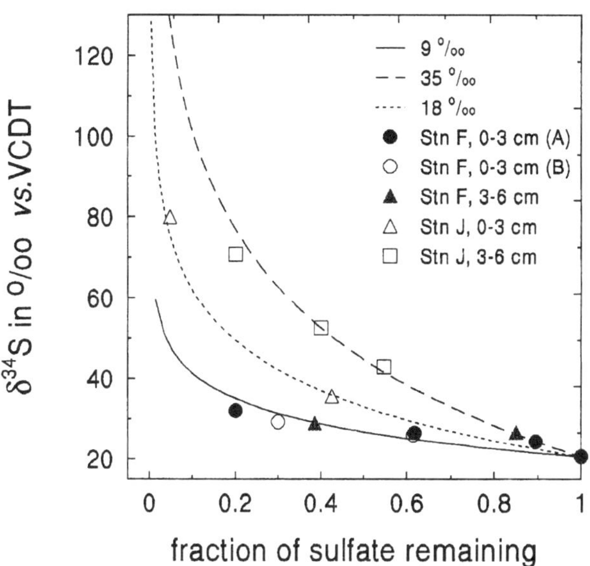

Figure 5. Sulfur isotope fractionation of whole-sediment sulfate-reducing community in Arctic sediments from Spitsbergen. In situ temperatures were 0.5 °C. Unamended sediment material from two stations from two different depth intervals was incubated in sealed bags to simulate closed system conditions. Isotope fractionations were calculated using the Rayleigh equations described in Brüchert et al. (2001). Different fractionations in the different depth intervals at Station J are consistent with FISH observation of different groups of sulfate-reducing bacteria and suggest the presence of more incomplete oxidizers in the 0–3 cm than the 3–6 cm depth.

by the narrow thermodynamic limits of the overall process. Specific sulfate reducers with high substrate affinity, potentially of the group *Desulfosarcina/Desulfococcus*, may exploit this ecological niche. Thermodynamically, the process only operates when the concentration of the reactants, sulfate and methane, are high enough to overcome the thermodynamic limits, while the intermediate electron donor (e.g., hydrogen) is kept at low concentration. This condition is maintained when the syntrophic partner (in this case, the sulfate-reducing bacterium) has abundant sulfate and readily consumes any available intermediate. This situation therefore reflects electron donor limitation for the sulfate reducer and could theoretically yield high fractionations. In marine sediments, however, this process has often been observed at diffusive interfaces between sulfate and methane, where sulfate concentrations are only in the micromolar range (Iversen and Jørgensen, 1985). In this case, low pore-water sulfate concentrations may be limiting cellular uptake, and overall low fractionations would be predicted.

Isotope Fractionation and Whole-Community Sulfate Reduction Rates

There has been a long discussion in the literature on whether sulfate reduction rates control isotope fractionation. If so, sediments with ^{34}S-enriched pyrite could be readily interpreted to reflect environments with high rates of bacterial sulfate reduction. Habicht and Canfield (1997) pointed out the important distinction between whole-community sulfate reduction rates and single-cell bacterial sulfate reduction rates. Whole-community sulfate reduction rates may be fast, but rates per cell can be slow if the population of sulfate-reducing bacteria is large. The whole-community sulfur isotope fractionation is the sum of the individual sulfur isotope fractionations produced by each species of sulfate reducers. Each bacterial group has a specific isotope fractionation and variable single-cell sulfate reduction rates. The single-cell sulfate reduction rates are controlled by the availability of substrate. Thus, the composition of the community and the cell-specific rates of each species regulate the whole-community sulfur isotope fractionation. The important distinction to be made is whether variations in isotope fractionations can be attributed either to differences in cell-specific rates or to differences in the sulfate-reducing bacterial community. In principle, this issue cannot be resolved until in situ cell-specific sulfate-reduction rates or other group-specific activity measures of phylogenetically identifiable sulfate reducers are available.

GEOLOGICAL IMPLICATIONS

Large and stratigraphically consistent depletions in the isotope composition of sedimentary pyrite (<−35‰ versus VCDT) have been used to support the existence of ancient euxinia (Beier and Hayes, 1989; Werne et al., 2003). The presence of ^{34}S-enriched pyrite (up to +20‰ versus VCDT) in ancient sediments has either been used to support the presence of low-sulfate (e.g., freshwater) environments or been explained by sedimentary sulfide formed from anaerobic methane oxidation (Jørgensen et al., 2004). However, between these two extreme compositions exist pyrites with isotope compositions ranging over 55‰ that are not readily amenable for unambiguous paleoenvironmental interpretations. The most common interpretation has been that pyrites with an intermediate isotope composition (e.g., between 0 and −30‰ versus VCDT) have formed in sediments with variable diffusion limitation. The compositional range has been explained by variable reservoir limitation (i.e., so-called partially open versus partially closed system effects). In essence, this argument implies that at high rates of bacterial sulfate reduction, there is extreme sulfate depletion, transport of sulfate becomes limiting, and the isotope effect would be small. In such systems, the instantaneous whole-community isotope fractionation may be larger than the net isotope fractionation. In pore-water profiles, such systems resemble a Rayleigh system and have been modeled accordingly (Mariotti et al., 1981; Hartmann and Nielsen, 1969; Aharon and Fu, 2000). There are several implications using a Rayleigh model or any approximations thereof that do not apply to sedimentary systems. Natural sedimentary systems, even sediments that are buried, are open, since dissolved species can be transported through each layer by diffusion. Therefore, only reaction-diffusion-advection models can be applied to isotope depth profiles to derive isotope fractionations (Jørgensen, 1979; Goldhaber and Kaplan, 1980; Wortmann et al., 2001; Rudnicki et al., 2001; Habicht et al., 2002). The choice of model has large implications for the determination of isotope fractionations because approximations using a Rayleigh model can yield significantly lower isotope fractionations than open system models.

Most geological interpretations implicitly assume that, irrespective of the sedimentary environment, the isotope fractionation of the natural sulfate-reducing community is constant or varies only as a function of sulfate reduction rate (Jørgensen, 1979; Habicht and Canfield, 1997; Habicht et al., 2002; Jørgensen et al., 2004). Community-related effects on isotope fractionation have not been taken into account. However, such effects can be significant considering that a community dominated by complete-oxidizing bacteria may produce pyrite after isotope fractionations of nearly 45‰, whereas whole-community isotope fractionations by a dominantly incomplete-oxidizing sulfate-reducing community may only fractionate sulfate by 5‰. The resulting pyrites from the two scenarios would have an isotope composition of −25‰ as opposed to +15‰. Both pyrites could have formed near the sediment-water interface from sulfide formed by the heterotrophic breakdown of organic matter by sulfate-reducing bacteria. In the sedimentary record, however, these widely differing isotope compositions could have been used for very different paleoenvironmental interpretations.

It would be premature to predict that changes in microbial sulfate-reducing community composition can be preserved in the isotope composition of sedimentary sulfides, in particular since the isotope composition of sulfur can be further modified during disproportionation of intermediate sulfur compounds (Habicht and Canfield, 2001). The new data presented require further testing in different recent environments. Additional supporting evidence is required, possibly in the form of bacterial phospholipids of specific sulfate reducers. It should also be kept in mind that the diversity of possible fermentation pathways derived from a complex carbon source yields a number of potentially usable substrates, which are simultaneously available for complete- and incomplete-oxidizing bacteria. Therefore, in most marine sediments, a composite isotope signal derived from the mixture of complete- and incomplete-oxidizing sulfate reducers should be expected. It would be a significant step forward in our understanding of the sulfur isotope fractionation by natural communities if experimentally determined in situ isotope fractionations can be ascribed to sulfate-reducing communities that vary as a function of organic carbon source.

Such data could provide supporting evidence for past anaerobic carbon mineralization processes in ancient sediments and point to the type of organic material that had been degraded by the sulfate-reducing bacteria.

ACKNOWLEDGMENTS

I would like to thank Ralf Rabus, Marcel Kuypers, Lev Neretin, Helle Ploug, Axel Schippers, two anonymous reviewers, and Tim Lyons for critically reading an earlier version of this manuscript. Discussions with Marga Bauer, Tim Ferdelman, Michael Böttcher, and Jan Kuever helped in shaping the scope of the manuscript. This study was funded by the Max-Planck Society.

REFERENCES CITED

Aharon, P., and Fu, B., 2000, Microbial sulfate reduction rates and sulfur and oxygen isotope fractionations at oil and gas seeps in deepwater Gulf of Mexico: Geochimica et Cosmochimica Acta, v. 64, p. 233–246, doi: 10.1016/S0016-7037(99)00292-6.

Albert, D.B., and Martens, C.S., 1997, Determination of low-molecular weight organic acid concentrations in seawater and pore-water samples via HPLC: Marine Chemistry, v. 56, p. 27–37, doi: 10.1016/S0304-4203(96)00083-7.

Amann, R., and Ludwig, W., 2000, Ribosomal RNA-targeted nucleic acid probes for studies in microbial ecology: FEMS Microbiology Reviews, v. 24, p. 555–565, doi: 10.1016/S0168-6445(00)00044-9.

Beier, J.A., and Hayes, J.M., 1989, Geochemical and isotopic evidence for paleoredox conditions during deposition of the Devonian-Mississippian New Albany Shale, southern Indiana: Geological Society of America Bulletin, v. 101, p. 774–782, doi: 10.1130/0016-7606(1989)1012.3.CO;2.

Berg, P., Petersen-Risgaard, N., and Rysgaard, S., 1998, Interpretation and measured concentration profiles in sediment pore water: Limnology and Oceanography, v. 43, p. 1500–1510.

Berner, R.A., and Petsch, S., 1998, The sulfur cycle and atmospheric oxygen: Science, v. 282, p. 1426–1427.

Boetius, A., Ravenschlag, K., Schubert, C.J., Rickert, D., Widdel, F., Gieseke, A., Amann, R., Jørgensen, B.B., Witte, U., and Pfannkuche, O., 2000, A marine microbial consortium apparently mediating anaerobic oxidation of methane: Nature, v. 407, p. 623–626, doi: 10.1038/35036572.

Bolliger, C., Schroth, M.H., Bernasconi, S.M., Kleikemper, J., and Zeyer, J., 2001, Sulfur isotope fractionation during microbial sulfate reduction by toluene-degrading bacteria: Geochimica et Cosmochimica Acta, v. 65, p. 3289–3298, doi: 10.1016/S0016-7037(01)00671-8.

Böttcher, M., Sievert, S., and Kuever, J., 1999, Fractionation of sulfur isotopes during dissimilatory reduction of sulfate by a thermophilic gram-negative bacterium at 60 °C: Archives of Microbiology, v. 172, p. 125–128, doi: 10.1007/S002030050749.

Brüchert, V., and Arnosti, C., 2003, Anaerobic carbon transformation: Experimental studies with flow-through cells: Marine Chemistry, v. 80, p. 171–183, doi: 10.1016/S0304-4203(02)00119-6.

Brüchert, V., Knoblauch, C., and Jørgensen, B.B., 2001, Microbial controls on the stable sulfur isotope fractionation during bacterial sulfate reduction in Arctic sediments: Geochimica et Cosmochimica Acta, v. 65, p. 753–766.

Brüchert, V., Jørgensen, B.B., Neumann, K., Riechmann, D., Schlösser, M., and Schulz, H., 2003, Regulation of bacterial sulfate reduction and hydrogen sulfide fluxes in the central Namibian coastal upwelling zone: Geochimica et Cosmochimica Acta, v. 67, p. 4505–4518, doi: 10.1016/S0016-7037(03)00275-8.

Brysch, K., Schneider, C., Fuchs, G., and Widdel, F., 1987, Lithoautotrophic growth of sulfate-reducing bacteria, and description of Desulfobacterium autotrophicum gen. nov. sp. nov: Archaeological Microbiology, v. 148, p. 264–274.

Canfield, D.E., 2001a, Biogeochemistry of sulfur isotopes, *in* Valley, J.W., and Cole, D.R., eds., Stable isotope geochemistry: Reviews in Mineralogy and Geochemistry, Mineralogical Society of America, p. 607–636.

Canfield, D.E., 2001b, Isotope fractionation by natural populations of sulfate-reducing bacteria: Geochimica et Cosmochimica Acta, v. 65, p. 1117–1124, doi: 10.1016/S0016-7037(00)00584-6.

Canfield, D.E., and Teske, A., 1996, Late Proterozoic rise in atmospheric oxygen concentration inferred from phylogenetic and sulphur-isotope studies: Nature, v. 382, p. 127–132, doi: 10.1038/382127A0.

Canfield, D.E., and Thamdrup, B.T., 1994, The production of ^{34}S-depleted sulfide during disproportionation of elemental sulfur: Science, v. 266, p. 1973–1975.

Chambers, L.A., Trudinger, P.A., Smith, J.W., and Burns, M.S., 1975, Fractionation of sulfur isotopes by continuous cultures of *Desulfovibrio desulfuricans*: Canadian Journal of Microbiology, v. 21, p. 1602–1607.

Chambers, L.A., Trudinger, P.A., Smith, J.W., and Burns, M.S., 1976, A possible boundary condition in bacterial sulfur isotope fractionation: Geochimica et Cosmochimica Acta, v. 40, p. 1191–1194, doi: 10.1016/0016-7037(76)90154-X.

Conrad, R., Schink, B., and Phelps, T.J., 1986, Thermodynamics of H_2-consuming and H_2-producing metabolic reactions in diverse methanogenic environments under in situ conditions: FEMS Microbial Ecology, v. 38, p. 353–360, doi: 10.1016/0378-1097(86)90013-3.

Crane, B.R., Siegel, L.M., and Getzoff, E.D., 1995, Sulfite reductase structure at 1.6A: Evolution and catalysis for reduction of inorganic anions: Science, v. 270, p. 59–67.

Cypionka, H., 1989, Characterization of sulfate transport in *Desulvovibrio desulfuricans*: Archives of Microbiology, v. 152, p. 237–243.

Cypionka, H., 1994, Sulfate transport: Methods in Enzymology, v. 243, p. 3–14.

Cypionka, H., 1995, Solute transport and cell energetics, *in* Barton, L.L., ed., Sulfate-reducing bacteria: New York, Plenum Press, p. 151–184.

Detmers, J., Brüchert, V., Habicht, K.S., and Kuever, J., 2001a, Diversity of sulfur isotope fractionations by sulfate-reducing prokaryotes: Applied and Environmental Microbiology, v. 67, p. 888–894, doi: 10.1128/AEM.67.2.888-894.2001.

Friedrich, M.W., 2002, Phylogenetic analysis reveals multiple lateral transfers of adenosine-5′-phosphosulfate reductase genes among sulfate-reducing microorganisms: Journal of Bacteriology, v. 184, p. 278–289, doi: 10.1128/JB.184.1.278-289.2002.

Fry, B., Gest, H., and Hayes, J.M., 1988, $^{34}S/^{32}S$ fractionation in sulfur cycles catalyzed by anaerobic bacteria: Applied and Environmental Microbiology, v. 54, p. 250–256.

Goldhaber, M.B., and Kaplan, I.R., 1980, Mechanisms of sulfur incorporation and isotope fractionation during early diagenesis in sediments of the Gulf of California: Marine Chemistry, v. 9, p. 95–143.

Gonfiantini, R., 1984, Report on Advisory Group Meeting on stable isotope reference samples for geochemical and hydrological investigations: Report to the Director General, International Atomic Energy Agency, Sept. 19–21, 1983.

Habicht, K.S., and Canfield, D.E., 2001, Isotope fractionation by sulfate-reducing natural populations and the isotope composition of sulfide in marine sediments: Geology, v. 29, p. 555–558, doi: 10.1130/0091-7613(2001)0292.0.CO;2.

Habicht, K.S., and Canfield, D.E., 1997, Sulfur isotope fractionation during bacterial sulfate reduction in organic-rich sediments: Geochimica et Cosmochimica Acta, v. 61, p. 5351–5361, doi: 10.1016/S0016-7037(97)00311-6.

Habicht, K., Gade, M., Thamdrup, B., Berg, P., and Canfield, D.E., 2002, Calibration of sulfate levels in the Archaean ocean: Science, v. 298, p. 2372–2374, doi: 10.1126/SCIENCE.1078265.

Hansen, J.W., Thamdrup, B., and Jørgensen, B.B., 2000, Anoxic incubation of sediment in gas-tight plastic bags: a method for biogeochemical process studies: Marine Ecology Progress Series, v. 208, p. 273–282.

Harrison, A.G., and Thode, H.G., 1958, Mechanism of the bacterial reduction of sulphate from isotope fractionation studies: Transactions of the Faraday Society, v. 54, p. 84–92.

Hartmann, M., and Nielsen, H., 1969, d34S Werte in rezenten Meeressedimenten und ihre Deutung am Beispiel einiger Sedimentprofile aus der westlichen Ostsee: Geologische Rundschau, v. 58, p. 621–655.

Hipp, W.M., Pott, A.S., Thum-Schmitz, N., Faath, I., Dahl, C., and Trüper, H.G., 1997, Towards the phylogeny of APS reductases and sirohaem sulfite reductases in sulfate-reducing and sulfur-oxidizing prokaryotes: Microbiology, v. 143, p. 2891–2902.

Hoehler, T.M., Alperin, M., Albert, D.B., and Martens, C.S., 1998, Thermodynamic control on hydrogen concentrations in anoxic sediments: Geochimica et Cosmochimica Acta, v. 62, p. 1745–1756, doi: 10.1016/S0016-7037(98)00106-9.

Holser, W.T., Schidlowski, M., Mackenzie, F.T., and Maynard, J.B., 1988, Geochemical cycles of carbon and sulfur, *in* Gregor, C.B., et al., eds., Chemical cycles in the evolution of the Earth, Wiley, p. 105–173.

Iversen, N., and Jørgensen, B.B., 1985, Anaerobic methane oxidation rates at the sulfate-methane transition in marine sediments from Kattegat and Skagerrak (Denmark): Limnology and Oceanography, v. 30, p. 944–955.

Joachimski, M.M., Ostertag-Henning, C., Pancost, R.D., Strauss, H., Freeman, K.H., Littke, R., Sinninghe Damsté, J.S., and Racki, G., 2001, Water column anoxia, enhanced productivity and concomitant changes in delta C-13 and delta S-34 across the Frasnian-Famennian boundary (Kowala Holy Cross Mountains/Poland): Chemical Geology, v. 175, p. 109–131, doi: 10.1016/S0009-2541(00)00365-X.

Jones, G.E., and Starkey, R.L., 1957, Fractionation of stable isotopes of sulfur by micro-organisms and their role in the deposition of native sulfur: Applied Microbiology, v. 5, p. 111.

Jørgensen, B.B., 1979, A theoretical model of the stable sulfur isotope distribution in marine sediments: Geochimica et Cosmochimica Acta, v. 43, p. 363–374, doi: 10.1016/0016-7037(79)90201-1.

Jørgensen, B.B., Böttcher, M.E., Lüschen, H., Neretin, L.N., and Volkov, I.I., 2004, Anaerobic methane oxidation and a deep H_2S sink generate isotopically heavy sulfides in Black Sea sediments: Geochimica et Cosmochimica Acta, v. 68, p. 2095–2118.

Kaplan, I.R., and Rittenberg, S.C., 1964, Microbiological fractionation of sulphur isotopes: Journal of General Microbiology, v. 34, p. 195–212.

Kemp, A.L.W., and Thode, H.G., 1968, The mechanism of the bacterial reduction of sulphate and sulphite from isotope fractionation studies: Geochimica et Cosmochimica Acta, v. 32, p. 71–91, doi: 10.1016/0016-7037(68)90088-4.

Klein, M., Friedrich, M., Roger, A.J., Hugenholtz, P., Fishbain, S., Abicht, H., Blackall, L.L., Stahl, D.A., and Wagner, M., 2001, Multiple lateral transfers of dissimilatory sulfite reductase genes between major lineages of sulfate-reducing prokaryotes: Journal of Bacteriology, v. 183, p. 6028–6035, doi: 10.1128/JB.183.20.6028-6035.2001.

Kuever, J., Rainey, F., and Hippe, H., 1999, Description of *Desulfotomaculum* sp. Groll as *Desulfotomaculum gibsoniae* sp. nov: International Journal of Systematic Bacteriology, v. 49, p. 1801–1808.

Llobet-Brossa, E., Rabus, R., Böttcher, M.E., Könneke, M., Finke, N., Schramm, A., Meyer, R.L., Grötzschel, S., Rosselló-Mora, R., and Amann, R., 2002, Community structure and activity of sulfate-reducing bacteria in an intertidal surface sediment: A multi-method approach: Aquatic Microbial Ecology, v. 29, p. 211–226.

Mariotti, A., Germon, J.C., Hubert, P., Kaiser, P., Letolle, R., Tardieux, A., and Tardieux, P., 1981, Experimental determination of nitrogen kinetic isotope fractionation: Some principles: Illustration for the denitrification and nitrification processes: Plant and Soil, v. 62, p. 413–430.

McCready, R.G.L., 1975, Sulphur isotope fractionation by *Desulvovibrio* and *Desulfotomaculum* species: Geochimica et Cosmochimica Acta, v. 39, p. 1395–1401, doi: 10.1016/0016-7037(75)90118-0.

Parkes, R.J., Gibson, G.R., Mueller-Harvey, I., Buckingham, W.J., and Herbert, R.A., 1989, Determination of the substrates for sulphate-reducing bacteria within marine and estuarine sediments with different rates of sulphate reduction: Journal of General Microbiology, v. 135, p. 175–187.

Parkes, R.J., Dowling, N.J.E., White, D.C., Herbert, R.A., and Gibson, G.R., 1993, Characterization of sulfate-reducing bacterial populations within marine and estuarine sediments with different rates of sulfate reduction: FEMS Microbiology Ecology, v. 102, p. 235–250.

Passier, H., Bosch, H.-J., Nijenhuis, I.A., Lourens, L.J., Böttcher, M.E., Leenders, A., Sinninghe Damsté, J.S., De Lange, G.J., and de Leeuw, J.W., 1999, Sulphidic Mediterranean surface waters during Pliocene sapropel formation: Nature, v. 397, p. 146–149, doi: 10.1038/16441.

Paytan, A., Kastner, M., Campbell, D., and Thiemens, M., 1998, Sulfur isotope composition of Cenozoic seawater sulfate: Science, v. 282, p. 1459–1462, doi: 10.1126/SCIENCE.282.5393.1459.

Rabus, R., Hansen, T., and Widdel, F., 2000, Dissimilatory sulfate- and sulfur-reducing prokaryotes, *in* Dworkin, M., Falkow, S., Rosenberg, E., Schleifer, K.H., and Stackebrandt, E., eds., The prokaryotes: An evolving electronic resource for the microbiological community: Heidelberg, Springer Science online, http://www.springerlink.com.

Rabus, R., Brüchert, V., Amann, J., and Könneke, M., 2002, Physiological response to temperature changes of the marine, sulfate-reducing bacterium *Desulfobacterium autotrophicum*: FEMS Microbiology Ecology, v. 42, p. 409–417, doi: 10.1016/S0168-6496(02)00358-6.

Ravenschlag, K., Sahm, K., Knoblauch, C., Jørgensen, B., and Amann, R., 2000, Community structure, cellular rRNA content and activity of sulfate-reducing bacteria in marine arctic sediments: Applied and Environmental Microbiology, v. 66, p. 3592–3602, doi: 10.1128/AEM.66.8.3592-3602.2000.

Ravenschlag, K., Sahm, K., and Amann, R., 2001, Quantitative molecular analysis of the microbial community in marine Arctic sediments (Svalbard): Applied and Environmental Microbiology, v. 67, p. 387–395, doi: 10.1128/AEM.67.1.387-395.2001.

Rees, C.E., 1973, A steady state model for sulphur isotope fractionation in bacterial reduction processes: Geochimica et Cosmochimica Acta, v. 37, p. 1141–1162, doi: 10.1016/0016-7037(73)90052-5.

Rosselló-Mora, R., Thamdrup, B., Schäfer, H., Weller, R., and Amann, R., 1999, The response of the microbial community of marine sediments to organic carbon input under anaerobic conditions: Systematic and Applied Microbiology, v. 22, p. 237–248.

Rudnicki, M.D., Elderfield, H., and Spiro, B., 2001, Fractionation of sulfur isotopes during bacterial sulfate reduction in deep ocean sediments at elevated temperatures: Geochimica et Cosmochimica Acta, v. 65, p. 777–789, doi: 10.1016/S0016-7037(00)00579-2.

Rueter, P., Rabus, R., Wilkes, H., Aeckersberg, F., Rainey, F., Jannasch, H., and Widdel, F., 1994, Anaerobic oxidation of hydrocarbons in crude oil by new types of sulphate-reducing bacteria: Nature, v. 372, p. 455–458, doi: 10.1038/372455A0.

Schauder, R., Eikmanns, B., Thauer, R.K., Widdel, F., and Fuchs, G., 1986, Acetate oxidation to CO_2 in anaerobic bacteria via a novel pathway not involving reactions of the citric acid cycle: Archives of Microbiology, v. 145, p. 162–172.

Shen, Y.A., Buick, R., and Canfield, D.E., 2001, Isotope evidence for microbial sulphate reduction in the early Archaean era: Nature, v. 410, p. 77–81, doi: 10.1038/35065071.

Sørensen, J., Christensen, D., and Jørgensen, B.B., 1981, Volatile fatty acids and hydrogen as substrates for sulfate-reducing bacteria in anerobic marine sediment: Applied and Environmental Microbiology, v. 42, p. 5–11.

Steuber, J., and Kroneck, P.M.H., 1998, Desulfoviridin, the dissimilatory sulfite reductase from *Desulfuvibrio desulfuricans* (Essex): New structural and functional aspects of the membranous enzyme: Inorganica Chimica Acta, v. 275-276, p. 52–57, doi: 10.1016/S0020-1693(97)06143-4.

Strauss, H., 1999, Geological evolution from isotope proxy signals—sulfur: Chemical Geology, v. 161, p. 89–101, doi: 10.1016/S0009-2541(99)00082-0.

Thamdrup, B., Finster, K., Hansen, J.W., Bak, F., 1993, Bacterial disproportionation of elemental sulfur coupled to chemical reduction of iron and manganese: Applied and Environmental Microbiology, v. 59, p. 101–108.

Valentine, D.L., and Reeburgh, W.S., 2000, New perspectives on anaerobic methane oxidation: Environmental Microbiology, v. 2, p. 477–484, doi: 10.1046/J.1462-2920.2000.00135.X.

Werne, J.P., Sageman, B.B., Lyons, T.W., and Hollander, D.J., 2002, An integrated assessment of a "type euxinic" deposit: Evidence for multiple controls on black shale deposition in the Middle Devonian Oatka Creek Formation: American Journal of Science, v. 302, p. 110–143.

Werne, J.P., Lyons, T.W., Hollander, D.J., Formolo, M.J., and Sinninghe Damsté, J.S., 2003, Reduced sulfur in euxinic sediments of the Cariaco Basin: Sulfur isotope constraints on organic sulfur formation: Chemical Geology, v. 195, p. 159–179.

Westermann, P., 1994, The effect of incubation-temperature on steady-state concentrations of hydrogen and volatile fatty-acids during anaerobic degradation in slurries from wetland sediments: FEMS Microbiology Ecology, v. 13, p. 295–302.

White, D., 1995, The physiology and biochemistry of prokaryotes: New York, Oxford University Press, 378 p.

Wortmann, U.G., Bernasconi, S.M., and Böttcher, M.E., 2001, Hypersulfidic deep biosphere indicates extreme sulfur isotope fractionation during single-step microbial sulfate reduction: Geology, v. 29, p. 647–650.

Manuscript Accepted by the Society March 28, 2004

Printed in the USA

Geological Society of America
Special Paper 379
2004

Microbially mediated sulfur-redox: Energetics in marine hydrothermal vent systems

Jan P. Amend*
Karyn L. Rogers
D'Arcy R. Meyer-Dombard
Department of Earth and Planetary Sciences, Washington University, St. Louis, Missouri 63130, USA

ABSTRACT

Many archaea and bacteria obtain metabolic energy by catalyzing the oxidation or reduction of sulfur. In marine hydrothermal systems, chemolithoautotrophs that oxidize hydrogen sulfide (H_2S) or elemental sulfur (S^0) account for much of the primary biomass synthesis. Under reducing conditions in these systems, both S^0 and sulfate can serve as terminal electron acceptors. The energetics of chemolithoautotrophy in marine hydrothermal systems are discussed, focusing principally on sulfur-redox, but also touching on methanogenesis and organic synthesis. Examples are given from deep- and shallow-sea hydrothermal environments, the early Earth, Mars, and Europa. In addition, we present a detailed analysis of the Gibbs free energies (ΔG_r) of 25 sulfur-redox reactions in a model shallow-marine hydrothermal ecosystem—the seeps, wells, and vents of Vulcano Island (Italy). A number of these reactions represent known metabolisms, but other reactions with no known microbial catalyst are also included to investigate their potential as possible energy sources. The reactions considered couple SO_4^{2-}, S^0, and H_2S with a variety of terminal electron acceptors (O_2, NO_3^-, Fe(III), CO_2) and electron donors (H_2, CH_4, formic acid, acetic acid, propanoic acid, NH_4^+, Fe^{2+}). At all seven study sites on Vulcano, which vary considerably in temperature, pH, and chemical composition, sulfate- and S^0-reduction reactions are energy-yielding where H_2, CH_4, or carboxylic acids serve as the electron donors, but energy-consuming with NH_4^+ or Fe^{2+} as the reductant. Elemental sulfur- and sulfide-oxidation reactions are energy-yielding at all sites when O_2, NO_3^-, or Fe(III) are the terminal electron acceptors, but energy-consuming with CO_2 as the oxidant.

Keywords: thermophile metabolism, reaction energetics, microbial sulfur redox, Vulcano Island (Italy), marine hydrothermal ecosystem.

INTRODUCTION

Sulfur[1] is the sixth most abundant element (by mass) in the Earth, residing naturally in minerals (e.g., elemental sulfur, metal sulfides), gases (e.g., SO_2, H_2S), and aqueous species (e.g., SO_4^{2-}, HS^-). Sulfur is also highly redox sensitive, occurring in a variety of oxidation states from −2 in sulfide to +6 in sulfate; common intermediate oxidation states are 0 in elemental sulfur (S^0), +2 in thiosulfate, and +4 in sulfite. Because of its abundance, range of oxidation states, and chemical reactivity, S plays an important role in biogeochemical processes, and S-redox serves as the central catabolism (Table 1) in a wide variety of archaea and bacteria. The cycling of S in the open ocean, marine sediments, and continental environments is well established, and the isotopic

*Corresponding author: amend@levee.wustl.edu

[1]We use "sulfur" and "S" generically to indicate the element; S^0 is used to indicate the crystalline form of elemental sulfur.

Amend, J.P., Rogers, K.L., and Meyer-Dombard, D.R., 2004, Microbially mediated sulfur-redox: Energetics in marine hydrothermal vent systems, *in* Amend, J.P., Edwards, K.J., and Lyons, T.W., eds., Sulfur Biogeochemistry—Past and Present: Geological Society of America Special Paper 379, p. 17–34. For permission to copy, contact editing@geosociety.org.

TABLE 1. GLOSSARY OF TERMS

Metabolism	All biochemical reactions in a cell, both energy-consuming and energy-yielding.
Catabolism	The biochemical conversion of organic and inorganic compounds, generally leading to the production of energy.
Mesophile	An organism that grows at ambient temperature, generally between 15 and 45 °C.
Thermophile	Any organism that grows at high temperatures.
Moderate thermophile	An organism with an optimum growth temperature between 45 and 80 °C.
Hyperthermophile	An organism with an optimum growth temperature ≥80 °C.
Chemolithoautotroph	An organism that obtains its metabolic energy from the oxidation of inorganic compounds and its carbon for biosynthesis from CO_2 or CO.
Chemoorganoheterotroph	An organism that obtains both its metabolic energy and its carbon for biosynthesis from organic compounds.
Terminal electron acceptor	The oxidant (e.g., O_2, NO_3^-, SO_4^{2-}, CO_2) in a biologically mediated overall redox process.
Exergonic	Refers to a chemical reaction that is energy-yielding, i.e., for which $\Delta G_r < 0$.
Endergonic	Refers to a chemical reaction that is energy-consuming, i.e., for which $\Delta G_r > 0$.

fractionations, redox intermediates, specific enzymes, organisms involved, and symbiotic relationships have largely been elucidated (e.g., Bates et al., 1994; Cutter and Kluckhohn, 1999; Simo and Alio, 1999; Kasten and Jørgensen, 2000; Brüchert et al., 2001; Grossman and Desrocher, 2001; Schippers and Jørgensen, 2002; this volume). In hydrothermal systems, however, the picture is less clear. This is despite the fact that among the cultured hyperthermophiles, S-redox is a very common metabolic strategy. In fact, most non-methanogenic hyperthermophiles show enhanced growth with or even require the presence of S^0 in the growth medium, and the majority of these hyperthermophiles facilitate the oxidation of sulfide and S^0 or the reduction of sulfate, sulfite, thiosulfate, and S^0.

Based in part on the physiology of bacterial and archaeal isolates deep in the tree of life, S-redox was likely a very early mode of metabolism. Sulfur isotope evidence forms the basis for the widely accepted age of microbial sulfate-reduction as early as ~2.7 Ga (Canfield and Raiswell, 1999). Shen et al. (2001) push the origin of microbial sulfate-reduction even further back in time; they report maximum ^{34}S fractionations of 21‰ in sulfides from the ca. 3.47 Ga barites from North Pole, Australia, and interpret these data as direct evidence of microbial sulfate-reduction. This interpretation, however, is not without its critics. Runnegar et al. (2002), for example, argue for an abiotic precipitation of these barites. For a recent review of early Earth sulfur geochemistry in a microbiological context, see Lyons et al. (2004).

Sulfur-oxidation in both deep- and shallow-marine hydrothermal environments is mediated for energy-gain by diverse populations of archaea and bacteria. Beginning with a study more than two decades ago at the Galápagos Rift vent system, it was noted that the primary synthesis of biomass at deep-sea hydrothermal sites is based largely on chemolithoautotrophic oxidation of sulfide and S^0 (Karl et al., 1980). Since then, many organisms that carry out S-oxidation have been identified at hydrothermal sites. For example, members of the genus *Thiomicrospira* appear to dominate the community of S-oxidizers at the Galápagos site (Jannasch and Mottl, 1985), at deep-sea vents of the Trans-Atlantic Geotraverse and Snake Pit sites on the Mid-Atlantic Ridge (Muyzer et al., 1995), and at shallow submarine hydrothermal vents near Milos, Greece (Brinkhoff et al., 1999). Close relatives of the S-oxidizer *Thiovolum* in the ε-subclass of the Proteobacteria dominate the bacterial clone library of a microbial mat at the Loihi Seamount (Hawaii) hydrothermal vent system (Moyer et al., 1995); physiological attributes, however, could not be unambiguously assigned to these organisms. In shallow submarine thermal waters, sediments, and solfataras of the Aeolian Islands (Italy), a number of acidophilic S-oxidizers have been found, including the mesophile *Thiobacillus prosperus* (Huber and Stetter, 1989), other *Thiobacillus*-like strains (Gugliandolo and Maugeri, 1993; Gugliandolo et al., 1999), and members of *Acidianus* (Segerer et al., 1986). The thermal vents on Vulcano (Aeolian Islands) and Iceland have yielded members of the hyperthermophilic bacterium *Aquifex*, which can oxidize sulfide or S^0 aerobically at temperatures up to ~95 °C (Huber et al., 1992; Deckert et al., 1998).

Sulfur-reduction, in particular S^0- and sulfate-reduction, also represents a common catabolic strategy of microorganisms in marine hydrothermal environments. For example, several thermophilic chemolithoautotrophs, including *Pyrodictium, Thermodiscus*, and *Acidianus* (Fischer et al., 1983; Stetter et al., 1983; Segerer et al., 1986), obtain energy from the reaction

$$S^0 + H_2 \rightarrow H_2S. \quad (1)$$

A variety of thermophilic heterotrophs, including members of *Thermococcus, Pyrococcus,* and *Palaeococcus* (Zillig et al., 1983; Fiala and Stetter, 1986; Takai et al., 2000; Amend et al., 2003a), use an array of monomeric and polymeric organic compounds as electron donors to drive the reduction of S^0 to sulfide. Fiala and Stetter (1986) and Huber et al. (1986) argued that thermophilic heterotrophs do not gain energy from the process represented by reaction 1, but merely mediate it to decrease the concentration of H_2, which is known to inhibit cellular growth. That organisms would catalyze exergonic reactions without exploiting the released energy seems counter-intuitive, and in fact, Kelly and Adams (1994) showed that the facultative S^0-reducing archaeon *Pyrococcus furiosus* obtains energy from reaction 1 while simultaneously disposing of the metabolic by-product H_2.

In addition to S^0-reduction, microbial sulfate-reduction has been documented in marine hydrothermal systems, though relatively little is known about the microbial populations that carry out this process in these habitats, especially at temperatures >80 °C. In heated deep-sea sediments of Guaymas Basin, Gulf of California, very high microbial sulfate-reduction rates (up to 61 μM SO_4^{2-} per day) were measured at temperatures ≥80 °C and in fact as high as 110 °C (Fossing and Jørgensen, 1990; Gundersen et al., 1992; Elsgaard et al., 1994; Weber and Jørgensen, 2002). In shallow-marine hydrothermal sediments at Vulcano, similarly high microbial sulfate reduction rates (72 μM SO_4^{2-} per day) were observed at 90 °C (Tor et al., 2003). Nonetheless, only a few sulfate reducers have been cultured from marine hydrothermal environments (Elsgaard et al., 1995; Sievert and Kuever, 2000). Among the >130 species of sulfate-reducing archaea and bacteria that have been described to date, comprising members of four bacterial phyla and one archaeal genus (Loy et al., 2002), only three marine hyperthermophiles are known. These species are exclusively members of the genus *Archaeoglobus* within the euryarchaeota[2] and include organisms from both Guaymas Basin (*A. profundus*; Burggraf et al., 1990b) and Vulcano (*A. fulgidus*; Stetter, 1988). In addition to these isolates, *Archaeoglobus* spp. have also been identified by 16S rRNA sequence analyses of Guaymas sediments (Teske et al., 2002) and in an in situ growth chamber deployed at a deep-sea hydrothermal vent on the Mid-Atlantic Ridge (Reysenbach et al., 2000). Finally, *A. fulgidus* and *A. profundus* were isolated from high-temperature oil field formation waters in the North Sea and Alaska (Stetter et al., 1993; Beeder et al., 1994; Nilsen et al., 1996).

The vast majority of crenarchaeota, a considerable number of hyperthermophilic euryarchaeota, and several high-temperature bacteria carry the biochemical machinery required to oxidize or reduce S-bearing compounds. A principal parameter that governs the presence of active S-oxidizers and S-reducers in a thermal environment is the availability of energy sources. In this regard, it is not only vital to determine whether S-redox is exergonic, but also to quantify the amount of energy that is released in situ from specific reactions. If a reaction is endergonic, the microbial catalysis of that reaction is a moot point. However, if the reaction of interest yields energy, it can be incorporated into a framework of microbial metabolism. The next level of complexity then includes a comparison of energy-yields, first from an array of S-redox reactions and ultimately from an even larger set of reactions that includes non-sulfur-bearing terminal electron acceptors (TEAs) and electron donors. Quantifying the energetics of this larger set is beyond the scope of this chapter; here, we concentrate on evaluating overall Gibbs free energies of reaction (ΔG_r) for S-redox processes in a model system of shallow-marine hydrothermal activity. To put these calculations into context, however, we first briefly review the energetics of chemolithoautotrophy in experimentally investigated or computationally modeled hydrothermal systems. Where possible, we focus on S-redox, but we also include some discussion of redox among C-, N-, and Fe-bearing compounds.

[2]Several moderately thermophilic, nonmarine sulfate reducers are known, which include the bacteria *Thermodesulfobacterium commune* (Zeikus et al., 1983), *T. hveragerdense* (Sonne-Hansen and Ahring, 1999), and *Thermodesulfovibrio yellowstonii* (Henry et al., 1994); *T. hydrogenophilum* is a moderately thermophilic, marine sulfate reducer isolated from Guaymas Basin (Jeanthon et al., 2002).

ENERGETICS OF CHEMOLITHOAUTOTROPHY IN MARINE HYDROTHERMAL SYSTEMS

In this section, we consider the energetics of chemolithoautotrophy in a variety of hydrothermal environments, including those in the present abyssal and shallow sea, those on early Earth, and in putative systems on Mars and Europa. We do not pretend to give an exhaustive review of chemolithoautotrophy in terrestrial or extraterrestrial hydrothermal systems—past or present. Rather, we highlight examples that consider microbial processes within a geochemical framework. In many cases, a paucity of compositional data has precluded energy calculations of S-dependent metabolisms or other simple chemolithoautotrophic reactions. Nevertheless, insight can be gleaned from models of abiotic organic synthesis and energetics of chemoorganoheterotrophy, and we review these processes in several hydrothermal environments.

Deep-Sea Hydrothermal Systems

Mixing of hot, chemically reduced, slightly acid hydrothermal fluid with cold, oxidized, slightly alkaline seawater provides geochemical energy in deep-sea vent systems (Jannasch, 1985; Karl, 1995). This chemical disequilibrium, coupled with sluggish reaction kinetics for redox reactions, allows certain microorganisms to harness this energy (McCollom and Shock, 1997). Thermodynamic calculations show that in such mixing environments at 21°N on the East Pacific Rise, for example, the aerobic oxidation of H_2S, CH_4, Fe^{2+}, and Mn^{2+} is exergonic at low temperatures (<40 °C), but endergonic at high (>40 °C) temperatures (McCollom and Shock, 1997). It also was shown that the reduction of SO_4^{2-}, S^0, and CO_2 with H_2 as the electron donor is energy-consuming at low temperatures, but energy-yielding at high temperatures. In vent plumes, however, the aerobic oxidation of S^0, metal sulfides, and H_2, as well as chemolithotrophic sulfate-reduction, methanogenesis, and aerobic methanotrophy are all exergonic (McCollom, 2000).

By comparison, the energetics of chemoorganoheterotrophy at deep-sea vents have not received much attention, largely because few studies have been published that give concentrations of aqueous organic compounds. However, several studies evaluated the energetics of abiotic organic and biomolecule synthesis. In these investigations, dissolved H_2, present in the hydrothermal fluid due to high-temperature water-rock interactions, serves as the electron donor in the reduction of CO_2 (or HCO_3^-). As an example, thermodynamic computations revealed that the synthesis of 11 of the 20 common amino acids from CO_2, NH_4^+, H_2S, and H_2 is exergonic at 100 °C in a deep-sea vent mixing zone, and the synthesis of all 20 amino acids is energetically favored in a 100 °C hydrothermal fluid relative to the synthesis in cold, oxidized seawater (Amend and Shock, 1998). Further, Shock and Schulte (1998) demonstrated that as hydrothermal fluids mix with seawater, it is thermodynamically feasible for much or all of the inorganic carbon to be converted to a mixture of carboxylic acids, alcohols, and ketones. The thermodynamic drive for organic synthesis in deep-sea hydrothermal systems indicates that carbon sources required for chemoorganoheterotrophy may be produced abiotically and be subsequently available for microbial metabolism.

Shallow-Sea Hydrothermal Systems

With respect to microbial metabolism, shallow marine hydrothermal systems differ from their deep counterparts in several ways. These include their proximity to the oxygen-rich atmosphere, significantly lower hydrostatic pressures, generally lower temperatures of the vent fluid, the likely presence of photosynthesizers and associated organic matter, the impact of subaerial landmasses and, hence, the transport of organic and inorganic compounds due to weathering and human activities. However, zones of mixing between vent fluids and seawater occur in both shallow and deep marine hydrothermal systems. In addition, in both types of systems, some of the same chemotrophic organisms are found, and energy-yielding, but kinetically inhibited redox reactions are required for organisms to thrive. Arguably the best-studied site of hyperthermophily is the shallow vent system of Vulcano, with approximately one-third of all known hyperthermophilic genera identified in culturing studies. It was shown recently that in the beach sediments and submarine vent fluids at Vulcano, exergonic reactions include a wide variety of TEAs (Amend et al., 2003b). Reactions with O_2, NO_3^-, and Fe(III) release by far the most energy per electron transferred, and large differences in ΔG_r (up to ~60 kJ/mol e^-) for Fe-redox reactions were noted. These differences were due predominantly to variations in the in situ concentrations of Fe^{2+}, H^+, and H_2, and not due to the differences in temperature (up to ~45 °C). The energetics of S-dependent chemolithotrophy and chemoorganotrophy at Vulcano are discussed in the section "Energetics of Sulfur-Redox at Vulcano: A Case Study of Shallow Marine Vents" below.

Early Earth

Models of Earth's history and evolution show that marine hydrothermal systems were likely more numerous and dynamic on early Earth than they are today. Such systems have probably been populated by microorganisms for billions of years (Rasmussen, 2000), making them the oldest continuously inhabited ecosystems on Earth (Reysenbach and Shock, 2002). Much of the attention regarding hydrothermal environments in the Hadean (>3.8 Ga) and Early Archean (3.8–3.4 Ga) is linked to origin of life hypotheses, and several recent publications deal at length with this topic. For a treatise on the emergence of chemolithoautotrophic life from abiotic geochemistry in Hadean hydrothermal systems, see Martin and Russell (2003). Shock et al. (2000) provide a comparison of two of the most influential origin of life theories—one that forces the emergence of heterotrophic life from an organic-rich soup under the cover of a reducing atmosphere (Oparin, 1924; Haldane, 1929; Oparin, 1936; Miller, 1957; Miller and Urey, 1959; Lazcano and Miller, 1996; Miller et al., 1997; Lazcano and Miller, 1999; Miyakawa et al., 2002), and the other that leads to the conclusion that the first organisms were autotrophic and a natural consequence of ordinary geologic forces and inescapable chemical disequilibrium in marine hydrothermal systems (Baross and Hoffman, 1985; Wächtershäuser, 1988, 1990, 1992; Shock et al., 1995; Russell and Hall, 1997; Huber and Wächtershäuser, 1998; Shock et al., 1998; Wächtershäuser, 1998; Huber et al., 2003). It is worth reiterating that not only is the chemical pathway uncertain that led from an abiotic planet to one that harbors a brilliant diversity of life forms, but the timing of this event remains in question. A window of time that seemed, until recently, to be securely bracketed by the age of the Earth at 4.55 Ga (Brown and Mussett, 1993) and the age of the earliest bona fide fossils at 3.465 Ga (Schopf et al., 2002), has now widened with the controversy regarding the evidence of these oldest fossils (Brasier et al., 2002; Pasteris and Wopenka, 2002).

Owing to the dearth of requisite compositional data in early Earth hydrothermal environments, our summary is limited in this section to studies of energy budgets for organic synthesis. An important control on the amount of chemical energy available for organic synthesis in mixing zones of hydrothermal systems, past or present, is the composition of the host rock (Shock and Schulte, 1998). However, the mineralogy of the Hadean and Early Archean oceanic crust is only poorly constrained. Sizable uncertainties also pertain for early Earth seawater and atmospheric compositions, both of which are incorporated in free energy calculations. A reasonable assumption is that the Hadean and Early Archean oceanic crust consisted of ultramafic rocks, which, when reacted with seawater, could produce strongly reduced hydrothermal vent fluids (Wetzel and Shock, 2000). The oxygen-deficient atmosphere would have further depressed the oxidation state of seawater and hydrothermal fluid. Under such postulated conditions in the Hadean, synthesis of simple organic compounds from the reduction of CO_2 would have been far more favorable than under present conditions, which feature predominantly basalt-hosted hydrothermal systems and an oxygen-rich atmosphere (Shock and Schulte, 1998). Similar conclusions were reached for amino acid synthesis energetics in putative hydrothermal systems on early Earth (Amend and Shock, 2000). It was shown that the synthesis of all 20 protein-forming amino acids (at concentrations of 10^{-5} M) was exergonic in a model Hadean hydrothermal system at 100 °C and 250 bar. Again, the reducing potential of the model hydrothermal fluid, due to seawater reactions with hot olivine gabbro (McCollom and Shock, 1998), was principally responsible for the favorable energetics of organic synthesis.

Mars and Europa

The Earth provides the only irrefutable evidence of life in our solar system. To date, the search for signs of extraterrestrial life has focused on Mars and the Jovian satellite Europa. On these extraterrestrial bodies, hydrothermal systems may have once existed (or may still exist) (Farmer, 1996; Newsom et al., 1999; Chyba, 2000; Greenberg and Geissler, 2002; Rathbun and Squyres, 2002), and, like on early Earth, the microbial catalysis of redox reactions among S-bearing compounds seems plausible. It is generally hypothesized that the putative extraterrestrial life is unicellular and carbon-based, requiring liquid water and geochemical energy sources. Evidence that liquid water existed on the surface of Mars some time in its history is mounting, as is the evidence for present or past subsurface ice. For example, NASA's Mars Orbital Laser Altimeter (MOLA) on the Mars Global Surveyor Mission (MGS) revealed high-resolution topographic data suggesting that the Martian highlands have undergone extensive fluvial resurfacing, particularly in the Margaritifer Sinus region (Hynek and Phillips, 2001). This area, located near the eastern end of Valles Marineris, features well-preserved valleys and channels, which provide strong evidence of past surface water, perhaps due to precipitation-recharged groundwater sapping (Carr and Chuang, 1997; Grant, 2000; Grant and Parker, 2002). Furthermore, the Thermal Emission Spectrometer (TES) on MGS detected gray crystalline hematite (Fe_2O_3) in Meridiani Planum as well as in several minor deposits in other regions (Christensen et al., 2000). The formation of hematite on Earth usually requires the presence of liquid water, and the Meridiani Planum formation is hypothesized to have accumulated in an ancient, subaqueous environment (Edgett and Parker, 1997).

Results from recent missions to Mars support the view that Mars was once wet. In particular, data obtained by NASA's Mars Exploration Rover *Opportunity* at Meridiani Planum have corroborated this hypothesis with analyses of Martian rocks with high sulfate salt contents and hematite nodules, which were almost certainly deposited in a shallow lake environment (Arvidson, 2004; Morris et al., 2004; Squyres, 2004). These new data indicate that the pertinent question is no longer *if* liquid water existed on the surface of Mars, but rather *how much* and *when*. Indeed, various precipitation events, groundwater, and surface water (both liquid and frozen) may have played a large role in shaping the surface of early Mars and in providing putative habitable environments.

In addition to apparent sources of surface and subsurface water, Mars exhibits morphological evidence of heat sources, many of which occur in association with evidence for liquid water (Brakenridge et al., 1985; Gulick and Baker, 1990; Farmer, 1996). A likely consequence of these concurrent events is the formation of hydrothermal systems, which on Mars could have resulted from the interaction of groundwater or subsurface ice with magmatic intrusions (Gulick, 1998), or due to hydrothermal convection in crater-lakes driven by the thermal anomaly produced by impact (Rathbun and Squyres, 2002).

Europa, the second Galilean satellite of Jupiter, has potential hydrothermal systems as well. Magnetometer data from NASA's Galileo probe have indicated the presence of a liquid water ocean beneath Europa's icy crust, and tidal dissipation in Europa's rocky core due to shared orbital resonance with its sister satellites Io and Ganymede may lead to hydrothermal heating at the water-rock interface (Chyba, 2000; Greenberg and Geissler, 2002). Fluid mixing in postulated hydrothermal systems may provide (or have provided) the geochemical energy sources for primary biomass synthesis and perhaps chemolithoautotrophy on Europa as well as Mars.

Both Mars and Europa have been the focus of geochemical energy modeling in recent years. McCollom (1999) identified potential energy sources for autotrophs in a postulated Europan hydrothermal system, showing that methanogenesis from CO_2 and H_2 would be exergonic regardless whether the Europan ocean is reduced and methane-rich or oxidized and sulfate- and bicarbonate-rich. In certain geochemical scenarios, sulfate-reduction would also supply sufficient energy to support microbial metabolism. This view, however, is counter to that of Gaidos et al. (1999), who argue that a lack of oxidants in the Europan ocean would severely minimize the chances of diverse life surrounding hydrothermal systems. They further note that Fe(III)-reduction might support a simple community of microorganisms, but methanogens, sulfate reducers, and aerobic chemolithoautotrophs are unlikely to thrive on Europa. It is worth reiterating that McCollom (1999) does not envision a dense biota surrounding the hydrothermal vents on Europa, nor a complex community structure, but merely concludes that geochemical energy sources could support the emergence and persistence of life in localized ecosystems. Similarly low, but nevertheless noteworthy energy yields were also computed by Jakosky and Shock (1998), who inventoried the amount of geochemical energy from volcanic activity and mineral weathering reactions in model Martian and Europan hydrothermal systems. They found that energy was sufficient on Mars for life to have emerged, but also concluded that life is not now, and probably never was, ubiquitous on Mars or Europa. More optimistic about the biological potential of Mars is a recent study by Varnes et al. (2003), which asserts that substantial geochemical energy may be available in Martian hydrothermal systems, depending on the mineral composition of the host rock.

ENERGETICS OF SULFUR-REDOX AT VULCANO: A CASE STUDY OF SHALLOW MARINE VENTS

Pyrodictium occultum emerged from a shallow-sea hydrothermal vent field at Vulcano as the first organism in pure culture to grow optimally at temperatures >100 °C (Stetter, 1982; Stetter et al., 1983). Since then, a number of other archaea that can grow at these temperatures have been cultured and characterized. They include *Aeropyrum pernix; Caldococcus litoralis; Hyperthermus butylicus; Methanopyrus kandleri;* several members of *Pyrobaculum, Pyrococcus, Pyrodictium*, and *Thermococcus;*

Pyrolobus fumarii; Stetteria hydrogenophila; Thermofilum pendens; Thermoproteus uzoniensis; and most recently, strain 121 with a maximum growth temperature of 121 °C (Stetter et al., 1983; Zillig et al., 1983; Fiala and Stetter, 1986; Huber et al., 1987; Svetlitshnyi et al., 1987; Zillig et al., 1987; Huber et al., 1989; Bonch-Osmolovskaya et al., 1990; Zillig et al., 1990; Kurr et al., 1991; Pledger and Baross, 1991; Pley et al., 1991; Erauso et al., 1993; Völkl et al., 1993; Sako et al., 1996; Blöchl et al., 1997; Jochimsen et al., 1997; Gonzalez et al., 1998; Kashefi and Lovley, 2003). Like *P. occultum*, several of these hyperthermophiles hail from the hydrothermal seeps and vents of Vulcano. In light of the hyperthermophile diversity documented there—including, by extension, the metabolic diversity—we chose to evaluate the energetics of a number of redox reactions at in situ geochemical conditions. We can regard the seeps, wells, and vents at Vulcano as a model system for shallow-sea hydrothermal sites. Other shallow marine vent environments are known off Ambitle and Lihir Islands, Papua New Guinea (Pichler and Dix, 1996; Pichler et al., 1999a; Pichler and Veizer, 1999; Pichler et al., 1999b); near Milos, Greece (Brinkhoff et al., 1999; Sievert et al., 1999; Stuben and Glasby, 1999; Sievert and Kuever, 2000; Wenzhofer et al., 2000); at Bahia Concepcion and Punta Mita, Mexico (Prol-Ledesma, 2003; Alfonso, et al., 2003); on the Mid-Atlantic Kolbeinsey Ridge, north of Iceland (Burggraf et al., 1990a; Kurr et al., 1991; Botz et al., 1999); and near the Aleutian Islands, Alaska (T. Pichler, 2003, personal commun.), to name only a few.

The energetics of 90 chemolithoautotrophic reactions in the H-O-N-S-C-Fe chemical system at Vulcano are discussed at length in Amend et al. (2003b); here, we reconsider several of the most important S-redox reactions and also compute values of ΔG_r for chemoorganoheterotrophic reactions in which carboxylic acids serve as the electron donors. Despite the ubiquity of thermophilic heterotrophs, few studies have focused on the composition of dissolved organic carbon in hydrothermal systems (Amend et al., 1998). Twenty-five different autotrophic and heterotrophic reactions, divided into four groups, are taken into account here: sulfate-reduction, S^0-reduction and -disproportionation, S^0-oxidation, and sulfide-oxidation. Nine of the 25 reactions are listed twice, once as the forward and once as the reverse reaction. Consequently, a total of 34 reactions are tabulated. The amount of energy yielded or consumed by a reaction (ΔG_r) can be computed from values of the standard Gibbs free energy of a reaction at the temperature and pressure of interest ($\Delta G_r°$) and activities derived from in situ chemical compositions. It should be pointed out that the thermodynamic calculations are based on the compositions of the mixed hydrothermal solutions and not on an end-member vent fluid that gets diluted by ambient seawater. As noted above, the mixing of two chemically distinct aqueous solutions with sluggish reaction kinetics commonly provides the chemical energy in marine hydrothermal systems, and it is in fact this stored energy that we are quantifying. The method to calculate ΔG_r for the S-redox reactions is discussed below, but values of $\Delta G_r°$ required in these calculations are obtained from Amend and Shock (2001).

Known and Unknown Microbial S-Redox Reactions

Numerous dissimilatory S-redox processes are known that provide metabolic energy to archaea and bacteria. A second group of S-redox reactions, which are currently not known to be utilized by any microorganisms, can also be considered. Below, we compute the energetics of both known and unknown reactions under the geochemical conditions that obtain at Vulcano. An evaluation of the energetics of the second group of reactions may aid geomicrobiologists in identifying other potential metabolisms and in designing culturing protocols to isolate novel S-reducers and S-oxidizers.

A variety of anaerobes use sulfate or S^0 as a TEA with low molecular weight organic compounds or H_2 as electron donors. For example, members of *Archaeoglobus, Desulfotomaculum, Desulfacinum,* and *Thermodesulfobacterium* can grow chemolithotrophically on H_2 plus sulfate; chemolithotrophic S^0-reduction with H_2 as electron donor is carried out, for example, by *Pyrodictium, Acidianus, Thermoproteus, Aquifex, Desulfurella, Hyperthermus,* and *Stetteria*. In addition, *Desulfocapsa* and *Desulfobulbus* can harness metabolic energy by disproportionating S^0. The majority of sulfate reducers are organotrophs, commonly utilizing carboxylic acids as electron donors. Examples of organisms that oxidize formic, acetic, or propanoic acid include *Desulfovibrio, Desulfotomaculum, Desulfococcus, Desulfobacterium,* and *Archaeoglobus*. It has also been shown that anaerobic methane oxidation is coupled to sulfate-reduction, catalyzed, most likely, by a microbial consortium that includes a methanogen operating in reverse (as a methanotroph) and a sulfate reducer (Hinrichs et al., 1999). Other microorganisms couple the oxidation of organic acids to S^0-reduction; these include members of the *Thermoproteales, Sulfurospirillum, Desulfuromonas, Geobacter,* and *Desulfurella*.

Aerobic as well as anaerobic S-oxidizers thrive in acidic and circumneutral waters, both in marine and nonmarine ecosystems. For example, members of *Thiobacillus, Acidianus, Aquifex, Metallosphaera, Sulfolobus, Sulfobacillus, Beggiatoa, Thiovolum,* and *Thiomicrospira* can oxidize H_2S and/or S^0 with O_2 as the TEA. Further, some members of *Thiobacillus, Thioploca, Aquifex, Ferroglobus,* and *Thermothrix* gain energy by coupling nitrate-reduction to S-oxidation. However, anaerobic S-oxidation is not limited to nitrate reducers; members of *Thiobacillus*, for example, catalyze the oxidation of S^0 with Fe(III) as the TEA.

In addition to known S-redox reactions just highlighted, we also investigate the energetics of as yet unknown sulfur metabolisms. For example, in the "Sulfate Reduction" section below, we compute values of ΔG_r for unknown incomplete sulfate-reduction reactions, ones that terminate in S^0 instead of H_2S. In the "Sulfate-Reduction" and "S^0-Reduction and S^0-Disproportionation" sections, we also calculate the energetics of sulfate- and S^0-reduction reactions in which NH_4^+ and Fe^{2+} serve as electron donors. Lastly, unknown S^0- and sulfide-oxidation reactions are discussed in the "S^0-Oxidation" and "Sulfide-Oxidation" sections, respectively, where we evaluate ΔG_r for reactions with CO_2, NO_3^-, and Fe(III) as TEAs.

Methods

Sampling and Chemical Analyses

For a detailed discussion of sampling procedures and water and gas analyses, see Amend et al. (2003b). Briefly, water samples from several submarine vents, sediment seeps, and geothermal wells on Vulcano were analyzed in the field or preserved for subsequent analysis. Temperature, pH, and conductivity were measured in situ with hand-held meters and probes. The redox-sensitive compounds Fe^{2+}, NO_3^-, NH_4^+, H_2S, and dissolved oxygen were analyzed by spectrophotometry in the field. Concentrations of major inorganic cations and anions as well as carboxylate anions were determined on an ion chromatograph equipped with an electrochemical detector. The chemical composition of free gases was determined by gas chromatography using hot wire and flame ionization detectors placed in series.

Dissolved Organic Carbon

Water samples for dissolved organic carbon (DOC) were collected in all-glass bottles, immediately poisoned with an aliquot of $HgCl_2$ (0.2 µM final concentration) to kill all microorganisms, and transported to the geochemistry laboratory at the Marcello Carapezza Center on Vulcano. There, samples were filtered under low vacuum in glass filtration units (Millipore) with GF/F filters. Each filtered sample was collected in a second all-glass bottle, refrigerated, and shipped cold (~4 °C) to the United States, where each sample was frozen (–20 °C) until analysis. All handling of samples, containers, and gear for DOC analysis was carried out with gloved hands. All glassware was muffled for 6 h at 500 °C; Whatman GF/F filters were muffled for 6 h at 450 °C.

Samples were analyzed with a nondispersive infrared detector on an Apollo 9000HS combustion TOC Analyzer (Tekmar). Prior to analysis, samples were acidified (pH <3) with phosphoric acid, converting (bi)carbonate to CO_2. Air was bubbled through the sample to purge the CO_2. The remaining carbon in the sample (nonpurgeable organic carbon) was converted to CO_2 by combustion at 680 °C with a platinum catalyst.

Calculating ΔG_r

Values of ΔG_r were computed with the expression

$$\Delta G_r = \Delta G_r^\circ + RT \ln Q_r, \quad (2)$$

where ΔG_r and ΔG_r° are as defined above, R and T stand for the gas constant and temperature in Kelvin, respectively, and Q_r denotes the activity product. The approach used to calculate values of ΔG_r° and Q_r for the reactions of interest have been published elsewhere (e.g., McCollom and Shock, 1997; Amend and Shock, 2001; Amend et al., 2003b). Briefly, values of ΔG_r° can be computed with equations of state and thermodynamic data for minerals, gases, and aqueous species given in Helgeson et al. (1978, 1981), Shock and Helgeson (1988, 1990), Tanger and Helgeson (1988), Shock et al. (1989, 1992), Shock and Koretsky (1993, 1995), Shock (1995), and McCollom and Shock (1997). Evaluating Q_r requires activities and/or fugacities of the reactants and products at the environmental conditions of interest, which can be computed from chemical analyses using the speciation program EQ3 (Wolery, 1992). This program explicitly accounts for activity coefficients and complex formation. In the present study, redox reactions among S-, C-, N-, Fe-, H-, and O-bearing compounds were suppressed, thus preserving geochemical disequilibria.

Concentrations and Activities of the Compounds of Interest

Field and analytical data of waters from seeps, vents, and wells at Vulcano are given in Table 2. A map depicting the locations of the seven sampling sites is shown in Figure 1. The temperature at five of the seven sites is >80 °C; at Grip and Pozzo Istmo, the temperature is ~55 °C. The pH is acid at all seven sites, but ranges considerably from 1.98 at Pozzo Vasca to 5.84 at Pozzo Istmo. It can be deduced from the conductivity measurements (Table 2) and the concentrations of major cations and anions (Amend et al., 2003b) that four of the hydrothermal solutions (at Stinky Surf Rock, Grip, Acque Calde 2, and Pozzo Istmo) are dominated by a marine end member; the other three solutions (Punto 1, Punto 7, Pozzo Vasca) are characterized by a substantially larger contribution from a fresh water end member. Particularly large variations among the sites are seen in concentrations of Fe^{2+}, NO_3^-, and H_2. For example, Fe^{2+} ranges from 0.02 ppm at Pozzo Istmo to ~300 ppm at Punto 7 and Pozzo Vasca. Nitrate concentrations vary from <2 ppm at Stinky Surf Rock and Grip to 200 ppm at Acque Calde 2, and H_2 is at a minimum of 8.1 ppmv at Pozzo Istmo and at a maximum of almost 20,000 ppmv at Acque Calde 2. Concentrations of carboxylate anions also vary significantly across the sites (0.05–1.14 ppm for formate, 0.09–1.86 ppm for acetate, and 0.06–1.26 ppm for propanoate). In general, the maximum concentrations of the organic compounds are slightly higher than previously reported (Amend et al., 1998). DOC concentrations were determined in five of the seven samples and ranged from <1 ppmC at Pozzo Istmo (0.41) and Acque Calde 2 (0.86) to >20 ppmC at Pozzo Vasca (21.36) and Punto 7 (24.0). For comparison, DOC in local surface seawater at Vulcano was 0.83 ppmC. As shown below, the differences in chemical composition, and hence the variations in activities (Table 3), have a significant effect on the range of ΔG_r values from one site to another.

Sulfate-Reduction

Fourteen sulfate-reduction reactions are given in Table 4A, including sulfate-reduction coupled to formate-oxidation (reaction 4.3), which is known to support growth of the Vulcano hyperthermophile *A. fulgidus* (Stetter et al., 1987; Stetter, 1988). In each of reactions 4.1–4.7, eight electrons are transferred from the electron donor to SO_4^{2-}, which is reduced to H_2S in the process; the electron donors are H_2, CH_4, formic, acetic, and

TABLE 2. FIELD AND ANALYTICAL DATA OF SEVEN HYDROTHERMAL FLUIDS AT VULCANO

	Punto 1	Punto 7	Stinky Surf Rock	Grip	Acque Calde 2	Pozzo Istmo	Pozzo Vasca
Sample type	S	S	S	V	V	W	W
Latitude (38°N)	25'6.66"	25'4.21"	25'6.39"	25'6.34"	24'59.03"	25'3.41"	24'59.37"
Longitude (14°E)	57'33.63"	57'33.17"	57'35.51"	57'35.68"	57'36.69"	57'31.42"	57'34.28"
T (°C)	88.9	81.0	84.8	54.0	87.5	56.4	81.8
pH	3.561	2.717	3.673	5.205	5.650	5.840	1.976
Conductivity	6.32	17.6	118	53.4	86.9	96.0	24.3
Fe^{2+}	4.0	309	7.7	12.1	0.7	0.02	293
NH_4^+	55	5	12	10.6	20.6	3.31	30
SO_4^{2-}	1546	3646	2901	2268	2624	3013	5976
NO_3^-	6	4	1.6	1.9	200	38	34
$HCOO^-$	0.14	1.14	0.18	0.05	0.06	0.07	0.18
CH_3COO^-	0.24	1.14	1.86	0.30	0.27	0.09	1.45
$C_2H_5COO^-$	0.07	0.06	0.11	0.11	–	1.26	0.41
$H_2S(aq)$	5.1	8.7	0.75	10.4	9.3	12.8	4.2
$O_2(aq)$	700	380	643	–	2100	382	406
$H_2(g)$	1812	221	1068	2374	19813	8.1	2909
$CH_4(g)$	1228	5568	1253	1497	1289	1104	1215
$CO_2(g)$	98.90	70.66	98.83	98.64	98.84	99.10	99.16
DOC	6.02	24.0	–	–	0.86	0.41	21.36

Note: Concentrations of aqueous species are in ppm (or ppmC for DOC), except for dissolved oxygen, which is in ppb. H_2 and CH_4 are in ppmv of the dry gas, and CO_2 is in vol% of the dry gas. Conductivity is in mS/cm. S—seep from heated sediment; V—shallow submarine vent; W—geothermal well; – indicates concentration not measured or not detected.

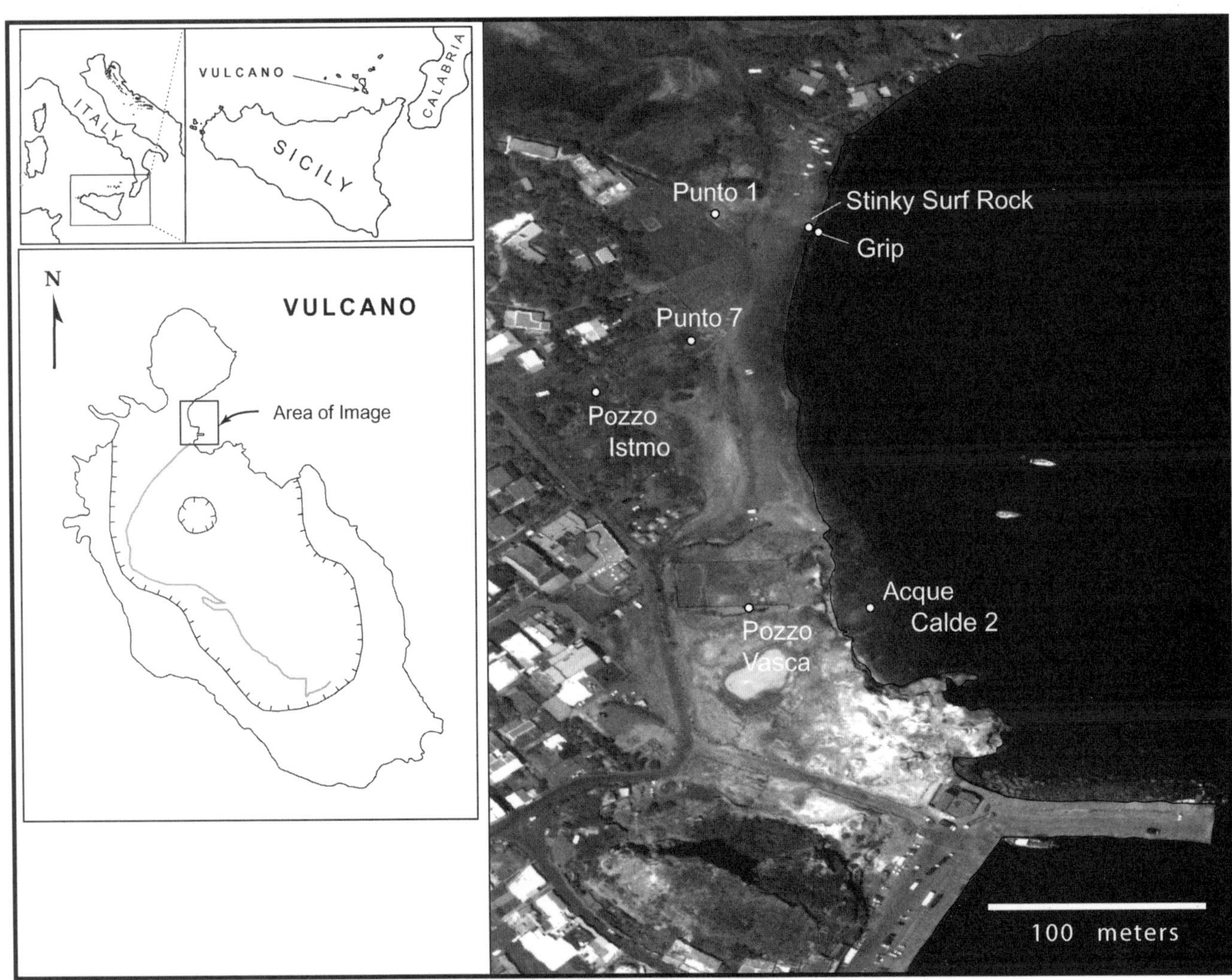

Figure 1. Aerial photograph of Baia di Levante on Vulcano (see inset map). Seven sampling sites are indicated by white circles, with latitudes and longitudes of all sites given in Table 2.

TABLE 3. LOG ACTIVITIES OF AQUEOUS SPECIES USED IN CALCULATIONS OF ΔG_r FOR REDOX REACTIONS AT SEVEN VULCANO SITES

	Punto 1	Punto 7	Stinky Surf Rock	Grip	Acque Calde 2	Pozzo Istmo	Pozzo Vasca
H^+	–3.561	–2.717	–3.673	–5.205	–5.650	–5.840	–1.976
$H_2O(l)$[a]	0.997	0.997	0.982	0.984	0.983	0.980	0.996
Fe^{2+}	–4.63	–2.78	–4.62	–4.35	–5.66	–7.18	–2.81
NH_4^+	–2.66	–3.72	–3.42	–3.45	–3.18	–3.97	–2.95
SO_4^{2-}	–2.52	–2.25[b]	–2.69	–2.65	–2.73	–2.61	–2.33[b]
NO_3^-	–4.14	–4.33	–4.80	–4.70	–2.70	–3.41	–3.40
HCOOH(aq)	–5.75	–4.65	–5.78				–5.42
$HCOO^-$				–6.28	–6.22	–6.22	
CH_3COOH(aq)	–5.42	–4.73	–4.57				–4.62
CH_3COO^-				–5.68	–5.72	–6.16	
C_2H_5COOH(aq)	–6.02	–6.11	–5.87				–5.26
$C_2H_5COO^-$				–6.21		–5.10	
H_2S(aq)	–3.81	–3.58	–4.65	–3.52	–3.64	–3.50	–3.90
O_2(aq)	–4.66	–4.92	–4.68	–4.67[c]	–4.17	–4.89	–4.89
H_2(aq)	–6.21	–7.01	–6.37	–5.82	–5.14	–8.30	–5.89
CH_4(aq)	–6.32	–5.53	–6.23	–5.89	–6.27	–6.04	–6.21
CO_2(aq)	–2.29	–2.13	–2.20	–1.82	–2.26	–1.83	–2.15

[a] Values represent the activities, not log values.

[b] HSO_4^- dominates, but in the energy calculations the activities of SO_4^{2-} were used.

[c] Computed assuming the same concentration as reported for Stinky Surf Rock.

propanoic acid, NH_4^+, and Fe^{2+}. In reactions 4.8–4.14, sulfate is reduced to S^0, a six-electron transfer. Values of ΔG_r for the reactions in Table 4A are given in Table 4B for each of the seven sites considered. In Figure 2, values of ΔG_r are depicted as a function of reaction number.

It can be seen in Table 4B and Figure 2 that all reactions in which H_2, CH_4, or carboxylic acids serve as electron donors (4.1–4.5, 4.8–4.12) are exergonic at all seven sites, with values of ΔG_r ranging from –9 to –208 kJ/mol SO_4^{2-}. Reactions in which NH_4^+ or Fe^{2+} act as electron donors (4.6, 4.7, 4.13, 4.14) are strongly endergonic, with values of ΔG_r between +112 and +540 kJ/mol SO_4^{2-}. In terms of per mole of sulfate reduced, reactions to H_2S are more exergonic than reactions to S^0. The largest energy yield (170–208 kJ/mol SO_4^{2-}) is observed from reaction (4.3) at Punto 1, Punto 7, Stinky Surf Rock, and Pozzo Vasca. In this reaction, formic acid serves as the electron donor. The range of ΔG_r values at the different sites for most of the sulfate-reduction reactions is relatively large; with only two exceptions (reactions 4.6 and 4.13), the range is >50 kJ/mol, commonly it is ~100 kJ/mol, and in one case (reaction 4.14), even >200 kJ/mol. At times, it is advantageous to normalize values of ΔG_r per electron transferred ($\Delta G_r/e^-$), because this enables a more direct comparison of redox reaction energetics (Amend et al., 2003b). When normalized in this way, analogous reactions in which SO_4^{2-} is reduced to H_2S or to S^0 (reactions 4.3 and 4.10) have very similar energy yields. For example, $\Delta G_{4.3}/e^-$ ranges from –8 to –26 kJ/mol e^-, and $\Delta G_{4.10}/e^-$ is between –6 and –25 kJ/mol e^-.

It should be pointed out that at sites where the pH is only slightly acid (5–6), the carboxylate anions formate, acetate, and propanoate dominate over formic, acetic, and propanoic acid. Therefore, at Grip, Acque Calde 2, and Pozzo Istmo, values of ΔG_r for reactions 4.3–4.5 and 4.10–4.12 in Table 4B are not for the reactions as written, but for their counterparts, which consider the carboxylate anion instead of the acid. This is perhaps best illustrated with an example. At Punto 1, Punto 7, Stinky Surf Rock, and Pozzo Vasca, where the pH is between 2.0 and 3.7, sulfate-reduction coupled to formic acid-oxidation can be represented by

$$SO_4^{2-} + 4HCOOH + 2H^+ \rightarrow H_2S + 4CO_2 + 4H_2O. \quad (3)$$

At Grip, Acque Calde 2, and Pozzo Istmo, the corresponding reaction is instead

TABLE 4A. SULFATE-REDUCTION REACTIONS

	Reaction
4.1	$SO_4^{2-} + 4H_2 + 2H^+ \rightarrow H_2S + 4H_2O$
4.2	$SO_4^{2-} + CH_4 + 2H^+ \rightarrow H_2S + CO_2 + 2H_2O$
4.3	$SO_4^{2-} + 4CHOOH + 2H^+ \rightarrow H_2S + 4CO_2 + 4H_2O$
4.4	$SO_4^{2-} + CH_3COOH + 2H^+ \rightarrow H_2S + 2CO_2 + 2H_2O$
4.5	$SO_4^{2-} + 4/7C_2H_5COOH + 2H^+ \rightarrow H_2S + 12/7CO_2 + 12/7H_2O$
4.6	$SO_4^{2-} + NH_4^+ \rightarrow H_2S + NO_3^- + H_2O$
4.7	$SO_4^{2-} + 12Fe^{2+} + 12H_2O \rightarrow H_2S + 4Fe_3O_4 + 22H^+$
4.8	$SO_4^{2-} + 3H_2 + 2H^+ \rightarrow S^0 + 4H_2O$
4.9	$SO_4^{2-} + 0.75CH_4 + 2H^+ \rightarrow S^0 + 0.75CO_2 + 2.5H_2O$
4.10	$SO_4^{2-} + 3CHOOH + 2H^+ \rightarrow S^0 + 3CO_2 + 4H_2O$
4.11	$SO_4^{2-} + 0.75CH_3COOH + 2H^+ \rightarrow S^0 + 1.5CO_2 + 2.5H_2O$
4.12	$SO_4^{2-} + 3/7C_2H_5COOH + 2H^+ \rightarrow S^0 + 9/7CO_2 + 16/7H_2O$
4.13	$SO_4^{2-} + 0.75NH_4^+ + 0.5H^+ \rightarrow S^0 + 0.75NO_3^- + 7/4H_2O$
4.14	$SO_4^{2-} + 9Fe^{2+} + 8H_2O \rightarrow S^0 + 3Fe_3O_4 + 16H^+$

TABLE 4B. VALUES OF ΔG_r (kJ/mol) FOR REACTIONS IN TABLE 4A AT SEVEN VULCANO HYDROTHERMAL SITES

Reaction	Punto 1	Punto 7	Stinky Surf Rock	Grip	Acque Calde 2	Pozzo Istmo	Pozzo Vasca
4.1	−101.09	−94.36	−101.52	−102.29	−99.74	−30.84	−135.92
4.2	−61.91	−76.49	−64.18	−32.97	−30.40	−24.39	−83.88
4.3	−170.85	−207.80	−170.83	−87.65*	−77.21*	−64.97*	−199.31
4.4	−111.53	−123.27	−117.99	−68.23*	−71.98*	−53.81*	−136.28
4.5	−115.78	−122.63	−117.07	−74.99*		−69.41*	−138.16
4.6	412.89	421.12	410.35	427.89	429.51	438.48	420.29
4.7	508.14	458.62	496.07	368.62	367.69	514.01	539.46
4.8	−69.80	−69.98	−65.58	−70.01	−62.30	−14.32	−101.87
4.9	−40.41	−56.58	−37.58	−18.02	−10.30	−9.49	−62.84
4.10	−115.20	−148.28	−110.69	−52.78*	−38.48*	−33.64*	−142.59
4.11	−70.71	−84.89	−71.07	−38.21*	−34.55*	−25.26*	−95.32
4.12	−73.90	−84.40	−70.37	−43.28*		−36.96*	−96.73
4.13	315.68	316.63	318.32	327.62	334.64	337.67	315.29
4.14	276.04	261.84	269.32	136.45	112.70	228.49	344.24

*Values of ΔG_r correspond to the reaction in which the carboxylate anion, not the carboxylic acid, is considered (see text).

$$SO_4^{2-} + 4HCOO^- + 6H^+ \rightarrow H_2S + 4CO_2 + 4H_2O. \quad (4)$$

By the same argument, S^0-reduction reactions (see below) at Grip, Acque Calde 2, and Pozzo Istmo consider the carboxylate anion, but at Punto 1, Punto 7, Stinky Surf Rock, and Pozzo Vasca, they consider the protonated form. In each case, the number of protons is adjusted accordingly on the left hand side of the reactions.

It should be pointed out that reactions 4.3–4.5 and 4.10–4.12 are less energy-yielding (or less energy-consuming) at Grip, Acque Calde 2, and Pozzo Istmo than at Punto 1, Punto 7, Stinky Surf Rock, and Pozzo Vasca. In other words, the chemoorganotrophic reactions considered here are thermodynamically more favorable at strongly acid sites than at weakly acid sites. If we again consider reaction 4.3 as an example, we see in Table 4B and Figure 2 that values of $\Delta G_{4.3}$ are ~110 kJ/mol less negative at Grip, Acque Calde 2, and Pozzo Istmo than at Punto 1, Punto 7, Stinky Surf Rock, and Pozzo Vasca. Although these thermodynamic calculations show that more energy is released from sulfate-reduction at lower pH, it has not yet been demonstrated whether microbial sulfate-reduction actually occurs under conditions of low pH and high temperature. Sulfate-reduction has been measured at low pH (Kühl et al., 1998), but little is known about the effect of pH on the composition of sulfate-reducing microbial populations and what species of sulfate reducers are active under acidic conditions. In a recent study (Küsel et al., 2001), a sulfate-reducing bacterium with a pH growth optimum of 5.5 was isolated from an acidic environment. This bacterium continued to reduce sulfate at a pH value as low as 4.9. Although this finding suggests that sulfate reducers may be present in low pH habitats at mesophilic temperatures, nothing is known about the presence and identity of sulfate reducers in low-pH/high-temperature environments. In addition, no sulfate reducer has been described to date that can be considered a true acidophile (i.e., with a pH optimum <4).

S^0-Reduction and S^0-Disproportionation

Pyrodictium occultum obtains energy by catalyzing the reduction of S^0 to H_2S; H_2 serves as the electron donor. This is one of seven S^0-reduction reactions, in addition to one S^0-dispropor-

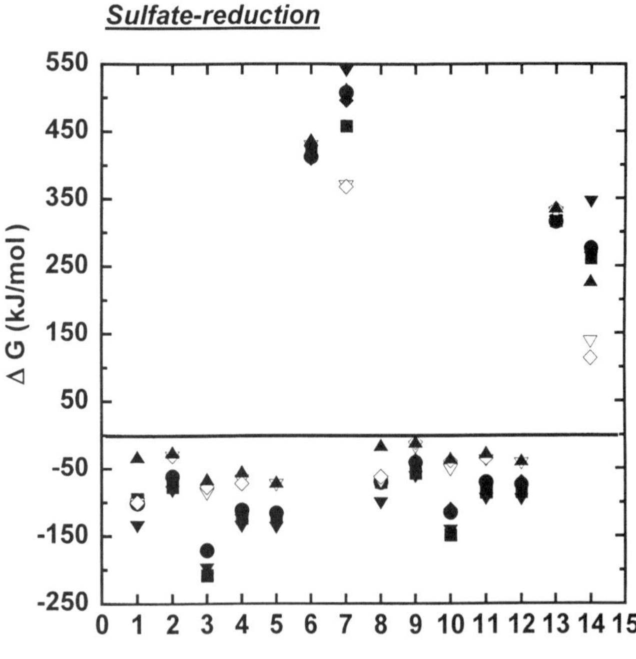

Figure 2. ΔG_r (kJ/mol) of the sulfate-reduction reactions listed in Table 4A (data given in Table 4B) plotted against reaction number. Symbols correspond to sampling sites as follows: ●, Punto 1; ■, Punto 7; ♦, Stinky Surf Rock; ▲, Pozzo Istmo; ▼, Pozzo Vasca; ◊, Acque Calde 2; ∇, Grip. Equilibrium ($\Delta G_r = 0$) is indicated by a solid horizontal line.

TABLE 5A. S^0-REDUCTION AND S^0-DISPROPORTIONATION REACTIONS

	Reaction
5.1	$S^0 + H_2 \rightarrow H_2S$
5.2	$S^0 + 0.25CH_4 + 0.5H_2O \rightarrow H_2S + 0.25CO_2$
5.3	$S^0 + HCOOH \rightarrow H_2S + CO_2$
5.4	$S^0 + 0.25CH_3COOH + 0.5H_2O \rightarrow H_2S + 0.5CO_2$
5.5	$S^0 + 1/7C_2H_5COOH + 4/7H_2O \rightarrow H_2S + 3/7CO_2$
5.6	$S^0 + 0.25NH_4^+ + 0.75H_2O \rightarrow H_2S + 0.25NO_3^- + 0.5H^+$
5.7	$S^0 + 3Fe^{2+} + 4H_2O \rightarrow H_2S + Fe_3O_4 + 6H^+$
5.8	$S^0 + H_2O \rightarrow 0.75H_2S + 0.25SO_4^{2-} + 0.5H^+$

TABLE 5B. VALUES OF ΔG_r (kJ/mol) FOR REACTIONS IN TABLE 5A AT SEVEN VULCANO HYDROTHERMAL SITES

Reac.	Punto 1	Punto 7	Stinky Surf Rock	Grip	Acque Calde 2	Pozzo Istmo	Pozzo Vasca
5.1	–31.29	–24.37	–35.93	–32.28	–37.44	–16.52	–34.06
5.2	–21.49	–19.91	–26.60	–14.95	–20.10	–14.91	–21.05
5.3	–55.65	–59.52	–60.14	–34.87*	–38.74*	–31.34*	–56.72
5.4	–40.82	–38.39	–46.93	–30.02*	–37.43*	–28.55*	–40.97
5.5	–41.88	–38.23	–46.70	–31.71*		–32.45*	–41.44
5.6	97.21	104.49	92.03	100.26	94.88	100.81	105.00
5.7	83.99	86.23	75.70	36.54	20.90	64.42	114.65
5.8	–6.02	–0.79	–10.56	–6.71	–12.50	–8.81	–0.07

*Values of ΔG_r correspond to the reaction in which the carboxylate anion, not the carboxylic acid, is considered (see text).

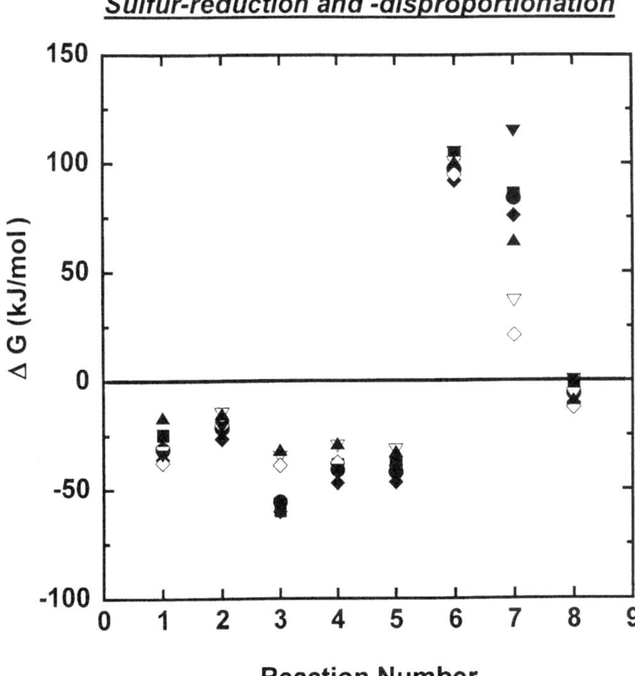

Figure 3. ΔG_r (kJ/mol) of the S^0-reduction and S^0-disproportionation reactions listed in Table 5A (data given in Table 5B) plotted against reaction number. Symbols are as in Figure 2. Equilibrium ($\Delta G_r = 0$) is indicated by a solid horizontal line.

tionation reaction, given in Table 5A. In each of the S^0-reduction reactions (5.1–5.7), two electrons are transferred from the electron donor to S^0, which, in turn, is reduced to H_2S. In the disproportionation reaction (5.8), S^0 is both oxidized to SO_4^{2-} and reduced to H_2S without an additional TEA or electron donor. Values of ΔG_r for reactions in Table 5A are listed in Table 5B for the seven Vulcano sites. Tabulated values of ΔG_r for reactions 5.1–5.8 are also plotted as a function of reaction number in Figure 3.

Note that in Table 5B and Figure 3, reactions 5.1–5.5 are exergonic at all sites considered, yielding between 14 and 61 kJ/mol S^0. The electron donors in these reactions are H_2, CH_4, and carboxylic acids. Again, at Grip, Acque Calde 2, and Pozzo Istmo, values of ΔG_r in Table 5B are calculated for reactions written with the carboxylate anion and not the protonated carboxylic acid. In general, values of ΔG_r are more negative for those reactions in which the carboxylic acids (or carboxylate anions) serve as the electron donors (5.3–5.5) than for those with inorganic reductants. Although all five reactions mentioned (5.1–5.5) are exergonic, they generally yield less energy per mole as written than most sulfate-reduction examples discussed above. It should be pointed out, however, that per mole of electrons transferred, these two sets of reactions yield comparable energy, 7–30 kJ/mol e^- for S^0-reduction and 1–26 kJ/mol e^- for sulfate-reduction. Analogous to reactions in Table 4, S^0-reduction reactions in which NH_4^+ and Fe^{2+} serve as the electron donor are endergonic, consuming 92–105 kJ/mol S^0 for reaction 5.6 and 20–115 kJ/mol S^0 for

TABLE 6A. S^0-OXIDATION REACTIONS

	Reaction
6.1	$S^0 + 1.5O_2 + H_2O \rightarrow SO_4^{2-} + 2H^+$
6.2	$S^0 + 0.75CO_2 + 2.5H_2O \rightarrow SO_4^{2-} + 0.75CH_4 + 2H^+$
6.3	$S^0 + 0.75NO_3^- + 7/4H_2O \rightarrow SO_4^{2-} + 0.75NH_4^+ + 0.5H^+$
6.4	$S^0 + 3Fe_3O_4 + 16H^+ \rightarrow SO_4^{2-} + 9Fe^{2+} + 8H_2O$

TABLE 6B. VALUES OF ΔG_r (kJ/mol) FOR REACTIONS IN TABLE 6A AT SEVEN VULCANO HYDROTHERMAL SITES

Reac.	Punto 1	Punto 7	Stinky Surf Rock	Grip	Acque Calde 2	Pozzo Istmo	Pozzo Vasca
6.1	–530.65	–517.13	–534.39	–561.91	–566.40	–567.17	–507.61
6.2	40.41	56.58	37.58	18.02	10.30	9.49	62.84
6.3	–315.68	–316.63	–318.32	–327.62	–334.64	–337.67	–315.29
6.4	–276.04	–261.84	–269.32	–136.45	–112.70	–228.49	–344.24

reaction 5.7. Reaction 5.6 consumes 46–53 kJ and reaction 5.7 consumes 10–58 kJ per mole of electrons transferred. S^0-disproportionation (reaction 5.8) is near equilibrium at all seven sites, yielding less than 13 kJ/mol S^0 at each site. Although reaction 5.1 is one of the most common metabolic strategies employed in the laboratory by hyperthermophilic chemolithoautotrophs, including archaea and bacteria isolated from Vulcano, it is perhaps surprising that its energy-yield at Vulcano is rather modest, releasing only 26 ± 10 kJ/mol S^0. To put this into perspective, the phosphorylation of one mole of adenosine diphosphate (ADP) to adenosine triphosphate (ATP) at standard temperature and pressure requires ~31 kJ.

S^0-Oxidation

As stated above, S^0-oxidation provides the chemical energy that drives much of the biomass synthesis in marine hydrothermal systems. It should thus come as no surprise that reaction 6.1 yields copious amounts of energy in the Vulcano vent system; values of $\Delta G_{6.1}$ range from –508 to –567 kJ/mol S^0. Four S^0-oxidation reactions, in which O_2, CO_2, NO_3^-, and Fe(III) serve as TEAs, are considered in Table 6A. Values of ΔG_r for these reactions are shown in Table 6B and Figure 4 for all seven Vulcano sites. Note that S^0-oxidation with NO_3^- or Fe(III) in magnetite also yield considerable amounts of energy, in fact, between 315 and 338 kJ/mol S^0 for reaction 6.3 and between 112 and 345 kJ/mol S^0 for reaction 6.4. As in earlier examples that consider Fe-redox reactions, the wide range of ΔG_r values in reaction 6.4 is due largely to considerable variations in pH and Fe^{2+} concentrations at the seven sites (see Table 2). Not all S^0-oxidation reactions considered here yield energy, however. Reaction 6.2, in which CO_2 serves as the TEA, is endergonic at all seven sites, consuming between 9 and 63 kJ/mol S^0. This is consistent with the fact that no methanogenic microorganism, thermophilic or not, is known to use S^0 as the sole electron donor.

Sulfide-Oxidation

Eight sulfide-oxidation reactions are given in Table 7A; in four of these (reactions 7.1–7.4), H_2S is converted in a two-electron transfer to S^0, and in the other four (7.5–7.8), eight electrons are transferred as H_2S is oxidized to SO_4^{2-}. Values of ΔG_r for

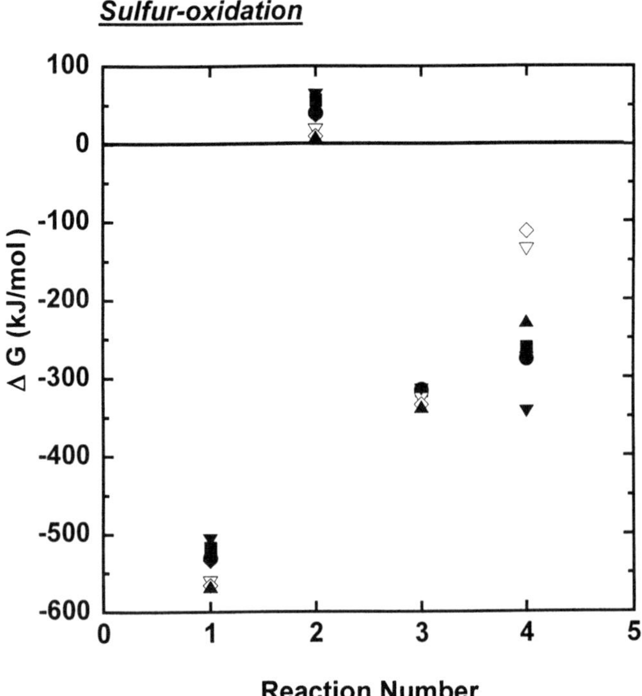

Figure 4. ΔG_r (kJ/mol) of the S^0-oxidation reactions listed in Table 6A (data given in Table 6B) plotted against reaction number. Symbols are as in Figure 2. Equilibrium ($\Delta G_r = 0$) is indicated by a solid horizontal line.

TABLE 7A. SULFIDE-OXIDATION REACTIONS

	Reaction
7.1	$H_2S + 0.5O_2 \rightarrow S^0 + H_2O$
7.2	$H_2S + 0.25CO_2 \rightarrow S^0 + 0.25CH_4 + 0.5H_2O$
7.3	$H_2S + 0.25NO_3^- + 0.5H^+ \rightarrow S^0 + 0.25NH_4^+ + 0.75H_2O$
7.4	$H_2S + Fe_3O_4 + 6H^+ \rightarrow S^0 + 3Fe^{2+} + 4H_2O$
7.5	$H_2S + 2O_2 \rightarrow SO_4^{2-} + 2H^+$
7.6	$H_2S + CO_2 + 2H_2O \rightarrow SO_4^{2-} + CH_4 + 2H^+$
7.7	$H_2S + NO_3^- + H_2O \rightarrow SO_4^{2-} + NH_4^+$
7.8	$H_2S + 4Fe_3O_4 + 22H^+ \rightarrow SO_4^{2-} + 12Fe^{2+} + 12H_2O$

TABLE 7B. VALUES OF ΔG_r (kJ/mol) FOR REACTIONS IN TABLE 7A AT SEVEN VULCANO HYDROTHERMAL SITES

Reac.	Punto 1	Punto 7	Stinky Surf Rock	Grip	Acque Calde 2	Pozzo Istmo	Pozzo Vasca
7.1	–168.86	–171.33	–164.05	–178.36	–172.13	–177.31	–169.10
7.2	21.49	19.91	26.60	14.95	20.10	14.91	21.05
7.3	–97.21	–104.49	–92.03	–100.26	–94.86	–100.81	–105.00
7.4	–83.99	–86.23	–75.70	–36.54	–20.90	–64.42	–114.65
7.5	–699.51	–688.46	–698.45	–740.27	–738.53	–744.49	–676.71
7.6	61.91	76.49	64.18	32.97	30.40	24.39	83.88
7.7	–412.89	–421.12	–410.35	–427.89	–429.51	–438.48	–420.29
7.8	–508.14	–458.62	–496.07	–368.62	–367.69	–514.01	–539.46

reactions 7.1–7.8 listed in Table 7B are also plotted as a function of reaction number in Figure 5. Note in this figure that, with the exception of reactions 7.2 and 7.6, in which CO_2 serves as the oxidant, all reactions are exergonic. However, the amount of energy released varies tremendously from only ~21 kJ/mol S^0 for reaction 7.4 at Acque Calde 2 to almost 750 kJ/mol S^0 for reaction 7.5 at Grip, Acque Calde 2, and Pozzo Istmo. Not surprisingly, reactions in which O_2 serves as the TEA are the most exergonic, yielding 171 ± 8 kJ/mol S^0 for reaction 7.1 and 710 ± 34 kJ/mol S^0 for reaction 7.5. Again, per mole of electrons transferred, these two reactions yield comparable amounts of energy (82–89 kJ for reaction 7.1 and 85–93 kJ for reaction 7.5).

A COMPARISON OF DEEP AND SHALLOW MARINE VENTS

McCollom and Shock (1997) showed that at temperatures >40 °C in the deep-sea hydrothermal system at 21°N on the East Pacific Rise, sulfate-reduction (reaction 4.1) and S^0-reduction (reaction 5.1) with H_2 as the electron donor are exergonic, but S^0-oxidation (reaction 6.1) and sulfide-oxidation (reaction 7.5) with O_2 as the TEA are endergonic. By comparison, these four reactions are all exergonic at the seven Vulcano sites investigated here. At temperatures between 50 and 100 °C, values of ΔG_r for sulfate- and S^0-reduction at the deep-sea site yield ~130 and ~30 kJ/mol, respectively. The energetics of these reactions at Vulcano are similar, yielding between 90 and 140 kJ/mol sulfate (except at Pozzo Istmo) and between 16 and 38 kJ/mol S^0. At temperatures between 50 and 100 °C, the oxidation of S^0 and sulfide consumes

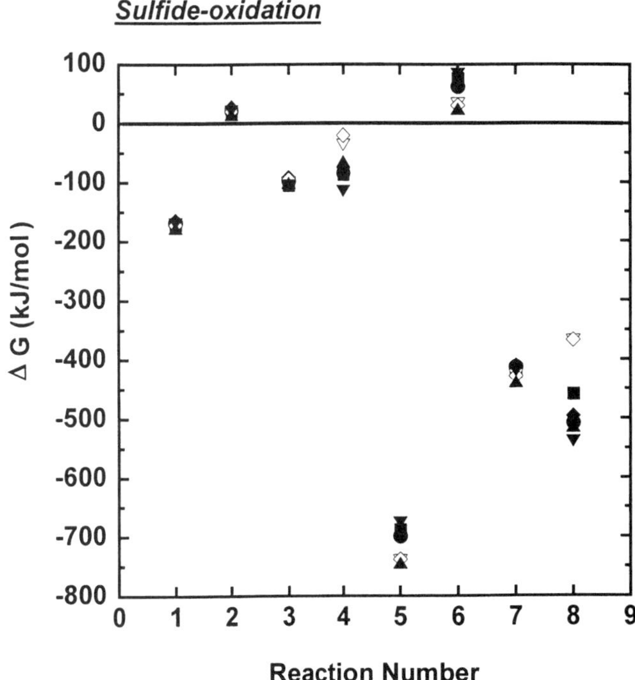

Figure 5. ΔG_r (kJ/mol) of the sulfide-oxidation reactions listed in Table 7A (data given in Table 7B) plotted against reaction number. Symbols are as in Figure 2. Equilibrium ($\Delta G_r = 0$) is indicated by a solid horizontal line.

~25 and ~30 kJ/mol, respectively, at the deep-sea site, but the same reactions yield 507–568 kJ/mol S^0 and 676–745 kJ/mol H_2S at Vulcano. These extreme differences in ΔG_r for the S-oxidation reactions are due almost exclusively to the large differences in oxygen concentration predicted for the deep-sea site and that measured at Vulcano. These results suggest that significantly larger amounts of geochemical energy for aerobic thermophiles and hyperthermophiles are available in the shallow vent environment at Vulcano than in the mixing zones of the deep-sea system. These thermodynamic results also show that anaerobic thermophiles can make a living from S-reduction in both the deep and shallow systems.

Microbial culturing and small subunit ribosomal RNA analyses at shallow and deep vents are largely consistent with these findings. At Vulcano, at other shallow-marine sites, and around acidic solfataras, aerobic S-oxidizers (e.g., *Aquifex aeolicus, Acidianus infernus*) and anaerobic S-reducers (e.g., *Archaeoglobus fulgidus, Pyrodictium occultum, P. brockii, Thermodiscus maritimus*) have been found (Fischer et al., 1983; Stetter et al., 1983; Segerer et al., 1986; Stetter, 1988; Deckert et al., 1998). At deep-sea vent sites, aerobic S-oxidizers have been observed, including members of *Thiobacillus* and *Thiomicrospira* (Jannasch et al., 1985; Durand et al., 1993; Taylor et al., 1999), but these are predominantly mesophiles with growth temperatures <50 °C. In addition, three microaerophilic, moderately thermophilic members of the genus *Persephonella* and several microaerophilic mesophiles and moderate thermophiles belonging to the ε-Proteobacteria were recently isolated from deep-sea hydrothermal vents (Götz et al., 2002; Takai et al., 2003). Anaerobic, moderately thermophilic and hyperthermophilic S-reducers, on the other hand, have been identified and/or isolated from deep-sea hydrothermal systems. These organisms include, for example, the sulfate- and sulfite-reducers *Archaeoglobus profundus* and *A. veneficus* from vent sites at Guaymas Basin and the Mid-Atlantic Ridge (Burggraf et al., 1990b; Huber et al., 1997) and the S^0-reducers *Ignicoccus pacificus* from Black Smoker samples at 9°N and 104°W (Huber et al., 2000) and *Desulfobacterium thermolithotrophum* from the Mid-Atlantic Ridge (L'Haridon et al., 1998).

CONCLUDING REMARKS

The sulfur isotope record of sedimentary rocks from the early Archean (3.4–2.8 Ga) suggests that seawater sulfate concentrations were low owing to low atmospheric O_2-concentrations (Canfield et al., 2000). It follows that microorganisms on early Earth relied neither on aerobic S-oxidation nor on sulfate-reduction as their primary metabolisms, except perhaps in localized sulfate- or oxygen-rich oases (Lyons et al., 2004). Until 2.7 Ga, which represents the first occurrence of robust evidence for oxygenic photosynthesis (Brocks et al., 1999), microbial metabolism on Earth probably relied on TEAs other than O_2 or sulfate. Elemental sulfur, for example, remains a very plausible candidate as TEA on early Earth. However, in a variety of present-day ecosystems that are characterized by low redox potentials, sulfate-reduction is pervasive. Although sulfate-reduction has been documented in high-temperature marine sediments, very few hyperthermophiles are known that mediate this process. This is despite the fact that compared with S^0-reduction, energy-yields are relatively high from both chemolithotrophic and chemo-organotrophic sulfate-reduction in marine hydrothermal systems. Nevertheless, S^0-reduction is a far more ubiquitous metabolic strategy among hyperthermophiles than is sulfate-reduction. This seems paradoxical. Perhaps hyperthermophilic sulfate reducers are more prevalent than is currently documented, but culturing methods for this group of organisms are inadequate. If this scenario is true, high-temperature sulfate reducers may be grossly underrepresented in culture collections. It is also plausible that because S^0 provides a solid substrate to which cells can attach, organisms "prefer" S^0 over aqueous sulfate, despite its lower energy yield. Lastly, assuming that competition among metabolically diverse groups of organisms is influenced considerably by energy-yield, it is perhaps less revealing to consider energy per mole of substrate (sulfate or S^0 in this case) than energy per mole of electrons transferred. In the latter reference frame, sulfate- and S^0-reduction at Vulcano yield nearly identical amounts of energy. In fact, this energy parity among sulfate- and S^0-reduction on a per electron basis is also observed in the deep-sea hydrothermal system considered in McCollom and Shock (1997).

ACKNOWLEDGMENTS

Financial support was provided by National Science Foundation grants OCE-0221417 and OCE-0234460, a grant (37032-G2) from the donors of the Petroleum Research Fund administered by the American Chemical Society, a Spencer T. Olin Fellowship for Women in Graduate Study at Washington University (to KLR), and a NASA-GSRP grant NGT5-50348 (to DRM-D). We thank Natasha Zolotova for carrying out the DOC and organic acid analyses and Everett Shock, Jason Tor, Sergio Gurrieri, and the research staff at the Istituto Nationale de Geofisica e Volcanologia in Palermo for their assistance in Italy.

REFERENCES CITED

Alfonso, P., Prol-Ledesma, R.M., Canet, C., Melgarejo, J.C., and Fallick, A.E., 2003, Sulfur isotope geochemistry of the submarine hydrothermal coastal vents of Punta Mita, Mexico: Journal of Geochemical Exploration, v. 78-79, p. 301–304.

Amend, J.P., and Shock, E.L., 1998, Energetics of amino acid synthesis in hydrothermal ecosystems: Science, v. 281, p. 1659–1662, doi:10.1126/SCIENCE.281.5383.1659.

Amend, J.P., and Shock, E.L., 2000, Thermodynamics of amino acid synthesis in hydrothermal systems on early Earth, *in* Goodfriend, G., Collins, M.J., Fogel, M.L., Macko, S.A., and Wehmiller, J. F., eds., Perspectives in amino acid and protein geochemistry: New York, Oxford University Press, p. 23–40.

Amend, J.P., and Shock, E.L., 2001, Energetics of overall metabolic reactions of thermophilic and hyperthermophilic Archaea and Bacteria: FEMS Microbiology Reviews, v. 25, p. 175–243, doi:10.1016/S0168-6445(00)00062-0.

Amend, J.P., Amend, A.C., and Valenza, M., 1998, Determination of volatile fatty acids in the hot springs of Vulcano, Aeolian Islands, Italy: Organic Geochemistry, v. 28, p. 699–705, doi:10.1016/S0146-6380(98)00023-0.

Amend, J.P., Meyer-Dombard, D.R., Sheth, S.N., Zolotova, N., and Amend, A.C., 2003a, *Palaeococcus helgesonii* sp. nov., a facultatively anaerobic, hyperthermophilic archaeon from a geothermal well on Vulcano Island, Italy: Archives of Microbiology, v. 179, p. 394–401.

Amend, J.P., Rogers, K.L., Shock, E.L., Gurrieri, S., and Inguaggiato, S., 2003b, Energetics of chemolithoautotrophy in the hydrothermal system of Vulcano Island, southern Italy: Geobiology, v. 1, p. 37–58.

Arvidson, R.E., 2004, Geology of Meridiani Planum as inferred from Mars Exploration Rover Observation: Lunar and Planetary Science Conference, 35th Annual Meeting, Houston, Texas, Abstract no. 2165.

Baross, J.A., and Hoffman, S.E., 1985, Submarine hydrothermal vents and associated gradient environments as sites for the origin and evolution of life: Origins of Life, v. 15, p. 327–345.

Bates, T.S., Kiene, R.P., Wolfe, G.V., Matrai, P.A., Chavez, F.P., Buck, K.R., Blomquist, B.W., and Cuhel, R.L., 1994, The cycling of sulfur in surface seawater of the northeast Pacific: Journal of Geophysical Research, C, Oceans, v. 99, p. 7835–7843, doi:10.1029/93JC02782.

Beeder, J., Nilsen, R.K., Rosnes, J.T., Torsvik, T., and Lien, T., 1994, *Archaeoglobus fulgidus* isolated from hot North Sea oil field waters: Applied and Environmental Microbiology, v. 60, p. 1227–1231.

Blöchl, E., Rachel, R., Burggraf, S., Hafenbradl, D., Jannasch, H.W., and Stetter, K.O., 1997, *Pyrolobus fumarii,* gen. and sp. nov., represents a novel group of archaea, extending the upper temperature limit for life to 113°C: Extremophiles, v. 1, p. 14–21, doi:10.1007/S007920050010.

Bonch-Osmolovskaya, E.A., Miroshnichenko, M.L., Kostrikina, N.A., Chernych, N.A., and Zavarzin, G.A., 1990, *Thermoproteus uzoniensis* sp. nov., a new extremely thermophilic archaebacterium from Kamchatka continental hot springs: Archives of Microbiology, v. 154, p. 556–559.

Botz, R., Winckler, G., Bayer, R., Schmitt, M., Schmidt, M., Garbe-Schonberg, D., Stoffers, P., and Kristjansson, J.K., 1999, Origin of trace gases in submarine hydrothermal vents of the Kolbeinsey Ridge, north Iceland: Earth and Planetary Science Letters, v. 171, p. 83–93, doi: 10.1016/S0012-821X(99)00128-4.

Brakenridge, G.R., Newsom, H.E., and Baker, V.R., 1985, Ancient hot springs on Mars: Origins and paleoenvironmental significance of small Martian valleys: Geology, v. 13, p. 859–862.

Brasier, M.D., Green, O.R., Jephcoat, A.P., Kleppe, A.K., Van Kranendonk, M.J., Lindsay, J.F., Steele, A., and Grassineau, N.V., 2002, Questioning the evidence for Earth's oldest fossils: Nature, v. 416, p. 76–81, doi:10.1038/416076A.

Brinkhoff, T., Sievert, S.M., Kuever, J., and Muyzer, G., 1999, Distribution and diversity of sulfur-oxidizing *Thiomicrospira* spp. at a shallow-water hydrothermal vent in the Aegean Sea, Milos, Greece: Applied and Environmental Microbiology, v. 65, p. 1127–1132.

Brocks, J.J., Logan, G.A., Buick, R., and Summons, R.E., 1999, Archean molecular fossils and the early rise of eukaryotes: Science, v. 285, p. 1033–1036, doi:10.1126/SCIENCE.285.5430.1033

Brown, G.C., and Mussett, A.E., 1993, The inaccessible Earth: An integrated view to its structure and composition: London, Chapman & Hall, 276 p.

Brüchert, V., Knoblauch, C., and Jørgensen, B.B., 2001, Controls on stable sulfur isotope fractionation during bacterial sulfate reduction in Arctic sediments: Geochimica et Cosmochimica Acta, v. 65, p. 763–776, doi: 10.1016/S0016-7037(00)00557-3.

Burggraf, S., Fricke, H., Neuner, A., Kristjansson, J., Rouvier, P., Mandelco, L., Woese, C.R., and Stetter, K.O., 1990a, *Methanococcus igneus* sp. nov., a novel hyperthermophilic methanogen from a shallow submarine hydrothermal vent: Systematic and Applied Microbiology, v. 13, p. 263–269.

Burggraf, S., Jannasch, H.W., Nicolaus, B., and Stetter, K.O., 1990b, *Archaeoglobus profundus* sp. nov. represents a new species within the sulfate-reducing archaebacteria: Systematic and Applied Microbiology, v. 13, p. 24–28.

Canfield, D.E., and Raiswell, R., 1999, The evolution of the sulfur cycle: American Journal of Science, v. 299, p. 697–723.

Canfield, D.E., Habicht, K.S., and Thamdrup, B., 2000, The Archean sulfur cycle and the early history of atmospheric oxygen: Science, v. 288, p. 658–661.

Carr, M.H., and Chuang, F.C., 1997, Martian drainage densities: Journal of Geophysical Research, v. 102, p. 9145–9152, doi:10.1029/97JE00113.

Christensen, P.R., Bandfield, J.L., Clark, R.N., Edgett, K.S., Hamilton, V.E., Hoefen, T., Kieffer, H.H., Kuzmin, R.O., Lane, M.D., Malin, M.C., Morris, R.V., Pearl, J.C., Pearson, R., Roush, T.L., Ruff, S.W., and Smith, M.D., 2000, Detection of crystalline hematite mineralization on Mars by the Thermal Emission Spectrometer: Evidence for near-surface water: Journal of Geophysical Research, v. 105, p. 9623–9642, doi:10.1029/1999JE001093.

Chyba, C.F., 2000, Energy for microbial life on Europa: Nature, v. 403, p. 381–382, doi:10.1038/35000281.

Cutter, G.A., and Kluckhohn, R.S., 1999, The cycling of particulate carbon, nitrogen, sulfur, and sulfur species (iron monosulfide, greigite, pyrite, and organic sulfur) in the water columns of Framvaren Fjord and the Black Sea: Marine Chemistry, v. 67, p. 149–160, doi:10.1016/S0304-4203(99)00056-0.

Deckert, G., Warren, P.V., Gaasterland, T., Young, W.G., Lenox, A.L., Graham, D.E., Overbeek, R., Snead, M.A., Keller, M., Aujay, M., Huber, R., Feldman, R.A., Short, J.M., Olsen, G.J., and Swanson, R.V., 1998, The complete genome of the hyperthermophilic bacterium *Aquifex aeolicus*: Nature, v. 392, p. 353–358, doi:10.1038/32831.

Durand, P., Reysenbach, A.-L., Prieur, D., and Pace, N., 1993, Isolation and characterization of *Thiobacillus hydrothermalis* sp. nov., a mesophilic obligately chemolithotrophic bacterium isolated from a deep-sea hydrothermal vent in Fiji Basin: Archives of Microbiology, v. 159, p. 39–44.

Edgett, K.S., and Parker, T.J., 1997, Water on early Mars: Possible subaqueous sedimentary deposits covering ancient cratered terrain in western Arabia and Sinus Meridiani: Geophysical Research Letters, v. 24, p. 2897–2900, doi:10.1029/97GL02840.

Elsgaard, L., Guezennec, J., Benbouzid-Rollet, N., and Prieur, D., 1995, Mesophilic sulfate reducing bacteria from three deep-sea hydrothermal vent sites: Oceanologica Acta, v. 18, p. 95–104.

Elsgaard, L., Isaksen, M.F., Jørgensen, B.B., Alayse, A.-M., and Jannasch, H.W., 1994, Microbial sulfate reduction in deep-sea sediments at the Guaymas basin hydrothermal vent area: Influence of temperature and substrates: Geochimica et Cosmochimica Acta, v. 58, p. 3335–3343, doi:10.1016/0016-7037(94)90089-2.

Erauso, G., Reysenbach, A.-L., Godfroy, A., Meunier, J.-R., Crump, B., Partensky, F., Baross, J.A., Marteinsson, V., Barbier, G., Pace, N.R., and Prieur, D., 1993, *Pyrococcus abyssii* sp. nov., a new hyperthermophilic archaeon isolated from a deep-sea hydrothermal vent: Archives of Microbiology, v. 160, p. 338–349.

Farmer, J.D., 1996, Hydrothermal systems on Mars: An assessment of present evidence, *in* Bock, G. R., and Goode, J. A., eds., Evolution of hydrothermal systems on Earth (and Mars?): Chichester, Wiley, p. 273–299.

Fiala, G., and Stetter, K.O., 1986, *Pyrococcus furiosus* sp. nov. represents a novel genus of marine heterotrophic archaebacteria growing optimally at 100 °C: Archives of Microbiology, v. 145, p. 56–61.

Fischer, F., Zillig, W., Stetter, K.O., and Schreiber, G., 1983, Chemolithoautotrophic metabolism of anaerobic extremely thermophilic archaebacteria: Nature, v. 301, p. 511–513.

Fossing, H., and Jørgensen, B.B., 1990, Oxidation and reduction of radiolabeled inorganic sulfur compounds in an estuarine sediment, Kysing Fjord, Denmark: Geochimica et Cosmochimica Acta, v. 54, p. 2731–2742, doi: 10.1016/0016-7037(90)90008-9.

Gaidos, E.J., Nealson, K.H., and Kirschvink, J.L., 1999, Life in ice-covered oceans: Science, v. 284, p. 1631–1633, doi:10.1126/SCIENCE.284.5420.1631.

Gonzalez, J.M., Masuchi, Y., Robb, F.T., Ammerman, J.W., Maeder, D.L., Yanagibayashi, M., Tamaoka, J., and Kato, C., 1998, *Pyrococcus horikoshii* sp. nov., a hyperthermophilic archaeon isolated from a hydrothermal vent at the Okinawa Trough: Extremophiles, v. 2, p. 123–130, doi:10.1007/S007920050051.

Götz, D., Banta, A., Beveridge, T.J., Rushdi, A.I., Simoneit, B.R.T., and Reysenbach, A.-L., 2002, *Persephonella marina* gen. nov., sp. nov. and *Persephonella guaymasensis* sp. nov., two novel, thermophilic, hydrogen-oxidizing microaerophiles from deep-sea hydrothermal vents: International Journal of Systematic and Evolutionary Microbiology, v. 52, p. 1349–1359, doi:10.1099/IJS.0.02126-0.

Grant, J.A., 2000, Valley formation in Margaritifer Sinus, Mars, by precipitation-recharged ground-water sapping: Geology, v. 28, p. 223–226, doi: 10.1130/0091-7613(2000)0282.3.CO;2.

Grant, J. A. and Parker, T. J., 2002, Drainage evolution in the Margaritifer Sinus region, Mars: Journal of Geophysical Research, v. 107, p. 4–1 to 4–19.

Greenberg, R., and Geissler, P., 2002, Europa's dynamic icy crust: Meteoritics & Planetary Science, v. 37, p. 1685–1710.

Grossman, E.L., and Desrocher, S., 2001, Microbial sulfur cycling in terrestrial subsurface environments, *in* Fredrickson, J.K., and Fletcher, M., eds., Subsurface microbiology and biogeochemistry: Wiley-Liss, Inc., p. 219–248.

Gugliandolo, C., and Maugeri, T.L., 1993, Chemolithotrophic sulfur-oxidizing bacteria from a marine shallow hydrothermal vent of Vulcano (Italy): Geomicrobiology Journal, v. 11, p. 109–120.

Gugliandolo, C., Italiano, F., Maugeri, T.L., Inguaggiato, S., Caccamo, D., and Amend, J.P., 1999, Submarine hydrothermal vents of the Aeolian islands: Relationship between microbial communities and thermal fluids: Geomicrobiology Journal, v. 16, p. 105–117, doi:10.1080/014904599270794.

Gulick, V.C., 1998, Magmatic intrusions and a hydrothermal origin for fluvial valleys on Mars: Journal of Geophysical Research, v. 103, p. 19365–19387, doi:10.1029/98JE01321.

Gulick, V.C., and Baker, V.R., 1990, Origin and evolution of valleys on Martian volcanoes: Journal of Geophysical Research, v. 95, p. 14325–14344.

Gundersen, J.K., Jørgensen, B.B., Larsen, E., and Jannasch, H.W., 1992, Mats of giant sulphur bacteria on deep-sea sediments due to fluctuating hydrothermal flow: Nature, v. 360, p. 454–455, doi:10.1038/360454A0.

Haldane, J.B.S., 1929, The origin of life: The Rationalist Annual.

Helgeson, H.C., Delany, J.M., Nesbitt, W.H., and Bird, D.K., 1978, Summary and critique of the thermodynamic properties of rock-forming minerals: American Journal of Science, v. 278A, p. 1–229.

Helgeson, H.C., Kirkham, D.H., and Flowers, G.C., 1981, Theoretical prediction of the thermodynamic behavior of aqueous electrolytes at high pressures and temperatures: IV. Calculation of activity coefficients, osmotic coefficients, and apparent molal and standard and relative partial molal properties to 600 °C and 5 kb: American Journal of Science, v. 281, p. 1249–1516.

Henry, E.A., Devereux, R., Maki, J.S., Gilmour, C.C., Woese, C.R., Mandelco, L., Schauder, R., Remsen, C.C., and Mitchell, R., 1994, Characterization of a new thermophilic sulfate-reducing bacterium, *Thermodesulfovibrio yellowstonii*, gen. nov. and sp. nov.: Its phylogenetic relationship to *Thermodesulfobacterium commune* and their origins deep within the bacterial domain: Archives of Microbiology, v. 161, p. 62–69, doi:10.1007/S002030050022.

Hinrichs, K.-U., Hayes, J.M., Sylva, S.P., Brewer, P.G., and DeLong, E.F., 1999, Methane-consuming archaebacteria in marine sediments: Nature, v. 398, p. 802–805, doi:10.1038/19751.

Huber, C., and Wächtershäuser, G., 1998, Peptides by activation of amino acids with CO on (Ni,Fe)S surfaces: Implications for the origin of life: Science, v. 281, p. 670–672, doi:10.1126/SCIENCE.281.5377.670.

Huber, C., Eisenreich, W., Hecht, S., and Wächtershäuser, G., 2003, A possible primordial peptide cycle: Science, v. 301, p. 938–940, doi:10.1126/SCIENCE.1086501.

Huber, H., Jannasch, H., Rachel, R., Fuchs, T., and Stetter, K.O., 1997, *Archaeoglobus veneficus* sp. nov., a novel facultative chemolithoautotrophic hyperthermophilic sulfur reducer, isolated from abyssal black smokers: Systematic and Applied Microbiology, v. 20, p. 374–380.

Huber, H., Burggraf, S., Mayer, T., Wyschkony, I., Biebl, M., Rachel, R., and Stetter, K.O., 2000, *Ignicococcus* gen. nov., a novel genus of hyperthermophilic, chemolithoautotrophic *Archaea*, represented by two new species *Ignicococcus islandicus* sp. nov. and *Ignicococcus pacificus* sp. nov: International Journal of Systematic and Evolutionary Microbiology, v. 50, p. 2093–2100.

Huber, R., and Stetter, K.O., 1989, *Thiobacillus prosperus* sp. nov., represents a new group of halotolerant metal-mobilizing bacteria isolated from a marine geothermal field: Archives of Microbiology, v. 151, p. 479–485.

Huber, R., Langworthy, T.A., König, H., Thomm, M., Woese, C.R., Sleytr, U.B., and Stetter, K.O., 1986, *Thermotoga maritima* sp. nov. represents a new genus of unique extremely thermophilic eubacteria growing up to 90 °C: Archives of Microbiology, v. 144, p. 324–333.

Huber, R., Kristjansson, J.K., and Stetter, K.O., 1987, *Pyrobaculum* gen. nov., a new genus group of neutrophilic, rod-shaped archaebacteria from continental solfataras growing optimally at 100 °C: Archives of Microbiology, v. 149, p. 95–101.

Huber, R., Kurr, M., Jannasch, H.W., and Stetter, K.O., 1989, A novel group of abyssal methanogenic archaebacteria (*Methanopyrus*) growing at 110 °C: Nature, v. 342, p. 833–834, doi:10.1038/342833A0.

Huber, R., Wilharm, T., Huber, D., Trincone, A., Burggraf, S., König, H., Rachel, R., Rockinger, I., Fricke, H., and Stetter, K.O., 1992, *Aquifex pyrophilus* gen. nov. sp. nov., represents a novel group of marine hyperthermophilic hydrogen-oxidizing Bacteria: Systematic and Applied Microbiology, v. 15, p. 340–351.

Hynek, B.M., and Phillips, R.J., 2001, Evidence for extensive denudation of the Martian highlands: Geology, v. 29, p. 407–410, doi:10.1130/0091-7613(2001)0292.0.CO;2.

Jakosky, B.M., and Shock, E.L., 1998, The biological potential of Mars, the early Earth, and Europa: Journal of Geophysical Research, v. 103, p. 19359–19364, doi:10.1029/98JE01892.

Jannasch, H.W., 1985, Chemosynthetic support for life and the microbial diversity at deep-sea hydrothermal vents: Proceedings of the Royal Society of London, v. 225, p. 277–297.

Jannasch, H.W., and Mottl, M.J., 1985, Geomicrobiology of deep-sea hydrothermal vents: Science, v. 229, p. 717–725.

Jannasch, H.W., Wirsen, C.O., Nelson, D.C., and Robertson, L.A., 1985, *Thiomicrospira crunogena* sp. nov., a colorless, sulfur-oxidizing bacterium from a deep-sea hydrothermal vent: International Journal of Systematic Bacteriology, v. 35, p. 422–424.

Jeanthon, C., L'Haridon, S., Cueff, V., Banta, A., Reysenbach, A.-L., and Prieur, D., 2002, *Thermodesulfobacterium hydrogenophilum* sp. nov., a thermophilic, chemolithoautotrophic, sulfate-reducing bacterium isolated from a deep-sea hydrothermal vent at Guaymas Basin, and emendation of the genus *Thermodesulfobacterium*: International Journal of Systematic and Evolutionary Microbiology, v. 52, p. 765–772, doi:10.1099/IJS.0.02025-0.

Jochimsen, B., Peinemann-Simon, S., Volker, H., Stuben, D., Botz, R., Stoffers, P., Dando, P.R., and Thomm, M., 1997, *Stetteria hydrogenophila* gen. nov. and sp. nov., a novel mixotrophic sulfur-dependent *crenarchaeote* isolated from Milos, Greece: Extremophiles, v. 1, p. 67–73, doi:10.1007/S007920050016.

Karl, D.M., 1995, Ecology of free-living, hydrothermal vent microbial communities, *in* Karl, D. M., ed., The microbiology of deep-sea hydrothermal vents: Boca Raton, CRC Press, p. 35–124.

Karl, D.M., Wirsen, C.O., and Jannasch, H.W., 1980, Deep-sea primary production at the Galapagos hydrothermal vents: Science, v. 207, p. 1345–1347.

Kashefi, K., and Lovley, D.R., 2003, Extending the upper temperature limit for life: Science, v. 301, p. 934, doi:10.1126/SCIENCE.1086823.

Kasten, S., and Jørgensen, B.B., 2000, Sulfate reduction in marine sediments, *in* Schulz, H.D., and Zabel, M., eds., Marine geochemistry: Berlin, Springer-Verlag, p. 263–281.

Kelly, R.M., and Adams, M.W.W., 1994, Metabolism in hyperthermophilic microorganisms: Antonie van Leeuwenhoek, v. 66, p. 247–270.

Kühl, M., Steuckert, C., Eickert, G., and Jeroschewski, P., 1998, A H_2S microsensor for profiling biofilms and sediments—application in an acidic lake sediment: Aquatic Microbial Ecology, v. 15, p. 201–209.

Kurr, M., Huber, R., König, H., Jannasch, H.W., Fricke, H., Trincone, A., Krisjansson, J.K., and Stetter, K.O., 1991, *Methanopyrus kandleri*, gen. and sp. nov. represents a novel group of hyperthermophilic methanogens, growing to 110 °C: Archives of Microbiology, v. 156, p. 239–247.

Küsel, K., Roth, U., Trinkwalter, T., and Pfeiffer, S., 2001, Effect of pH on the anaerobic microbial cycling of sulfur in mining-impacted freshwater lake sediments: Environmental and Experimental Microbiology, v. 46, p. 213–223.

Lazcano, A., and Miller, S.L., 1996, The origin and early evolution of life: prebiotic chemistry, the pre-RNA world, and time: Cell, v. 85, p. 793–798.

Lazcano, A., and Miller, S.L., 1999, On the origin of metabolic pathways: Journal of Molecular Evolution, v. 49, p. 424–431.

L'Haridon, S., Cilia, V., Messner, P., Raguenes, G., Gambacorta, A., Sleytr, U.B., Prieur, D., and Jeanthon, C., 1998, *Desulfobacterium thermolithotrophum* gen. nov. sp. nov., a novel autotrophic, sulfur-reducing bacterium isolated from a deep-sea hydrothermal vent: International Journal of Systematic Bacteriology, v. 48, p. 701–711.

Loy, A., Lehner, A., Lee, N., Adamczyk, J., Meier, H., Ernst, J., Schleifer, K.-H., and Wagner, M., 2002, Oligonucleotide microarray for 16S rRNA gene-based detection of all recognized lineages of sulfate-reducing prokaryotes in the environment: Applied and Environmental Microbiology, v. 68, p. 5064–5081, doi:10.1128/AEM.68.10.5064-5081.2002.

Lyons, T.W., Kah, L.C., and Gellatly, A.M., 2004, The Precambrian sulfur isotope record of evolving atmospheric oxygen, *in* Eriksson, P.G., et al., eds., The Precambrian Earth: Tempos and events: Developments in Precambrian geology: Amsterdam, Elsevier, p. 421–439.

Martin, W., and Russell, M.J., 2003, On the origins of cells: A hypothesis for the evolutionary transitions from abiotic geochemistry to chemoautotrophic prokaryotes, and from prokaryotes to nucleated cells: Philosophical Transactions of the Royal Society of London, v. 358, p. 59–85.

McCollom, T.M., 1999, Methanogenesis as a potential source of chemical energy for primary biomass production by autotrophic organisms in hydrothermal systems on Europa: Journal of Geophysical Research, v. 104, p. 30729–30742, doi:10.1029/1999JE001126.

McCollom, T.M., 2000, Geochemical constraints on primary productivity in submarine hydrothermal vent plumes: Deep-Sea Research I, v. 47, p. 85–101.

McCollom, T.M., and Shock, E.L., 1997, Geochemical constraints on chemolithoautotrophic metabolism by microorganisms in seafloor hydrothermal systems: Geochimica et Cosmochimica Acta, v. 61, p. 4375–4391, doi:10.1016/S0016-7037(97)00241-X.

McCollom, T.M., and Shock, E.L., 1998, Fluid-rock interactions in the lower oceanic crust: Thermodynamic models of hydrothermal alteration: Journal of Geophysical Research, v. 103, p. 547–575, doi:10.1029/97JB02603.

Miller, S.L., 1957, The formation of organic compounds on the primitive Earth: Annals of the New York Academy of Sciences, v. 69, p. 260–275.

Miller, S.L., and Urey, H.C., 1959, Organic compound synthesis on the primitive Earth: Science, v. 130, p. 245–251.

Miller, S.L., Schopf, J.W., and Lazcano, A., 1997, Oparin's "origin of life": Sixty years later: Journal of Molecular Evolution, v. 44, p. 351–353.

Miyakawa, S., Yamanashi, H., Kobayashi, K., Cleaves, H.J., and Miller, S.L., 2002, Prebiotic synthesis from CO atmospheres: implications for the origins of life: Proceedings of the National Academy of Sciences of the United States of America, v. 99, p. 14628–14631, doi:10.1073/PNAS.192568299.

Morris, R.V., Squyres, S.W., Arvidson, R.E., Bell, J.F., III, Christensen, P.E., Gorevan, S., Herkenhoff, K., Klingelhofer, G., Rieder, R., Farrand, W., Ghosh, A., Glotch, T., Johnson, J.R., Lemmon, M., McSween, H.Y., Ming, D.W., Schroeder, C., de Souza, P., and Wyatt, M., 2004, A first look at the mineralogy and geochemistry of the MER-B landing site in Meridiani Planum: Lunar and Planetary Science Conference, 35th Annual Meeting, Houston, Texas, Abstract no. 2179.

Moyer, C.L., Dobbs, F.C., and Karl, D.M., 1995, Phylogenetic diversity of the bacterial community from a microbial mat in an active, hydrothermal vent system, Loihi Seamount, Hawaii: Applied and Environmental Microbiology, v. 61, p. 1555–1562.

Muyzer, G., Teske, A., Wirsen, C.O., and Jannasch, H.W., 1995, Phylogenetic relationships of *Thiomicrospira* species and their identification in deep-sea hydrothermal vent samples by denaturing gradient gel electrophoresis of 16S rDNA fragments: Archives of Microbiology, v. 164, p. 165–172, doi:10.1007/S002030050250.

Newsom, H.E., Hagerty, J.J., and Goff, F., 1999, Mixed hydrothermal fluids and the origin of the Martian soil: Journal of Geophysical Research, E, Planets, v. 104, p. 8717–8728, doi:10.1029/1998JE900043.

Nilsen, R.K., Beeder, J., Thorstenson, T., and Torsvik, T., 1996, Distribution of thermophilic marine sulfate reducers in North Sea oil field waters and oil reservoirs: Applied and Environmental Microbiology, v. 62, p. 1793–1798.

Oparin, A.I., 1924, Proiskhozhdenie zhizny: Moscow, Moskovskii Rabochii.

Oparin, A.I., 1936, Origin of Life: Dover, Translated from the Russian by S. Margolis.

Pasteris, J.D., and Wopenka, B., 2002, Images of the Earth's earliest fossils?: Nature, v. 420, p. 476–477, doi:10.1038/420476B

Pichler, T., and Dix, G.R., 1996, Hydrothermal venting within a coral reef ecosystem, Ambitle Island, Papua New Guinea: Geology, v. 24, p. 435–438, doi:10.1130/0091-7613(1996)0242.3.CO;2.

Pichler, T., and Veizer, J., 1999, Precipitation of Fe(III) oxyhydroxide deposits from shallow-water hydrothermal fluids in Tutum Bay, Ambitle Island, Papua New Guinea: Chemical Geology, v. 162, p. 15–31, doi:10.1016/S0009-2541(99)00068-6.

Pichler, T., Giggenbach, W.F., McInnes, B.I.A., Buhl, D., and Duck, B., 1999a, Fe sulfide formation due to seawater-gas-sediment interaction in a shallow-water hydrothermal system at Lihir Island, Papua New Guinea: Economic Geology and the Bulletin of the Society of Economic Geologists, v. 94, p. 281–288.

Pichler, T., Veizer, J., and Hall, G.E.M., 1999b, The chemical composition of shallow-water hydrothermal fluids in Tutum Bay, Ambitle Island, Papua New Guinea and their effect on ambient seawater: Marine Chemistry, v. 64, p. 229–252, doi:10.1016/S0304-4203(98)00076-0.

Pledger, R.J., and Baross, J.A., 1991, Preliminary description and nutritional characterization of a chemoorganotrophic archaebacterium growing at temperatures of up to 110 °C isolated from a submarine hydrothermal vent environment: Journal of General Microbiology, v. 137, p. 203–211.

Pley, U., Schipka, J., Gambacorta, A., Jannasch, H.W., Fricke, H., Rachel, R., and Stetter, K.O., 1991, *Pyrodictium abyssi* sp. nov. represents a novel heterotrophic marine archaeal hyperthermophile growing at 110 °C: Systematic and Applied Microbiology, v. 14, p. 245–253.

Prol-Ledesma, R.M., 2003, Similarities in the chemistry of shallow submarine hydrothermal vents: Geothermics, v. 32, p. 639–644.

Rasmussen, B., 2000, Filamentous microfossils in a 3,235-million-year-old volcanogenic massive sulphide deposit: Nature, v. 405, p. 676–679, doi:10.1038/35015063.

Rathbun, J.A., and Squyres, S.W., 2002, Hydrothermal systems associated with Martian impact craters: Icarus, v. 157, p. 362–372, doi:10.1006/ICAR.2002.6838.

Reysenbach, A.-L., and Shock, E.L., 2002, Merging genomes with geochemistry in hydrothermal ecosystems: Science, v. 296, p. 1077–1082, doi:10.1126/SCIENCE.1072483.

Reysenbach, A.-L., Longnecker, K., and Kirshtein, J., 2000, Novel bacterial and archaeal lineages from an in situ growth chamber deployed at a Mid-Atlantic Ridge hydrothermal vent: Applied and Environmental Microbiology, v. 66, p. 3798–3806, doi:10.1128/AEM.66.9.3798-3806.2000.

Runnegar, B., Coath, C.D., Lyons, J.R., and McKeegan, K.D., 2002, Mass-independent and mass-dependent sulfur processing throughout the Archean: Geochimica et Cosmochimica Acta, v. 66, p. A655.

Russell, M.J., and Hall, A.J., 1997, The emergence of life from iron monosulphide bubbles at a submarine hydrothermal redox and pH front: Journal of the Geological Society (London), v. 154, p. 377–402.

Sako, Y., Nomura, N., Uchida, A., Ishida, Y., Morii, H., Koga, Y., Hoaki, T., and Maruyama, T., 1996, *Aeropyrum pernix* gen. nov., sp. nov., a novel aerobic hyperthermophilic archaeon growing at temperatures up to 100 °C: International Journal of Systematic Bacteriology, v. 46, p. 1070–1077.

Schippers, A., and Jørgensen, B.B., 2002, Biogeochemistry of pyrite and iron sulfide oxidation in marine sediments: Geochimica et Cosmochimica Acta, v. 66, p. 85–92, doi:10.1016/S0016-7037(01)00745-1.

Schopf, J.W., Kudryavtsev, A.B., Agresti, D.G., Wdowiak, T.J., and Czaja, A.D., 2002, Laser-Raman imagery of Earth's earliest fossils: Nature, v. 416, p. 73–76, doi:10.1038/416073A.

Segerer, A., Neuner, A., Kristjansson, J.K., and Stetter, K.O., 1986, *Acidianus infernus* gen. nov., sp. nov., and *Acidianus brierleyi* comb. nov.: Facultatively aerobic, extremely acidophilic thermophilic sulfur-metabolizing archaebacteria: International Journal of Systematic Bacteriology, v. 36, p. 559–564.

Shen, Y., Buick, R., and Canfield, D.E., 2001, Isotopic evidence for microbial sulphate reduction in the early Archaean era: Nature, v. 410, p. 77–81, doi:10.1038/35065071.

Shock, E.L., 1995, Organic acids in hydrothermal solutions: Standard molal thermodynamic properties of carboxylic acids, and estimates of dissociation constants at high temperatures and pressures: American Journal of Science, v. 295, p. 496–580.

Shock, E.L., and Helgeson, H.C., 1988, Calculation of the thermodynamic and transport properties of aqueous species at high pressures and temperatures: Correlation algorithms for ionic species and equation of state predictions to 5 kb and 1000 °C: Geochimica et Cosmochimica Acta, v. 52, p. 2009–2036, doi:10.1016/0016-7037(88)90181-0.

Shock, E.L., and Helgeson, H.C., 1990, Calculation of the thermodynamic and transport properties of aqueous species at high pressures and temperatures: Standard partial molal properties of organic species: Geochimica et Cosmochimica Acta, v. 54, p. 915–945, doi:10.1016/0016-7037(90)90429-O.

Shock, E.L., and Koretsky, C.M., 1993, Metal-organic complexes in geochemical processes—calculation of standard partial molal thermodynamic properties of aqueous acetate complexes at high pressures and temperatures: Geochimica et Cosmochimica Acta, v. 57, p. 4899–4922, doi:10.1016/0016-7037(93)90128-J.

Shock, E.L., and Koretsky, C.M., 1995, Metal-organic complexes in geochemical processes: Estimation of standard partial molal thermodynamic properties of aqueous complexes between metal cations and monovalent organic acid ligands at high pressures and temperatures: Geochimica et Cosmochimica Acta, v. 59, p. 1497–1532, doi:10.1016/0016-7037(95)00058-8.

Shock, E.L., Helgeson, H.C., and Sverjensky, D.A., 1989, Calculation of the thermodynamic and transport properties of aqueous species at high pressures and temperatures: Standard partial molal properties of inorganic neutral species: Geochimica et Cosmochimica Acta, v. 53, no. 9, p. 2157–2183, doi:10.1016/0016-7037(89)90341-4.

Shock, E.L., and Schulte, M.D., 1998, Organic synthesis during fluid mixing in hydrothermal systems: Journal of Geophysical Research, v. 103, p. 28513–28527, doi:10.1029/98JE02142.

Shock, E.L., Oelkers, E.H., Johnson, J.W., Sverjensky, D.A., and Helgeson, H.C., 1992, Calculation of the thermodynamic properties of aqueous species at high pressures and temperatures: Effective electrostatic radii, dissociation constants and standard partial molal properties to 1000 °C and 5 kbar: Journal of the Chemical Society, Faraday Transactions 1: Physical Chemistry in Condensed Phases, v. 88, no. 6, p. 803–826.

Shock, E.L., McCollom, T., and Schulte, M.D., 1995, Geochemical constraints on chemolithoautotrophic reactions in hydrothermal systems: Origins of Life and Evolution of the Biosphere, v. 25, p. 141–159.

Shock, E.L., McCollom, T., and Schulte, M.D., 1998, The emergence of metabolism from within hydrothermal systems, *in* Wiegel, J., and Adams, M. W. W., eds., Thermophiles: the keys to molecular evolution and the origin of life?: London, Taylor & Francis, p. 59–76.

Shock, E.L., Amend, J.P., and Zolotov, M.Y., 2000, The early Earth vs. the origin of life, *in* Canup, R., and Righter, K., eds., Origin of the Earth and Moon: Tucson, University of Arizona Press, p. 527–543.

Sievert, S.M., and Kuever, J., 2000, *Desulfacinum hydrothermale*, sp. nov., a thermophilic sulfate-reducing bacterium from geothermally heated sediments near Milos island, Greece: International Journal of Systematic and Evolutionary Microbiology, v. 50, p. 1239–1246.

Sievert, S.M., Brinkhoff, T., Muyzer, G., Ziebis, W., and Kuever, J., 1999, Spatial heterogeneity of bacterial population along an environmental gradient at a shallow submarine hydrothermal vent near Milos Island, Greece: Applied and Environmental Microbiology, v. 65, p. 3834–3842.

Simo, R., and Alio, C.P., 1999, Short-term variability in the open ocean cycle of dimethylsulfide: Global Biogeochemical Cycles, v. 13, p. 1173–1181, doi:10.1029/1999GB900081.

Sonne-Hansen, J., and Ahring, B.K., 1999, *Thermodesulfobacterium hveragerdense* sp. nov., and *Thermodesulfovibrio islandicus* sp. nov., two thermophilic sulfate reducing bacteria isolated from an Icelandic hot spring: Systematic and Applied Microbiology, v. 22, p. 559–564.

Squyres, S.W., 2004, Initial results from the MER Athena Science Investigation at Gusev Crater and Meridiani Planum: Lunar and Planetary Science Conference, 35th Annual Meeting, Houston, Texas, Abstract no. 2187.

Stetter, K.O., 1982, Ultrathin mycelia-forming organisms from submarine volcanic areas having an optimum growth temperature of 105 °C: Nature, v. 300, p. 258–260.

Stetter, K.O., 1988, *Archaeoglobus fulgidus* gen. nov., sp. nov.: a new taxon of extremely thermophilic arachaebacteria: Systematic and Applied Microbiology, v. 10, p. 172–173.

Stetter, K.O., König, H., and Stackebrandt, E., 1983, *Pyrodictium* gen. nov., a new genus of submarine disc-shaped sulphur reducing archaebacteria growing optimally at 105 °C: Systematic and Applied Microbiology, v. 4, p. 535–551.

Stetter, K.O., Lauerer, G., Thomm, M., and Neuner, A., 1987, Isolation of extremely thermophilic sulfate reducers: Evidence for a novel branch of archaebacteria: Science, v. 236, p. 822–824.

Stetter, K.O., Huber, R., Blöchl, E., Kurr, M., Eden, R.D., Fielder, M., Cash, H., and Vance, I., 1993, Hyperthermophilic archaea are thriving in deep North Sea and Alaskan oil reservoirs: Nature, v. 365, p. 743–745, doi:10.1038/365743A0.

Stuben, D., and Glasby, G.P., 1999, Geochemistry of shallow submarine hydrothermal fluids from Paleohori Bay, Milos, Aegean Sea: Exploration and Mining Geology, v. 8, p. 273–287.

Svetlitshnyi, V.A., Slesarev, A.I., Svetlichnaya, T.P., and Zavarzin, G.A., 1987, *Caldococcus litoralis* gen. nov., sp. nov., a new marine extremely thermophilic archaebacterium reducing elemental sulfur: Mikrobiologiya, v. 56, p. 831–838.

Takai, K., Sugai, A., Itoh, T., and Horikoshi, K., 2000, *Palaeococcus ferrophilus* gen. nov., sp. nov., a barophilic, hyperthermophilic archaeon from a deep-sea hydrothermal vent chimney: International Journal of Systematic and Evolutionary Microbiology, v. 50, p. 489–500.

Takai, K., Inagaki, F., Nakagawa, S., Hirayama, H., Nunoura, T., Sako, Y., Nealson, K.H., and Horikoshi, K., 2003, Isolation and phylogenetic diversity of members of previously uncultivated ε-Proteobacteria in deep-sea hydrothermal fields: FEMS Microbiology Letters, v. 218, p. 167–174, doi:10.1016/S0378-1097(02)01147-3.

Tanger, J.C., and Helgeson, H.C., 1988, Calculation of the thermodynamic and transport properties of aqueous species at high pressures and temperatures: Revised equations of state for the standard partial molal properties of ions and electrolytes: American Journal of Science, v. 288, p. 19–98.

Taylor, C.D., Wirsen, C.O., and Gaill, F., 1999, Rapid microbial production of filamentous sulfur mats at hydrothermal vents: Applied and Environmental Microbiology, v. 65, p. 2253–2255.

Teske, A., Hinrichs, K.-U., Edgcomb, V., Gomez, A.d.V., Kysela, D., Sylva, S.P., Sogin, M.L., and Jannasch, H.W., 2002, Microbial diversity of hydrothermal sediments in the Guaymas Basin: Evidence for anaerobic methanotrophic communities: Applied and Environmental Microbiology, v. 68, p. 1994–2007.

Tor, J.M., Amend, J.P., and Lovley, D.R., 2003, Metabolism of organic compounds in anaerobic, hydrothermal sulfate-reducing marine sediments: Environmental Microbiology, v. 5, p. 583–591, doi:10.1046/J.1462-2920.2003.00441.X.

Varnes, E.S., Jakosky, B.M., and McCollom, T.M., 2003, Biological potential of Martian hydrothermal systems: Astrobiology, v. 3, p. 407–414, doi:10.1089/153110703769016479.

Völkl, P., Huber, R., Drobner, E., Rachel, R., Burggraf, S., Trincone, A., and Stetter, K.O., 1993, *Pyrobaculum aerophilum* sp. nov., a novel nitrate-reducing hyperthermophilic archaeum: Applied and Environmental Microbiology, v. 59, p. 2918–2926.

Wächtershäuser, G., 1988, Pyrite formation, the first energy source for life: A hypothesis: Systematic and Applied Microbiology, v. 10, p. 207–210.

Wächtershäuser, G., 1990, Evolution of the first metabolic cycles: Proceedings of the National Academy of Sciences of the United States of America, v. 87, p. 200–204.

Wächtershäuser, G., 1992, Groundworks for an evolutionary biochemistry: The iron-sulphur world: Progress in Biophysics and Molecular Biology, v. 58, p. 85–201, doi:10.1016/0079-6107(92)90022-X.

Wächtershäuser, G., 1998, Origin of life in an iron-sulfur world, *in* Brack, A., ed., The Molecular Origins of Life: Assembling Pieces of the Puzzle: Cambridge, Cambridge University Press, p. 206–218.

Weber, A. and Jørgensen, B.B., 2002, Bacterial sulfate reduction in hydrothermal sediments of the Guaymas Basin, Gulf of California, Mexico: Deep-Sea Research I, v. 49, p. 827–841.

Wenzhofer, F., Holby, O., Glud, R.N., Nielsen, H.K., and Gundersen, J.K., 2000, In situ microsensor studies of a shallow water hydrothermal vent at Milos, Greece: Marine Chemistry, v. 69, p. 43–54, doi:10.1016/S0304-4203(99)00091-2.

Wetzel, L.R., and Shock, E.L., 2000, Distinguishing ultramafic- from basalt-hosted submarine hydrothermal systems by comparing calculated vent fluid compositions: Journal of Geophysical Research, v. 105, p. 8319–8346, doi:10.1029/1999JB900382.

Wolery, T., 1992, EQ3NR, A computer program for geochemical aqueous speciation-solubility calculations: Theoretical manual, user's guide, and related documentation: Lawrence Livermore National Laboratory.

Zeikus, J.G., Dawson, M.A., Thompson, T.E., Ingvorsen, K., and Hatchikian, E.C., 1983, Microbial ecology of volcanic sulfidogenesis: isolation and characterization of *Thermodesulfobacterium commune* gen. nov. and sp. nov: Journal of General Microbiology, v. 129, p. 1159–1169.

Zillig, W., Holz, I., Janekovic, D., Schäfer, W., and Reiter, W.D., 1983, The archaebacterium *Thermococcus celer* represents a novel genus within the thermophilic branch of the Archaebacteria: Systematic and Applied Microbiology, v. 4, p. 88–94.

Zillig, W., Holz, I., Klenk, H.-P., Trent, J., Wunderl, S., Janekovic, D., Imsel, E., and Haas, B., 1987, *Pyrococcus woesei*, sp. nov., an ultra-thermophilic marine archaebacterium, representing a nover order, *Thermococcales*: Systematic and Applied Microbiology, v. 9, p. 62–70.

Zillig, W., Holz, I., Janekovic, D., Klenk, H.-P., Imsel, E., Trent, J., Wunderl, S., Forjaz, V.H., Coutinho, R., and Ferreira, T., 1990, *Hyperthermus butylicus*, a hyperthermophilic sulfur-reducing archaebacterium that ferments peptides: Journal of Bacteriology, v. 172, p. 3959–3965.

Manuscript Accepted by the Society March 28, 2004

Printed in the USA

Geological Society of America
Special Paper 379
2004

Eukaryotes of the Cariaco, Soledad, and Santa Barbara Basins: Protists and metazoans associated with deep-water marine sulfide-oxidizing microbial mats and their possible effects on the geologic record

Joan M. Bernhard*
Department of Environmental Health Sciences, Arnold School of Public Health, University of South Carolina, Columbia, South Carolina 29208, USA

Kurt R. Buck
Monterey Bay Aquarium Research Institute, 7700 Sandholdt Road, Moss Landing, California 95039, USA

ABSTRACT

Sulfide-enriched environments are not typically considered to be sites that support abundant eukaryotes, yet it is known that plentiful and relatively diverse protistan and metazoan fauna inhabit at least one modern bathyal sulfidic site (Santa Barbara Basin, California). This contribution adds to our knowledge of eukaryotic communities inhabiting sulfide-enriched deep-water sediments by presenting data from Soledad Basin (off the western coast of Baja California, Mexico) and Cariaco Basin (off Venezuela). Results indicate that, when considered at the appropriate scale, the density of eukaryotes in Soledad Basin was comparable to that of Santa Barbara Basin. Eukaryotic biovolume and abundance were dominated by foraminifera at all three sites. Unlike the Santa Barbara Basin assemblage, Soledad eukaryotic abundance and biovolume were not dominated by eukaryotes with associated putative symbionts. An undescribed polychaete found in Cariaco *Beggiatoa*-laden sediments had bacterial ectobionts. Sub-millimeter life-position analysis indicated that Soledad eukaryotes concentrated within the top 2 mm even when the bottom-water oxygen concentration was relatively high (2.7 μM). Observations suggest that the eukaryotic fauna of a *Thioploca*-dominated site (Soledad) varied substantially in taxonomic composition and sub-millimeter life positions from *Beggiatoa*-dominated sites (Cariaco and Santa Barbara).

Keywords: *Beggiatoa*, Cariaco Basin, ciliate, flagellate, foraminifera, nematode, polychaete, Santa Barbara Basin, Soledad Basin, symbiosis, *Thioploca*.

* Present address: Department of Geology & Geophysics, Woods Hole Oceanographic Institution, Woods Hole, Massachusetts 02543, USA, jbernhard@whoi.edu.

Bernhard, J.M., and Buck, K.R., 2004, Eukaryotes of the Cariaco, Soledad, and Santa Barbara Basins: Protists and metazoans associated with deep-water marine sulfide-oxidizing microbial mats and their possible effects on the geologic record, *in* Amend, J.P., Edwards, K.J., and Lyons, T.W., eds., Sulfur Biogeochemistry—Past and Present: Geological Society of America Special Paper 379, p. 35–47. For permission to copy, contact editing@geosociety.org.

INTRODUCTION

Although it has long been known that certain modern-day marine basins such as the Black Sea are sulfide enriched, the use of manned submersibles and remotely operated vehicles for deep-ocean exploration has revealed that sulfidic environments in the modern ocean are relatively common. The discovery of deep-water hydrothermal vents in the 1970s prompted extensive oceanic exploration to determine their extent. In the process of exploring these environments, extensive chemoautotrophic communities were discovered, the energy sources for which are reduced compounds such as hydrogen sulfide and methane, rather than sunlight. Original biological studies concentrated on the larger fauna and the symbiotic prokaryotes (i.e., cells lacking nuclei: bacteria and archaea) of these chemosynthetic communities. Only now are studies elucidating the smaller eukaryotic (cells with nuclei: single-celled protists and multi-celled metazoans) fauna of sulfide-enriched deep-sea habitats. This contribution presents a synopsis of current knowledge of deep-water, sulfide-tolerant protistan and metazoan meiofauna, new ecological data from selected sulfide-enriched sites, and a discourse on the possible effects that these eukaryotes have on the geologic record. Because Earth's early evolution and subsequent events of oceanic anoxia were likely to have included sulfide enrichment (e.g., Canfield 1998), understanding present-day "thiobiotic" communities can help unravel past episodes in Earth's history.

Sulfide-enriched habitats in today's oceans can occur anywhere there is organic enrichment. Besides the hard-substrate hydrothermal vents and the soft-sediment associated "cold" seeps, sulfidic conditions occur in fjords, silled basins (e.g., Santa Barbara Basin, Cariaco Basin, Black Sea), and along the open ocean margins with well-developed oxygen minimum zones (e.g., Monterey Bay, California; off Mazatlan, Mexico). In addition, sulfide enrichment has been observed at large food falls, which are areas on the seafloor where the carcass of a large mammal sank to the seafloor (e.g., whale falls; Bennett et al., 1994; Deming et al., 1997).

The megafaunal (e.g., bivalves, tube worms) and macrofaunal (e.g., polychaetes) chemoautotrophic communities of hydrothermal vents have been extensively studied for their physiology (e.g., Childress and Fisher, 1992), ecology (Van Dover, 2000, and references therein) and biogeography (e.g., Van Dover et al., 2002). Much information is also available regarding cold seep chemoautotrophic communities (e.g., Sibuet and Olu, 1998). Recent studies of whale-fall carcasses indicate interesting taxonomic and gene-flow patterns between these stepping stones and seeps and vents (e.g., Smith and Baco, 1998). Less is known about the eukaryotic fauna of silled basins and deep-water fjords because many of these environments were thought to support only prokaryotes, given that they typically lack larger fauna. Recent studies show that high densities of protists and meiofaunal metazoans occur in at least one bathyal oxygen-depleted, sulfide-enriched silled basin (Santa Barbara Basin, off California; Bernhard et al., 2000). Few studies have addressed the similarly small fauna of hydrothermal vents, but results suggest the presence of ciliates (Small and Gross, 1985; Edgcomb et al., 2002) and flagellates (Edgcomb et al., 2002) at the sediment-covered Guaymas Basin hydrothermal vent. Studies on water column samples collected in proximity to hydrothermal vents and the Guaymas Basin indicated flagellates capable of withstanding high concentrations of hydrogen sulfide (30 mM), suggesting that these taxa may be important components of deep-water hydrothermal vent communities (Atkins et al., 2000, 2002).

Because hydrogen sulfide at micromolar concentrations inhibits respiration, the aerobic eukaryotes of sulfide-enriched environments must have physiological adaptations to allow them to survive such conditions. The majority of physiological studies on chemoautotrophic communities have been devoted to macrofauna and megafauna; little is known about the physiology of meiofauna (e.g., foraminifera, nematodes) and nanobiota (i.e., ciliates, flagellates; see Gage and Tyler [1991] for more discussion on organism size classes) in any sulfide-enriched environment, and even less is known about those inhabiting deep-water sulfidic environments. In general, of the eukaryotes inhabiting sulfidic environments that have been studied, many have prokaryotic associates (e.g., Fenchel and Finlay, 1995; Gaill 1993). For example, nematodes from shallow-water environments are known to harbor putative symbionts (e.g., Ott et al., 1991), as do oligochaetes (Giere et al., 1991).

These putative symbionts presumably provide their host with some metabolic byproduct(s) to promote their survival, but demonstrating metabolic exchange between host and prokaryote is difficult, especially in small eukaryotes such as nanobiota and meiofauna. Prokaryotic associates can be endobionts (living inside the host) or ectobionts (living on the host). In some cases, both endobionts and ectobionts occur (e.g., species of the ciliate *Metopus*; Esteban et al., 1995). Symbionts of metazoan (aerobic) hosts are typically sulfide oxidizers (e.g., nematodes, Polz et al., 1994; Hentschel et al., 1999) or methanotrophs (e.g., bivalves; Vetter and Fry, 1998). In some vent mollusks, both of these types of endosymbionts can be present (e.g., Cavanaugh et al., 1992). In addition, an oligochaete species is known to harbor two types of endosymbionts: sulfate-reducers and sulfide-oxidizers (Dubilier et al., 2001). Symbionts of anaerobic flagellate hosts are known to be methanogens (e.g., Fenchel and Finlay, 1992) or sulfate reducers (Fenchel and Ramsing, 1992).

A number of eukaryotes inhabiting sulfide-enriched habitats lack symbionts; their adaptations to sulfide exposure are varied. For example, in some animals, sulfide oxidation occurs in mitochondria. The hemoglobin of some metazoans binds sulfide to prevent or minimize its detrimental effects on respiration. When respiration is inhibited, a sulfur-dependent anaerobic energy metabolism can be invoked. For details regarding these adaptations, the reader is directed to reviews by Somero et al. (1989), Vismann (1991), Childress (1995), Grieshaber and Völkel (1998), and Hagerman (1998).

From the geological perspective, laminated sediments are invaluable to studies of climate change on annual, decadal, and millennial scales (e.g., Schimmelmann et al., 1992; Behl and Kennett, 1996; Schaaf and Thurow, 1997; Cannariato et al., 1999; Kennett et al., 2000). Considering the fact that sulfide-enriched, laminated sediments support considerable abundances of metazoans (Bernhard et al., 2000; Pike et al., 2001), it is crucial to determine the millimeter-scale vertical distributions of those possible bioturbators. One would not expect disruption of laminated sediments if the metazoans lived exclusively in the surface millimeter, which is the approximate thickness of a typical lamina. However, if metazoans occur throughout the top centimeter of laminated sediments, it is possible that they migrate actively throughout this centimeter given their mobility and need for dissolved oxygen. Protists may also migrate actively throughout surface sediments and thus should also be considered potential "microbioturbators." Furthermore, because the origin and early diversification of eukaryotes coincided with the existence of oxygen-depletion and possible sulfide-enrichment in the deep oceans of the Proterozoic (Canfield, 1998; Canfield and Raiswell, 1999; Condie et al., 2001; Shen et al., 2002), it is important to understand eukaryotic survival capabilities to better understand the Proterozoic ecosystem and thus its geologic record. Finally, given that either sulfate-reducing or sulfide-oxidizing symbionts are associated with extant protists inhabiting sulfidic environments (e.g., Fenchel and Ramsing, 1992; Buck et al., 2000; Vopel et al., 2002), it is likely that the Proterozoic supported similar fauna that was potentially dynamic in terms of sulfur cycling.

HABITAT DESCRIPTION AND METHODS

Samples for this study were collected from three silled basins: Cariaco, Soledad, and Santa Barbara. The Cariaco Basin, which is located off Venezuela, has a maximum depth of ~1400 m and a sill depth of 146 m (Richards, 1975). The anoxic-oxic interface intersected the seafloor in Cariaco at 244 m in May 2001 (Fig. 1; ~10°49.45′N, 64°42.36′W), which is when multicore samples were obtained. The Soledad Basin, which is located off the western coast of Baja California (Mexico; van Geen et al., 2001), has a maximum depth of ~540 m and a sill depth of ~290 m (van Geen et al., 2003). The dissolved oxygen concentration, which was determined by microwinkler analysis (Broenkow and Cline, 1969) of CTD rosette samples, in Soledad's bottom waters at the time of sampling was ~2.7 μM. Multicores from Soledad were collected in November 1999 at 25°12.03′N, 112°43.00′W. The Santa Barbara Basin, which is located off California (United States), has a maximum depth of ~600 m and sill depth of ~425 m. The Santa Barbara sample described here was collected in September 1999 when bottom-water oxygen was 0.1 μM at the site (34°13.5′N, 120°02′W; see Bernhard et al. [2003] for more details). It is important to note that both bottom-water oxygen and hydrogen sulfide concentrations vary considerably in Santa Barbara (e.g., Reimers et al., 1990, 1996; Kuwabara et al., 1999; Bernhard et al., 2003) and, presumably, Soledad Basins. The location of the anoxic-oxic interface and concomitant sulfidic zone in Cariaco also varies (Scranton et al., 1987).

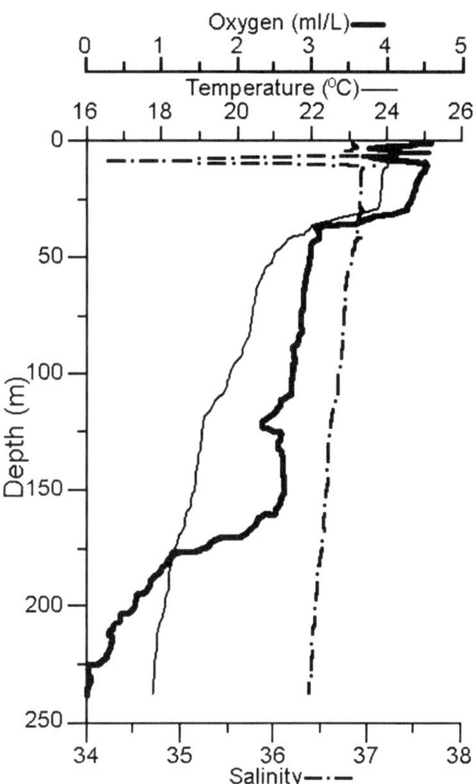

Figure 1. Profiles of salinity, temperature, and dissolved oxygen in the water column above the Cariaco Basin sampling site at the time of collection (May 2001).

The specific environment addressed in this study is that of microbial mats of sulfide-oxidizing bacteria. The bacteria in mats from Cariaco and Santa Barbara were morphologically similar to *Beggiatoa*, while the bacteria in Soledad sediments were morphologically similar to *Thioploca*. *Thioploca* and *Beggiatoa* are both comprised of filaments, but *Beggiatoa* are individual filaments while *Thioploca* filaments are bundled inside polysaccharide sheaths (Jørgensen and Gallardo, 1999). Because of the close phylogenetic affinities of various species of the two genera (Teske et al., 1999), it is possible that the two morphologies merely represent different life habits. Evidence suggests that both *Thioploca* and *Beggiatoa* can use nitrate as their electron acceptor (e.g., Jørgensen and Gallardo, 1999; McHatton et al., 1996). Thus, both types require sulfide and little or no oxygen. Although sulfide concentrations have not been measured in Soledad Basin, the sediments must be sulfidic, at least enough to support copious populations of the sulfide-oxidizer *Thioploca*.

The methods employed follow our standard procedures. Nanobiotic and meiofaunal quantification were executed according to the density gradient extraction technique and DAPI staining method described in Bernhard et al. (2000). A second sample was quantified for foraminiferal species counts using the method of Bernhard et al. (1997). Ultrastructural procedures were those described in Bernhard et al. (2000). Sub-millimeter life positions were determined using the recently described Fluorescently Labeled Embedded Core (FLEC) method (Bernhard et al., 2003). FLEC material was imaged with an Olympus Fluoview personal laser scanning confocal microscope (LSCM).

RESULTS

Abundance and Taxonomic Composition

The eukaryotic fauna of Soledad basin was abundant in the top 1 cm, but was not as dense as that of the laminated sediments of Santa Barbara Basin (Fig. 2A). The Soledad community was, however, denser than a comparable, more aerated site located off Southern California (Fig. 2A).

Foraminifera were the dominant eukaryotic biovolume contributor in Soledad Basin (Fig. 2B), as they are in Santa Barbara Basin (Bernhard et al., 2000). Although quantitative samples are not available for Cariaco, this site also appeared to be dominated in terms of biovolume by foraminifera. In all three sites, the foraminiferal fauna was dominated by a single species: *Bolivina subadvena* in Soledad (59%, Table 1), *Virgulinella fragilis* in Cariaco (Bernhard, 2003), and *Nonionella stella* in Santa Barbara (Bernhard et al., 1997, 2000).

A total of 18 foraminiferal species occurred in the *Thioploca*-laden Soledad sediments (Table 1). In addition to the dominant *B. subadvena*, Soledad had low densities of species that also occur in Santa Barbara Basin (i.e., *Bolivina argentea*, *Chilostomella oolina*, *Fursenkoina rotundata*, *Nonionella stella*; Bernhard et al., 1997). *Buliminella* species occurred in Soledad, although they differ from those of Santa Barbara Basin. A single specimen of *Virgulinella* was observed; this specimen appears to be a species other than *V. fragilis*, which dominates Cariaco sediments. Although allogromids (tectinous foraminifera) are relatively rare in oxygen-depleted environments (Gooday et al., 2000), they occurred in relatively high abundances in Soledad sediments (Table 1).

Most species of flagellates and ciliates from deep-sea sedimentary environments are unidentified. Of the taxa we can confidently identify in these two groups, it does not appear that any species occurred in all three sites (Soledad, Cariaco, Santa Barbara). *Parablepharisma* ciliates were common to all three basins, but additional studies are needed to determine if these *Parablepharisma* are conspecifics. The ciliate genus *Metopus* also occurred at all three sites, sometimes in multiple morphotypes. Soledad and Santa Barbara both had the ciliate *Metopus verrucosus* (Bernhard et al., 2000; Buck and Bernhard, 2001) and the flagellates *Calkinsia aureus* (Bernhard et al., 2000; Buck and Bernhard, 2001) and *Sphenomonas* sp.

The only metazoans observed in Soledad samples were nematodes. Nematodes were also the dominant metazoan taxon in the Cariaco *Beggiatoa* sample. However, as in Santa Barbara Basin, gastrotrichs and polychaetes were also present in the *Beggiatoa*-

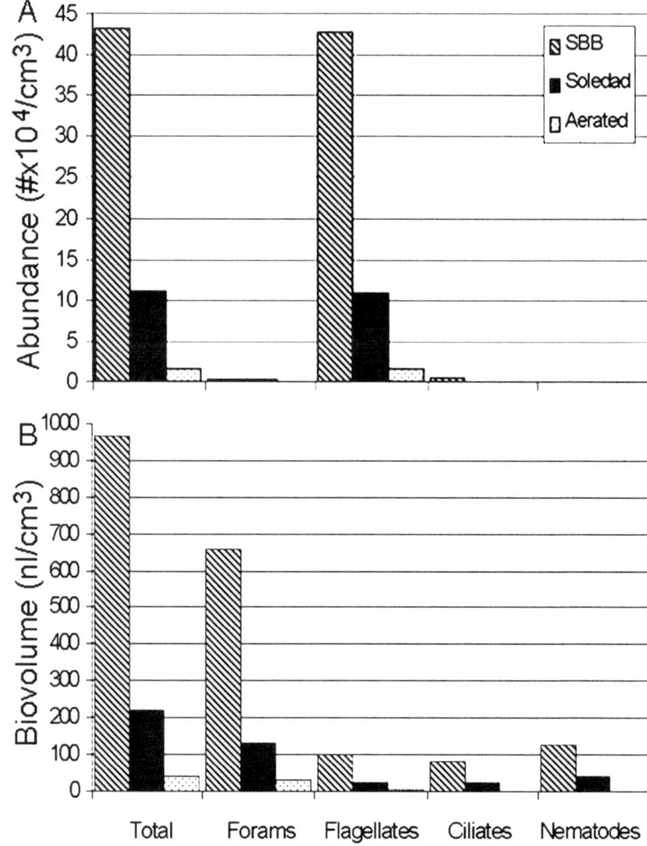

Figure 2. Histograms of eukaryote abundance (A) and biovolume (B) from Santa Barbara Basin (SBB), Soledad Basin, and comparative aerated, sulfide-free samples collected off Southern California, presented by major taxa. Data for Santa Barbara Basin and comparable aerated site are from Bernhard et al. (2000).

TABLE 1. ABUNDANCES OF ROSE BENGAL STAINED BENTHIC FORAMINIFERA FROM AN ALIQUOT OF SOLEDAD BASIN SEDIMENTS

Species	n	# cm^{-3}	% of total
Adercotryma sp.?	70	92.3	3.8
Bolivina argentea	8	10.5	0.4
Bolivina cf *seminuda*	1	1.32	0.1
Bolivina cf *subadvena*	1097	1450	59.0
Buliminella sp. 1	16	21.1	0.9
Buliminella sp. 2	82	108	4.4
Chilostomella oolina	6	7.91	0.3
Dorothia pseudofiliformis?	84	111	4.5
Epistominella sp.	9	11.9	0.5
Fursenkoina rotundata	10	13.2	0.5
Lagenammina sp.	125	165	6.7
Leptohalysis sp.	107	141	5.8
Nonionella stella	31	40.9	1.7
Stainforthia fusiformis	87	115	4.7
Textularia cf *agglutinans*	88	116	4.7
Virgulinella sp.	1	1.32	0.1
Unidentified rotalid	2	2.64	0.1
Allogromid spp.	35	46.1	1.9
Total foraminifera	1859	2450	–

Note: n—number picked from 0.759 cm^3 of sediment.

laden Cariaco sediments. The single gastrotrich morphotype is most likely a new *Urodasys* species because it is smaller than its Santa Barbara Basin congener *U. anorektoxys* (Todaro et al., 2000). The Cariaco polychaete is a nerillid, but not the species present in Santa Barbara Basin (*Xenonerilla bactericola*, Müller et al., 2001). Most likely, this Cariaco polychaete is new to science (M. Müller, 2002, personal commun.). A single oligochaete was also observed in the Cariaco *Beggiatoa* sample; additional specimens are required for its identification.

Prevalence of Putative Symbionts

Each of the major eukaryotic biomass contributors in Santa Barbara (*Nonionella stella*) and Cariaco (*Virgulinella fragilis*) exhibit the unusual characteristic of sequestering chloroplasts (Bernhard and Bowser, 1999; Bernhard, 2003). Although these foraminifers live in darkness, the sequestered plastids are presumably crucial to their dominance in these sulfide-enriched environments (Grzymski et al., 2002). The ultrastructure of foraminifers from Soledad has not been analyzed, so it is unclear if its dominant species, *Bolivina subadvena,* similarly sequesters chloroplasts.

Neither the Soledad flagellate nor ciliate community was dominated by specimens with associated putative symbionts (in terms of either abundance or biomass; Fig. 3). This contrasts with those communities of Santa Barbara Basin (Fig. 3; Bernhard et al., 2000). The flagellate and ciliate community of Cariaco's *Beggiatoa*-laden sediments also appeared to be dominated (in terms of abundance and biovolume) by eukaryotes with prokaryotic associates. Although the available non-quantitative Cariaco sample only allows relative comparisons, ciliates with putative symbionts comprised 96.5% of the ciliate population while 57.7% of the flagellates similarly had putative symbionts.

As in Santa Barbara's *Xenonerilla bactericola* (Bernhard et al., 2000; Müller et al., 2001), the Cariaco polychaete had rod-shaped bacterial ectobionts (Fig. 4). The epibionts had well-developed attachments to the polychaete epidermis (Fig. 4B). A specimen of a Cariaco gastrotrich appeared to have bacterial associates when examined using DAPI to identify nuclei and prokaryotes. Ultrastructural analysis of this species is pending. In the few specimens examined ultrastructurally to date, putative symbionts were not observed in either of the other Cariaco metazoans (i.e., nematode, oligochaete). Although the ultrastructure of Soledad nematodes has not yet been examined, DAPI results suggest the presence of bacterial ectobionts.

Sub-Millimeter Life Positions in Laminated Soledad and Santa Barbara Sediments

Eukaryotes in Soledad sediments were concentrated in the surface ~2 mm, even when bottom water O_2 was relatively high (2.7 µM; Fig. 5). When such "high" oxygen concentrations occurred in Santa Barbara, the eukaryotes were also concentrated in the top few mm, but they also occurred in relatively high abun-

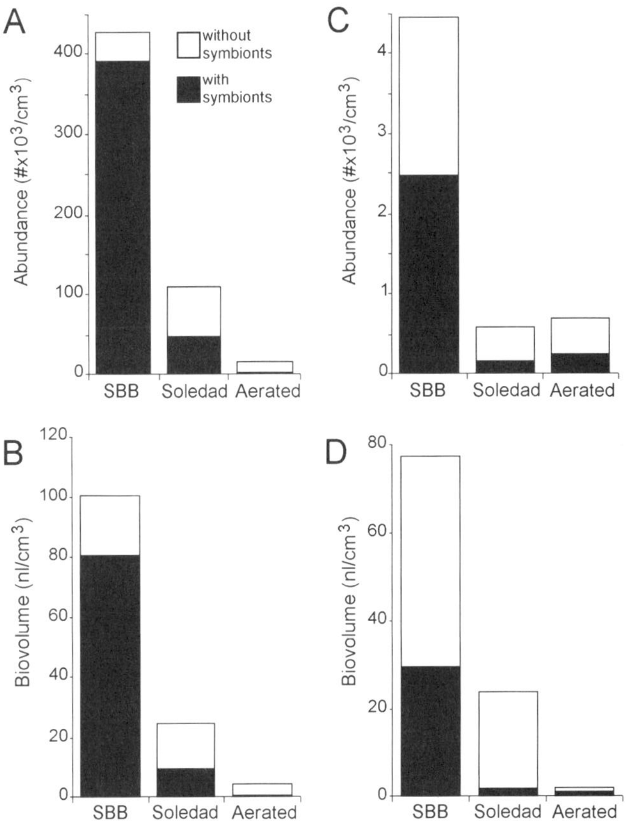

Figure 3. Histograms of flagellate (A, B) and ciliate (C, D) abundance and biovolume data, presented for specimens with and without putative symbionts. SBB—Santa Barbara Basin.

dances to depths of at least 8 mm (2.4 µM O_2; Bernhard et al., 2003). In the Soledad FLEC material examined to date, very few protists (foraminifera, flagellates, ciliates) were observed at a depth >2mm, although nematodes were noted to depths of ~1.5 cm. Life-position data is not available for Cariaco sediments.

Soledad FLEC sections show that the laminae in the top 2 mm were substantially disrupted (Fig. 6). The disruptions were not caused by sampling because prokaryotes appeared to selectively inhabit lighter-colored layers rather than darker layers (Fig. 6A). Because gravity cores and piston cores show that Soledad subsurface sediments are laminated to a depth of >5 m (van Geen et al., 2001), it is possible that disrupted laminae "realign" during compaction due to sediment burial or that modern environmental conditions differ from those that produced the well-preserved laminae.

Higher magnification examination of Soledad FLEC material shows some of the abundant foraminifera, flagellates, ciliates, and filamentous bacteria other than *Thioploca* (Fig. 7). Specimens of the agglutinated foraminifer *Leptohalysis* sp. and the tectinous foraminifer *Nodellum* sp. were easily identified (Fig. 7A). The presence of *Nodellum* sp. in Soledad samples is,

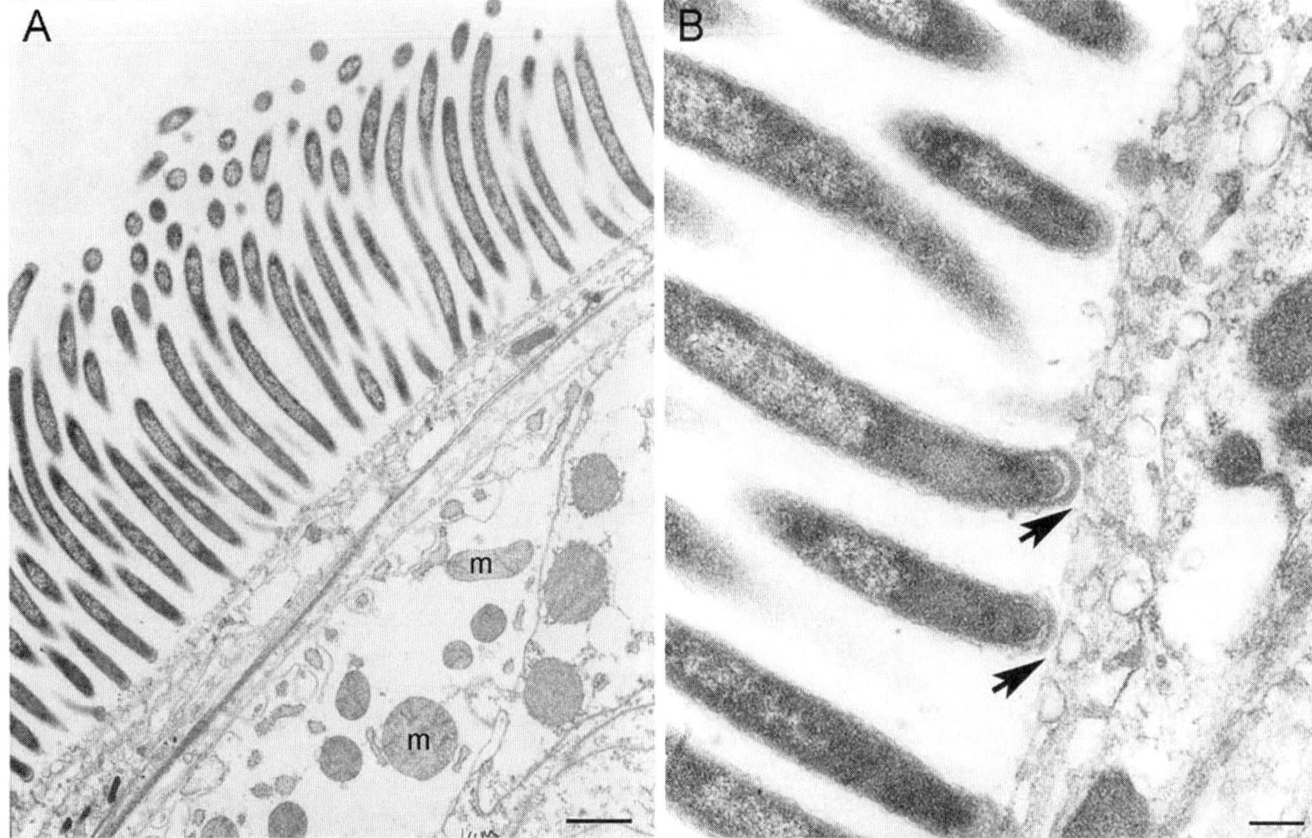

Figure 4. Transmission electron micrographs of the undescribed Cariaco polychaete. (A) View including the polychaete epidermis and attached bacterial ectobionts. (B) Higher-magnification view showing morphological modifications at attachment points (arrows). m—mitochondria. Scale bars: A =1 µm; B = 200 nm.

to our knowledge, the first recorded occurrence of this tectinous genus in laminated sediments.

Sub-millimeter life positions of Santa Barbara Basin eukaryotes have recently been described (Bernhard et al., 2003). Examination of additional material indicates that metazoans did not occur exclusively at the sediment-water interface even when oxygen was nearly undetectable (Fig. 8; 0.1 µM O_2; Bernhard et al., 2003). More specifically, polychaetes and nematodes were both observed at depths of at least 6 mm (Fig. 8) in laminated sediments (see Figure 2D in Bernhard et al., 2003).

Given the apparently strict occurrence of Soledad eukaryotes in the surface 1–2 mm (Fig. 5), abundance data illustrated in Figure 2 was recalculated to determine the density in the top 2 mm of Soledad sediments. When considered in this manner, Soledad eukaryote densities and biovolume in Soledad exceed those in the surface centimeter of Santa Barbara Basin (i.e., Soledad density ~55.9×10^4 cm^{-3} versus Santa Barbara ~43.3×10^4 cm^{-3}; Soledad biovolume ~1101 nl cm^{-3} versus Santa Barbara ~967 nl cm^{-3}). Such a perspective suggests that ecological approaches at the centimeter scale are not always representative of in situ conditions at scales relevant to the microorganisms.

DISCUSSION

Our observations suggest that, in general, the eukaryotic fauna of deep-water sulfidic habitats is relatively diverse, with a high dominance of few or one species, and a high total abundance, if considered at the appropriate scale. In addition, protists typically dominate abundance and biovolume, and most eukaryotic taxa have bacterial associates. Taxonomic comparisons indicate similarities in metazoan taxa, with nematodes in all three basins and gastrotrichs and primitive polychaetes occurring in two of the basins (Table 2). In at least two of the three basins, the dominant foraminifer sequesters chloroplasts, even though the basin seafloor lies far below the maximum extent of the photic zone.

Although little is known regarding the taxonomy of flagellates and ciliates from oxygen-depleted deep-water sediments (or their associated putative symbionts), much is known about the comparable fauna from shallow-water environments (e.g., Fenchel and Finlay, 1995). Given the recent and controversial assertion that protistan species are cosmopolitan and, by extrapolation, of limited diversity (Finlay, 2002), deep-water nanobiota are crucial communities with which to test this hypothesis. Our

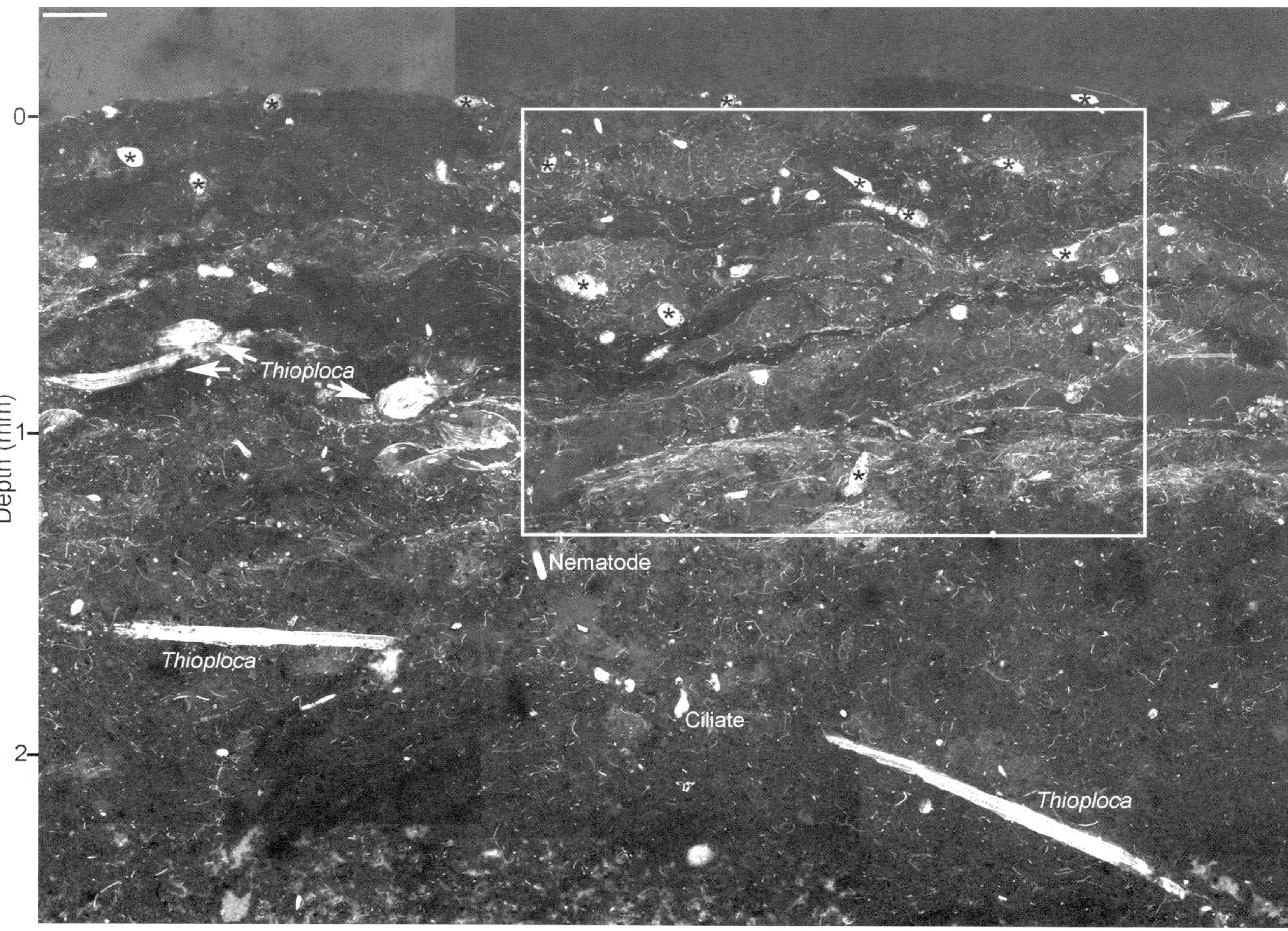

Figure 5. Laser scanning confocal microscope montage of Fluorescently Labeled Embedded Core section from Soledad Basin surface sediments. *—foraminifera. Flagellates and most ciliates are not labeled because they are too small to be seen at this magnification. Scale bar = 200 μm.

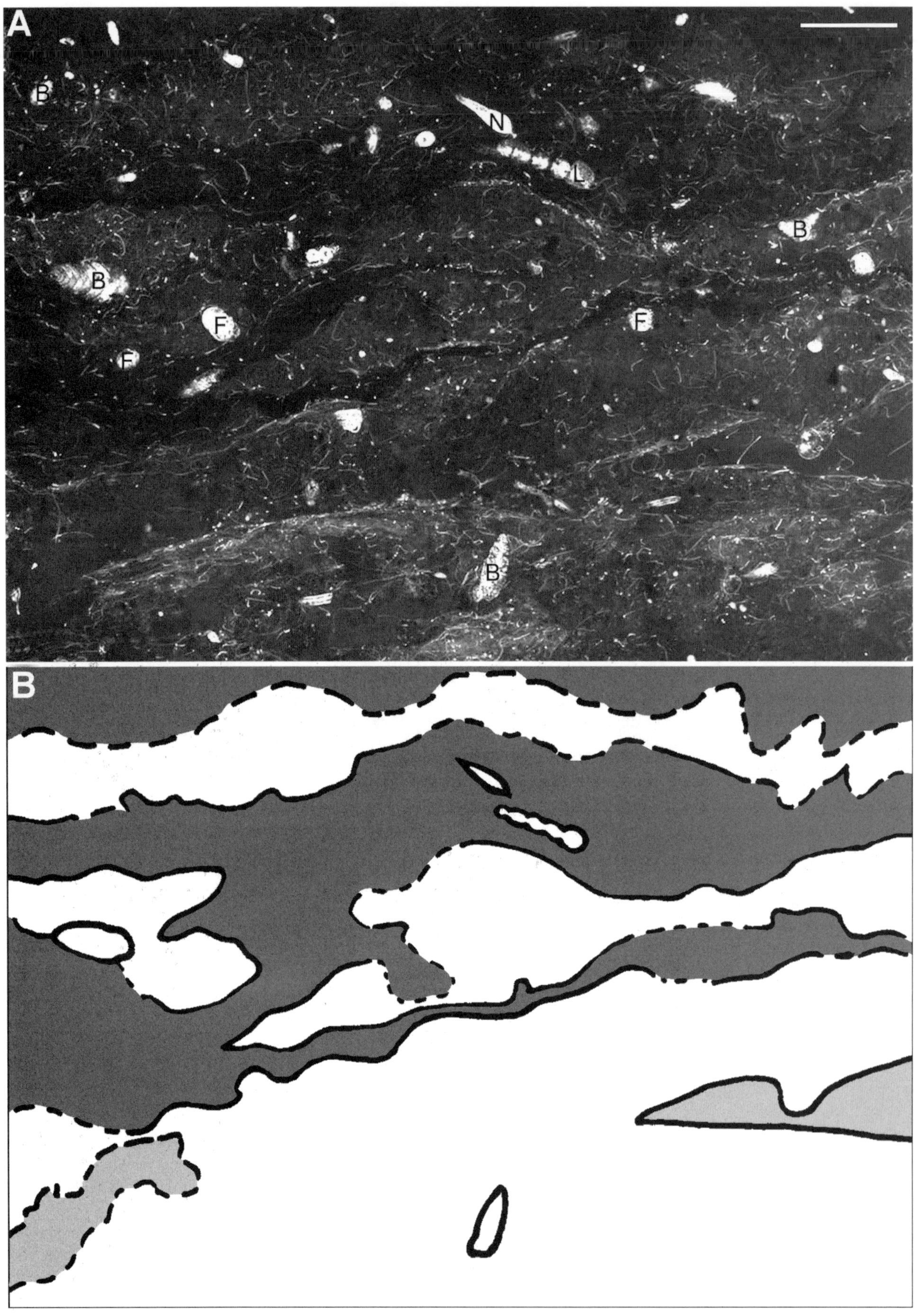

Figure 6. (A) Higher magnification laser scanning confocal microscope montage illustrating selected area shown in Figure 5. (B) Schematic of laser scanning confocal microscope montage shown in A, with some larger eukaryotes outlined for orientation. Darker shaded areas represent dark lamina. Note that prokaryotes appear to occur in lighter laminae (unshaded regions). Also note pore-water voids (light-gray shading). Dashed lines represent indistinct boundaries. Scale bar = 200 μm.

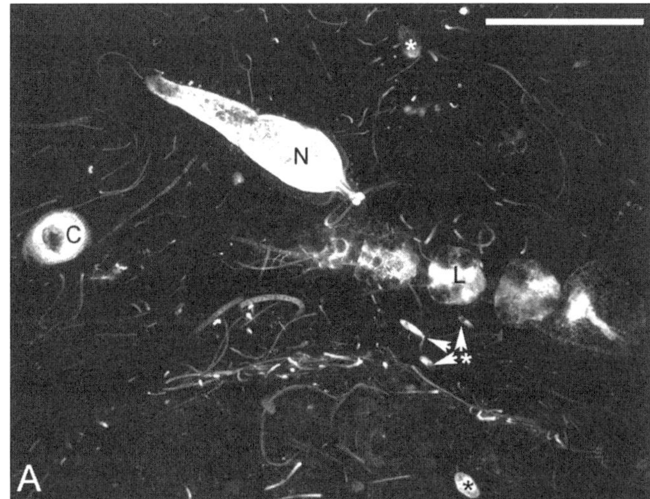

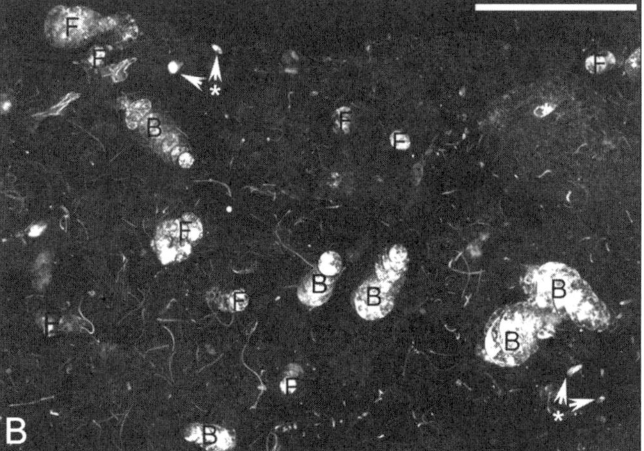

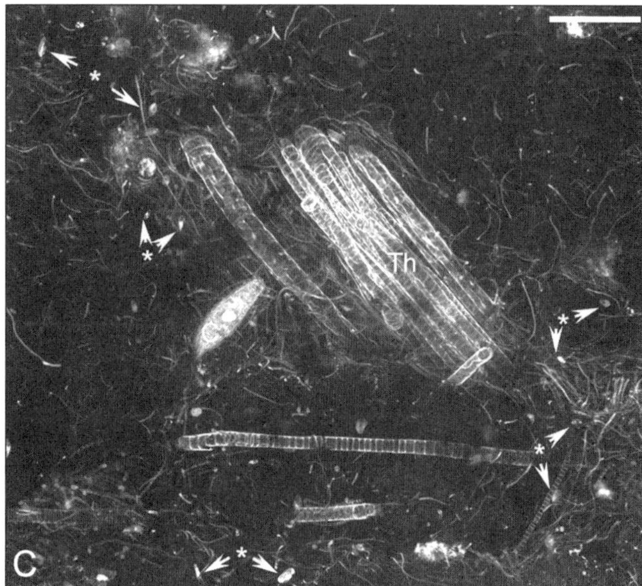

Figure 7. Higher magnification laser scanning confocal microscope images of Soledad Fluorescently Labeled Embedded Core sections. (A) Allogromid foraminifer *Nodellum* sp. (N), the agglutinated foraminifer *Leptohalysis* (L), a ciliate (C), and flagellates (*) from small area included in Figures 5 and 6. (B) Numerous foraminifers near sediment surface of the same section as that shown in Figures 5 and 6. Sediment-water interface occurs near top of image. B—*Bolivina subadvena*, F—unidentified foraminifer, *—flagellates. (C) *Thioploca* (Th) bundle along with other filamentous prokaryotes, a ciliate (C), and flagellates (*). Not all flagellates are labeled. Image was taken from a different multicore from that imaged in Figures 5 and 6. *Thioploca* bundle was ~1.5 mm below the sediment-water interface. Number of images compiled/distance between images (μm): A—36/0.7; B—83/0.8; C—186/0.7. Scale bars, A, C—100 μm; B—200 μm.

data to date show some consistencies between nanobiotic taxa, but the communities from the three basins certainly are not identical. Molecular analyses of these communities would yield additional insights into protistan biogeography as well as their diversity. It is equally important to establish that the eukaryotes present in these environments are actually viable and active.

Soledad and Santa Barbara Basin had different sulfide-oxidizing bacteria dominating their laminated sediments even when bottom water O_2 was similar: Soledad was dominated by *Thioploca* while Santa Barbara Basin supported *Beggiatoa*. While sheaths of *Thioploca* impart the ability to access hydrogen sulfide from deep within the sediments, *Beggiatoa* is thought to occur only where sulfide gradients are particularly steep (Jørgensen and Revsbech, 1983; Schulz et al., 1996). Given these generalizations, it may be expected that the H_2S in Soledad surface sediments was less than that in Santa Barbara Basin. Thus, it may also be expected that eukaryotes, which are generally negatively impacted by hydrogen sulfide, would be distributed deeper in Soledad sediments compared to eukaryotic distributions in Santa Barbara. This trend, however, was not evident from the material examined to date. Besides pore-water sulfide concentrations, additional driving forces causing the observed different sub-millimeter distributions and microbial composition are unknown but are probably linked to variations in sediment porosity, sediment fabric, and eukaryote physiology. Dedicated geochemical and microbiological studies will help reveal the driving forces for such faunal differences.

Although *Thioploca*- and *Beggiatoa*-dominated sites showed differences in sub-millimeter life positions, the taxonomic composition of those communities was somewhat similar. For example, Soledad (*Thioploca*) and Santa Barbara Basin (*Beggiatoa*) both supported the foraminifer *Nonionella stella*, the flagellates *Calkinsia aureus* and *Sphenomonas* sp., and the ciliate *Metopus verrucosus*. Given the relatively close proximity, however, between Santa Barbara and Soledad, one might expect more similarities in taxonomic composition.

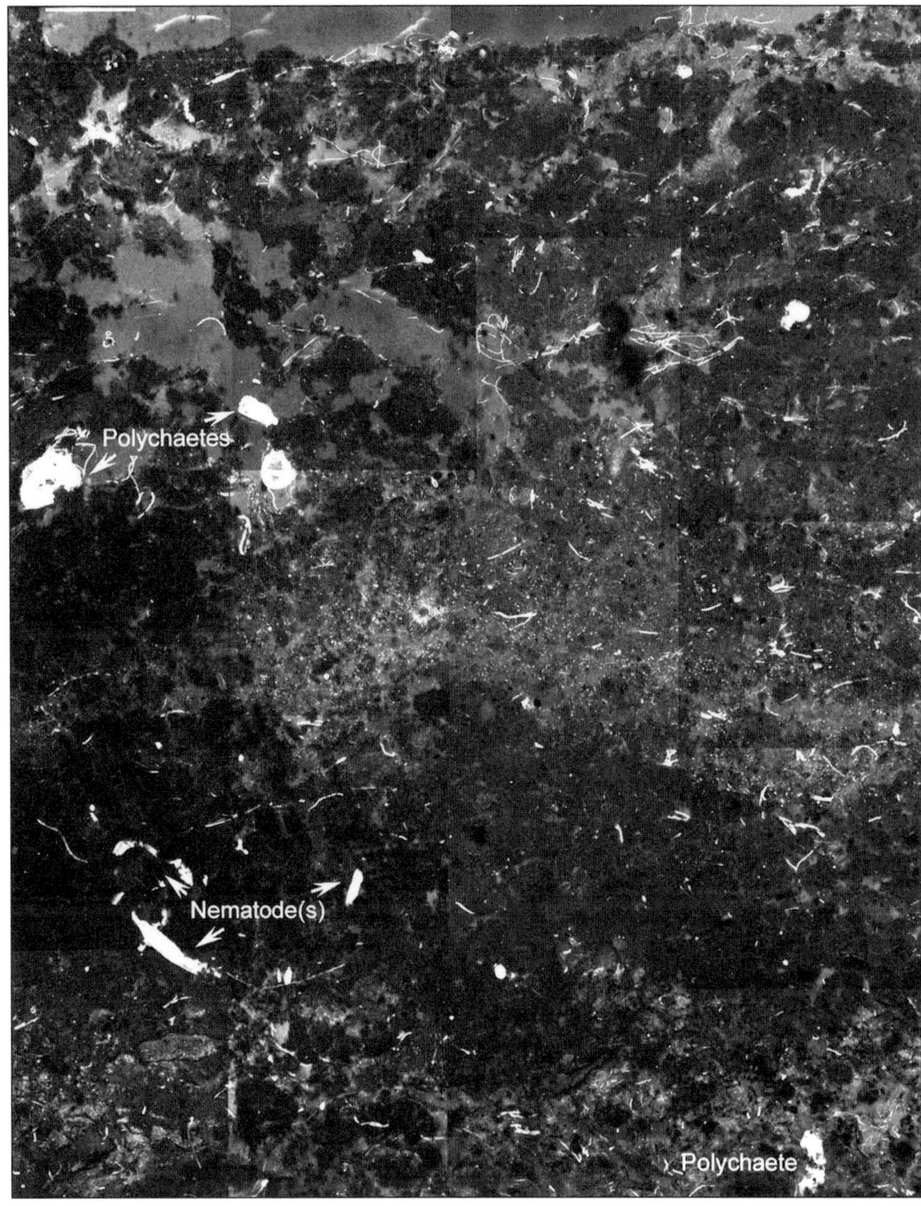

Figure 8. Laser scanning confocal microscope montage of Fluorescently Labeled Embedded Core section from Santa Barbara Basin surface sediments showing metazoans to ~6.7 mm depth. Unlabeled fluorescent shapes are protists or prokaryotes. Scale bar = 0.5 mm.

Effects on the Geologic Record

Geochemistry

Because certain meiofauna and ciliates can increase solute transport by greater than twice that of diffusion (Aller and Aller, 1992; Glud and Fenchel, 1999), eukaryotes inhabiting sulfide-enriched sediments must promote geochemical reactions. For example, due to the infiltration of oxygen into sediments, caused by protistan and metazoan activities and the enhanced nucleation of pyrite formation in slightly oxidized settings (Wilkin and Barnes, 1996; Benning et al., 2000), pyrite is probably formed at depth at a higher rate than appreciated (Pike et al., 2001). The oxidation rate of organic matter is also likely to be increased due to eukaryote activities. Even if oxygen is present only in trace concentrations, it will be drawn down into the sediments, promoting oxidation.

In areas with fluctuating redox boundaries, the mobility of eukaryotes could promote symbiont growth and longevity compared to that of free-living prokaryotes because the eukaryotes presumably track particular geochemical conditions, thereby continually exposing their prokaryotic associates to ideal conditions. Compared to metazoans, free-living prokaryotes are likely to move slowly and would thus be less likely to successfully track quickly migrating geochemical cues. When a migrating redox boundary shoals and, therefore, steepens, eukaryotes will be congregated in a smaller space, and thus rates of some geochemical processes may be increased. Fluctuating redox

TABLE 2. GENERAL QUALITATIVE COMPARISON BETWEEN THE THREE SULFIDE-ENRICHED BASINS STUDIED

	Santa Barbara	Cariaco	Soledad
Dominant colorless S bacterium	*Beggiatoa*	*Beggiatoa*	*Thioploca*
Eukaryote abundance	High	High	High
Foraminifera with plastids	Yes	Yes	ND
Nerillid polychaete with epibionts	Yes	Yes	?
Urodasys gastrotrich	Yes	Yes	?
Nematodes with epibionts	Yes	?	Yes

Note: Yes—presence of that organism/taxon; ND—not yet investigated; ?—not yet observed.

boundaries are not only expected in laminated sediments but also in sediments associated with cold seeps that have variable flow rates. Bacterial mats associated with cold seeps are also known to support numerous eukaryotes (Buck and Barry, 1998), often with symbionts (Buck et al., 2000).

Recent life-position studies of the Santa Barbara Basin microbial community indicate that laminated sediments are a vertically and horizontally heterogeneous mosaic of organism distributions and inferred chemical regimes (Bernhard et al., 2003). Thus, physiochemical conditions along any given lamina should not be expected to be consistent. Although pore-water geochemistry is an important force in structuring microorganism distributions in Santa Barbara Basin, it is likely that the distributions of the microorganisms also modify the localized geochemical environment (Bernhard et al., 2003). Soledad FLEC analyses suggest, however, that life positions of both eukaryotes and prokaryotes in laminated sediments from different sites are not necessarily identical. For example, initial observations suggest Soledad protists are concentrated in the surface 2 mm. Additional analyses of more material are required before confident conclusions can be asserted about the effects of eukaryotes on laminites in general. In the sulfide-enriched sediments examined to date, however, the high eukaryotic abundance and the varied physiologies of their epibionts and endobionts suggests that these sites are hotspots of carbon and sulfur cycling.

Knowledge of the benthic foraminiferal assemblages inhabiting sulfur-oxidizing microbial mats will aid paleoecology. If species' threshold tolerances to sulfide are determined, geochemical reconstructions will also benefit. In addition, at least one foraminiferal species is known to be strongly depleted in ^{13}C (*Virgulinella fragilis* from Cariaco; $\delta^{13}C = -6.4‰$; Bernhard, 2003), indicating that this species, and possibly others, are good indicators of environments with high rates of sulfate reduction (Bernhard, 2003).

Sedimentology

Early sediment fabric of laminated sediments is also probably affected by eukaryotes. The likely movement of metazoans throughout the surface millimeters to centimeters of laminated sediments could cause stratigraphic slurring by transporting material between layers without disrupting laminae boundaries (Pike et al., 2001). Indeed, the presence of both endosymbiotic sulfate reducers and sulfide oxidizers in oligochaete individuals (*Olavius algarvensis*; Dubilier et al., 2001) suggests active oligochaete migration between oxidized and reduced sediments to provide symbionts with required oxidants and reductants. Although metazoans from deep-water laminated sediments are not known to support these two types of symbionts, such a possibility cannot be discounted.

In sum, given the high abundances of metazoans to depths of at least 3 cm in Santa Barbara Basin sediments (Todaro et al., 2000; Müller et al., 2001; Pike et al., 2001), the need for oxygen by all metazoans in at least part of their life cycle (Fenchel and Finlay, 1995), and life-position observations presented here, they must migrate through sediments and thus have significant impacts on the geochemistry and sedimentology of laminated sediments. Assuming a sedimentation rate of 1 cm/yr (before compaction; Reimers et al., 1990), the top 3 cm in Santa Barbara Basin sediments equates to up to 36 months of exposure to microbioturbation. Observations suggest that deep-water laminated sulfidic sediments are not necessarily postdepositionally pristine in terms of disturbance or geochemistry. The high abundances of microbiota of various physiologies could have substantial impacts on the rates of carbon and sulfur cycling. Only integrated biogeochemical studies will elucidate the magnitude and significance that the eukaryotic community has on the geological record.

ACKNOWLEDGMENTS

We thank the captains and crews of the RV *Melville* and RV *Hermano Gines*; Lex van Geen for providing the opportunity to join the cruise off Baja; Ron Comer, Shad Baiz, Eric Tappa, and Mark Woodworth for sampling assistance; Rob Bourgeois for field and laboratory assistance; Andy Gooday for sharing his knowledge on *Nodellum*; Tom Chandler for LSCM access; Pam Murphy for FLEC assistance; and Jessica Blanks and Christie Robinson for Soledad foraminiferal counts. We also thank K. Edwards and J. Amend for organizing the truly multidisciplinary "Sulfur Session" and two anonymous reviewers for their helpful comments. Soledad ship time was provided by National Science Foundation (NSF) grant OCE-9809026 to A. van Geen. Support for KRB came from a grant from the Packard Foundation to MBARI (to J.P. Barry). Funded by NSF grant OCE-0095564 and NSF grant OCE-9711812, both to JMB.

REFERENCES CITED

Aller, R.C., and Aller, J.Y., 1992, Meiofauna and solute transport in marine muds: Limnology and Oceanography, v. 37, p. 1018–1033.

Atkins, M.S., Teske, A.P., and Anderson, O.R., 2000, A survey of flagellate diversity at four deep-sea hydrothermal vents in the eastern Pacific Ocean using structural and molecular approaches: The Journal of Eukaryotic Microbiology, v. 47, p. 400–411.

Atkins, M.S., Hanna, M.A., Kupetsky, E.A., Saito, M.A., Taylor, C.D., and Wirsen, C.O., 2002, Tolerance of flagellated protists to high sulfide and metal concentrations potentially encountered at deep-sea hydrothermal vents: Marine Ecology Progress Series, v. 226, p. 63–75.

Behl, R.J., and Kennett, J.P., 1996, Brief interstadial events in the Santa Barbara Basin, NE Pacific, during the past 60 kyr: Nature, v. 379, p. 243–246, doi: 10.1038/379243A0.

Bennett, B.A., Smith, C.R., Glaser, B., and Maybaum, H.L., 1994, Faunal community structure of a chemoautotrophic assemblage on whale bones in the deep northeast Pacific Ocean: Marine Ecology Progress Series, v. 108, p. 205–223.

Benning, L.G., Wilkin, R.T., and Barnes, H.L., 2000, Reaction pathways in the Fe-S system below 100°C: Chemical Geology, v. 167, p. 25–51, doi: 10.1016/S0009-2541(99)00198-9.

Bernhard, J.M., 2003, Potential symbionts in bathyal foraminifera: Science, v. 299, p. 861, doi: 10.1126/SCIENCE.1077314.

Bernhard, J.M., and Bowser, S.S., 1999, Benthic foraminifera of dysoxic sediments: Chloroplast sequestration and functional morphology: Earth Science Reviews, v. 46, p. 149–165, doi: 10.1016/S0012-8252(99)00017-3.

Bernhard, J.M., Sen Gupta, B.K., and Borne, P.F., 1997, Benthic foraminiferal proxy to estimate dysoxic bottom-water oxygen concentrations: Santa Barbara Basin, U.S. Pacific Continental Margin: Journal of Foraminiferal Research, v. 27, p. 301–310.

Bernhard, J.M., Buck, K.R., Farmer, M.A., and Bowser, S.S., 2000, The Santa Barbara Basin is a symbiosis oasis: Nature, v. 403, p. 77–80, doi: 10.1038/47476.

Bernhard, J.M., Visscher, P.T., and Bowser, S.S., 2003, Sub-millimeter life positions of bacteria, protists, and metazoans in laminated sediments of the Santa Barbara Basin: Limnology and Oceanography, v. 48, p. 813–828.

Broenkow, W.W., and Cline, J.D., 1969, Colorimetric determination of dissolved oxygen at low concentrations: Limnology and Oceanography, v. 14, p. 450–454.

Buck, K.R., and Barry, J.P., 1998, Monterey Bay cold seep infauna: Quantitative comparison of bacterial mat meiofauna with non-seep control sites: Cahiers de Biologie Marine, v. 39, p. 333–335.

Buck, K.R., and Bernhard, J.M., 2001, Protistan-Prokaryotic symbioses in deep-sea sulfidic sediments, *in* Seckbach, J., ed., Symbiosis: Amsterdam, Kluwer Academic, p. 507–517.

Buck, K.R., Barry, J.P., and Simpson, A.G.B., 2000, Monterey Bay cold seep biota: Euglenozoa with chemoautotrophic bacterial epibionts: European Journal of Protistology, v. 36, p. 117–126.

Canfield, D.E., 1998, A new model for Proterozoic ocean chemistry: Nature, v. 396, p. 450–453, doi: 10.1038/24839.

Canfield, D.E., and Raiswell, R., 1999, The evolution of the sulfur cycle: American Journal of Science, v. 299, p. 697–723.

Cannariato, K.G., Kennett, J.P., and Behl, J.P., 1999, Biotic response to late Quaternary rapid climate switches in Santa Barbara Basin: Ecological and evolutionary implications: Geology, v. 27, p. 63–66, doi: 10.1130/0091-7613(1999)0272.3.CO;2.

Cavanaugh, C.M., Wirsen, C.O., and Jannasch, H.W., 1992, Evidence for methylotrophic symbionts in a hydrothermal vent mussel (Bivalvia: Mytilidae) from the Mid-Atlantic Ridge: Applied and Environmental Microbiology, v. 58, p. 3799–3803.

Childress, J.J., 1995, Life in sulfidic environments: historical perspective and current research trends: American Zoologist, v. 35, p. 83–90.

Childress, J.J., and Fisher, C.R., 1992, The biology of hydrothermal vent animals—physiology, biochemistry, and autotrophic symbioses: Oceanography and Marine Biology, v. 30, p. 337–441.

Condie, K.C., Des Marais, D.J., and Abbott, D., 2001, Precambrian superplumes and supercontinents: a record in black shales, carbon isotopes, and paleoclimates?: Precambrian Research, v. 106, p. 239–260, doi: 10.1016/S0301-9268(00)00097-8.

Deming, J.W., Reysenbach, A.L., Macko, S.A., and Smith, C.R., 1997, Evidence for the microbial basis of a chemoautotrophic invertebrate community at a whale fall on the deep seafloor: Bone colonizing bacteria and invertebrate symbionts: Microscopy Research and Technique, v. 37, p. 162–170, doi: 10.1002/(SICI)1097-0029(19970415)37:23.0.CO;2-Q.

Dubilier, N., Mülders, C., Ferdelman, T., de Beer, D., Pernthaler, A., Klein, M., Wagner, M., Erséus, C., Thiermann, F., Krieger, J., Giere, O., and Amann, R., 2001, Endosymbiotic sulphate-reducing and sulphide-oxidizing bacteria in an oligochaete worm: Nature, v. 411, p. 298–302, doi: 10.1038/35077067.

Edgcomb, V.P., Kysela, D.T., Teske, A., de Vera Gomez, A., and Sogin, M.L., 2002, Benthic eukaryotic diversity in the Guaymas Basin hydrothermal vent environment: Proceedings of the National Academy of Sciences of the United States of America, v. 99, p. 7658–7662, doi: 10.1073/PNAS.062186399.

Esteban, G., Fenchel, T., and Finlay, B., 1995, Diversity of free-living morphospecies in the ciliate genus *Metopus*: Archiv fur Protistenkunde, v. 146, p. 137–164.

Fenchel, T., and Finlay, B.J., 1992, Production of methane and hydrogen by anaerobic ciliates containing symbiotic methanogens: Archives of Microbiology, v. 157, p. 475–480.

Fenchel, T., and Finlay, B.J., 1995, Ecology and evolution in anoxic worlds: Oxford, Oxford University Press, 276 p.

Fenchel, T., and Ramsing, N.B., 1992, Identification of sulphate-reducing ectosymbiotic bacteria from anaerobic ciliates using 16S rRNA binding oligonucleotide probes: Archives of Microbiology, v. 158, p. 394–397.

Finlay, B.J., 2002, Global dispersal of free-living microbial eukaryotic species: Science, v. 296, p. 1061–1063, doi: 10.1126/SCIENCE.1070710.

Gage, J.D., and Tyler, P.A., 1991, Deep-sea biology: A natural history of organisms living at the deep-sea floor: Cambridge, Cambridge University Press, 504 p.

Gaill, F., 1993, Aspects of life development at deep-sea hydrothermal vents: The FASEB Journal, v. 7, p. 558–565.

Giere, O., Conway, N.M., Gastrock, G., and Schmidt, C., 1991, Regulation of gutless annelid ecology by endosymbiotic bacteria: Marine Ecology Progress Series, v. 68, p. 287–299.

Glud, R.N., and Fenchel, T., 1999, The importance of ciliates for interstitial solute transport in benthic communities: Marine Ecology Progress Series, v. 186, p. 87–93.

Gooday, A.J., Bernhard, J.M., Levin, L.A., and Suhr, S., 2000, Foraminifera in the Arabian Sea oxygen minimum zone and other oxygen-deficient settings: taxonomic composition, diversity, and relation to metazoan faunas: Deep-Sea Research II, v. 47, p. 25–54, doi: 10.1016/S0967-0645(99)00099-5.

Grieshaber, M.K., and Völkel, S., 1998, Animal adaptations for tolerance and exploitation of poisonous sulfide: Annual Review of Physiology, v. 60, p. 33–53, doi: 10.1146/ANNUREV.PHYSIOL.60.1.33.

Grzymski, J., Schofield, O.M., Falkowski, P.G., and Bernhard, J.M., 2002, The function of plastids in the deep-sea foraminifer *Nonionella stella*: Limnology and Oceanography, v. 47, p. 1569–1580.

Hagerman, L., 1998, Physiological flexibility: A necessity for life in anoxic and sulphidic habitats: Hydrobiologia, v. 375-376, p. 241–254, doi: 10.1023/A:1017033711985.

Hentschel, U., Berger, E.C., Bright, M., Felbeck, H., and Ott, J.A., 1999, Metabolism of nitrogen and sulfur in ectosymbiotic bacteria of marine nematodes (Nematoda, Stilbonematinae): Marine Ecology Progress Series, v. 183, p. 149–158.

Jørgensen, B.B., and Gallardo, A., 1999, *Thioploca* spp.: Filamentous sulfur bacteria with nitrate vacuoles: FEMS Microbiology Ecology, v. 28, p. 301–313, doi: 10.1016/S0168-6496(98)00122-6.

Jørgensen, B.B., and Revsbech, N.P., 1983, Colorless sulfur bacteria, *Beggiatoa* spp. and *Thioploca* spp., in O_2 and H_2S microgradients: Applied and Environmental Microbiology, v. 45, p. 1261–1270.

Kennett, J.P., Cannariato, K.G., Hendy, I.L., and Behl, R.J., 2000, Carbon isotopic evidence for methane hydrate instability during Quaternary interstadials: Science, v. 288, p. 128–133, doi: 10.1126/SCIENCE.288.5463.128.

Kuwabara, J.S., van Geen, A., McCorkle, D.C., and Bernhard, J.M., 1999, Dissolved sulfide distributions in the water column and sediment pore waters of the Santa Barbara Basin: Geochimica et Cosmochimica Acta, v. 63, p. 2199–2209, doi: 10.1016/S0016-7037(99)00084-8.

McHatton, S.C., Barry, J.P., Jannasch, H.W., and Nelson, D.C., 1996, High nitrate concentrations in vacuolated, autotrophic marine *Beggiatoa* spp: Applied and Environmental Microbiology, v. 62, p. 954–958.

Müller, M.C., Bernhard, J.M., and Jouin-Toulmond, C., 2001, A new member of Nerillidae (Annelida: Polychaeta), *Xenonerilla bactericola* nov. gen., nov. sp., collected off California: USA: Cahiers de Biologie Marine, v. 42, p. 203–217.

Ott, J.A., Novak, R., Schiemer, F., Hentschel, U., Nebelsick, M., and Polz, M., 1991, Tackling the sulfide gradient: A novel strategy involving marine nematodes and chemoautotrophic ectosymbionts: Marine Ecology—Pubblicazioni della Stazione Zoologica de Napoli, v. 12, p. 261–279.

Pike, J., Bernhard, J.M., Moreton, S.G., and Butler, I.B., 2001, Microbioirrigation of marine sediments in dysoxic environments: Implications for early sediment fabric formation and diagenetic processes: Geology, v. 29, p. 923–926, doi: 10.1130/0091-7613(2001)0292.0.CO;2.

Polz, M.F., Distel, D.L., Zarda, B., Amann, R., Felbeck, H., Ott, J.A., and Cavanaugh, C.M., 1994, Phylogenetic analysis of a highly specific association between ectosymbiotic, sulfur-oxidizing bacteria and a marine nematode: Applied and Environmental Microbiology, v. 60, p. 4461–4467.

Reimers, C.E., Lange, C.B., Tabak, M., and Bernhard, J.M., 1990, Seasonal spillover and varve formation in the Santa Barbara Basin, California: Limnology and Oceanography, v. 35, p. 1577–1585.

Reimers, C.E., Ruttenberg, K.C., Canfield, D.E., Christiansen, M.B., and Martin, J.B., 1996, Porewater pH and authigenic phases formed in the uppermost sediments of the Santa Barbara Basin: Geochimica et Cosmochimica Acta, v. 60, p. 4037–4057, doi: 10.1016/S0016-7037(96)00231-1.

Richards, F.A., 1975, The Cariaco Basin (Trench): Oceanography and Marine Biology Annual Review, v. 13, p. 11–67.

Schaaf, M., and Thurow, J., 1997, Tracing short cycles in long records: the study of inter-annual to inter-centennial climate change from long sediment records, examples from the Santa Barbara Basin: Journal of the Geological Society of London, v. 154, p. 613–622.

Schimmelmann, A., Lange, C.B., Berger, W.H., Simon, A., Burke, S.K., and Dunbar, R.B., 1992, Extreme climatic conditions recorded in Santa Barbara Basin laminated sediments—The 1835–1840 *Macoma* Event: Marine Geology, v. 106, p. 279–299, doi: 10.1016/0025-3227(92)90134-4.

Schulz, H.N., Jørgensen, B.B., Fossing, H.A., and Ramsing, N.B., 1996, Community structure of filamentous, sheath-building sulfur bacteria, *Thioploca* spp., off the coast of Chile: Applied and Environmental Microbiology, v. 62, p. 1855–1862.

Scranton, M.I., Sayles, F.A., Bacon, M.P., and Brewer, P.G., 1987, Temporal changes in the hydrography and chemistry of the Cariaco Trench: Deep-Sea Research, v. 34, p. 945–963, doi: 10.1016/0198-0149(87)90047-1.

Shen, Y., Canfield, D.E., and Knoll, A.H., 2002, Middle Proterozoic ocean chemistry: Evidence from the McArthur Basin, Northern Australia: American Journal of Science, v. 302, p. 81–109.

Sibuet, M., and Olu, K., 1998, Biogeography, biodiversity and fluid dependence of deep-sea cold-seep communities at active and passive margins: Deep-Sea Research Part II, v. 45, p. 517–567, doi: 10.1016/S0967-0645(97)00074-X.

Small, E.B., and Gross, M.E., 1985, Preliminary observations of protistan organisms, especially ciliates, from the 21°N hydrothermal vent site: Biological Society of Washington Bulletin, v. 6, p. 401–410.

Smith, C.R., and Baco, A.R., 1998, Phylogenetic and functional affinities between whale-fall, seep and vent chemoautotrophic communities: Cahiers de Biologie Marine, v. 39, p. 345–346.

Somero, G.N., Childress, J.J., and Anderson, A.E., 1989, Transport, metabolism, and detoxification of hydrogen sulfide in animals from sulfide-rich marine environments: CRC Critical Reviews in Aquatic Sciences, v. 1, p. 591–614.

Teske, A., Sogin, M.L., Nielsen, L.P., and Jannasch, H.W., 1999, Phylogenetic relationships of a large marine *Beggiatoa*: Systematic and Applied Microbiology, v. 22, p. 39–44.

Todaro, M.A., Bernhard, J.M., and Hummon, W.D., 2000, A new species of *Urodasys* (Gastrotricha, Macrodasyida) from dysoxic sediments of the Santa Barbara Basin: Bulletin of Marine Science, v. 66, p. 467–476.

Van Dover, C.L., 2000, The ecology of deep-sea hydrothermal vents: Princeton, New Jersey, Princeton University Press, 424 p.

Van Dover, C.L., German, C.R., Speer, K.G., Parson, L.M., and Vrijenhoek, R.C., 2002, Marine biology, evolution, and biogeography of seep-sea vent and seep invertebrates: Science, v. 295, p. 1253–1257, doi: 10.1126/SCIENCE.1067361.

van Geen, A., and the Scientific Party of RV *Melville*, 2001, Baja California coring cruise OXMZ01MV, Core descriptions and CTD/Rosette data: Lamont-Doherty Observatory Technical Report, LDEO 2001–01.

van Geen, A., Zheng, Y., Bernhard, J.M., Cannariato, K.G., Carriquiry, J., Dean, W.E., Eakins, B.W., Ortiz, J.D., and Pike, J., 2003, On the preservation on laminated sediments along the western margin of North America: Paleoceanography, v. 18, 1098, doi: 10.1029/2003PA000911.

Vetter, R.D., and Fry, B., 1998, Sulfur contents and sulfur isotope compositions of thiotrophic symbioses in bivalve mollusks and vestimentiferan worms: Marine Biology, v. 132, p. 453–460, doi: 10.1007/S002270050411.

Vismann, B., 1991, Sulfide tolerance: physiological mechanisms and ecological implications: Ophelia, v. 34, p. 1–27.

Vopel, K., Reick, C.H., Arlt, G., Pohn, M., and Ott, J.A., 2002, Flow microenvironment of two marine peritrich ciliates with ectobiotic chemoautotrophic bacteria: Aquatic Microbial Ecology, v. 29, p. 19–28.

Wilkin, R.T., and Barnes, H.L., 1996, Pyrite formation by reactions of iron monosulfide with dissolved inorganic and organic sulfur species: Geochimica et Cosmochimica Acta, v. 60, p. 4167–4179, doi: 10.1016/S0016-7037(97)81466-4.

Manuscript Accepted by the Society March 28, 2004

Geological Society of America
Special Paper 379
2004

Biogeochemistry of metal sulfide oxidation in mining environments, sediments, and soils

Axel Schippers*
Section Geomicrobiology, Federal Institute for Geosciences and Natural Resources (BGR), Stilleweg 2, D-30655 Hannover, Germany

ABSTRACT

Metal sulfide oxidation is an important process in the past and present global biogeochemical sulfur cycles. In this process, various sulfur compounds, namely elemental sulfur, polysulfides, thiosulfate, polythionates, sulfite, and sulfate, are generated in different environments. The formation of the sulfur compounds depends on the mineralogy of the metal sulfide and the geochemical conditions in the environment, mainly the pH and the presence of different oxidants. Metal sulfide oxidation can be described by two different pathways: the thiosulfate mechanism and the polysulfide mechanism. Microorganisms play a crucial role in the oxidation of intermediate sulfur compounds, which are formed by the chemical dissolution of the metal sulfides. Under oxic and acidic conditions (e.g., in sulfidic mine waste or in acid sulfate soils), microorganisms oxidize Fe(II) to Fe(III), which serves as an oxidant for the metal sulfides and for most of the intermediate sulfur compounds. Additionally, microorganisms may catalyze the oxidation of intermediate sulfur compounds to sulfate. Under oxic and pH-neutral conditions (e.g., in carbonate-buffered sulfidic mine waste or at the surface of marine sediments) the metal sulfides are chemically oxidized by molecular oxygen via a Fe(II)/Fe(III) shuttle to the metal (hydr)oxide, intermediate sulfur compounds, and sulfate. Microorganisms oxidize the intermediate sulfur compounds to sulfate and, at low partial pressure of molecular oxygen, may catalyze Fe(II) oxidation. Under anoxic and pH-neutral conditions (e.g., in marine sediments), metal sulfides and intermediate sulfur compounds are oxidized either chemically by MnO_2 or by microorganisms using nitrate as an electron acceptor.

Keywords: metal sulfide, pyrite, mining environment, sulfur compounds, microorganisms, oxidation mechanism.

INTRODUCTION

Metal sulfide oxidation is the only major sulfate-generating biogeochemical process on Earth. It is a process of major environmental impact, causing acid rock drainage (ARD) or acid mine drainage (AMD), the development of acid sulfate soils, and aquifer contamination. Metal sulfides are formed and oxidized in sediments. Metal sulfide oxidation is also important for processing ores for metal recovery (e.g., in bioleaching applications).

Chemical and biological processes interact in metal sulfide oxidation, and metal sulfides are oxidized via several inorganic sulfur compounds. The occurrence of inorganic sulfur compounds has been documented for different metal sulfide–containing environments, as shown in Table 1.

*a.schippers@bgr.de

Schippers, A., 2004, Biogeochemistry of metal sulfide oxidation in mining environments, sediments, and soils, *in* Amend, J.P., Edwards, K.J., and Lyons, T.W., eds., Sulfur biogeochemistry—Past and present: Boulder, Colorado, Geological Society of America Special Paper 379, p. 49–62. For permission to copy, contact editing@geosociety.org.

TABLE 1. DETECTION OF INORGANIC SULFUR COMPOUNDS IN METAL SULFIDE CONTAINING ENVIRONMENTS

Environment	Inorganic sulfur compound	References
Pyritic mine waste heaps and tailings	sulfate, elemental sulfur, thiosulfate, trithionate, tetrathionate, pentathionate, pyrite	Elberling et al., 2000; Schippers, 1998; Schippers et al., 2000
Marine and freshwater sediments	sulfate, sulfite, sulfide, polysulfide, elemental sulfur, thiosulfate, tetrathionate, iron sulfide, pyrite	Troelsen and Jørgensen, 1982; Herlihy et al., 1988; Jørgensen, 1990a, 1990b; Jørgensen and Bak, 1991; Thamdrup et al., 1994a, 1994b; Podgorsek and Imhoff, 1999; Luther et al., 2001
Salt marshes and paddy soil	sulfate, sulfite, sulfide, polysulfide, elemental sulfur, thiosulfate, tetrathionate (polythionates), iron sulfide, pyrite	Boulegue et al., 1982; Howarth et al., 1983; Luther et al., 1986; 1991; 2001; Wind and Conrad, 1995

In this chapter it will be shown that the occurrence of these sulfur compounds can be explained by metal sulfide oxidation mechanisms. After explaining these mechanisms, the importance of microorganisms for metal sulfide oxidation at low and circumneutral pH as well as at oxic and anoxic conditions will be discussed. The biogeochemical coupling of metal sulfide oxidation with the reduction of molecular oxygen, Fe(III), Mn(IV), nitrate, and CO_2 will be addressed. Finally, I will report on metal sulfide oxidation in different environments.

METAL SULFIDE OXIDATION MECHANISMS

The sulfur moiety of metal sulfides has an oxidation state of −2 (e.g., ZnS, sphalerite) or −1 (e.g., FeS_2, pyrite). The oxidation state of the sulfate S is +6, which means that for a complete oxidation of ZnS to $ZnSO_4$, eight electrons have to be transferred, and for a complete oxidation of FeS_2 to 0.5 $Fe_2(SO_4)_3$ and 0.5 H_2SO_4, 15 electrons have to be transferred (1 for Fe and 14 for S_2). Since redox reactions occur in steps of one or two electrons only, these 15 electrons are transferred in multiple steps, which means that various intermediate inorganic sulfur compounds are formed in the course of the multi-step oxidation process (Moses et al., 1987; Luther, 1987, 1990).

Metal sulfides are conductors, semiconductors, or insulators, and their metal and sulfur atoms are bound in the crystal lattice (Vaughan and Craig, 1978; Xu and Schoonen, 2000). According to molecular orbital and valence band theory, the orbitals of single atoms or molecules form electron bands with different energy levels. The metal sulfides FeS_2, MoS_2 (molybdenite), and WS_2 (tungstenite) consist of pairs of sulfur atoms (Vaughan and Craig, 1978) that form nonbonding orbitals. Consequently, the valence bands of these metal sulfides are only derived from orbitals of metal atoms, whereas the valence bands of all other metal sulfides are derived from both metal and sulfur orbitals (Borg and Dienes, 1992). Thus, the valence bands of FeS_2, MoS_2, and WS_2 do not contribute to the bonding between the metal and the sulfur moiety of the metal sulfide, which explains the resistance of these metal sulfides against a proton attack. The bonds can only be broken via multi-step electron transfers with an oxidant like Fe(III). For the other metal sulfides, in addition to an oxidant like Fe(III), protons can remove electrons from the valence band, causing a cleavage of the bonds between the metal and the sulfur moiety of the metal sulfide. Consequently, these metal sulfides are more or less soluble in acid, whereas FeS_2, MoS_2, and WS_2 are insoluble (Singer and Stumm, 1970; Tributsch and Bennett, 1981a; Crundwell, 1988; Rossi, 1993; Sand et al. 2001).

Because two different groups of metal sulfides exist, two different metal sulfide oxidation mechanisms have been proposed (Schippers and Sand, 1999; Schippers et al., 1996a, 1999; Sand et al., 2001). These mechanisms explain the occurrence of all inorganic sulfur compounds that have been documented for different metal sulfide–containing environments, as shown in Table 1. The pH and the availability of oxygen and other oxidants determine the accumulation of intermediate sulfur compounds and the importance of microorganisms for the oxidation process. These mechanisms will be summarized in the following sections.

Metal sulfide oxidation mechanisms may depend on surface-controlled processes (Evangelou, 1995; Vaughan et al., 1997; Bebie et al., 1998; Eggelston et al., 1996, Guevremont et al., 1998; De Giudici and Zuddas, 2001; Becker et al., 2001, 2003; Elsetinow et al., 2001, 2003; Gerson and O'Dea, 2003; Nesbitt et al., 2003; Todd et al., 2003), metal sulfide structure, impurities, dislocations and stacking faults (Martello et al., 1994; Cruz et al., 2001; Rimstidt and Vaughan, 2003; Thomas et al., 2003), photochemical reactions (El-Halim et al., 1995; Schoonen et al., 2000; Giannetti et al., 2001), or galvanic interactions (Rossi, 1990; Lizama and Suzuki, 1991; Gantayat et al., 2000), but since little information about the effect of these aspects on metal sulfide oxidation mechanisms is available, they will not be discussed in this review. Furthermore, the kinetics of metal sulfide oxidation will not be targeted, and the reader is referred to other publications on this topic (Wiersma and Rimstidt, 1984; Nicholson et al., 1988, 1990; Morse, 1991; Nakamura et al., 1994; Sasaki, 1994; Williamson and Rimstidt, 1994; De Giudici and Zuddas, 2001; Lengke and Tempel, 2001, 2003).

Oxidation Mechanism for the Acid Insoluble Metal Sulfides FeS_2, MoS_2, and WS_2

FeS_2 is the most widespread sulfide mineral in nature, whereas MoS_2 and WS_2 only rarely occur. Since FeS_2 oxidation is also the most studied among metal sulfides (for reviews, see

Dutrizac and MacDonald, 1974; Lowson, 1982; Nordstrom, 1982; Evangelou, 1995; Evangelou and Zhang, 1995; Nordstrom and Southam, 1997; Nordstrom and Alpers, 1999a; Rimstidt and Vaughan, 2003), FeS_2 will be used as an example for the three metal sulfides FeS_2, MoS_2, and WS_2 in the following text.

In nature, molecular oxygen and Fe(III) may serve as oxidants for FeS_2 (Lowson, 1982). Luther (1987) used molecular orbital theory to explain why Fe(III), rather than molecular oxygen, reacts with the FeS_2 surface. In contrast to molecular oxygen, hydrated Fe(III) ions are connected to the pyrite surface via σ-bonds. These bonds can facilitate an electron transfer from the sulfur moiety of FeS_2 to the Fe(III) ions. Moses and Herman (1991) showed that even at neutral pH, Fe(III) is a FeS_2 oxidizing agent. The Fe(II) remains adsorbed to the FeS_2 surface and is oxidized by molecular oxygen to Fe(III), again attacking FeS_2. The abiotic Fe(II) oxidation by molecular oxygen is promoted by Fe(II)-CO_2 complexes on the FeS_2 surface (Evangelou et al., 1998). Additionally, Fe(III) complexing organic substances can influence the FeS_2 oxidation rate (Peiffer and Stubert, 1999). An adsorbed Fe(II)/Fe(III) shuttle has been suggested for FeS_2 oxidation by molecular oxygen (Moses and Herman, 1991; Eggleston et al., 1996) and by MnO_2 under anoxic conditions (Schippers and Jørgensen, 2001). Consequently, Fe(III) seems to be the most important oxidant for FeS_2 in nature. Molecular oxygen or other oxidants are important to provide the FeS_2 attacking agent Fe(III) via the oxidation of Fe(II). Depending on the geochemical conditions, this process may be efficiently catalyzed by Fe(II)-oxidizing microorganisms, which will be discussed below.

After initial attack of the oxidant Fe(III), the sulfur moiety of FeS_2 is oxidized to soluble sulfur intermediates. Moses et al. (1987) and Luther (1987) presented a detailed reaction mechanism for FeS_2 dissolution by Fe(III) in which thiosulfate is the first soluble sulfur intermediate. According to this mechanism, hydrated Fe(III) ions oxidize the S_2 of FeS_2 to a sulfonic acid group by several electron transfers. Due to this transformation, the bonds between Fe and the two sulfur atoms are cleaved, and hydrated Fe(II) ions and thiosulfate are formed (Luther, 1987):

$$FeS_2 + 6\,[Fe(H_2O)_6]^{3+} + 9\,H_2O \rightarrow S_2O_3^{2-} + 7\,[Fe(H_2O)_6]^{2+} + 6\,H^+. \quad (1)$$

At circumneutral pH, thiosulfate could be detected in the presence of molecular oxygen (Steger and Desjardins, 1978; Goldhaber, 1983; Moses et al., 1987; Nesbitt and Muir, 1994; Schippers et al., 1996a; Bonnissel-Gissinger et al., 1998; Descostes et al., 2001) or MnO_2 as oxidant for FeS_2 (Schippers and Jørgensen, 2001).

To confirm the formation of thiosulfate during FeS_2 oxidation at pH 2, an experiment with silver(I) ions was carried out (Schippers et al., 1996a). At low pH, silver(I) ions react with thiosulfate to silver sulfide, which prevents the quick oxidation of thiosulfate by Fe(III) (see below). In the experiment in which FeS_2 was oxidized by Fe(III) in the presence of silver(I) ions at pH 2, silver sulfide could be detected. This result shows that thiosulfate is the product of FeS_2 oxidation at low pH.

In addition, an electrochemical study supported the occurrence of thiosulfate in the course of FeS_2 oxidation at pH 2 (Mishra and Osseo-Asare, 1988). Cyclic voltagrams revealed only one anodic peak and two cathodic peaks in the return sweep. These peaks were attributed to the electroadsorption/desorption of OH groups on pyrite surfaces. It is proposed that the electrocatalytic electroadsorption of OH groups on FeS_2 is due to the presence of Fe 3d electrons in the upper portions of the valence band. Thus, OH^- ions are oxidized by holes on Fe 3d states in the first step. These groups are transferred to S_2^{2-} sites in the second step. A mechanism for the anodic dissolution of FeS_2 has been proposed, according to which elemental sulfur is not an intermediate product, but rather is a product that forms due to the decomposition of thiosulfate (Mishra and Osseo-Asare, 1988).

Apart from thiosulfate, sulfite, and sulfate, high amounts of polythionates, namely trithionate, tetrathionate, and pentathionate, were detected in the course of FeS_2 oxidation at circumneutral pH, when either molecular oxygen (Goldhaber, 1983; Moses et al., 1987; Pichtel and Dick, 1991; Schippers et al., 1996a) or MnO_2 (Schippers and Jørgensen, 2001) were present. At pH 2, in the presence of Fe(III), mainly sulfate and high amounts of elemental sulfur were detected, along with low amounts of tetrathionate and pentathionate (Schippers et al., 1996a, 1999).

Tetrathionate formation by thiosulfate oxidation has been shown for pH 2.9 – 8.6 in the presence of molecular oxygen and FeS_2 (Xu and Schoonen, 1995) or ZnS (Xu et al., 1996), at pH 8 in the presence of MnO_2 (Schippers and Jørgensen, 2001), and at pH 2 in the presence of Fe(III) (Schippers et al., 1996a; Williamson and Rimstidt, 1994):

$$2\,S_2O_3^{2-} + 2\,Fe^{3+} \rightarrow S_4O_6^{2-} + 2\,Fe^{2+}. \quad (2)$$

This reaction is much faster than the acid decomposition of thiosulfate to sulfite and elemental sulfur that occurs in the absence of Fe(III) (Williamson and Rimstidt, 1994). Thus, tetrathionate is the main product of thiosulfate degradation in the course of FeS_2 oxidation.

The degradation of tetrathionate strongly depends on pH and on the availability of catalysts. Tetrathionate quickly decomposes in alkaline solution (Zhang and Dreisinger, 2002) but is quite stable in acid solution even in the presence of Fe(III) ions (Schippers et al., 1996a; Druschel et al., 2003). In contrast, tetrathionate degrades at low pH in the presence of FeS_2 if the suspension is vigorously shaken (Moses et al., 1987; Schippers et al., 1996a). Obviously, the FeS_2 surface acts as a catalyst for tetrathionate degradation. The kinetics of tetrathionate degradation in presence of FeS_2 is unknown and may depend on the type and surface area of the FeS_2, the shaking or stirring rate, the temperature, and/or the pH. However, according to Steudel et al. (1987) and Schippers et al. (1996a), tetrathionate is hydrolyzed on the FeS_2 surface to disulfane-monosulfonic acid:

$$S_4O_6^{2-} + H_2O \rightarrow HS_3O_3^- + SO_4^{2-} + H^+. \qquad (3)$$

Disulfane-monosulfonic acid is very unstable and reacts to trithionate, pentathionate, elemental sulfur, and sulfite according to the following equations:

$$S_3O_3^{2-} + 1.5\ O_2 \rightarrow S_3O_6^{2-}, \qquad (4)$$

$$S_3O_3^{2-} + S_2O_3^{2-} + 0.5\ O_2 + 2\ H^+ \rightarrow S_5O_6^{2-} + H_2O, \qquad (5)$$

$$S_3O_3^{2-} + S_4O_6^{2-} \rightarrow S_2O_3^{2-} + S_5O_6^{2-}, \qquad (6)$$

$$4\ S_3O_3^{2-} \rightarrow S_8 + 4\ SO_3^{2-}. \qquad (7)$$

Instead of molecular oxygen (Equations 4 and 5), Fe(III) might again serve as the alternative oxidant.

In analogy to Equation 3, trithionate might be hydrolyzed to thiosulfate and sulfate (Steudel et al., 1987; Schippers et al., 1996a):

$$S_3O_6^{2-} + H_2O \rightarrow S_2O_3^{2-} + SO_4^{2-} + 2\ H^+. \qquad (8)$$

Thiosulfate might again be oxidized to tetrationate (Equation 2). This series of reactions results in a cyclic degradation of thiosulfate via polythionates to sulfate (Fig. 1; Schippers et al., 1996a, 1999). Since thiosulfate is a key compound in the reaction series, this FeS_2 oxidation mechanism has been named the "thiosulfate mechanism" (Schippers and Sand, 1999). Elemental sulfur, sulfite, and pentathionate only occur as side products. Elemental sulfur may accumulate during FeS_2 oxidation because it is quite stable and only degradable in the presence of sulfur-oxidizing microorganisms (de Donato, 1993; Sasaki et al., 1995; Schippers et al., 1996a, 1999; McGuire et al., 2001a, 2001b).

In contrast to the Equations 3 and 4, Druschel et al. (2003) proposed that tetrathionate is degraded to trithionate via the following equations:

$$S_4O_6^{2-} + Fe^{3+} \rightarrow S_3O_3^{0} + SO_3^{*-} + Fe^{2+}, \qquad (9)$$

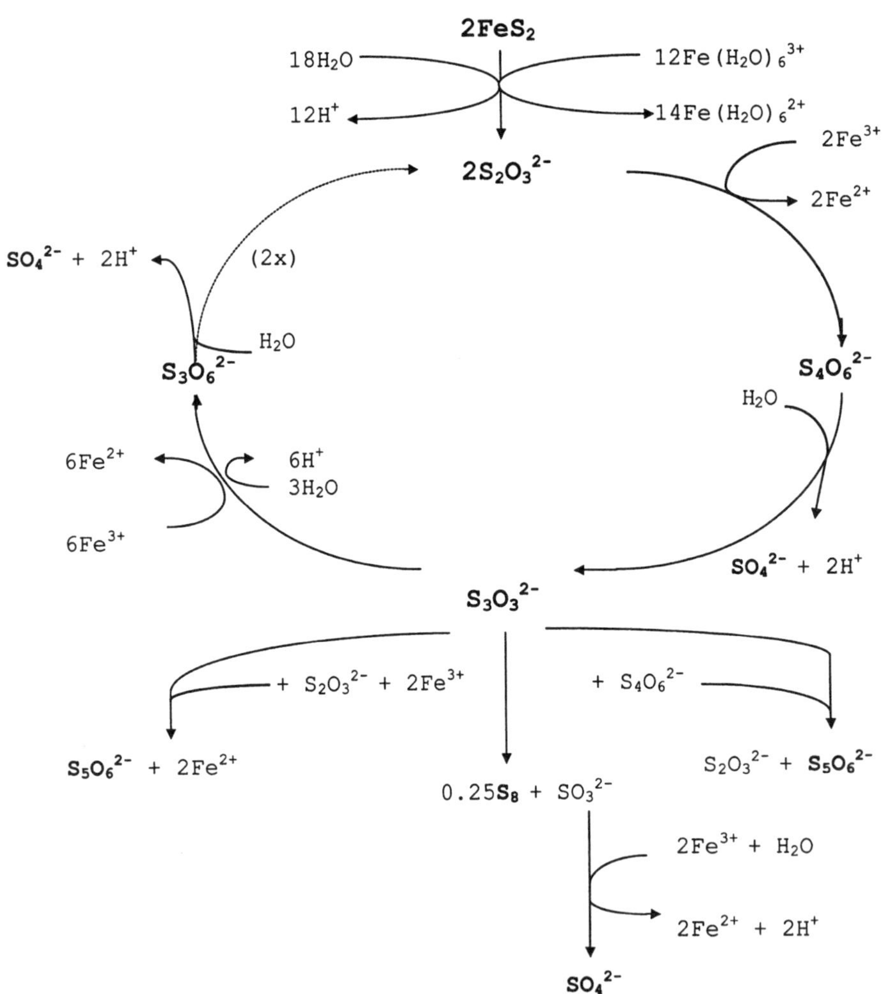

Figure 1. Thiosulfate mechanism of pyrite oxidation (modified from Schippers et al., 1996a, 1999). Pyrite is attacked by Fe(III) hexahydrate ions. Thiosulfate, as the first intermediary sulfur compound, is degraded via tetrathionate, disulfane-monosufonic acid, and trithionate to sulfate in the cycle. In side reactions, elemental sulfur, pentathionate, and sulfate occur. Oxidation of intermediary sulfur compounds proceeds with Fe(III) ions, NO_3^-, MnO_2, or O_2 as oxidant.

$$S_3O_3^0 + H_2O \rightarrow S_3O_4^{2-} + 2\ H^+, \quad (10)$$

$$S_3O_4^{2-} + O_2 \rightarrow S_3O_6^{2-}. \quad (11)$$

The product trithionate would be degraded to sulfate via analogous reactions to those given for tetrathionate degradation (Equation 9–11). An overall equation for tetrathionate degradation to sulfate by Fe(III) is given by Druschel et al. (2003):

$$S_4O_6^{2-} + 3\ Fe^{3+} + 2.75\ O_2 + 4.5\ H_2O \rightarrow 4\ SO_4^{2-} + 3\ Fe^{2+} + 9\ H^+. \quad (12)$$

Since neither the intermediate $S_3O_3^{2-}$ (Equation 3) nor the intermediates $S_3O_3^0$ and SO_3^{*-} (Equation 9) and $S_3O_4^{2-}$ (Equation 10) have been detected during FeS_2 oxidation experiments, the detailed mechanism of tetrathionate degradation remains to be resolved.

However, the occurrence of all sulfur compounds that have been detected in FeS_2 oxidation experiments or in FeS_2-containing environments, as listed in Table 1, can be explained by the thiosulfate mechanism. As well, Reedy et al. (1991) used ^{18}O-labeled molecular oxygen or water and studied the isotopic composition of the FeS_2 oxidation products. Their results can be explained by the different oxidation reactions of the thiosulfate mechanism in which both Fe(III) and molecular oxygen can be oxidants. Overall, the thiosulfate mechanism described for FeS_2 is also valid for MoS_2 and WS_2, and can be summarized by the following equations:

$$FeS_2 + 6\ Fe^{3+} + 3\ H_2O \rightarrow S_2O_3^{2-} + 7\ Fe^{2+} + 6\ H^+, \quad (13)$$

$$S_2O_3^{2-} + 8\ Fe^{3+} + 5\ H_2O \rightarrow 2\ SO_4^{2-} + 8\ Fe^{2+} + 10\ H^+. \quad (14)$$

Oxidation Mechanism for Acid-Soluble Metal Sulfides

The previously discussed metal sulfides FeS_2, MoS_2, and WS_2 can only be degraded by oxidation in the environment. Most other metal sulfides, like As_2S_3 (orpiment), As_4S_4 (realgar), $CuFeS_2$ (chalcopyrite), FeS (troilite), Fe_7S_8 (pyrrhotite), MnS_2 (hauerite), PbS (galena), and ZnS (sphalerite), can also be dissolved by protons. At pH 2, the sulfur moiety of these metal sulfides is chemically oxidized by Fe(III) ions mainly to elemental sulfur (Dutrizac and MacDonald, 1974; Schippers and Sand, 1999; McGuire et al., 2001b). At circumneutral pH with MnO_2 as oxidant, elemental sulfur was the main product of FeS oxidation as well (Schippers and Jørgensen, 2001). By contrast, in the case of FeS_2 and MoS_2 with Fe(III) as oxidant, sulfate was the dominant oxidation product (>90% yield), in addition to ~1–2% of polythionates (Schippers and Sand, 1999).

Due to the acid solubility of most of the metal sulfides (MS), the first reaction is assumed to be:

$$MS + 2\ H^+ \rightarrow M^{2+} + H_2S. \quad (15)$$

In contrast to FeS_2 oxidation, the M-S bonds in the acid-soluble metal sulfides can be cleaved before the sulfidic sulfur is oxidized.

A mechanism of aqueous sulfide oxidation by molecular oxygen in seawater in which sulfite, thiosulfate, and sulfate and not elemental sulfur are the major oxidation products has been described by Zhang and Millero (1993). This mechanism does not seem to be relevant for metal sulfide oxidation; however, aqueous sulfide oxidation via free radicals to elemental sulfur by molecular oxygen has been proposed by Chen and Morris (1972):

$$HS^- + O_2 \rightarrow HS^* + O_2^-, \quad (16)$$

$$HS^* + O_2^- \rightarrow S + HO_2^-. \quad (17)$$

By contrast, elemental sulfur was the dominant product of aqueous sulfide oxidation if Fe(III) or Mn(IV) was the oxidant (Yao and Millero, 1996). A mechanism for aqueous sulfide oxidation by Fe(III) has been described in detail by Steudel (1996). According to his work, the H_2S is subjected to a single electron oxidation by an Fe(III) ion:

$$H_2S + Fe^{3+} \rightarrow H_2S^{*+} + Fe^{2+}. \quad (18)$$

The cation radical H_2S^{*+} may also be formed directly by an attack of Fe(III) ions on a metal sulfide (Schippers and Sand, 1999):

$$MS + Fe^{3+} + 2\ H^+ \rightarrow M^{2+} + H_2S^{*+} + Fe^{2+}. \quad (19)$$

By dissociation of the strong acid H_2S^{*+}, the radical HS^* occurs:

$$H_2S^{*+} + H_2O \rightarrow H_3O^+ + HS^*. \quad (20)$$

Two of these radicals may react to a disulfide ion:

$$2\ HS^* \rightarrow HS_2^- + H^+. \quad (21)$$

The disulfide ion can be oxidized further by an Fe(III) ion (Equation 12) or a HS^* radical:

$$HS_2^- + HS^* \rightarrow HS_2^* + HS^-. \quad (22)$$

Tetrasulfide can occur by dimerization of two HS_2^* or trisulfide by reaction of HS_2^* with HS^* radicals. Chain elongation of the polysulfides may proceed by analogous reactions.

In acidic solutions, polysulfides decompose to rings of elemental sulfur, mainly S_8, with yields exceeding 99%:

$$HS_9^- \rightarrow HS^- + S_8. \quad (23)$$

This mechanism does not necessarily function only in the presence of Fe(III) ions. In cases of molecular oxygen as oxidant, the oxygen molecule is reduced via a superoxide radical and a peroxide molecule to water (Tributsch and Gerischer, 1976; Zhang and Millero, 1993). However, Fe(III) ions are generally much more efficient in extracting electrons from a metal sulfide lattice than molecular oxygen (Tributsch and Bennett, 1981a, 1981b).

The series of reactions for acid-soluble metal sulfides inherently explains the formation of elemental sulfur via polysulfides, which have been detected during dissolution of Fe_7S_8 (Thomas et al., 1998, 2001), PbS (Smart et al., 2000), and $CuFeS_2$ (Hackl et al., 1995). Consequently, this oxidation mechanism for acid soluble metal sulfides has been named the "polysulfide" mechanism (Schippers and Sand, 1999). The polysulfide mechanism is summarized in Figure 2.

Elemental sulfur is the end product in the reaction scheme. However, although elemental sulfur is chemically inert in natural environments, it can be biologically oxidized to sulfuric acid (Equation 26). Overall, the polysulfide mechanism can be described by the following equations (Schippers and Sand, 1999):

$$MS + Fe^{3+} + H^+ \rightarrow M^{2+} + 0.5\ H_2S_n + Fe^{2+},\ (n \geq 2) \quad (24)$$

$$0.5\ H_2S_n + Fe^{3+} \rightarrow 0.125\ S_8 + Fe^{2+} + H^+, \quad (25)$$

$$0.125\ S_8 + 1.5\ O_2 + H_2O \rightarrow SO_4^{2-} + 2\ H^+. \quad (26)$$

In cases of the oxidation of acid soluble metal sulfides, minor amounts of sulfate and polythionates can be reaction products that may be formed via reactions with thiosulfate (see the thiosulfate mechanism). The thiosulfate may arise by side reactions of the polysulfide mechanism, and the following reactions have been proposed (Schippers and Sand, 1999):

$$HS_n^- + 1.5\ O_2 \rightarrow HS_2O_3^- + [0.125\ (n-2)]S_8, \quad (27)$$

$$HS_n^- + 6\ Fe^{3+} + 3\ H_2O \rightarrow HS_2O_3^- + [0.125\ (n-2)]S_8 + 6\ Fe^{2+} + 6\ H^+. \quad (28)$$

Thiosulfate may also be formed in the following reaction:

$$0.125\ S_8 + HSO_3^- \rightarrow HS_2O_3^-. \quad (29)$$

As well, thiosulfate may be formed as a product from the oxidation of aqueous sulfide (formed in Equation 15) in a series of reactions (Chen and Morris, 1972; Zhang and Millero, 1993).

Summarizing, thiosulfate and polythionates play a minor role in the polysulfide mechanism, but a major role in the thiosulfate mechanism. Polysulfides and elemental sulfur play a key role in the polysulfide mechanism, while elemental sulfur is only a side product in the thiosulfate mechanism. The occurrence of all the sulfur compounds that have been detected in metal sulfide oxidation experiments or in metal sulfide containing environments, as listed in Table 1, can be explained by the thiosulfate mechanism or the polysulfide mechanism. Metal sulfide oxidation and formation of inorganic sulfur compounds in different environments will be discussed after highlighting the importance of microorganisms for metal sulfide oxidation in the following section.

$$MS \xrightarrow[Fe^{3+}\ \rightarrow\ Fe^{2+}]{2\,H^+\ \rightarrow\ M^{2+}} [\ H_2S^{*+} \xrightarrow{H^+} HS^* \longrightarrow H_2S_n\ ] \xrightarrow[Fe^{3+}\ \rightarrow\ Fe^{2+}]{H^+} S_8$$

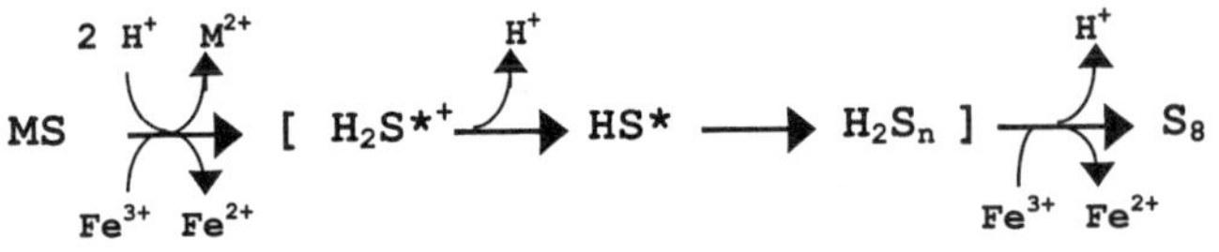

Figure 2. Simplified scheme of the polysulfide mechanism for acid soluble metal sulfides (Schippers and Sand, 1999).

IMPORTANCE OF MICROORGANISMS FOR METAL SULFIDE OXIDATION

Microorganisms are strongly involved in metal sulfide oxidation. Most relevant are Fe- and S-oxidizing microorganisms. Their contribution to the overall oxidation process depends on the geochemical conditions in the environment. The metal sulfide type, the availability of molecular oxygen or other oxidants, and the pH determine which species of different Fe- and S-oxidizing microorganisms contribute to metal sulfide oxidation. Table 2 shows the main compounds of chemical and biological oxidation of FeS_2 and acid soluble metal sulfides (MS) at pH 2 and pH 7–8 in the presence of oxygen or other oxidants, as well as the types of microorganisms involved. The importance of microorganisms for metal sulfide oxidation is discussed in detail in the following sections.

Oxic Biological Metal Sulfide Oxidation at Low pH

Oxic biological metal sulfide oxidation at low pH (around 2) is well documented in the literature as bioleaching or microbial catalyzed weathering of metal sulfides. Bioleaching is increasingly used by the mining industry to extract metals from ore, mainly for gold recovery in huge tank bioreactors and for copper recovery in mining heaps (Ehrlich and Brierley, 1990; Rossi, 1990; Bosecker, 1997; Brierley and Rawlings, 1997; Brandl, 2001; Ehrlich, 2002; Rawlings, 2002; Rawlings et al., 2003; Rohwerder et al., 2002). Microbial catalyzed weathering of metal sulfides in mine waste produces hazardous acid mine drainage (Colmer and Hinkle, 1947; Schippers et al., 1995; Schrenk et al., 1998; Edwards et al., 1999a, 1999b, 2000a, 2000b). Metal sulfide oxidizing organisms are chemolithoautotrophic, acidophilic Fe(II), and/or sulfur-compound oxidizing bacteria or archaea. Well-known organisms are *Acidithiobacillus* (formerly *Thiobacillus*) *ferrooxidans*, *Acidithiobacillus thiooxidans*, *Acidithiobacillus caldus*, *Leptospirillum ferrooxidans*, *Acidianus brierleyi*, and *Ferroplasma acidarmanus*. All these organisms grow well at pH 2; *Ferroplasma acidarmanus* grows even at pH 0 (Edwards et al., 2000a).

Most important for metal sulfide oxidation at low pH are acidophilic Fe(II)-oxidizing organisms like *Acidithiobacillus ferrooxidans*, *Leptospirillum ferrooxidans*, *Acidianus brierleyi*,

TABLE 2. MAIN SULFUR COMPOUND PRODUCTS OF CHEMICAL AND BIOLOGICAL OXIDATION OF FeS_2 AND ACID SOLUBLE METAL SULFIDES (MS) AT pH 2 AND pH 7–8 IN THE PRESENCE OF OXYGEN OR OTHER OXIDANTS, AND TYPES OF MICROORGANISMS INVOLVED

	Chemical oxidation	Biological oxidation
FeS_2, pH 2, oxic, O_2 as oxidant	{Sulfuric acid, elemental sulfur}	Sulfuric acid; aerobic strongly acidophilic Fe(II) (and S) oxidizer; e.g., *Acidithiobacillus ferrooxidans*
MS, pH 2, oxic, O_2 as oxidant	Elemental sulfur	Sulfate; aerobic, strongly acidophilic Fe(II) and S oxidizer; e.g., *Acidithiobacillus ferrooxidans*
FeS_2, pH 7–8, oxic, O_2 as oxidant	Trithionate, tetrathionate, sulfuric acid, thiosulfate	Sulfuric acid; {aerobic moderately acidophilic S oxidizer; e.g., *Thiomonas intermedia*}, microaerophilic Fe(II) oxidizer
MS, pH 7–8, oxic, O_2 as oxidant	Elemental sulfur, thiosulfate	Sulfate; aerobic moderately acidophilic S oxidizer; e.g., *Thiomicrospira frisia*, and microaerophilic Fe(II) oxidizer
FeS_2, pH 2, anoxic, Fe(III) ions as oxidant	Sulfuric acid, elemental sulfur	Sulfuric acid; {anaerobic, strongly acidophilic S oxidizer and Fe(III) reducer; e.g., *Acidithiobacillus ferrooxidans*}
MS, pH 2, anoxic, Fe(III) ions as oxidant	Elemental sulfur, sulfate	Sulfuric acid; {anaerobic, strongly acidophilic S oxidizer and Fe(III) reducer; e.g., *Acidithiobacillus ferrooxidans*}
FeS_2, pH 7–8, anoxic, Mn(IV) oxide as oxidant	Sulfate, tetrathionate, trithionate, thiosulfate	No oxidation
MS, pH 7–8, anoxic, Mn(IV) oxide as oxidant	Elemental sulfur, sulfate	Sulfate; {Sulfur disproportionating bacterium; e.g., *Desulfocapsa sulfoexigens*}
MS, pH 7–8, anoxic, Nitrate as oxidant	No oxidation	Sulfate; moderately acidophilic S [or Fe(II)] oxidizer; e.g., *Thiobacillus denitrificans*, and anaerobic Fe(II) oxidizer
MS, pH 7–8, anoxic, CO_2 as oxidant, light	No oxidation	Sulfate; anaerobic, phototrophic Fe(II) oxidizer; e.g., *Rhodovulum iodosum*

Note: Sulfur compounds in braces indicate that the oxidation rate is very low. Organisms in braces indicate that the organism only oxidizes intermediary sulfur compounds and does not increase the chemical metal sulfide dissolution rate (after Schippers, 1998; Schippers and Sand, 1999; Schippers and Jørgensen 2001, 2002; Schippers et al., 1996a, 1996b, 1999).

and *Ferroplasma acidarmanus* because they provide Fe(III), the most important oxidant for metal sulfides at low pH. As outlined in the previous section, Fe(III) attacks and dissolves metal sulfides and oxidizes intermediate sulfur compounds like thiosulfate or polysulfide, but not elemental sulfur. Elemental sulfur is exclusively oxidized biologically by acidophilic sulfur-compound oxidizing organisms like *Acidithiobacillus ferrooxidans*, *Acidithiobacillus thiooxidans, Acidithiobacillus caldus,* and *Acidianus brierleyi.* These organisms are also involved in the oxidation of other intermediates of metal sulfide oxidation like tetrathionate (Schippers et al., 1999; McGuire et al., 2001a).

In the literature about bioleaching it is regularly stated that bioleaching organisms oxidize metal sulfides in two different ways, "directly" and "indirectly." "Directly" indicates that organisms are attached to the metal sulfide surface and dissolve the metal sulfide without a soluble electron shuttle. "Indirectly" indicates that organisms are not attached to the mineral surface and that the metal sulfide is oxidized via the electron shuttle Fe(II)/Fe(III). So far, it has not been shown how organisms oxidize metal sulfides in a "direct" way. Gehrke et al. (1998) and Sand et al. (2001) detected high amounts of Fe bound in a layer of extracellular polymeric substances (EPS) of *Acidithiobacillus ferrooxidans* and of *Leptospirillum ferrooxidans*. Recently, Ehrlich (2002) suggested that this EPS-bound Fe may serve as an electron shuttle, as does Fe in the "indirect" way. Consequently, Fe(III) is generally the oxidant for biological metal sulfide dissolution, irrespective if cells are attached ("direct") or not attached ("indirect") to the mineral surface. This statement is supported by a SEM study of Edwards et al. (2001), who detected similar leaching patterns on metal sulfide surfaces in cases of bioleaching and abiotic Fe(III) leaching. Rawlings (2002) also emphasized the role of EPS-bound Fe for bioleaching and concluded that the bioleaching mechanism is indirect. Thus, a close contact of a cell to the mineral surface is not essential for bioleaching but increases the rate of bioleaching. He suggested replacing the term "direct leaching" with the term "contact leaching."

Irrespective of the different terms, bioleaching of metal sulfides is carried out by acidophilic Fe(II)-oxidizing organisms providing Fe(III) to oxidize metal sulfides most likely via the thiosulfate or the polysulfide mechanisms. The intermediate sulfur compounds are either oxidized chemically by Fe(III) or biologically by acidophilic sulfur-compound oxidizing organisms (Schippers et al., 1999; Schippers and Sand, 1999; McGuire et al., 2001a).

Oxic Biological Metal Sulfide Oxidation at Neutral to Alkaline pH

Oxic biological metal sulfide oxidation at neutral to alkaline pH is less well studied. At this pH, Fe(III) is insoluble and the above-mentioned bioleaching organisms cannot live, which prohibits biological oxidation by a pathway similar to the one at low pH. Biological dissolution of the acid-soluble metal sulfide FeS at circumneutral pH has been shown for moderately acidophilic sulfur compound-oxidizing organisms like *Thiomicrospira frisia* (Kuenen et al., 1992; Brinkhoff et al., 1999). These organisms produce protons by sulfur oxidation, which dissolves the acid-soluble metal sulfide. According to the polysulfide mechanism, intermediate sulfur compounds like elemental sulfur are formed and are subsequently biologically oxidized by moderately acidophilic sulfur compound-oxidizing bacteria. In the case of the acid-insoluble FeS_2, moderately acidophilic sulfur compound oxidizing organisms like *Thiomonas intermedia* only oxidize intermediate sulfur compounds formed by the chemical FeS_2 oxidation and do not increase the chemical FeS_2 dissolution rate (Arkesteyn, 1980; Schippers et al., 1996b). At low partial pressure of molecular oxygen, growth of microaerophilic, neutrophilic Fe(II)-oxidizing organisms with FeS and FeS_2 as substrate has been reported (Emerson and Moyer, 1997, 2002; Edwards et al., 2003), but it is not known how efficiently these organisms increase the metal sulfide dissolution rate or which sulfur compounds are formed.

Anoxic Biological Metal Sulfide Oxidation at Low pH

At low pH (around 2), Fe(III) is much more soluble than at neutral pH and efficiently oxidizes metal sulfides including FeS_2. According to the thiosulfate and the polysulfide mechanisms, elemental sulfur may accumulate in the course of the chemical metal sulfide oxidation. *Acidithiobacillus ferrooxidans, Acidithiobacillus thiooxidans*, and *Sulfolobus acidocaldarius* are able to oxidize elemental sulfur by reduction of Fe(III) (Brock and Gustafson, 1976; Pronk and Johnson, 1992). Because the regeneration of the oxidant Fe(III) at low pH depends on the presence of both aerobic, acidophilic Fe(II)-oxidizers and molecular oxygen (Singer and Stumm, 1970), Fe(III) has to be transported from an oxic zone to an anoxic zone (e.g., by diffusion or percolation in mine waste tailings or in salt marshes).

Anoxic Biological Metal Sulfide Oxidation at Neutral to Alkaline pH

Several experiments have been done to determine whether metal sulfides can be oxidized under anaerobic conditions at neutral to alkaline pH. In chemical experiments, FeS_2 and FeS were oxidized with MnO_2, according to the thiosulfate and the polysulfide mechanisms, respectively, but not with nitrate or amorphous Fe(III) oxide (Bonnissel-Gissinger et al., 1998; Schippers and Jørgensen, 2001, 2002). Bacteria could be enriched from anaerobic marine sediments, which oxidize FeS, but not FeS_2, anaerobically, using nitrate as the electron acceptor (Schippers and Jørgensen, 2002). Bacteria could not be isolated with amorphous Fe(III) oxide as electron acceptor. Similarly, no oxidation of FeS_2 was observed using ^{55}Fe-labeled FeS_2 (Schippers and Jørgensen, 2002).

The anaerobic FeS oxidation with nitrate as the electron acceptor can be catalyzed by anaerobic sulfur-oxidizing and nitrate-reducing bacteria like *Thiobacillus denitrificans* (Garcia-Gil and Golterman, 1993) and by anaerobic Fe(II)-oxidizing and nitrate-reducing bacteria (Straub et al., 1996; Edwards et al., 2003). Anaerobic, phototrophic Fe(II)-oxidizing bacteria like *Rhodovulum iodosum* (Ehrenreich and Widdel, 1994; Straub et al., 1999) can also oxidize FeS to Fe(III) and sulfate with CO_2 as the electron acceptor.

In cases of the chemical oxidation of FeS by MnO_2, elemental sulfur accumulates according to the polysulfide mechanism. The oxidation of elemental sulfur to sulfate in the presence of MnO_2 may be catalyzed by sulfur-disproportionating bacteria like *Desulfocapsa sulfoexigens* (Thamdrup et al., 1993; Finster et al., 1998).

METAL SULFIDE OXIDATION IN DIFFERENT ENVIRONMENTS

Metal sulfide oxidation has been studied in many different environments, but relatively few studies include the analyses of intermediate sulfur compounds and the microbiology of participating organisms. Results from such studies are summarized in this section to define general rules for metal sulfide oxidation in the environment. The following environments are considered: mining environments, sediments, and soils. This book also contains a chapter about metal sulfide formation and weathering at seafloor hydrothermal vent sites and in the ocean crust, written by K.J. Edwards (Chapter 6).

Mining Environments

Acid rock drainage (ARD), or acid mine drainage (AMD), is generated by the mining of metal sulfide deposits. In the most extreme case reported, the Richmond Mine of Iron Mountain, California, ARD/AMD contained metal concentrations as high as 200 g/L, sulfate concentrations as high as 760 g/L, and a pH as low as −3.6 (Nordstrom and Alpers, 1999b). The extremely acidophilic Fe(II)-oxidizing *Ferroplasma acidarmanus* was isolated from the Iron Mountain site. This microorganism dominated the microbial community at sites with a temperature of around 40 °C and a pH of 0–1, and constituted up to 85% of all microorganisms (Edwards et al., 2000a). At other sites with comparable temperature and pH, the Fe(II)-oxidizing *Leptospirillum ferrooxidans* was a dominant member of the microbial community. At sites with a temperature below 30 °C and a pH above 1.3, *Acidithiobacillus ferrooxidans* was the most abundant Fe(II)-oxidizing organism (Schrenk et al. 1998; Edwards et al., 1999a, 1999b, 2000a, 2000b).

In a Romanian sulfidic mine with less extreme conditions (temperature around 20 °C and pH around 3) the three bioleaching organisms *Leptospirillum ferrooxidans, Acidithiobacillus ferrooxidans*, and *Acidithiobacillus thiooxidans* were almost equally abundant (Sand et al., 1992).

In two German pyritic uranium mine waste heaps consisting of crushed carbonaceous black schist ore, the pH values fluctuated between 4 and 8 and the oxygen concentration decreased with increasing depth in the waste heap. Biological FeS_2

oxidation was confirmed, and the microbial activity in the solid waste material was measured by microcalorimetry (Schippers et al., 1995). *Acidithiobacillus ferrooxidans* was the dominant acidophilic Fe(II)-oxidizing organism and was most abundant in the top 2 m of the heaps. The acidophilic sulfur oxidizer *Acidithiobacillus thiooxidans* was less abundant. At neutral pH, moderately acidophilic sulfur compound-oxidizing organisms like *Thiomonas intermedia* dominated (Schippers et al., 1995). Intermediate sulfur compounds of pyrite oxidation, the substrate for the latter organisms, could be detected (Schippers, 1998). The occurrence of sulfur compounds and of chemolithoautotrophic bacteria in selected samples from these two heaps is shown in Table 3. Tetrathionate and pentathionate could only be detected in samples with circumneutral pH from heap 1, in which sulfur-oxidizers were not detected. Conversely, sulfur compound oxidizing organisms, but not tetrathionate and pentathionate, were detected in samples from heap 2.

Mine tailings consist of fine-grained waste material from mineral processing. In a pyritic mine tailings heap in Romania, high amounts of elemental sulfur as well as *Acidithiobacillus ferrooxidans, Acidithiobacillus thiooxidans*, and *Thiomonas intermedia* were detected in samples with a pH around 4, whereas no elemental sulfur and only *Thiomonas intermedia* were found in samples with circumneutral pH. In the latter case, *Thiomonas intermedia* does not increase the FeS_2 dissolution rate but produces acidity by consuming intermediate sulfur compounds (Schippers et al., 2000).

Similar microbiological results were obtained for three different sites in a Canadian mine tailings impoundment. At the first site, with a pH between 6.5 and 7.5, the moderately acidophilic sulfur oxidizer *Thiobacillus thioparus* and related species were predominant. At the second site, with a pH of 5.5, the moderately acidophilic sulfur oxidizer *Thiobacillus thioparus* and related species were predominant as well. At the third site, the pH dropped below 4, and the acidophilic Fe(II) and sulfur oxidizer *Acidithiobacillus ferrooxidans* and the acidophilic sulfur oxidizer *Acidithiobacillus thiooxidans* were most abundant (Blowes et al., 1995).

In an Arctic Canadian pyritic mine tailings pond that consisted of 75–95% FeS_2 with the remainder made up of dolomite and residual amounts of ZnS and PbS, thiosulfate, trithionate, tetrathionate, and pentathionate were measured in particularly high amounts in samples with circumneutral pH. These compounds were not detected in samples with a pH below 7, enabling growth of acidophilic bioleaching organisms. Instead, higher amounts of elemental sulfur were found in these samples (Table 4). Only approximately one third of the FeS_2 oxidation was biological, as measured by microcalorimetric FeS_2 oxidation rates (Elberling et al., 2000).

In mine waste heaps, heat is produced as a consequence of FeS_2 oxidation. A complete oxidation of FeS_2 to Fe(III) and sulfate produces a reaction energy of −1546 kJ/mol (Rohwerder et al., 1998). High FeS_2 oxidation rates may cause documented elevated temperatures in mine waste heaps. For example, temperatures up to 100 °C have been measured in the two German pyritic uranium mine waste heaps mentioned earlier (Schippers et al., 1995). Harries and Ritchie (1980) measured vertical temperature profiles in mine waste heaps and used the data to calculate the heat flow and consequently the rate of FeS_2 oxidation.

Low rates of FeS_2 oxidation can also be detected by microcalorimetry (Sand et al. 1993, 2001; Schippers et al., 1995, 2000; Rohwerder et al., 1998; Elberling et al., 2000). With this sensitive laboratory method, heat production of a few μW/g samples can

TABLE 3. SULFUR COMPOUNDS AND CHEMOLITHOAUTOTROPHIC BACTERIA IN SELECTED SAMPLES FROM TWO DIFFERENT PYRITIC URANIUM MINE WASTE HEAPS NEAR RONNEBURG, THURINGIA, GERMANY, 1993

Sampling site	Depth (m)	T (°C)	pH	Sulfate (mg/kg)	Elemental sulfur (mg/kg)	Tetra-thionate (mg/kg)	Penta-thionate (mg/kg)	Acid. Fe(II) oxidizer (N/g)	Acid. S^0 oxidizer (N/g)	Mod. Acid. $S_2O_3^{2-}$ oxidizer (N/g)
Heap 1, covered										
Core 1	1.0–1.3	32	5.9	8640	8	n.d.	n.d.	n.d.	n.d.	n.d.
	1.6–1.7	39	7.5	7680	19	36	8	n.d.	n.d.	n.d.
	1.9–2.2	40	7.7	6720	3	16	10	n.d.	n.d.	n.d.
	2.6–2.9	50	7.6	16320	115	n.d.	n.d.	n.d.	n.d.	n.d.
	3.2–3.6	62	7.0	8640	19	7	n.d.	n.d.	n.d.	n.d.
Core 2	0.9–1.4	60	7.1	27840	n.d.	152	20	n.d.	n.d.	n.d.
	1.7–1.8	86	5.8	10560	38	13	26	n.d.	n.d.	n.d.
Heap 2, not covered										
Core 1	0.0–0.1	28	4.4	24960	n.d.	n.d.	n.d.	n.d.	n.d.	n.d.
	0.1–0.7	19	5.0	32640	3	n.d.	n.d.	1,300	9	23
	0.9–1.5	22	2.9	21120	13	n.d.	n.d.	15,000	4	n.d.
	1.9–2.6	35	3.8	26880	234	n.d.	n.d.	n.d.	n.d.	n.d.
	2.9–3.6	43	5.3	31680	96	n.d.	n.d.	43	n.d.	43
	3.9–4.6	45	3.9	28320	n.d.	n.d.	n.d.	n.d.	n.d.	n.d.

Note: after Schippers, 1998; Schippers et al., 1995. Acid.—Acidophilic, Mod.—Moderate, n.d.—not detectable.

TABLE 4. SULFUR COMPOUNDS AND CHEMOLITHOAUTOTROPHIC BACTERIA IN SELECTED SAMPLES FROM ARTIC PYRITIC MINE TAILINGS, NANISIVIK MINE, CANADA, 1998

Sample	Depth (cm)	T (°C)	pH	Sulfate (mg/kg)	Elemental sulfur (mg/kg)	Thio-sulfate (mg/kg)	Tri-thionate (mg/kg)	Tetra-thionate (mg/kg)	Penta-thionate (mg/kg)	Acid. Fe(II) oxidizer (N/g)	Acid. S^0 oxidizer (N/g)	Mod. Acid. $S_2O_3^{2-}$ oxidizer (N/g)
17	n.d.	n.d.	2.9	7286	293	n.d.	n.d.	n.d.	n.d.	90	2,000	2,000
18	n.d.	n.d.	6.8	7488	779	n.d.	n.d.	n.d.	n.d.	5,000	20,000	500
19	n.d.	n.d.	6.5	5714	1539	n.d.	n.d.	n.d.	n.d.	n.d.	90	n.d.
Core I												
20	2–4	13	7.4	4510	62	2	4	312	55	200	8	n.d.
21	7–9	12	7.4	4380	45	1	4	275	55	n.d.	n.d.	500,000
22	12–14	10	7.3	3520	59	1	4	195	36	n.d.	n.d.	n.d.
23	17–19	9	7.8	3623	22	1	4	155	30	n.d.	n.d.	50,000
24	22–24	8	7.7	3755	31	1	18	183	32	n.d.	n.d.	n.d.
25	27–29	8	7.6	3307	167	1	5	109	24	n.d.	n.d.	n.d.
26	32–34	7	7.3	2896	33	1	9	150	36	n.d.	n.d.	n.d.
27	37–39	7	7.5	1452	59	3	11	162	33	n.d.	n.d.	n.d.
28	42–44	6	7.2	1681	35	2	7	161	35	n.d.	n.d.	n.d.
29	47–49	5	7.5	2193	69	1	13	190	38	n.d.	n.d.	n.d.
30	52–54	5	7.3	1645	37	3	13	104	22	n.d.	n.d.	n.d.
31	57–59	5	7.7	741	8	2	17	57	6	n.d.	n.d.	n.d.

Note: after Elberling et al., 2000; Acid.—Acidophilic, Mod.—Moderate, n.d.—not detectable.

be detected. For measurement with a thermal activity monitor, only a few grams of a sample are needed. The chemical oxidation rate can be distinguished from the biological oxidation rate if the heat production of the same sample is measured again after treating the sample with chloroform or heating the sample to stop bacterial activity.

For Arctic pyritic mine tailings, a good correlation between microcalorimetric FeS_2 oxidation rates and in situ O_2 uptake rates has been found (Elberling et al., 2000). In situ O_2 uptake rates were measured as changes in O_2 concentration over time within a gas chamber (Elberling et al., 1994; Elberling and Nicholson, 1996; Elberling, 2001). O_2 profiles were also measured in columns filled with undisturbed tailings in the laboratory. From the laboratory results it was possible to evaluate the in situ rates of pyrite oxidation (Elberling and Damgaard, 2001).

Sediments

In contrast to mining environments with different pH levels, marine and freshwater sediments usually have circumneutral pH. Furthermore, the oxygen concentration drops to zero within a few mm below the surface of marine coastal sediments, and therefore, bacterial sulfate reduction forming H_2S is a dominant process in marine sediments (Jørgensen, 1982). H_2S can react with other sulfur compounds like elemental sulfur or be oxidized by O_2, nitrate, Fe(III), or Mn(IV) to elemental sulfur, polysulfides, thiosulfate, or sulfate (Millero, 1986; Zhang and Millero, 1993; Yao and Millero, 1996; Otte et al., 1999). Thamdrup et al. (1994a) showed that for coastal marine sediments most of the H_2S precipitated as iron sulfides and elemental sulfur. Both Fe(III) and a nonsulfur-bound Fe(II) pool reacted efficiently with H_2S.

The reactivity of different sedimentary iron minerals toward sulfide is variable over several orders of magnitude (Canfield et al., 1992). Fossing and Jørgensen (1990), using $H_2{}^{35}S$ tracer experiments, showed that most of the injected tracer occurred immediately in the AVS (FeS) and FeS_2 pools, and that significant amounts of tracer were detected in the sulfate pool only several hours after injection. This result indicates that H_2S is oxidized via FeS_2 and FeS to sulfate, which also emphasizes the importance of thiosulfate as a FeS_2 oxidation product in the sulfur cycle of marine and freshwater sediments (Jørgensen, 1990a, 1990b; Jørgensen and Bak, 1991).

The iron sulfides FeS_2 and FeS can be transported by bioturbation to the sediment surface, where a chemical oxidation by O_2 can occur (Thamdrup et al., 1994a). Aerobic bacteria like *Thiomicrospira frisia* (Brinkhoff et al., 1999) oxidize intermediates of FeS_2 and FeS oxidation, such as thiosulfate, polythionates, and elemental sulfur, which have all been detected in marine sediments (Table 1), to sulfate. In the anoxic sediment, FeS_2 and FeS can be oxidized by MnO_2 (Aller and Rude, 1988; Schippers and Jørgensen, 2001). Sulfur intermediates might be oxidized by sulfur disproportionating bacteria like *Desulfocapsa sulfoexigens* (Thamdrup et al., 1993; Finster et al., 1998). Presumably, because of the low solubility of Fe(III) or the low concentration of Fe(III) complexes at circumneutral pH, Fe(III) oxide is not an oxidant for FeS_2 or FeS in marine sediment (Schippers and Jørgensen, 2002). However, FeS can be biologically oxidized in anoxic sediments by Fe(II)-oxidizing and nitrate-reducing bacteria or H_2S-oxidizing and nitrate-reducing bacteria like *Thiobacillus denitrificans* due to the acid solubility of FeS. Lithotrophic bacteria produce extracellular polymeric substances (EPS) to create a microenvironment that favors their metabolisms (Sand et al., 2001). In such a microenvironment, the pH might be much lower than eight, enabling FeS dissolution. Since FeS_2 is resistant against proton attack, these bacteria do not dissolve FeS_2. Thus, nitrate cannot be used by bacteria to oxidize FeS_2 in marine sediments (Schippers and Jørgensen, 2002). Isotopic evidence for anoxic FeS_2 oxidation has been given by Bottrell et al. (2000).

To quantify metal sulfide oxidation in sediments, the degradation of radioactively labeled metal sulfides may be used in laboratory experiments. So far, $^{55}FeS_2$, $Fe^{35}S_2$, and $Fe^{35}S$ have been used for marine sediments (Fossing and Jørgensen, 1990; Schippers and Jørgensen, 2001, 2002). For the quantification of metal sulfide oxidation in aquifer sediments, samples were incubated in gas impermeable, polymer laminate bags, and the gas composition in the bags was monitored over a period of nearly two months. Depletion of the O_2 and enrichment of CO_2 and N_2 was interpreted as due to FeS_2 oxidation in combination with calcite dissolution (Andersen et al., 2001).

A review of current knowledge on the chemical and microbiological oxidation processes in marine sediments is given in this book by B.B. Jørgensen (Chapter 5). In addition, Chapter 7 of this book, written by J. Zopfi, T.G. Ferdelman, and H. Fossing, specifically explores the distribution and fate of sulfur intermediates in marine sediments.

Soils

Inorganic sulfur compounds have been detected in hydromorphic soils, such as salt marshes (Boulegue et al., 1982; Howarth et al., 1983; Luther et al., 1986, 1991, 2001) and paddy soils (Wind and Conrad, 1995). As in marine sediments, bacterial sulfate reduction is the dominant anaerobic degradation process in the anoxic zone of these soils. Both FeS and FeS_2 are formed from H_2S. Thus, the biogeochemistry of metal sulfide oxidation in salt marshes and paddy soils is comparable to that in marine sediments. However, periodically, depending on the water level, O_2 penetrates these soils and oxidizes the metal sulfides. Consequently, the metal sulfide oxidation products polysulfides, elemental sulfur, thiosulfate, and polythionates have been detected (Table 1). Due to FeS_2 oxidation, the pH can fall dramatically. Values below pH 3 have been reported (Schachtschabel et al., 1989). At an acidic pH, besides O_2, complexed Fe(III) might be an oxidant for FeS and FeS_2 (Luther et al., 1992), enabling metal sulfide oxidation in the anoxic zone.

Soils with a permanent low pH are the so-called "acid sulfate soils" (FAO soil taxonomy: Orthi-Thionic-Fluvisol; Schachtschabel et al., 1989). Arkesteyn (1980) studied the pH drop in pyritic marine muds during aeration. He isolated the moderately acidophilic sulfur oxidizer *Thiobacillus thioparus* and the acidophilic sulfur oxidizer *Acidithiobacillus thiooxidans* from the acidifying soil material and suggested that these bacteria utilized sulfur compounds formed by chemical FeS_2 oxidation. Biological FeS_2 oxidation by the acidophilic Fe(II) and sulfur oxidizer *Acidithiobacillus ferrooxidans* became relevant when the pH dropped below 4, as it is in mining environments.

ACKNOWLEDGMENTS

I thank Volker Brüchert, Gregory K. Druschel, Katrina Edwards, Lev N. Neretin, and an anonymous reviewer for valuable comments to improve the manuscript.

REFERENCES CITED

Aller, R.C., and Rude, P.D., 1988, Complete oxidation of solid phase sulfides by manganese and bacteria in anoxic marine sediments: Geochimica et Cosmochimica Acta, v. 52, p. 751–765, doi: 10.1016/0016-7037(88)90335-3.

Andersen, M.S., Larsen, F., and Postma, D., 2001, Pyrite oxidation in unsaturated aquifer sediments. Reaction stoichiometry and rate of oxidation: Environmental Science & Technology, v. 35, p. 4074–4079, doi: 10.1021/ES0105919.

Arkesteyn, G.J.M.W., 1980, Pyrite oxidation in acid sulphate soils: Plant and Soil, v. 54, p. 119–134.

Bebie, J., Schoonen, M.A.A., Fuhrmann, M., and Strongin, D.R., 1998, Surface charge development on transition metal sulfides: An electrokinetic study: Geochimica et Cosmochimica Acta, v. 62, p. 633–642, doi: 10.1016/S0016-7037(98)00058-1.

Becker, U., Rosso, K.M., and Hochella, M.F., Jr, 2001, The proximity effect on semiconducting mineral surfaces: A new aspect of mineral surface reactivity and surface complexation theory?: Geochimica et Cosmochimica Acta, v. 65, p. 2641–2649, doi: 10.1016/S0016-7037(01)00624-X.

Becker, U., Rosso, K.M., Weaver, R., Warren, M., and Hochella, M.F., Jr, 2003, Metal island growth and dynamics on molybdenite surfaces: Geochimica et Cosmochimica Acta, v. 67, p. 923–934, doi: 10.1016/S0016-7037(02)01144-4.

Blowes, D.W., Al, T., Lortie, L., Gould, W.D., and Jambor, J.L., 1995, Microbiological, chemical, and mineralogical characterization of the Kidd Creek mine tailings impoundment, Timmins area, Ontario: Geomicrobiological Journal, v. 13, p. 13–31.

Bonnissel-Gissinger, P., Alnot, M., Ehrhardt, J.-J., and Behra, P., 1998, Surface oxidation of pyrite as a function of pH: Environmental Science & Technology, v. 32, p. 2839–2845, doi: 10.1021/ES980213C.

Borg, R.J., and Dienes, G.J., 1992, The physical chemistry of solids: Boston, Academic Press, 584 p.

Bosecker, K., 1997, Bioleaching: Metal solubilization by microorganisms: FEMS Microbiology Reviews, v. 20, p. 591–604, doi: 10.1016/S0168-6445(97)00036-3.

Bottrell, S.H., Parkes, R.J., Cragg, B.A., and Raiswell, R., 2000, Isotopic evidence for anoxic pyrite oxidation and stimulation of bacterial sulphate reduction in marine sediments: Journal of the Geological Society [London], v. 157, p. 711–714.

Boulegue, J., Lord, C.J., III, and Church, T.M., 1982, Sulfur speciation and associated trace metals (Fe, Cu) in the pore waters of Great Marsh, Delaware: Geochimica et Cosmochimica Acta, v. 46, p. 453–464, doi: 10.1016/0016-7037(82)90236-8.

Brandl, H., 2001, Microbial leaching of metals, *in* Rehm, H.-J., and Reed, G., in cooperation with Pühler, A., and Stadler, P., eds., Biotechnology, Volume 10: Weinheim, Germany, Wiley-VCH, p. 191–224.

Brierley, C.L., and Rawlings, D.E., editors, 1997, Biomining: Theory, microbes and industrial processes: New York, Springer, 302 p.

Brinkhoff, T., Muyzer, G., Wirsen, C.O., and Kuever, J., 1999, *Thiomicrospira kuenenii* sp. nov. and *Thiomicrospira frisia* sp. nov., two mesophilic obligately chemolithoautotrophic sulfur oxidizing bacteria isolated from an intertidal mud flat: International Journal of Systematic Bacteriology, v. 49, p. 385–392.

Brock, T.D., and Gustafson, J., 1976, Ferric iron reduction by sulfur- and iron-oxidizing bacteria: Applied and Environmental Microbiology, v. 32, p. 567–571.

Canfield, D.E., Raiswell, R., and Bottrell, S.H., 1992, The reactivity of sedimentary iron minerals towards sulfide: American Journal of Science, v. 292, p. 659–683.

Chen, K.Y., and Morris, J.C., 1972, Kinetics of oxidation of aqueous sulfide by O_2: Environmental Science & Technology, v. 6, p. 529–537.

Colmer, A.R., and Hinkle, M.E., 1947, The role of microorganisms in acid mine drainage: Science, v. 106, p. 253–256.

Crundwell, F.K., 1988, The influence of the electronic structure of solids on the anodic dissolution and leaching of semiconducting sulphide minerals: Hydrometallurgy, v. 21, p. 155–190, doi: 10.1016/0304-386X(88)90003-5.

Cruz, R., Bertrand, V., Monroy, M., and González, I., 2001, Effect of sulfide impurities on the reactivity of pyrite and pyritic concentrates: A multi-tool approach: Applied Geochemistry, v. 16, p. 803–819, doi: 10.1016/S0883-2927(00)00054-8.

De Donato, Ph., Mustin, C., Benoit, R., and Erre, R., 1993, Spatial distribution of iron and sulphur species on the surface of pyrite: Applied Surface Science, v. 68, p. 81–93, doi: 10.1016/0169-4332(93)90217-Y.

De Giudici, G., and Zuddas, P., 2001, In situ investigation of galena dissolution in oxygen saturated solution: Evolution of surface features and kinetic rate:

Geochimica et Cosmochimica Acta, v. 65, p. 1381–1389, doi: 10.1016/S0016-7037(00)00605-0.
Descostes, M., Mercier, F., Beaucaire, C., Zuddas, P., and Trocellier, P., 2001, Nature and distribution of chemical species on oxidized pyrite surface: complementary of XPS and nuclear microprobe analysis: Nuclear Instruments & Methods in Physics Research. Section B, Beam Interactions with Materials and Atoms, v. 181, p. 603–609, doi: 10.1016/S0168-583X(01)00627-9.
Druschel, G.K., Hamers, R.J., and Banfield, J.F., 2003, Kinetics and mechanism of polythionate oxidation of sulfate at low pH by O_2 and Fe^{3+}: Geochimica and Cosmochimica Acta, v. 67, p. 4457–4469, doi: 10.1016/S0016-7037(03)00388-0.
Dutrizac, J.E., and MacDonald, R.J.C., 1974, Ferric ion as a leaching medium: Minerals Science Engineering, v. 6, p. 59–100.
Edwards, K.J., Gihring, T.M., and Banfield, J.F., 1999a, Seasonal variations in microbial populations and environmental conditions in an extreme acid mine drainage environment: Applied and Environmental Microbiology, v. 65, p. 3627–3632.
Edwards, K.J., Goebel, B.M., Rodgers, T.M., Schrenk, M.O., Gihring, T.M., Cardona, M.M., Hu, B., McGuire, M.M., Hamers, R.J., and Pace, N.R., 1999b, Geomicrobiology of pyrite (FeS_2) dissolution: Case study at Iron Mountain, California: Geomicrobiology Journal, v. 16, p. 155–179, doi: 10.1080/014904599270668.
Edwards, K.J., Bond, P.L., Gihring, T.M., and Banfield, J.F., 2000a, An archaeal iron-oxidizing extreme acidophile important in acid mine drainage: Science, v. 287, p 1796–1799.
Edwards, K.J., Bond, P.L., Druschel, G.K., McGuire, M.M., Hamers, R.J., and Banfield, J.F., 2000b, Geochemical and biological aspects of sulfide mineral dissolution: Lessons from Iron Mountain, California: Chemical Geology, v. 169, p. 383–397, doi: 10.1016/S0009-2541(00)00216-3.
Edwards, K.J., Hu, B., Hamers, R.J., and Banfield, J.F., 2001, A new look at microbial leaching patterns on sulfide minerals: FEMS Microbiology Ecology, v. 34, p. 197–206, doi: 10.1016/S0168-6496(00)00094-5.
Edwards, K.J., Rogers, D.R., Wirsen, C.O., and McCollom, T.M., 2003, Isolation and characterization of novel psychrophilic, neutrophilic, Fe-oxidizing, chemolithoautotrophic α- and γ-*Proteobacteria* from the deep sea: Applied and Environmental Microbiology, v. 69, p. 2906–2913, doi: 10.1128/AEM.69.5.2906-2913.2003.
Eggelston, C.M., Ehrhardt, J.-J., and Stumm, W., 1996, Surface structural controls on pyrite oxidation kinetics: An XPS-UPS, STM, and modeling study: American Mineralogist, v. 81, p. 1036–1056.
Ehrenreich, A., and Widdel, F., 1994, Anaerobic oxidation of ferrous iron by purple bacteria, a new type of phototrophic metabolism: Applied and Environmental Microbiology, v. 60, p. 4517–4526.
Ehrlich, H.L., 2002, Geomicrobiology: New York, Marcel Dekker, Inc., 768 p.
Ehrlich, H.L., and Brierley, C.L., 1990, Microbial Mineral Recovery: New York, McGraw-Hill, 454 p.
Elberling, B., 2001, Environmental controls of the seasonal variation in oxygen uptake in sulfidic tailings deposited in a permafrost-affected area: Water Resources Research, v. 37, p. 99–107, doi: 10.1029/2000WR900259.
Elberling, B., and Damgaard, L.R., 2001, Microscale measurements of oxygen diffusion and consumption in subaqeous sulfide tailings: Geochimica et Cosmochimica Acta, v. 65, p. 1897–1905, doi: 10.1016/S0016-7037(01)00574-9.
Elberling, B., and Nicholson, R.V., 1996, Field determination of sulphide oxidation rates in mine tailings: Water Resources Research, v. 32, p. 1773–1784, doi: 10.1029/96WR00487.
Elberling, B., Nicholson, R.V., Reardon, E.J., and Tibble, P., 1994, Evaluation of sulphide oxidation rates—a laboratory study comparing oxygen fluxes and rates of oxidation product release: Canadian Geotechnology Journal, v. 31, p. 375–383.
Elberling, B., Schippers, A., and Sand, W., 2000, Bacterial and chemical oxidation of pyritic mine tailings at low temperatures: Journal of Contaminant Hydrology, v. 41, p. 225–238, doi: 10.1016/S0169-7722(99)00085-6.
El-Halim, A.M., Alonso-Vante, N., and Tributsch, H., 1995, Iron/sulphur centre mediated photoinduced charge transfer at (100) oriented pyrite surfaces: Journal of Electroanalytical Chemistry, v. 399, p. 29–39, doi: 10.1016/0022-0728(95)04223-7.
Elsetinow, A.R., Schoonen, M.A.A., and Strongin, D.R., 2001, Aqueous geochemical and surface science investigations of the effect of phosphate on pyrite oxidation: Environmental Science & Technology, v. 35, p. 2252–2257, doi: 10.1021/ES0016809.
Elsetinow, A.R., Strongin, D.R., Borda, M.J., Schoonen, M.A.A., and Rosso, K.M., 2003, Characterization of the structure and the surface reactivity of a marcasite thin film: Geochimica et Cosmochimica Acta, v. 67, p. 807–812, doi: 10.1016/S0016-7037(02)00923-7.
Emerson, D., and Moyer, C.L., 1997, Isolation and characterisation of novel iron-oxidizing bacteria that grow at circumneutral pH: Applied and Environmental Microbiology, v. 63, p. 4784–4792.
Emerson, D., and Moyer, C.L., 2002, Neutrophilic Fe-oxidizing bacteria are abundant at the Loihi seamount hydrothermal vents and play a major role in Fe oxide deposition: Applied and Environmental Microbiology, v. 68, p. 3085–3093, doi: 10.1128/AEM.68.6.3085-3093.2002.
Evangelou, V.P.B., 1995, Pyrite oxidation and its control: Boca Raton, Florida, CRC Press, 293 p.
Evangelou, V.P.B., and Zhang, Y.L., 1995, A review: pyrite oxidation mechanisms and acid mine drainage prevention: Critical Reviews in Environmental Science and Technology, v. 25, p. 141–199.
Evangelou, V.P., Seta, A.K., and Holt, A., 1998, Potential role of bicarbonate during pyrite oxidation: Environmental Science & Technology, v. 32, p. 2084–2091, doi: 10.1021/ES970829M.
Finster, K., Liesack, W., and Thamdrup, B., 1998, Elemental sulfur and thiosulfate disproportionation by *Desulfocapsa sulfoexigens* sp. nov., a new anaerobic bacterium isolated from marine surface sediment: Applied and Environmental Microbiology, v. 64, p. 119–125.
Fossing, H., and Jørgensen, B.B., 1990, Oxidation and reduction of radiolabelled inorganic sulfur compounds in an estuarine sediment, Kysing Fjord, Denmark: Geochimica et Cosmochimica Acta, v. 54, p. 2731–2742, doi: 10.1016/0016-7037(90)90008-9.
Gantayat, B.P., Rath, P.C., Paramguru, R.K., and Rao, S.B., 2000, Galvanic interaction between chalcopyrite and manganese dioxide in sulfuric acid medium: Metallurgical and Materials Transactions B, v. 31, p. 55–61.
Garcia-Gil, L.J., and Golterman, H.L., 1993, Kinetics of FeS-mediated denitrification in sediments from the Camargue (Rhone delta, southern France): FEMS Microbiology Ecology, v. 13, p. 85–92, doi: 10.1016/0168-6496(93)90026-4.
Gehrke, T., Telegdi, J., Thierry, D., and Sand, W., 1998, Importance of extracellular polymeric substances from *Thiobacillus ferrooxidans* for bioleaching: Applied and Environmental Microbiology, v. 64, p. 2743–2747.
Gerson, A.R., and O'Dea, A.R., 2003, A quantum chemical investigation of the oxidation and dissolution mechanisms of galena: Geochimica et Cosmochimica Acta, v. 67, p. 813–822, doi: 10.1016/S0016-7037(02)01147-X.
Giannetti, B.F., Bonilla, S.H., Zinola, C.F., and Rabóczkay, T., 2001, A study of the main oxidation products of natural pyrite by voltammetric and photoelectrochemical responses: Hydrometallurgy, v. 60, p. 41–53, doi: 10.1016/S0304-386X(00)00158-4.
Goldhaber, M.B., 1983, Experimental study of metastable sulfur oxyanion formation during pyrite oxidation at pH 6–9 and 30°C: American Journal of Science, v. 283, p. 193–217.
Guevremont, J.M., Elsetinow, A.R., Strongin, D.R., Bebie, J., and Schoonen, M.A.A., 1998, Structure sensitivity of pyrite oxidation: Comparison of the (100) and (111) planes: American Mineralogist, v. 83, p. 1353–1356.
Hackl, R.P., Dreisinger, D.B., Peters, E., and King, J.A., 1995, Passivation of chalcopyrite during oxidative leaching in sulfate media: Hydrometallurgy, v. 39, p. 25–48, doi: 10.1016/0304-386X(95)00023-A.
Harries, J.R., and Ritchie, A.I.M., 1980, The use of temperature profiles to estimate the pyritic oxidation rate in a waste rock dump from an opencut mine: Water, Air, and Soil Pollution, v. 15, p 405–423.
Herlihy, A.T., Mills, A.L., and Herman, J.S., 1988, Distribution of reduced inorganic sulfur compounds in lake sediments receiving acid mine drainage: Applied Geochemistry, v. 3, p. 333–344, doi: 10.1016/0883-2927(88)90110-2.
Howarth, R.W., Giblin, A., Gale, J., Petersen, B.J., and Luther, G.W., III, 1983, Reduced sulfur compounds in the pore waters of a New England salt marsh, *in* Hallberg, R., ed., Environmental Biogeochemistry: Ecological Bulletin (Stockholm), v. 35, p. 135–152.
Jørgensen, B.B., 1982, Mineralization of organic matter in the sea bed—the role of sulphate reduction: Nature, v. 296, p. 643–645.
Jørgensen, B.B., 1990a, A thiosulfate shunt in the sulfur cycle of marine sediments: Science, v. 249, p. 152–154.
Jørgensen, B.B., 1990b, The sulfur cycle of freshwater sediments: Role of thiosulfate: Limnology and Oceanography, v. 35, p. 1329–1342.
Jørgensen, B.B., and Bak, F., 1991, Pathways and microbiology of thiosulfate transformations and sulfate reduction in a marine sediment (Kattegat, Denmark): Applied and Environmental Microbiology, v. 57, p. 847–856.
Kuenen, J.G., Robertson, L.A., and Tuovinen, O.H., 1992, The genera *Thiobacillus, Thiomicrospira*, and *Thiosphera*, *in* Balows, A., Trüper, H.G., Dworkin, M., Harder W., and Schleifer, K.-H., eds., The Prokaryotes: Berlin, Springer Verlag, p. 2638–2657.

Lengke, M.F., and Tempel, R.N., 2001, Kinetic rates of amorphous As_2S_3 oxidation at 25 to 40 °C and initial pH of 7.3 to 9.4: Geochimica et Cosmochimica Acta, v. 65, p. 2241–2255, doi: 10.1016/S0016-7037(01)00592-0.

Lengke, M.F., and Tempel, R.N., 2003, Natural realgar and amorphous AsS oxidation kinetics: Geochimica et Cosmochimica Acta, v. 67, p. 859–871, doi: 10.1016/S0016-7037(02)01227-9.

Lizama, H.M., and Suzuki, I., 1991, Interaction of chalcopyrite and sphalerite with pyrite during leaching by *Thiobacillus ferrooxidans* and *Thiobacillus thiooxidans*: Canadian Journal of Microbiology, v. 37, p. 304–311.

Lowson, R.T., 1982, Aqueous oxidation of pyrite by molecular oxygen: Chemical Reviews, v. 82, p. 461–497.

Luther, G.W., III, 1987, Pyrite oxidation and reduction: molecular orbital theory considerations: Geochimica et Cosmochimica Acta, v. 51, p. 3193–3199, doi: 10.1016/0016-7037(87)90127-X.

Luther, G.W., III, 1990, The frontier-molecular-orbital theory approach in geochemical processes, *in* Stumm, W., ed., Aquatic Chemical Kinetics: New York, John Wiley & Sons, Inc., p. 173–198.

Luther, G.W., III, Church, T.M., Scudlark, J.R., and Cosman, M., 1986, Inorganic and organic sulfur cycling in salt-marsh pore waters: Science, v. 232, p. 746–749.

Luther, G.W., III, Ferdelman, T.G., Kostka, J.E., Tsamakis, E.J., and Church, T.M., 1991, Temporal and spatial variability of reduced species (FeS_2, $S_2O_3^{2-}$) and porewater parameters in salt marsh sediments: Biogeochemistry, v. 14, p. 57–88.

Luther, G.W., III, Kostka, J.E., Church, T.M., Sulzberger, B., and Stumm, W., 1992, Seasonal iron cycling in the salt-marsh sedimentary environment: The importance of ligand complexes with Fe(II) and Fe(III) in the dissolution of Fe(III) minerals and pyrite, respectively: Marine Chemistry, v. 40, p. 81–103, doi: 10.1016/0304-4203(92)90049-G.

Luther, G.W., III, Glazer, B.T., Hohmann, L., Popp, J.I., Taillefert, M., Rozan, T.F., Brendel, P.J., Theberge, S.M., and Nuzzio, D.B., 2001, Sulfur speciation monitored in situ with solid state gold amalgam voltammetric microelectrodes: Polysulfides as a special case in sediments, microbial mats and hydrothermal vent waters: Journal of Environmental Monitoring, v. 3, p. 61–66, doi: 10.1039/B006499H.

Martello, D.V., Vecchio, K.S., Diehl, J.R., Graham, R.A., Tamilia, J.P., and Pollack, S.S., 1994, Do dislocations and stacking faults increase the oxidation rate of pyrites?: Geochimica et Cosmochimica Acta, v. 58, p. 4657–4665, doi: 10.1016/0016-7037(94)90198-8.

McGuire, M.M., Edwards, K.J., Banfield, J.F., and Hamers, R.J., 2001a, Kinetics, surface chemistry, and structural evolution of microbially mediated sulfide mineral dissolution: Geochimica et Cosmochimica Acta, v. 65, p. 1243–1258, doi: 10.1016/S0016-7037(00)00601-3.

McGuire, M.M., Jallad, K.N., Ben-Amotz, D., and Hamers, R.J., 2001b, Chemical mapping of elemental sulfur on pyrite and arsenopyrite surfaces using near-infrared Raman imaging microscopy: Applied Surface Science, v. 178, p. 105–115, doi: 10.1016/S0169-4332(01)00303-8.

Millero, F.J., 1986, The thermodynamics and kinetics of the hydrogen sulfide system in natural waters: Marine Chemistry, v. 18, p. 121–147, doi: 10.1016/0304-4203(86)90003-4.

Mishra, K.K., and Osseo-Asare, K., 1988, Aspects of the interfacial electrochemistry of semiconductor pyrite (FeS_2): Journal of Electrochemical Society, v. 135, p. 2502–2509.

Morse, J.W., 1991, Oxidation kinetics of sedimentary pyrite in seawater: Geochimica et Cosmochimica Acta, v. 55, p. 3665–3667, doi: 10.1016/0016-7037(91)90064-C.

Moses, C.O., and Herman, J.S., 1991, Pyrite oxidation at circumneutral pH: Geochimica et Cosmochimica Acta, v. 55, p. 471–482, doi: 10.1016/0016-7037(91)90005-P.

Moses, C.O., Nordstrom, D.K., Herman, J.S., and Mills, A.L., 1987, Aqueous pyrite oxidation by dissolved oxygen and by ferric iron: Geochimica et Cosmochimica Acta, v. 51, p. 1561–1571, doi: 10.1016/0016-7037(87)90337-1.

Nakamura, H., Sato, S., and Hara, Y., 1994, The oxidation of pyrite: Journal of Hazardous Materials, v. 37, p. 253–263, doi: 10.1016/0304-3894(93)E0095-J.

Nesbitt, H.W., and Muir, I.J., 1994, X-ray photoelectron spectroscopic study of a pristine pyrite surface reacted with water vapour and air: Geochimica et Cosmochimica Acta, v. 58, p. 4667–4679, doi: 10.1016/0016-7037(94)90199-6.

Nesbitt, H.W., Schaufuß, A., Sciani, M., Höchst, H., Bancroft, G.M., and Szargan, R., 2003, Monitoring fundamental reactions at NiAsS surfaces by synchroton radiation X-ray photoelectron spectroscopy: As and S air oxidation by consecutive reaction schemes: Geochimica et Cosmochimica Acta, v. 67, p. 845–858, doi: 10.1016/S0016-7037(02)00944-4.

Nicholson, R.V., Gillham, R.W., and Reardon, E.J., 1988, Pyrite oxidation in carbonate-buffered solution: 1. Experimental kinetics: Geochimica et Cosmochimica Acta, v. 52, p. 1077–1085, doi: 10.1016/0016-7037(88)90262-1.

Nicholson, R.V., Gillham, R.W., and Reardon, E.J., 1990, Pyrite oxidation in carbonate-buffered solution: 2. Rate control by oxide coatings: Geochimica et Cosmochimica Acta, v. 54, p. 395–402, doi: 10.1016/0016-7037(90)90328-I.

Nordstrom, D.K., 1982, Aqueous pyrite oxidation and the consequent formation of secondary iron minerals, *in* Hossner, L.R., Kittrick, J.A., and Fanning, D.F., eds., Acid sulfate weathering, pedogeochemistry and relationship to manipulation of soil minerals: Madison, Wisconsin, Soil Science Society of America Press, p. 37–55.

Nordstrom, D.K., and Southam, G., 1997, Geomicrobiology of sulfide mineral oxidation, *in* Banfield, J.F., and Nealson, K.H., eds., Geomicrobiology: Interactions between microbes and minerals, Reviews in Mineralogy, Volume 35: Washington D.C., Mineralogical Society of America, p. 361–390.

Nordstrom, D.K., and Alpers, C.N., 1999a, Geochemistry of acid mine waters, *in* Plumlee, G. S., and Logsdon, M.J., eds., The environmental geochemistry of mineral deposits, Reviews in Economic Geology, Volume 6A: Littleton, Colorado, Society of Economic Geologists, Inc., p. 133–160.

Nordstrom, D.K., and Alpers, C.N., 1999b, Negative pH, efflorescent mineralogy, and consequences for environmental restoration at the Iron Mountain, superfund site, California: Proceedings of the National Academy of Sciences of the United States of America, v. 96, p. 3455–3462, doi: 10.1073/PNAS.96.7.3455.

Otte, S., Kuenen, J.G., Nielsen, L.P., Paerl, H.W., Zopfi, J., Schulz, H.N., Teske, A., Strotmann, B., Gallardo, V.A., and Jørgensen, B.B., 1999, Nitrogen, carbon, and sulfur metabolism in natural *Thioploca* samples: Applied and Environmental Microbiology, v. 65, p. 3148–3157.

Peiffer, S., and Stubert, I., 1999, The oxidation of pyrite at pH 7 in the presence of reducing and nonreducing Fe(III)-chelators: Geochimica et Cosmochimica Acta, v. 63, p. 3171–3182, doi: 10.1016/S0016-7037(99)00224-0.

Pichtel, J.R., and Dick, W.A., 1991, Sulfur, iron and solid phase transformations during biological oxidation of pyritic mine spoil: Soil Biology & Biochemistry, v. 23, p. 101–107, doi: 10.1016/0038-0717(91)90120-9.

Podgorsek, L., and Imhoff, J.F., 1999, Tetrathionate production by sulfur oxidizing bacteria and the role of tetrathionate in the sulfur cycle of Baltic sediments: Aquatic Microbial Ecology, v. 17, p. 255–265.

Pronk, J.T., and Johnson, D.B., 1992, Oxidation and reduction of iron by acidophilic bacteria: Geomicrobiological Journal, v. 10, p. 153–171.

Rawlings, D.E., 2002, Heavy metal mining using microbes: Annual Review of Microbiology, v. 56, p. 65–91, doi: 10.1146/ANNUREV.MICRO.56.012302.161052.

Rawlings, D.E., Dew, D., and du Plessis, C., 2003, Biomineralization of metal-containing ores and concentrates: Trends in Biotechnology, v. 21, p. 38–44, doi: 10.1016/S0167-7799(02)00004-5.

Reedy, B.J., Beattie, J.K., and Lowson, R.T., 1991, A vibrational spectroscopic ^{18}O tracer study of pyrite oxidation: Geochimica et Cosmochimica Acta, v. 55, p. 1609–1614, doi: 10.1016/0016-7037(91)90132-O.

Rimstidt, J.D., and Vaughan, D.J., 2003, Pyrite oxidation: A state-of-the-art assessment of the reaction mechanism: Geochimica et Cosmochimica Acta, v. 67, p. 873–880, doi: 10.1016/S0016-7037(02)01165-1.

Rohwerder, T., Schippers, A., and Sand, W., 1998, Determination of reaction energy values for biological pyrite oxidation by calorimetry: Thermochimica Acta, v. 309, p. 79–85, doi: 10.1016/S0040-6031(97)00352-3.

Rohwerder, T., Jozsa, P.-G., Gehrke, T., and Sand, W., 2002, Bioleaching, *in* Bitton, G., ed., Encyclopedia of Environmental Microbiology, Volume 2: New York, Wiley, p. 632–641.

Rossi, G., 1990, Biohydrometallurgy: Hamburg, McGraw-Hill, 609 p.

Rossi, G., 1993, Biodepyritization of coal: achievements and problems: Fuel, v. 72, p. 1581–1592, doi: 10.1016/0016-2361(93)90340-8.

Sand, W., Rohde, K., Sobotke, B., and Zenneck, C., 1992, Evaluation of *Leptospirillum ferrooxidans* for leaching: Applied and Environmental Microbiology, v. 58, p. 85–92.

Sand, W., Hallmann, R., Rohde, K., Sobotke, B., and Wentzien, S., 1993, Controlled microbiological in-situ stope leaching of a sulphidic ore: Applied Microbiology and Biotechnology, v. 40, p. 421–426.

Sand, W., Gehrke, T., Jozsa, P.-G., and Schippers, A., 2001, (Bio)chemistry of bacterial leaching—direct vs. indirect bioleaching: Hydrometallurgy, v. 59, p. 159–175, doi: 10.1016/S0304-386X(00)00180-8.

Sasaki, K., 1994, Effect of grinding on the rate of oxidation of pyrite by oxygen in acid solutions: Geochimica et Cosmochimica Acta, v. 58, p. 4649–4655, doi: 10.1016/0016-7037(94)90197-X.

Sasaki, K., Tsunekawa, M., Ohtsuka, T., and Konno, H., 1995, Confirmation of a sulfur-rich layer on pyrite after oxidative dissolution by Fe(III) ions around pH 2: Geochimica et Cosmochimica Acta, v. 59, p. 3155–3158, doi: 10.1016/0016-7037(95)00203-C.

Schachtschabel, P., Blume, H.-P., Brümmer, G., Hartge, K.-H., and Schwertmann, U., in cooperation with Fischer, W.R., Renger, M., and Strebel, O., 1989, Scheffer/Schachtschabel, Lehrbuch der Bodenkunde: Stuttgart, Germany, Enke-Verlag, 491 p.

Schippers, A., 1998, Untersuchungen zur Schwefelchemie der biologischen Laugung von Metallsulfiden [PhD-thesis]: Aachen, Germany, University of Hamburg, Shaker-Verlag, 102 p.

Schippers, A., and Sand, W., 1999, Bacterial leaching of metal sulfides proceeds by two indirect mechanisms via thiosulfate or via polysulfides and sulfur: Applied and Environmental Microbiology, v. 65, p. 319–321.

Schippers, A., and Jørgensen, B.B., 2001, Oxidation of pyrite and iron sulfide by manganese dioxide in marine sediments: Geochimica et Cosmochimica Acta, v. 65, p. 915–922, doi: 10.1016/S0016-7037(00)00589-5.

Schippers, A., and Jørgensen, B.B., 2002, Biogeochemistry of pyrite and iron sulfide oxidation in marine sediments: Geochimica et Cosmochimica Acta, v. 66, p. 85–92, doi: 10.1016/S0016-7037(01)00745-1.

Schippers, A., Hallmann, R., Wentzien, S., and Sand, W., 1995, Microbial diversity in uranium mine waste heaps: Applied and Environmental Microbiology, v. 61, p. 2930–2935.

Schippers, A., Jozsa, P.-G., and Sand, W., 1996a, Sulfur chemistry in bacterial leaching of pyrite: Applied and Environmental Microbiology, v. 62, p. 3424–3431.

Schippers, A., von Rège, H., and Sand, W., 1996b, Impact of microbial diversity and sulfur chemistry on safeguarding sulfidic mine waste: Minerals Engineering, v. 9, p. 1069–1079, doi: 10.1016/0892-6875(96)00099-4.

Schippers, A., Rohwerder, T., and Sand, W., 1999, Intermediary sulfur compounds in pyrite oxidation: implications for bioleaching and biodepyritization of coal: Applied Microbiology and Biotechnology, v. 52, p. 104–110, doi: 10.1007/S002530051495.

Schippers, A., Jozsa, P.-G., Sand, W., Kovacs, Zs.M., and Jelea, M., 2000, Microbiological pyrite oxidation in a mine tailings heap and its relevance to the death of vegetation: Geomicrobiology Journal, v. 17, p. 151–162, doi: 10.1080/01490450050023827.

Schoonen, M., Elsetinow, A., Borda, M., and Strongin, D., 2000, Effect of temperature and illumination on pyrite oxidation between pH 2 and 6: Geochemical Transactions, v. 4, p. 000–001.

Schrenk, M.O., Edwards, K.J., Goodman, R.M., Hamers, R.J., and Banfield, J.F., 1998, Distribution of *Thiobacillus ferrooxidans* and *Leptospirillum ferrooxidans* for generation of acid mine drainage: Science, v. 279, p. 1519–1522, doi: 10.1126/SCIENCE.279.5356.1519.

Singer, P.C., and Stumm, W., 1970, Acidic mine drainage: the rate-determining step: Science, v. 167, p. 1121–1123.

Smart, R.St.C., Jasieniak, M., Prince, K.E., and Skinner, W.M., 2000, SIMS studies of oxidation mechanisms and polysulfide formation in reacted sulfide surfaces: Minerals Engineering, v. 13, p. 857–870, doi: 10.1016/S0892-6875(00)00074-1.

Steger, H.F., and Desjardins, L.E., 1978, Oxidation of sulfide minerals, 4. Pyrite, chalcopyrite and pyrrhotite: Chemical Geology, v. 23, p. 225–237, doi: 10.1016/0009-2541(78)90079-7.

Steudel, R., 1996, Mechanism for the formation of elemental sulfur from aqueous sulfide in chemical and microbiological desulfurization processes: Industrial & Engineering Chemistry Research, v. 35, p. 1417–1423, doi: 10.1021/IE950558T.

Steudel, R., Holdt, G., Goebel, T., and Hazeu, W., 1987, Chromatographic separation of higher polythionates $S_nO_6^{2-}$ (n = 3...22) and their detection in cultures of *Thiobacillus ferrooxidans*; molecular composition of bacterial sulfur excretions: Angewandte Chemie International Edition in English, v. 26, p. 151–153, doi: 10.1002/ANIE.198701511.

Straub, K.L., Benz, M., Schink, B., and Widdel, F., 1996, Anaerobic, nitrate-dependent microbial oxidation of ferrous iron: Applied and Environmental Microbiology, v. 62, p. 1458–1460.

Straub, K.L., Rainey, F.A., and Widdel, F., 1999, *Rhodovulum iodosum* sp. nov. and *Rhodovulum robiginosum* sp. nov., two new marine phototrophic ferrous-iron-oxidizing purple bacteria: International Journal of Systematic Bacteriology, v. 49, p. 729–735.

Thamdrup, B., Finster, K., Würgler-Hansen, J., and Bak, F., 1993, Bacterial disproportionation of elemental sulfur coupled to chemical reduction of iron or manganese: Applied and Environmental Microbiology, v. 59, p. 101–108.

Thamdrup, B., Fossing, H., and Jørgensen, B.B., 1994a, Manganese, iron, and sulfur cycling in a coastal marine sediment, Aarhus Bay, Denmark: Geochimica et Cosmochimica Acta, v. 58, p. 5115–5129, doi: 10.1016/0016-7037(94)90298-4.

Thamdrup, B., Finster, K., Fossing, H., Würgler-Hansen, J., and Jørgensen, B.B., 1994b, Thiosulfate and sulfite distributions in porewater of marine sediments related to manganese, iron, and sulfur geochemistry: Geochimica et Cosmochimica Acta, v. 58, p. 57–73, doi: 10.1016/0016-7037(94)90445-6.

Thomas, J.E., Jones, C.F., Skinner, W.M., Smart, R.St.C., 1998, The role of surface sulfur species in the inhibition of pyrrhotite dissolution in acid conditions: Geochimica et Cosmochimica Acta, v. 62, p. 1555–1565, doi: 10.1016/S0016-7037(98)00087-8.

Thomas, J.E., Skinner, W.M., Smart, R.St.C., 2001, A mechanism to explain sudden changes in rates and products for pyrrhotite dissolution in acid solution: Geochimica et Cosmochimica Acta, v. 65, p. 1–12, doi: 10.1016/S0016-7037(00)00503-2.

Thomas, J.E., Skinner, W.M., Smart, R.St.C., 2003, A comparison of the dissolution behavior of troilite with other iron(II) sulfides; implications of structure: Geochimica et Cosmochimica Acta, v. 67, p. 831–843, doi: 10.1016/S0016-7037(02)01146-8.

Todd, E.C., Sherman, D.M., and Purton, J.A., 2003, Surface oxidation of pyrite under ambient atmospheric and aqueous (pH = 2 to 10) conditions: Electronic structure and mineralogy from X-ray absorption spectroscopy: Geochimica et Cosmochimica Acta, v. 67, p. 881–893, doi: 10.1016/S0016-7037(02)00957-2.

Tributsch, H., and Bennett, J.C., 1981a, Semiconductor-electrochemical aspects of bacterial leaching. 1. Oxidation of metal sulphides with large energy gaps: Journal of Chemical Technology and Biotechnology (Oxford, Oxfordshire : 1986), v. 31, p. 565–577.

Tributsch, H., and Bennett, J.C., 1981b, Semiconductor-electrochemical aspects of bacterial leaching. Part 2. Survey of rate-controlling sulphide properties: Journal of Chemical Technology and Biotechnology (Oxford, Oxfordshire: 1986), v. 31, p. 627–635.

Tributsch, H., and Gerischer, H., 1976, The oxidation and self-heating of metal sulphides as an electrochemical corrosion phenomenon: Journal of Applied Chemistry and Biotechnology, v. 26, p. 747–761.

Troelsen, H., and Jørgensen, B.B., 1982, Seasonal dynamics of elemental sulfur in two coastal sediments: Estuarine, Coastal and Shelf Science, v. 15, p. 255–266.

Vaughan, D.J., and Craig, J.R., 1978, Mineral chemistry of metal sulfides: Cambridge, Cambridge University Press, 512 p.

Vaughan, D.J., Becker, U., and Wright, K., 1997, Sulphide mineral surfaces: Theory and experiment: International Journal of Mineral Processing, v. 51, p. 1–14, doi: 10.1016/S0301-7516(97)00035-5.

Wiersma, C.L., and Rimstidt, J.D., 1984, Rates of reaction of pyrite and marcasite with ferric iron at pH 2: Geochimica et Cosmochimica Acta, v. 48, p. 85–92, doi: 10.1016/0016-7037(84)90351-X.

Williamson, M.A., and Rimstidt, J.D., 1994, The kinetics and electrochemical rate-determining step of aqueous pyrite oxidation: Geochimica et Cosmochimica Acta, v. 58, p. 5443–5454, doi: 10.1016/0016-7037(94)90241-0.

Wind, T., and Conrad, R., 1995, Sulfur compounds, potential turnover of sulfate and thiosulfate, and numbers of sulfate-reducing bacteria in planted and unplanted paddy soil: FEMS Microbiology Ecology, v. 18, p. 257–266, doi: 10.1016/0168-6496(95)00066-6.

Xu, Y., and Schoonen, M.A.A., 1995, The stability of thiosulfate in the presence of pyrite in low-temperature aqueous solutions: Geochimica et Cosmochimica Acta, v. 59, p. 4605–4622, doi: 10.1016/0016-7037(95)00331-2.

Xu, Y., and Schoonen, M.A.A., 2000, The absolute energy positions of conduction and valence bands of selected semiconducting minerals: American Mineralogist, v. 85, p. 543–556.

Xu, Y., Schoonen, M.A.A., and Strongin, D.R., 1996, Thiosulfate oxidation: Catalysis of synthetic sphalerite doped with transition metals: Geochimica et Cosmochimica Acta, v. 60, p. 4701–4710, doi: 10.1016/S0016-7037(96)00279-7.

Yao, W., and Millero, F.J., 1996, Oxidation of hydrogen sulfide by hydrous Fe(III) oxides in seawater: Marine Chemistry, v. 52, p. 1–16, doi: 10.1016/0304-4203(95)00072-0.

Zhang, H., and Dreisinger, D.B., 2002, The kinetics for the decomposition of tetrathionate in alkaline solutions: Hydrometallurgy, v. 66, p. 59–65, doi: 10.1016/S0304-386X(02)00078-6.

Zhang, J.-Z., and Millero, F.J., 1993, The products from the oxidation of H_2S in seawater: Geochimica et Cosmochimica Acta, v. 57, p. 1705–1718, doi: 10.1016/0016-7037(93)90108-9.

Manuscript Accepted by the Society March 28, 2004

Printed in the USA

Geological Society of America
Special Paper 379
2004

Sulfide oxidation in marine sediments: Geochemistry meets microbiology

Bo Barker Jørgensen*
Max Planck Institute for Marine Microbiology, Celsiusstrasse 1, D-28359 Bremen, Germany

Douglas C. Nelson
Section of Microbiology, University of California, Davis, California 95616, USA

ABSTRACT

The main pathways of sulfide oxidation in marine sediments involve complex interactions of chemical reaction and microbial metabolism. Sulfide becomes partly oxidized and bound by Fe(III), and the resulting iron-sulfur minerals are transported toward the oxic sediment-water interface by bioturbating and irrigating fauna. Although oxygen is the main oxidant for pyrite or amorphous iron sulfide, oxidation reactions may also take place with nitrate or manganese oxide. Intermediate oxidation products such as elemental sulfur or thiosulfate undergo disproportionation reactions and thereby provide shunts in the sedimentary sulfur cycle. Although of widespread occurrence, chemolithoautotrophic sulfide oxidizing bacteria, such as *Thiobacillus* spp. or *Thiomicrospira* spp., appear to be of minor significance relative to heterotrophic or mixotrophic sulfide oxidizers of diverse phylogenetic lineages. As a unique group, the large sulfur bacteria of the genera *Beggiatoa*, *Thioploca,* and *Thiomargarita* have developed specialized modes of sulfide oxidation using nitrate stored in intracellular vacuoles. By commuting between electron acceptor and donor, or by temporally bridging their occurrences in the environment through a great storage potential for both nitrate and elemental sulfur, these bacteria compete efficiently with other microbial pathways of sulfide oxidation. Dissimilatory nitrate reduction in these bacteria leads preferentially to ammonium rather than to dinitrogen, as in the denitrifying bacteria. *Beggiatoa* appears to be widely distributed in coastal sediments with a high organic load. In such sediments where *Beggiatoa* often occurs unnoticed in the anoxic, oxidized zone rather than growing as a visible mat on the sediment surface, dissimilatory nitrate reduction to ammonium may dominate over denitrification.

Keywords: sulfate reduction, disproportionation, sulfur cycle, sulfur bacteria, chemoautotrophy, *Beggiatoa*.

---

*bjoergen@mpi-bremen.de.

Jørgensen, B.B., and Nelson, D.C., 2004, Sulfide oxidation in marine sediments: Geochemistry meets microbiology, *in* Amend, J.P., Edwards, K.J., and Lyons, T.W., eds., Sulfur Biogeochemistry—Past and Present: Geological Society of America Special Paper 379, p. 63–81, For permission to copy, contact editing@geosociety.org.

INTRODUCTION

In the geochemist's view, sulfide oxidation in marine sediments takes place by heterogeneous reactions with oxidized forms of iron or manganese or by rapid reaction with oxygen. Iron-sulfur minerals, such as amorphous FeS or pyrite, are quantitatively the dominant forms of reduced sulfur, and an understanding of sulfide oxidation must consequently include their transformations. In the microbiologist's view, sulfide oxidation is a metabolic pathway in a range of chemoautotrophic or heterotrophic bacteria that are widespread in marine sediments. Although which of these bacteria predominate in any given sediment remains poorly understood, their activity is expectedly a prerequisite for the complete oxidation of sulfide to sulfate and thus for the continuous function of the sulfur cycle.

With the aim of reconciling these views, we will address the following questions in this paper:

- What is the role of microbiology versus geochemistry for the overall sulfide oxidation in marine sediments and how do they interact?
- Does sulfide oxidation mainly take place in the oxic or in the anoxic part of the sediment?
- Which are the important types of bacteria in the oxidative pathways of the sulfur cycle?
- What is the role of nitrate-accumulating sulfur bacteria in coupling the nitrogen and sulfur cycles?

In discussing these aspects, we will consider mostly shelf sediments that have a relatively high turnover of organic material and where sulfate reduction and sulfide oxidation are important in the mineralization processes. The discussion will have an emphasis on the coupling of sulfide oxidation with nitrate reduction, and the examples chosen will have a bias toward results from the authors' research groups.

SULFIDE PRODUCTION IN MARINE SEDIMENTS

Marine shelf sediments have redox zonations that are often recognizable from their color. The brown upper layer comprises the oxidized zone in which manganese and iron occur mostly as oxidized mineral phases. Oxygen is present here only in the top few millimeters or centimeters, and most of the oxidized zone is thus anoxic (i.e., without molecular oxygen) (Fig. 1A). In this suboxic zone, the metal oxides serve as oxidants in mineralization of organic matter, either directly by heterotrophic iron or manganese-reducing bacteria, or indirectly by reaction with sulfide formed by sulfate-reducing bacteria. The depth and intensity of metal oxide reduction are apparent from the maxima and steepness of Mn^{2+} and Fe^{2+} gradients. The pore water profiles in Figure 1A show that manganese is mainly reduced in the uppermost 0–1 cm, whereas iron reduction mainly takes place below 1 cm and down to at least 4 cm. Below 4 cm, the gradient of H_2S demonstrates an upward diffusive flux of H_2S originating from sulfate reduction deeper in

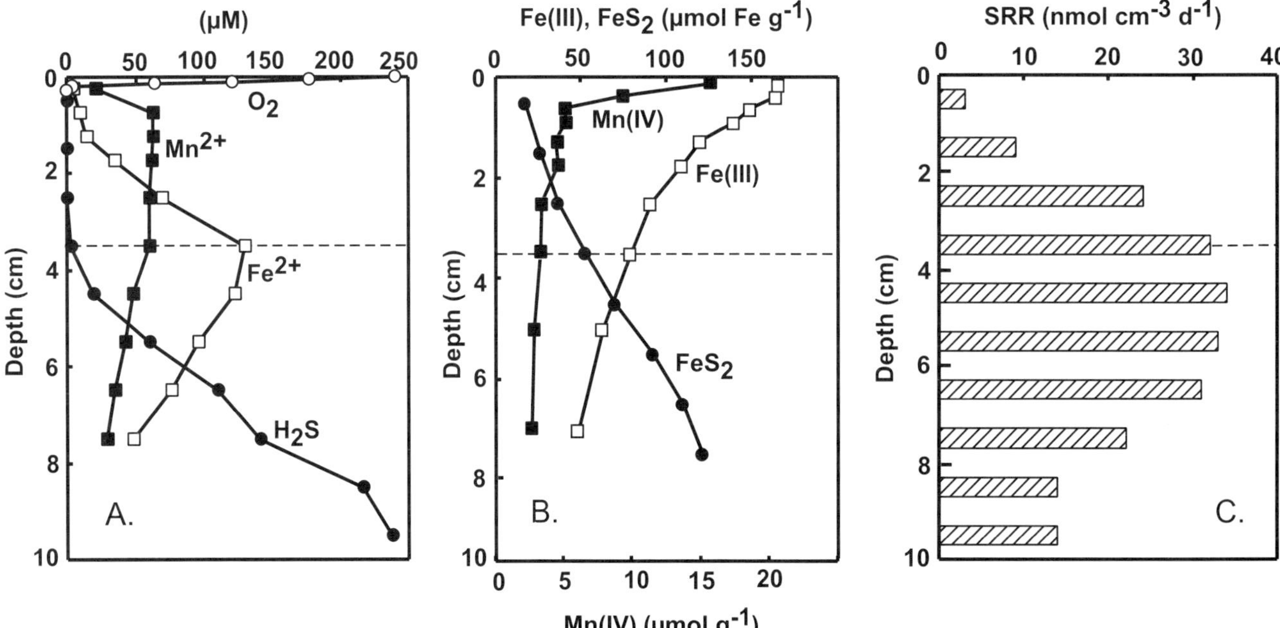

Figure 1. Chemical zonations typical for a coastal marine sediment (Aarhus Bay, Denmark, 16 m water depth). The brown oxidized zone (above the broken line) comprised the top 3–4 cm below which the color turned to gray or black. A. Pore water gradients of oxygen, metal ions, and hydrogen sulfide. Oxygen is consumed within the top 3 mm, below which manganese reduction, iron reduction, and sulfate reduction are sequentially the dominant redox processes. B. Solid phase distribution of reactive manganese and iron oxides (oxalate extracts) and pyrite. C. Sulfate reduction rates measured by ^{35}S-technique showing activity throughout all zones. (Redrawn from Thamdrup et al., 1994a.)

the sediment. Around 4 cm depth, the H_2S disappears, probably by reaction with iron oxides and by trapping in FeS and FeS_2.

Starting at a few centimeters or decimeters subsurface in organic-rich sediments, sulfate is the dominant electron acceptor down to the bottom of the sulfate zone at several meters depth. Even in the partially oxidized zone, however, sulfate reducing bacteria are present, and sulfate reduction is active concurrent with manganese and iron reduction (Fig. 1C; Jørgensen and Bak, 1991). Sulfate reduction is altogether the main pathway of anaerobic mineralization of organic matter in most continental shelf sediments, where it may account for 25%–50% of the overall carbon oxidation on an areal basis (Jørgensen, 1982a).

Vast amounts of H_2S are formed through sulfate reduction in shelf sediments, typically 0.1–1 mol S m^{-2} yr^{-1}. Only a small fraction of the H_2S, generally in the range of 5%–20%, is permanently buried within the sediment after being trapped as iron sulfide and pyrite (Jørgensen et al., 1990; Canfield and Teske, 1996). The remaining 80%–95% of the H_2S is recycled within the sediment and gradually oxidized back to sulfate. The reoxidation takes place at all depths and zones of the sediment, most rapidly in the upper oxidized layer but also in the deeper and sulfidic part (Elsgaard and Jørgensen, 1992).

OXIDATION OF SULFIDE

Marine sediments generally have a distinct separation of O_2 and H_2S by an intermediate zone where neither is present in detectable concentration. In this zone, iron and manganese oxides constitute an efficient barrier that oxidizes and binds H_2S diffusing up from below. A direct H_2S oxidation with O_2 is the exception and probably plays a role only under special conditions in which the metal oxide barrier is exhausted or is penetrated by advective transport. The latter may, for example, be due to bioirrigation, whereby oxygen is injected directly into the sulfide zone or vice versa. It may also result from current-induced advective pore water transport in porous, sandy sediments (Huettel et al., 1998), or from oxygen transport down into the root zone of sea grass beds (Ballbjerg et al., 1998).

In some coastal environments, where the organic sedimentation is so high that the reactive metal oxides are all reduced, the H_2S may diffuse freely up to the sediment surface. In this extreme situation, a diffusional reaction zone of oxygen and sulfide develops at the sediment-water interface, and the gradient-type of colorless sulfur bacteria, such as *Beggiatoa*, may flourish. Such hotspots of sulfide oxidation are recognizable from the black coloration of the sediment surface due to iron sulfide ("black spots"; Rusch et al., 1998) or from a film of white filamentous sulfur bacteria containing light refracting sulfur globules. Similar white mats are typical of hydrothermal vents or cold seeps that bring H_2S from the deep subsurface in direct contact with oxygenated seawater and support rich communities of chemoautotrophic sulfide oxidizing bacteria (Jannasch et al., 1989).

In most marine sediments, reoxidation of H_2S takes place without direct interaction with oxygen, but rather by reaction with iron oxides, manganese oxide, nitrate, or other potential oxidants. Evidence for this anoxic sulfide oxidation comes from a wide range of studies, including analyses of chemical pore water gradients (Fig. 1A), solid phase distributions of metal oxides and metal sulfides (Fig. 1B), mass balance calculations, and direct experimental determination of the processes involved. Experiments using radiolabeled H_2S added to anoxic sediment cores or slurries show a rapid transfer of the label into sulfur fractions defined as acid volatile sulfide (mostly FeS), chromium reducible sulfide (mostly FeS_2), elemental sulfur, and sulfate (Fossing and Jørgensen, 1990; Elsgaard and Jørgensen, 1992). Also, radiolabeled FeS and S^0 are oxidized to sulfate in anoxic sediments, whereas pyrite, FeS_2, is more stable and is not significantly oxidized over short experimental periods of up to a day (Fossing and Jørgensen, 1990). Yet, pyrite comprises the main sulfur pool in marine sediments, and, as shown below, undergoes slow transport and oxidation, which are critical for the sulfur cycle.

The oxidation of pyrite with oxygen under sediment conditions is a rather fast process that is well-described in the literature (e.g., Lowson, 1982; Luther, 1987; Moses and Herman, 1991; Morse, 1991). The oxidation may be purely abiotic, catalyzed by an electron shuttle between adsorbed Fe(II) and Fe(III) ions transferring electrons from pyrite to O_2. Sulfate is the end product of the sulfur oxidation and iron oxides coat the surface of the oxidizing pyrite grains.

Pyrite oxidation is not restricted to the oxic surface sediment. In anoxic sediments, it may also take place by reaction with manganese oxide, as suggested from chemical profiles by Canfield et al. (1993) and recently shown experimentally by Schippers and Jørgensen (2001, 2002). By the use of radiolabeled $^{55}FeS_2$ added to MnO_2-rich marine sediments, a slow dissolution of the ^{55}Fe was observed. The degree of pyrite dissolution and oxidation was directly related to the amount of MnO_2 in the sediment and was not detected below 0.15 wt% of total Mn (Fig. 2). It is interesting that two solid-phase minerals are able to react. The pyrite oxidation is purely chemical and has been proposed to occur by a Fe(II)/Fe(III)-shuttle in the pore fluid between the mineral surfaces of FeS_2 and MnO_2 (Schippers and Jørgensen, 2001; Fig. 3). Accordingly, the dissolution of pyrite is not affected by bacterial inhibitors and also takes place in sterile sediment. The immediate products of the oxidation are thiosulfate and polythionates. These can be further oxidized to sulfate by manganese-reducing bacteria, thus making the complete pyrite oxidation to sulfate dependent on microbial catalysis (Schippers and Jørgensen, 2001):

$$FeS_2 + 7.5\ MnO_2 + 11\ H^+ \rightarrow Fe(OH)_3 + 2\ SO_4^{2-} + 7.5\ Mn^{2+} + 4\ H_2O \quad (1)$$

FeS is also readily oxidized by MnO_2, but in this case, the immediate product is polysulfides, which are subsequently transformed into elemental sulfur:

$$FeS + 1.5\ MnO_2 + 3\ H^+ \rightarrow Fe(OH)_3 + S^0 + 1.5\ Mn^{2+} \quad (2)$$

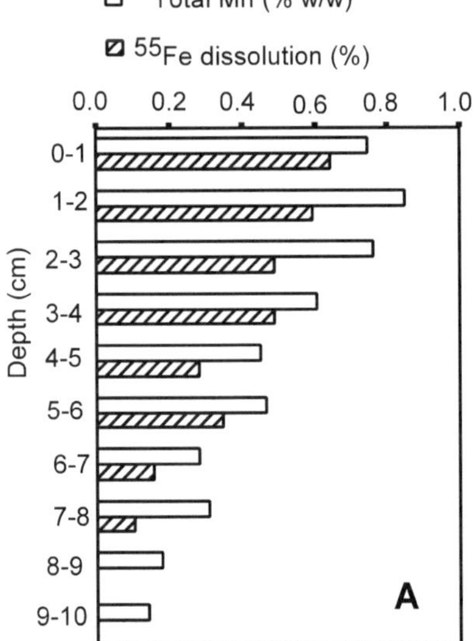

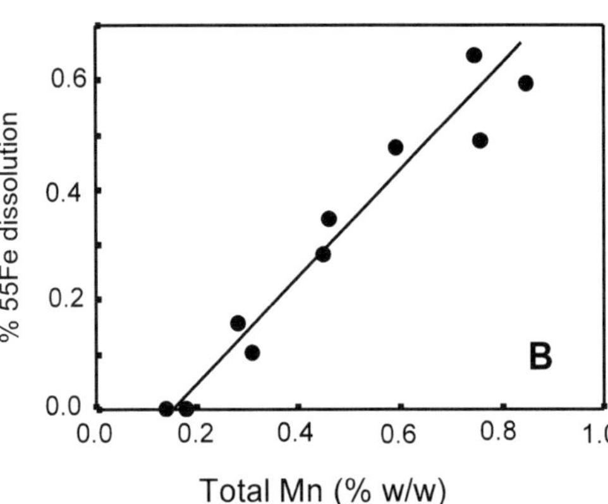

Figure 2. Oxidative dissolution of pyrite by manganese oxide in a continental shelf sediment from 700 m water depth in Skagerrak (Denmark). The ^{55}Fe-labeled pyrite was added to slurries from different sediment depths (A) and the fraction of FeS_2 dissolved was related to the total MnO_2 pool in the manganese-rich sediment (B). (Redrawn from Schippers and Jørgensen, 2001a.)

Thiosulfate and sulfite are important intermediates of sulfide oxidation but generally occur only in sub-micromolar concentration in the sediment pore water, which makes it difficult to measure their rapid turnover (Thamdrup et al., 1994b). If, however, the sediment is spiked with nonlabeled thiosulfate in incubations with $H_2{}^{35}S$, the thiosulfate also becomes transiently radiolabeled, which indicates that it is indeed an intermediate in the oxidation pathway to sulfate (Jørgensen, 1990).

Elemental sulfur is a dynamic sulfur constituent of marine sediments (Troelsen and Jørgensen, 1982) and is a substrate of diverse physiological types of bacteria that reduce, oxidize, or disproportionate it. The oxidation of H_2S to elemental sulfur is also difficult to demonstrate by radiotracer experiments. This, however, is due to rapid isotope exchange of the ^{35}S among the reduced inorganic sulfur pools of H_2S, S^0, polysulfide, and iron sulfide, which blurs the actual reaction pathways (Fossing et al., 1992).

DISPROPORTIONATION REACTIONS

Microbial disproportionation of intermediate oxidation products of sulfide was first discovered by Bak and Pfennig (1987), who were studying thiosulfate metabolism in sulfate-reducing bacteria, and it has since turned out to play an important role for sulfide oxidation in sediments. The disproportionation of thiosulfate, sulfite, and elemental sulfur has the following stoichiometry:

$$S_2O_3^{2-} + H_2O \rightarrow H_2S + SO_4^{2-} \quad (3a)$$

$$4SO_3^{2-} + 2H^+ \rightarrow H_2S + 3SO_4^{2-} \quad (3b)$$

$$4S^0 + 4H_2O \rightarrow 3H_2S + SO_4^{2-} + 2H^+ \quad (3c)$$

Important for the bacterial disproportionation processes is that the sulfur species are concurrently reduced to sulfide and oxidized to sulfate and that this reaction is independent of external reductants or oxidants. These processes can thus be considered a unique type of inorganic fermentation. By thiosulfate disproportionation (Equation 3a), the inner (sulfonate) sulfur atom changes oxidation step from +5 in $S_2O_3^{2-}$ to +6 in SO_4^{2-}, while the outer (sulfane) sulfur atom changes from −1 in $S_2O_3^{2-}$ to −2 in H_2S. The change in free energy by the internal electron transfer associated with thiosulfate disproportionation may be sufficient to cover the energy requirement of heterotrophic anaerobes and can even support an autotrophic metabolism. The disproportionating bacteria include both organisms such as *Desulfocapsa sulfoexigens*,

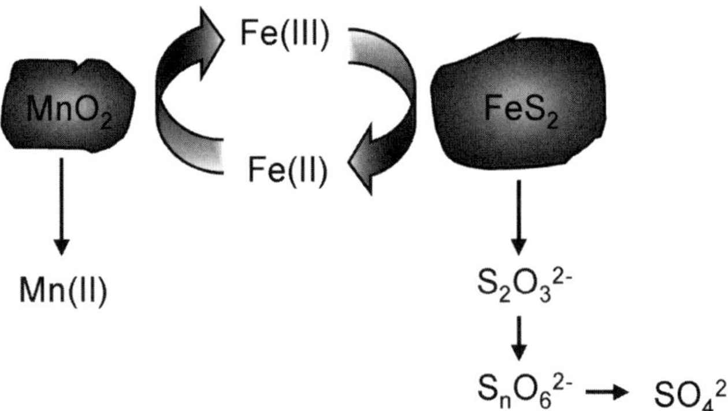

Figure 3. Model of anoxic FeS_2 oxidation by MnO_2 via the Fe(II)/Fe(III) shuttle. Thiosulfate and polythionates that form as immediate products may be further oxidized to sulfate by bacteria. (Redrawn from Schippers and Jørgensen, 2001.)

which are highly specialized for the process (Finster et al., 1998), and normal sulfate-reducing bacteria such as *Desulfovibrio desulfodismutans* and *Desulfocapsa thiozymogenes*, which are also able to carry out the process (Bak and Pfennig, 1987; Krämer and Cypionka, 1989; Janssen et al., 1996).

Although disproportionation reactions do not cause a net oxidation of the sulfur species, they have a key function in sulfide oxidation. Disproportionation provides a shunt in the sulfur cycle whereby the H_2S formed by this reaction may again be oxidized to the sulfur intermediate by metal oxides present in the sediment. For example, manganese oxides react rapidly with H_2S, which is oxidized quantitatively to S^0 (Burdige, 1993). This sulfide oxidation is a purely chemical process (Equation 4). Subsequently, the elemental sulfur may be disproportionated (Equation 5), which requires bacterial catalysis and is thus a purely biological process at normal environmental temperatures. By repeated cycling between H_2S and S^0, each time transferring a fourth of the sulfur into sulfate by disproportionation, a complete oxidation of H_2S to sulfate may result (Equation 6):

$$4\ H_2S + 4\ MnO_2 \rightarrow 4\ S^0 + 4\ Mn^{2+} + 8\ OH^- \quad (4)$$

$$4\ S^0 + 4\ H_2O \rightarrow 3\ H_2S + SO_4^{2-} + 2\ H^+ \quad (5)$$

$$H_2S + 4\ MnO_2 + 2\ H_2O \rightarrow SO_4^{2-} + 4\ Mn^{2+} + 6\ OH^- \quad (6)$$

The removal of H_2S, e.g., by oxidation with manganese (4), is important to drive the elemental sulfur disproportionation (5) as this bacterial process is only exergonic at low H_2S concentration and therefore requires a sulfide sink (Thamdrup et al., 1993). Iron sulfide in marine sediment may also be completely oxidized to sulfate by MnO_2 (Aller and Rude, 1988):

$$FeS + 4\ MnO_2 + 8\ H^+ \rightarrow Fe^{2+} + SO_4^{2-} + 4\ Mn^{2+} + 4\ H_2O \quad (7)$$

The process has been demonstrated in shelf sediments that are rich in manganese oxides (Schippers and Jørgensen, 2001). Because the immediate product of FeS oxidation by manganese oxide is elemental sulfur (Schippers and Jørgensen, 2001), it is plausible that Equation 7 also proceeds via elemental sulfur disproportionation.

Iron oxides are less efficient than manganese oxide in the anaerobic oxidation of sulfide to sulfate (Aller and Rude, 1988), although iron oxides are generally much more abundant than manganese oxide in marine sediments and may also enhance sulfur disproportionation by scavenging of the H_2S. The reason is probably that iron sulfide has a much lower solubility product than manganese sulfide. In contrast to Mn^{2+}, the Fe^{2+} formed by the oxidation of H_2S consequently precipitates as FeS with the H_2S that either remains in the pore water or is formed from the initial elemental sulfur disproportionation. In laboratory experiments, FeS does not react further with iron oxides. The low solubility product of iron sulfide thus prevents a more complete oxidation of sulfide with Fe(III):

$$3\ H_2S + 2\ FeOOH \rightarrow S^0 + 2\ FeS + 4\ H_2O \quad (8)$$

The elemental sulfur may, however, after reaction with HS^- and conversion to polysulfide, HS_n^-, combine with iron sulfide to form pyrite, giving the net reaction (e.g., Luther, 1991; Thamdrup et al., 1993):

$$3\ H_2S + 2\ FeOOH \rightarrow FeS + FeS_2 + 4\ H_2O \quad (9)$$

These and other reactions involving metal oxides and reduced sulfur species lead to complex pathways of anaerobic sulfur cycling that are difficult to quantify and for which reliable budgets are still needed. Many processes take place simultaneously, with the predominance depending on the overall redox chemistry and the availability of electron donors and acceptors. For example, thiosulfate is concurrently reduced, oxidized, and disproportionated in sediments, the main pathway changing gradually with depth in the sediment from predominantly oxidation near the surface to reduction at depth, but with disproportionation as the overall dominant process in the whole sediment (Jørgensen, 1990; Jørgensen and Bak, 1991).

The formation of intermediate products of sulfide oxidation and their further transformation by disproportionation is thus critical for the entire electron flow through the sulfur cycle. These intermediates may be products of chemical reactions with metal oxides or they may be formed by incomplete bacterial sulfide oxidation under conditions of limiting electron acceptors. As an example, *Thiobacillus thioparus*, isolated from marine sediment and grown in aerobic chemostat culture, carried out complete sulfide oxidation to sulfate under high oxygen availability but incomplete oxidation to thiosulfate, tetrathionate, and polysulfide under oxygen limitation (van den Ende and van Gemerden, 1993).

The important role of disproportionation as a pathway in sulfide oxidation is also indicated by the sulfur isotope geochemistry of marine sediments. A fractionation between the light and heavy sulfur isotopes, ^{32}S and ^{34}S, during sulfate reduction leads to H_2S enriched in ^{32}S relative to the sulfate. The direct fractionation during bacterial sulfate reduction varies greatly but is typically 20‰–40‰ (Habicht and Canfield, 1997; Canfield, 2001). The isotopic difference between pore water sulfate and sulfides in marine sediments is, however, 40‰–70‰, which implies an additional fractionation step in the sulfur cycle. Whereas sulfide oxidation itself is not associated with significant isotopic discrimination, the disproportionation pathways are, producing relatively light sulfide and heavy sulfate (Canfield and Thamdrup, 1994). In thiosulfate, the inner sulfur atom is already enriched in ^{34}S relative to the outer sulfur atom, whereas by elemental sulfur and sulfite disproportionation, the enzymatic conversion causes fractionation between the produced SO_4^{2-} and H_2S (Canfield et al., 1998; Habicht et al., 1998). The additional fractionation by partial sulfide reoxidation and subsequent disproportionation may explain the large isotopic difference between sulfate and sulfides observed in marine sediments (Habicht and Canfield, 2001). This mechanism still needs to be confirmed by direct quantification of disproportionation under in situ conditions in marine sediments.

THE SULFUR CYCLE

The sulfide transformations in anoxic sediments lead to the accumulation of FeS and FeS_2. In sediments with free H_2S in the pore water, the total accumulation of iron sulfides is primarily limited by the amount of reactive iron since the H_2S is produced in excess of the metal sulfide precipitation capacity. The further oxidation of the solid phase species, FeS and FeS_2, is therefore critical to explain the 80%–95% recycling of sulfide back into sulfate. Within the anoxic sediment, the oxidation may take place with MnO_2, as discussed above. Also, nitrate may oxidize FeS, but apparently it does not oxidize FeS_2, neither chemically nor biologically, at least not on a time scale of weeks to months (Schippers and Jørgensen, 2001a). The FeS-mediated denitrification may be catalyzed by bacteria such as *Thiobacillus denitrificans* (Garcia-Gil and Golterman, 1993) or by anaerobic Fe(II)-oxidizing NO_3^- reducing bacteria (Straub et al., 1996; Benz et al., 1998).

According to detailed flux and process studies in marine sediments, the main terminal oxidant for pyrite must be oxygen, since no other electron acceptor has a sufficiently high flux into the sediment to balance the total electron flow via sulfur cycling (e.g., Jørgensen, 1977a; Thamdrup et al., 1994a). A direct pyrite oxidation with O_2 or an indirect one via MnO_2, however, requires a mass transport of pyrite-containing sediment from the suboxic zone up to the oxic surface layer. Data from the coastal marine sediment shown in Figure 1 illustrate this coupling of iron, manganese, and sulfur transformations (Thamdrup et al., 1994a). The pyrite concentration increased between the sediment surface and 8 cm depth by 110 µmol Fe g^{-1} (dry weight), whereas the total iron oxide concentration conversely decreased by 121 µmol Fe g^{-1}. These opposite gradients of Fe(III) and FeS_2 provide balanced fluxes because they are affected by the same vertical transport coefficients: (*a*) downward transport of iron oxides supplies the iron required for partial sulfide oxidation and for the trapping of sulfide as pyrite (Equation 9); and (*b*) upward transport and oxidation of pyrite with oxygen or manganese oxide regenerates the iron oxide pool at the sediment surface. Thamdrup et al. (1994a) showed, for the coastal sediment studied, that bioturbation (i.e., the mixing of sediment due to the burrowing and sediment-feeding activity of benthic macrofauna) could realistically generate the mixing coefficient needed to balance the sulfur budget through vertical iron oxide and pyrite fluxes.

In conclusion, the sulfur cycling in the upper centimeters to decimeters of marine sediments is dependent on a conveyer-belt function of the benthic infauna, which brings pyrite upward to become reoxidized at the sediment-water interface and at the same time transports iron oxides downward and, thereby, recharges and maintains the sulfur cycle in the suboxic zone (Fig. 4). The H_2S formed below the suboxic zone accumulates in the pore water and diffuses upwards to become oxidized or trapped by metal oxides in the suboxic zone. Deeper in the sediment, a slow reaction with iron bound in sheet silicates may bind sulfide formed by very slow sulfate reduction. This iron may also cause a partial reoxidation of sulfide and regeneration of sulfate, which is particularly important on a long time scale of hundreds to thousands of years (Canfield and Raiswell, 1992). Thus, a reoxidation of sulfide may take place throughout the sediment column through slow reaction with Fe(III) and by further disproportionation into sulfide and sulfate, a combination that may potentially lead to the oxidation of sulfide completely to sulfate. The evidence for such a slow oxidation is seen in a number of sediments from downward-directed H_2S gradients below the sulfate zone and even from the complete disappearance of H_2S at depth (e.g., Pruysers, 1998; Jørgensen et al., 2004a).

THE SULFIDE OXIDIZING BACTERIA

The sulfide oxidation pathways discussed above raise the question of what role remains for the "classical" sulfide oxidizers such as *Thiobacillus* spp. or *Thiomicrospira* spp. Much of

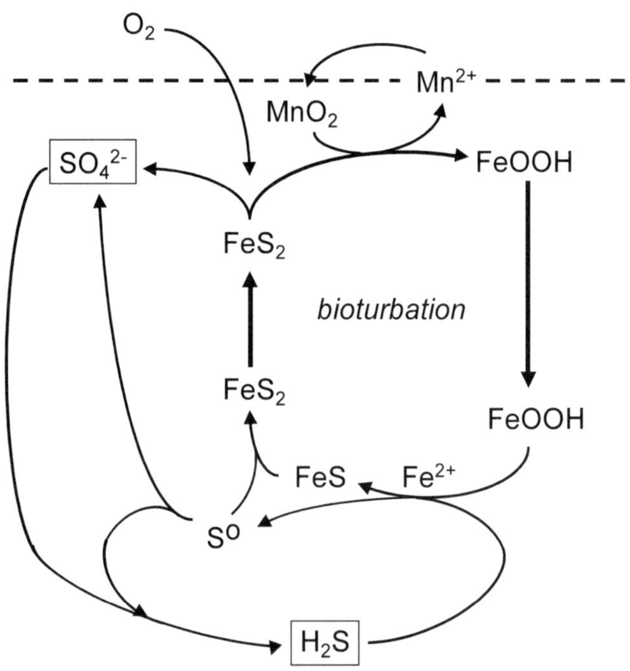

Figure 4. Principle of sulfur cycling in marine sediments. The H_2S generated from sulfate reduction reacts with iron oxides to form FeS, FeS_2, S^0 and other intermediate oxidation products such as $S_2O_3^{2-}$. These intermediate products (only S^0 is shown) may be disproportionated by anaerobic bacteria into H_2S and SO_4^{2-}, thereby generating a shunt in the sulfur cycle through which a complete oxidation of H_2S to SO_4^{2-} is possible by repeated cycling. The main pool of reduced sulfur is bound in pyrite, which is slowly transported up to become reoxidized (e.g., to FeOOH) near the sediment surface. The FeS_2 oxidation may partly take place in the suboxic zone by reaction with MnO_2, but overall the terminal oxidant in the sulfur cycle is O_2.

what we know about bacterial sulfide oxidation pathways and biochemistry originates from pure culture studies of these organisms. In reality, are they insignificant for the marine sulfur cycle, although they can be isolated from all types of marine sediments? Is it important for the coupling of the carbon and sulfur cycles whether sulfide oxidation is carried out by such chemoautotrophic bacteria?

The sulfide oxidizing bacteria comprise a broad physiological spectrum of chemolithoautotrophs, chemolithoheterotrophs, and mixotrophs (Robertson and Kuenen, 1992; Kuenen et al., 1992), depending on whether their main external source of cell carbon is CO_2, organic carbon, or both. Due to the chemical heterogeneity and temporal variability of bioturbated surface sediments, where most of the sulfide oxidation takes place, versatile sulfide oxidizers able to switch between different energy substrates and carbon sources presumably have a selective advantage (Kuenen et al., 1985). The availability of organic substrates determines to what extent sulfide oxidation is heterotrophic or is associated with the autotrophic fixation of CO_2 and, thereby, the formation of new biomass. Whereas this is highly interesting from a microbiological perspective, it does not significantly affect the overall organic carbon budget of the sediment, as the following calculation shows (Fig. 5).

In coastal marine sediments, about half of the deposited organic carbon may be oxidized directly by oxygen through aerobic organisms. Bacterial sulfate reduction is the second most important mineralization pathway and may account for nearly as much organic carbon oxidation as oxygen. Of all the H_2S formed from this sulfate reduction, on the order of 10% is trapped in pyrite, while the remaining 90% is reoxidized. Consequently, up to half of the oxygen uptake in such sediments may be directly or indirectly consumed for the reoxidation of sulfide. The resulting total oxygen uptake would thus be twice the sulfate reduction when calculated in oxidation equivalents (Jørgensen, 1982a).

Similar to other autotrophic organisms that fix CO_2 via the Calvin cycle, sulfide oxidizing bacteria have a rather low growth yield that may vary according to growth conditions (Kuenen, 1979). Based on culture data from thiobacilli and gradient-living *Beggiatoa*, yields of up to 6.7 g dry weight biomass per mol sulfide have been calculated (Kelly, 1982; Nelson et al., 1986a). This is equivalent to the use of ~15% of the electrons from sulfide for the reduction of CO_2 to cell biomass. The remaining 85% of the electron flow is transferred to oxygen in the adenosine triphosphate (ATP)–generating respiratory metabolism of the sulfide oxidizers. Whereas the complete oxidation of H_2S with oxygen has the following stoichiometry:

$$H_2S + 2O_2 \rightarrow SO_4^{2-} + 2H^+, \quad (10)$$

the net reaction of aerobic H_2S oxidation by chemoautotrophic bacteria is approximately:

$$4\,H_2S + 7\,O_2 + CO_2 + H_2O \rightarrow 4\,SO_4^{2-} + [CH_2O] + 8\,H^+. \quad (11)$$

In conclusion, the maximum autotrophic CO_2 fixation corresponds to $(0.5 \times 0.9 \times 0.15 \times 100 =)$ 7% of the organic carbon mineralization in the sediment (Fig. 5). This is at the limit of detectability in most studies aiming at a budget of the carbon cycle in marine sediments. If the sulfide oxidation mainly takes place through reactions with metal oxides and via disproportionation reactions, the contribution of autotrophic CO_2 fixation by sulfur bacteria is much less than 7%.

Population studies of sulfide oxidizing bacteria in sediments have been done by viable counting methods using either plate counts or most probable number estimates from dilution series. Such viable counts tend to underestimate the total cell numbers because of insufficient cell dispersion, inability to grow on the medium offered, cell death, or other factors. In the case of

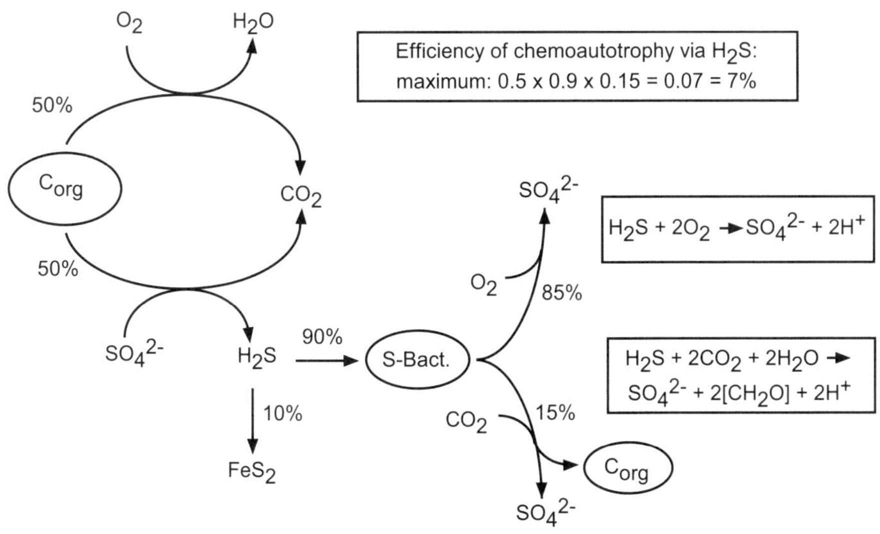

Figure 5. Coupling of carbon and sulfur cycles in marine sediments and the potential role of chemoautotrophic CO_2 assimilation for the carbon budget (see text).

sulfate reducers in marine sediments, a comparison of sulfate reduction rates and viable counts have demonstrated that the calculated metabolic rates per cell exceeded those of pure cultures by orders of magnitude (i.e., the true cell numbers must be underestimated by at least similar orders of magnitude; e.g., Jørgensen and Bak, 1991). In recent years, molecular methods based on DNA and RNA sequence analyses and quantification as well as fluorescence in situ hybridization (FISH) have provided more quantitative and detailed data on sulfide oxidizing bacteria. However, bacteria oxidizing sulfide and other reduced inorganic sulfur compounds constitute highly diverse phylogenetic lineages among the proteobacteria, and it has therefore been difficult to design suitable probes for this physiological group. Some recent studies in sediments of the German Wadden Sea illustrate both the progress and the problems.

Llobet-Brossa et al. (1998) made a comprehensive analysis of the phylogenetic groups of bacteria inhabiting the muddy and sandy sediments along the Wadden Sea coast. Out of the total bacteria, $2–4 \times 10^9$ cells per cm^3, identified by counting in the fluorescence microscope after DAPI staining, up to 73% hybridized with a general eubacterial FISH probe. This shows that FISH methods were able to generate quantitative data on the sediment bacteria. The eubacteria were further classified by a range of group-specific probes. The gamma-*Proteobacteria*, including the classical sulfide oxidizers, accounted for a rather constant 10% fraction of the *Eubacteria* in all samples, a sufficiently high fraction to indicate a real quantitative importance of this group.

The genus *Thiomicrospira* was originally defined from an isolate, *T. pelophila*, obtained from the Dutch Wadden Sea by Kuenen and Veldkamp (1972). More recently, *Thiomicrospira* species have been found to be common inhabitants of marine sediments (Brinkhoff and Muyzer, 1997; Brinkhoff et al., 1998; Sievert et al., 2000). In the absence of a general molecular probe for sulfide oxidizers, Brinkhoff et al. (1998) made most probable number (MPN) counts of the chemolithotrophic sulfide oxidizers in the Wadden Sea and found numbers of $10^5–10^6$ per g. This corresponds to ~0.01% of the total bacterial counts and 0.1% of the gamma-*Proteobacteria* counts. When testing the positive MPN tubes with a newly designed molecular probe specific for *Thiomicrospira*, only 1% or $10^3–10^4$ cells g^{-1} of the enriched sulfide oxidizers turned out to belong to the *Thiomicrospira* (Fig. 6). The vast majority of the rest were aerobic heterotrophs with the ability to oxidize sulfide. Whereas the *Thiomicrospira* were rather evenly distributed in the top 0–4 cm of the sediment, their rRNA content decreased significantly with depth, thus indicating that their metabolic activity was highest near the sediment surface and that below the oxidized zone, cells could be partly dormant. These results show the power of the molecular methods for quantitative population studies but also illustrate the difficulty in determining the functional role of sulfide oxidizers in the sulfur cycle.

If the relative scarcity of *Thiobacillus* and *Thiomicrospira* cells in MPN counts is representative of their scarcity in the sediment samples studied, then who are the main sulfide oxidizing bacteria in the seabed? In recent years, most studies of marine sulfide oxidizing bacteria have been done on hydrothermal vent systems, and less progress has been made on those bacteria inhabiting "normal" marine sediments. Tuttle and Jannasch (1972) isolated a large number of sulfur and thiosulfate oxidizing bacterial strains from marine sediments and waters. They concluded that the obligately chemoautotrophic thiobacilli are rare and that oxidation of reduced sulfur compounds is rather carried out by facultatively autotrophic bacteria of uncertain taxonomic affiliation. More recently, Podgorsek and Imhoff (1999) found very large populations of heterotrophic thiosulfate oxidizing bacteria, up to 10^7 cells cm^{-3}, by MPN counts in Baltic Sea sediments, with highest numbers in the oxidized zone near the sediment surface. The bacterial numbers correlated well with the high potential for thiosulfate oxidation during sediment incubations. The isolated thiosulfate oxidizing bacteria belonged to the gamma-*Proteobacteria*, similar to the *Thiobacillus* and *Thiomicrospira* species. Teske et al. (2000) similarly isolated thiosulfate oxidizers from sediments of the North Atlantic continental slope and abyssal plain and found the strains to belong to either the alpha- or the gamma-*Proteobacteria*. The alpha-*Proteobacteria*, and in particular the *Roseobacter* cluster, are very abundant in coastal seawater and sediments, and many of these heterotrophic organisms are able to oxidize reduced sulfur species (González et al., 1999).

Many heterotrophic bacteria produce tetrathionate as an intermediate or final product from the oxidation of inorganic

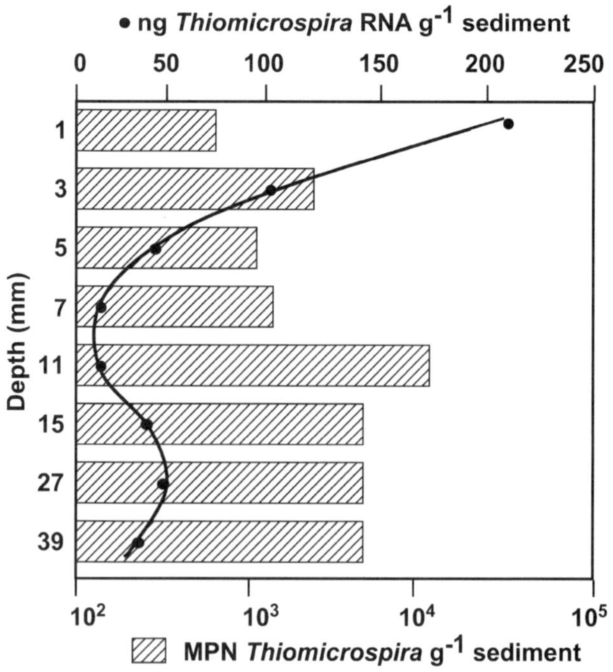

Figure 6. Depth distribution of sulfide oxidizing bacteria of the genus *Thiomicrospira*, based on viable counts, and of 16S rRNA of this group in near-shore sediments of the German Wadden Sea. (Redrawn from Brinkhoff et al., 1998).

sulfur species (Mason and Kelly, 1988; Sorokin, 1996). Podgorsek and Imhoff (1999) proposed a tetrathionate cycle in coastal sediments, whereby the thiosulfate ($S_2O_3^{2-}$) is oxidized to tetrathionate ($S_4O_6^{2-}$) by bacteria and tetrathionate is in turn reduced back to thiosulfate by chemical reaction with hydrogen sulfide. The sulfide becomes oxidized to elemental sulfur, which transiently accumulates in the sediment. The thiosulfate-tetrathionate cycle would in this way play a catalytic role in the oxidation of sulfide to elemental sulfur and might function in a system in which bacteria are unable to perform a more direct sulfide oxidation.

In conclusion, sulfide oxidation in sediments involves interactions between highly diverse, autotrophic, or heterotrophic bacteria and complex solid and liquid phase chemical reactions. The role of the classical aerobic, chemolithoautotrophic sulfur bacteria is not understood but could be minor.

THE NITRATE-STORING SULFUR BACTERIA

In addition to aerobic sulfide oxidation, denitrifying species of *Thiobacillus* and *Thiomicrospira* may oxidize sulfide according to

$$5\ H_2S + 8\ NO_3^- \rightarrow 5\ SO_4^{2-} + 4\ N_2 + 4\ H_2O + 2\ H^+ \quad (12)$$

A novel pathway of anaerobic sulfide oxidation by filamentous sulfur bacteria was realized in 1994 during a study of massive communities of filamentous sulfur bacteria, *Thioploca* spp., on the continental shelf of central Chile. These communities had been observed here by biologists already in the early 1960s but only became widely known fifteen years later through a publication by the Chilean biologist, V.A. Gallardo (Gallardo, 1977). The large thioplocas and their gelatinous sheaths form slimy masses of up to 800 g/m^2 (Schulz et al., 1996), enough to clog up the bottom trawl of local fishermen. Scientists who extracted the pore water of these sediments by a whole-core squeezing method discovered that the nitrate concentration rose with increasing squeezing pressure from the ambient 30–40 µM to an extreme of 5 mM (Thamdrup and Canfield, 1996). The source of this nitrate was the *Thioploca* inhabiting the sediments, and nitrate analyses in individual filaments subsequently revealed intracellular concentrations of up to 500 mM (Fossing et al., 1995).

This discovery has highly stimulated the interest in *Thioploca* and *Beggiatoa* and other large sulfide oxidizing bacteria, partly because of their fascinating biology and partly because of their potential role in the nitrogen and sulfur cycles of marine sediments. The nitrate accumulation explains why these bacteria grow to giant sizes and why they appear hollow, filled by a liquid vacuole in which nitrate is stored. Several earlier observations on the distribution of these bacteria and on nitrate reduction in sediments now make sense for the first time. The following discussion will, therefore, concentrate on their physiology and ecology and on their biogeochemical significance. Recent reviews on these organisms were published by Jørgensen and Gallardo (1999) and Schulz and Jørgensen (2001).

Marine thioplocas are abundant on the seafloor along the Pacific coast of South America. They occur primarily where coastal upwelling provides high nitrate enrichment, high primary productivity, seasonal anoxia of the lower water column, and high sulfide production from bacterial sulfate reduction in the underlying sediment (Ferdelman et al., 1997). The two dominant species, *T. chileae* and *T. araucae*, have diameters of 12–20 and 30–43 µm, respectively, but narrower and wider forms of undescribed taxa are also common. The filaments are many centimeters long and grow in bundles surrounded by a common sheath that penetrates 5–10 cm down into the muddy sediment (Fig. 7A). The bacteria are anaerobic sulfide oxidizers that use nitrate to oxidize sulfide to sulfate. As a solid intermediate in this oxidation, elemental sulfur globules are stored in the cells, and the gliding bacteria thereby transport large quantities of both sulfur and nitrate as reserves for their energy metabolism. Sulfide oxidizing *Thioploca* filaments have been shown experimentally to assimilate ^{14}C-labeled bicarbonate and acetate and can probably grow autotrophically or mixotrophically according to the availability of carbon and energy sources (Maier and Gallardo, 1984; Otte et al., 1999).

Their adaptation to sulfide oxidation in an anoxic environment is that of commuters between the electron acceptor and the electron donor. Up at the sediment surface, they stretch the long filaments into the flowing seawater and take up nitrate into the vacuoles. Down in the sediment, they oxidize the ambient sulfide to elemental sulfur, which they store as an energy reserve. In this manner, they may continuously carry out their chemoautotrophic metabolism even though they do not have simultaneous access to both sulfide and nitrate (Huettel et al., 1996). Due to their efficient sulfide oxidation, the H_2S concentration in the surrounding

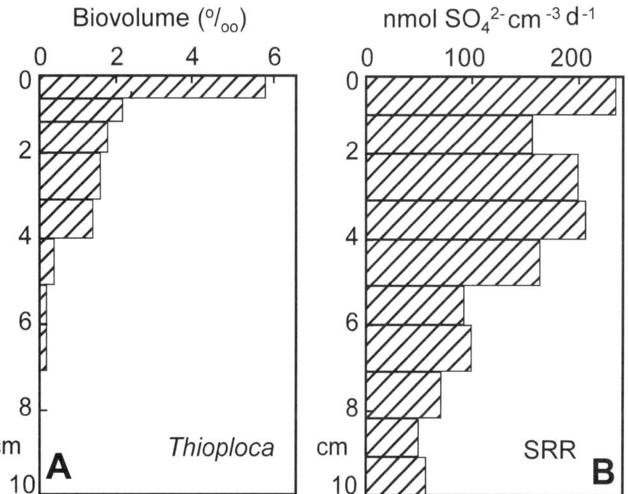

Figure 7. A. Distribution of *Thioploca* in shelf sediments off the Chilean coast. The biovolume is expressed in per mil of the total sediment volume, i.e., 1‰ = 1 mm^3 *Thioploca* biovolume per cm^3 sediment. B. Sulfate reduction rates in the same sediment measured by ^{35}S-tracer technique. (Redrawn from Schulz et al. [1996] and Thamdrup and Canfield [1996].)

pore water may be kept below detection limit, <1 μM, in spite of extremely high sulfate reduction rates (Fig. 7B).

The intracellular nitrate pool of *Thioploca* is so high that it vastly exceeds the free nitrate in the sediment. In sediment from the Chilean shelf, free nitrate penetrated only a few millimeters below the sediment surface by diffusion through the pore water (Fig. 8A), whereas intracellular nitrate penetrated down to 10 cm and near the surface reached 20-fold higher bulk sediment concentration than free nitrate (Fig. 8B). The intracellular elemental sulfur peaked near the sediment surface but was much lower than the extracellular S^0 concentration in the bulk sediment (Fig. 8C). A subsurface peak of extracellular S^0, although with lower peak concentration, has also been observed in other shelf sediments without such bacterial populations (Troelsen and Jørgensen, 1982).

Dense populations of *Thioploca* were recently discovered in sediments under the oxygen-minimum zone of the Arabian Sea, and *Thioploca* communities also occur in many other marine and freshwater habitats (Schmaljohann et al., 2001; overview in Jørgensen and Gallardo, 1999). The Benguela upwelling system along southwest Africa provides a hydrographic situation similar to that of the Humboldt system along Chile and, thus, similar benthic communities might be expected. During a research cruise in 1997 aimed to search for *Thioploca* along the coast of Namibia, giant sulfur bacteria with a similar physiology but different morphology and behavior were discovered. The organisms are spherical and form chains embedded in a gelatinous sheath. They are up to 750 μm in diameter (mean diameter 150–200 μm) and are thus the largest prokaryotic cells known. The cells appeared shining white in the black mud and were given the name *Thiomargarita namibiensis*, the sulfur pearls of Namibia (Schulz et al., 1999). They have so far only been found in the diatomaceous ooze off the coast of Namibia. However, free-living cells of similar morphology but even greater cell size were recently discovered at seep environments in the Gulf of Mexico (M. Joye, 2003, personal commun.).

The Namibian *Thiomargarita* also reach large population densities, although not quite as high as the Chilean *Thioploca*: up to 47 g living biomass per m^2 compared to a maximum of 160 g/m^2 for *Thioploca* (Schulz et al., 1999, 2000). To an even greater extreme, they have developed into liquid storage tanks for nitrate. About 98% of the cell volume consists of a large central vacuole in which the bacteria accumulate nitrate to several hundred mM concentration. As the cells are immotile, they apparently have access to nitrate from the overlying water only during transient episodes of resuspension or pore water advection. Calculations of their metabolic rate compared to the storage capacity of nitrate indicate, however, that they may endure nitrate starvation for months without running out of electron acceptor (Schulz and Jørgensen, 2001).

Recently, large diameter filaments of marine strains of *Beggiatoa* have been collected from a variety of hydrothermal vent and cold seep habitats. In all cases examined, they have been shown to possess internal vacuoles and to store nitrate (Table 1). The internal nitrate concentrations achieved by vacuolate *Beggiatoa*, *Thioploca*, and *Thiomargarita* are comparable, and in all cases examined, the accumulated concentrations are 4,000–10,000-fold above the extracellular concentrations (Table 1). Although there is no published record that any of these conspicuous vacuolate sulfur bacteria have been grown in pure culture, their phylogenetic relationships have nonetheless been analyzed from environmental samples. The techniques applied for this include amplification and sequencing of 16S rRNA, the portion of bacterial DNA that encodes the complex and functionally conserved structural ribosomal RNA molecule. By combining sequencing with FISH technique, it was possible to assign specific 16S rRNA sequences to each of several morphologically identifiable *Beggiatoa*, *Thioploca*, and *Thiomargarita* strains. Vacuoles are a rare feature in bacteria, and the analysis of rDNA sequences makes it clear that all of these morphologically distinct sulfur bacteria form a tight evolutionary cluster (Fig. 9; Teske et al., 1995, Jørgensen et al., 2004b). They are a distinct portion of the slightly larger cluster in the gamma-*Proteobacteria* defined by the inclusion of the sequences from non-vacuolate, narrow marine and freshwater *Beggiatoa* strains that can be grown in pure culture.

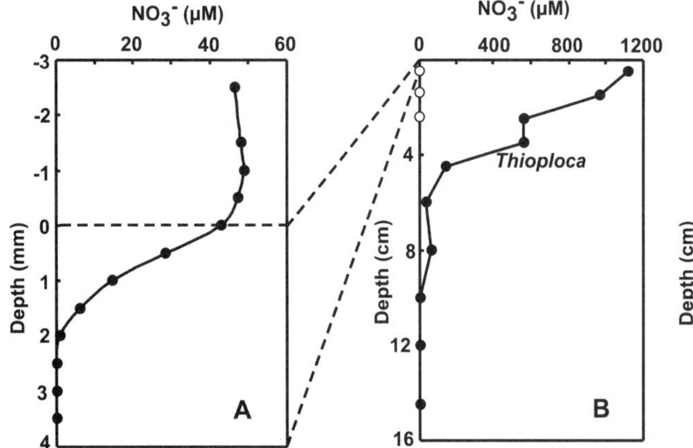

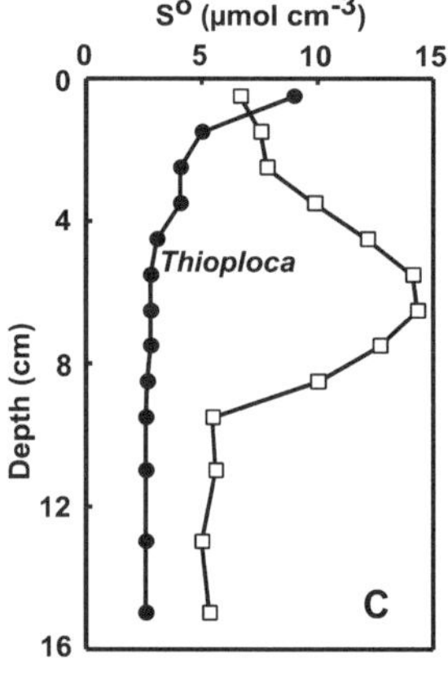

Figure 8. Nitrate and elemental sulfur in sediment and bacteria of *Thioploca*-inhabited shelf sediments off the Chilean coast. A. Nitrate in the pore water measured by a nitrate microsensor. Penetration depth 2 mm. B. Intracellular nitrate in the vacuoles of *Thioploca* (closed circles, penetration depth 10 cm) compared to the extracellular nitrate (open circles). C. Intracellular elemental sulfur in *Thioploca* filaments (filled circles) and extracellular sulfur in the bulk sediment (open squares). (Redrawn from Zopfi et al., 2001.)

TABLE 1. COLORLESS SULFUR BACTERIA AND NITRATE ACCUMULATION IN VACUOLES

Site and bacterium	Filament width (µm)	Vacuole present?	Protein/biovolume (mg protein cm^{-3})	Nitrate intracellular	Nitrate ambient
Culture—narrow control					
Beggiatoa 81–6	4	No; see next column	121 ± 17*	≤0.3 µM	1000 µM
Guaymas Basin HTV					
Beggiatoa sp.	88–140	Yes; electron microscopy	9.5	160 mM	0–40 µM
Monterey Canyon seeps					
Beggiatoa sp.	65–85	Yes; light microscopy	24	130 mM	40–45 µM
Juan de Fuca HTV					
Beggiatoa sp.	20–40	Yes; light microscopy	37	130–240 mM	N.D.
Gulf of Mexico seeps					
Beggiatoa sp.	29–46	Yes; light microscopy	≈25	10–110 mM	N.D.
Peru and Chile OMZ					
Thioploca araucae	12–22	Yes; electron microscopy	N.D.	150–500 mM	≈25 µM
Thioploca chileae	28–42	67%–84% vacuole			
Bay of Concepcion OMZ					
Beggiatoa sp.	35–40	Yes; light microscopy	N.D.	15–120 mM	N.D.
Namibia OMZ					
Thiomargarita namibiensis	100–750	Yes; electron microscopy 98% vacuole	N.D.	300 mM	5–28 µM

Note: HTV—hydrothermal vent; OMZ—oxygen minimum zone, N.D—no data.
*Protein/biovolume ratio for typical bacterial cytoplasm is predicted to be ~125 mg cm^{-3}.

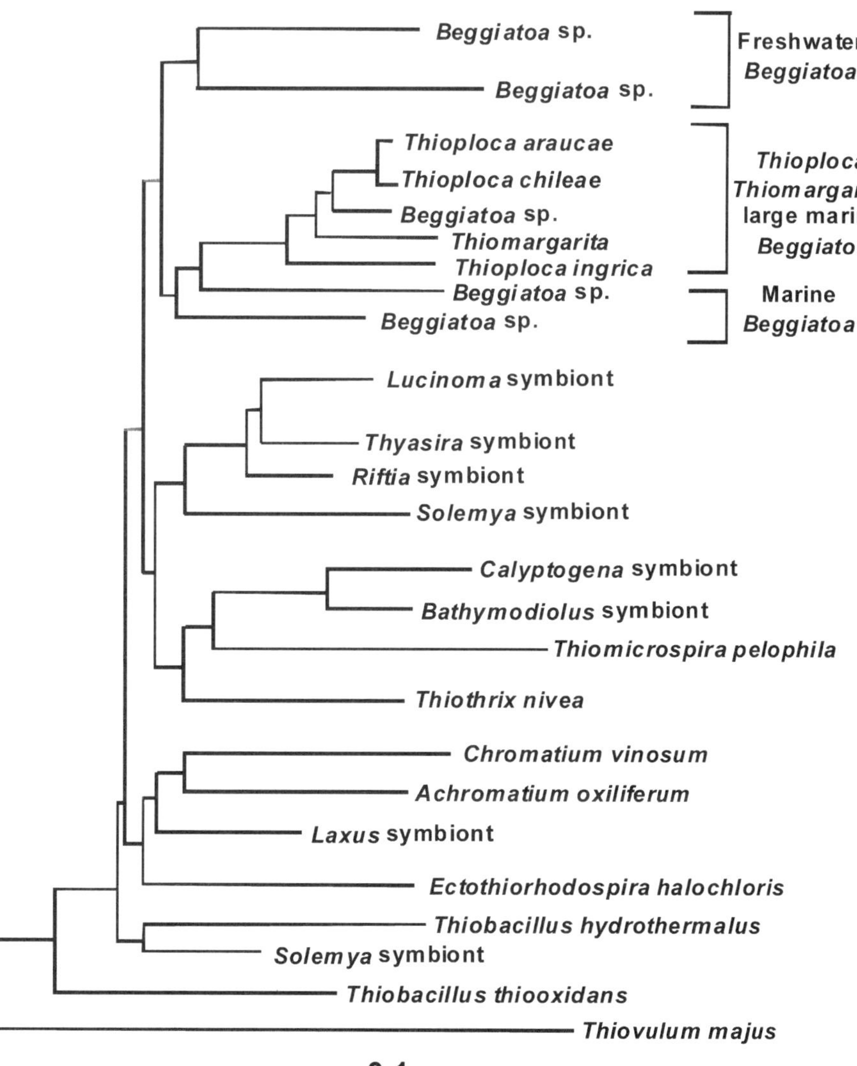

Figure 9. Phylogenetic distance tree for near-complete sequences of gamma-*Proteobacteria* showing the position of the genera *Beggiatoa*, *Thioploca* and *Thiomargarita*. The sequenced strains form three clusters: (A) the freshwater *Beggiatoa*; (B) the nitrate accumulating *Thioploca*, *Thiomargarita* and large marine *Beggiatoa*; and (C) the narrow marine *Beggiatoa*. The morphologically rather similar *Thiothrix* as well as the *Thiomicrospira* and *Thiobacillus* belong to a different lineage within the gamma-*Proteobacteria*. (Redrawn from Jørgensen et al., 2004b.)

PHYSIOLOGY, BIOCHEMISTRY, AND ENERGETICS OF NITRATE-STORING SULFUR BACTERIA

The conclusion from 16S rRNA-based evolutionary studies (Fig. 9) is that adaptive radiation from a single common non-vacuolate filamentous ancestor gave rise to the diversity of vacuolate, nitrate-accumulating sulfur bacteria that we currently recognize. It seems reasonable to speculate that nitrate is accumulated for the same purpose in these closely related bacteria, and that, given the significant energetic cost of concentrating it for storage (Table 2, footnote), it must be of central importance to cellular metabolism. In addition to denitrification (Equation 12), there is precedent (Eisenmann et al., 1995) among non-vacuolate sulfur-oxidizers for energy-conserving, dissimilatory reduction of nitrate to ammonia (DNRA):

$$HS^- + NO_3^- + H^+ + H_2O \rightarrow SO_4^{2-} + NH_4^+ \qquad (13)$$

The majority of evidence suggests that the energy-generating metabolic pathway employed by vacuolate sulfur bacteria produces ammonium as the major nitrogenous product of nitrate respiration. Cleaned bundles of *Thioploca* used internal stores of elemental sulfur to respire stored nitrate to ammonium at modest basal rates (1.0 nmol min^{-1} mg $protein^{-1}$). The rates were enhanced twofold to threefold by addition of exogenous nitrate or sulfide but abolished by mechanical disruption of the filaments (Otte et al., 1999). Addition of $^{15}NO_3^-$ to similarly washed *Thioploca* bundles resulted in its rapid uptake and the subsequent appearance of the majority of the label in NH_4^+ (Fig. 10). Regardless of the final product, the initial step in nitrate respiration is the conversion of nitrate to nitrite by a membrane-associated enzyme, which was detected at very high levels (200 nmol min^{-1} mg $protein^{-1}$) in very pure natural populations of vacuolate *Beggiatoa* from Monterey Canyon (McHatton et al., 1996). Importantly, related assays with this material demonstrated an even higher activity of a nitrite reductase enzyme that produced ammonium as the waste product (McHatton, 1998). The conclusions of these assays were supported by pore water profiles from sediment cores that contained vacuolate Monterey *Beggiatoa* distributed over the upper 10–15 cm. These sediments showed distinct ammonium peaks (2–3 mM maximum) in the main migratory zone of these gliding bacteria (McHatton, 1998).

Close evolutionary relationships make it attractive to envision a single physiological role and fate for nitrate accumulated by all vacuolate sulfur bacteria. Yet, the possibility of metabolic versatility among different species with respect to products of nitrate reduction must still be considered. The ^{15}N-nitrate studies of *Thioploca* (Otte et al., 1999) did show that ~15% of the label ended up in dinitrogen. The authors speculated that this activity might be ascribed to epibiotic bacteria on the *Thioploca* or its sheaths, yet a denitrifying activity of Thioploca cannot be excluded. Additional caution derives from conflicting observations of others on whether the final product of nitrate reduction in non-vacuolate freshwater *Beggiatoa* spp. is dinitrogen or ammonium (Sweerts et al., 1990; Vargas and Strohl, 1985).

In a search for additional insights regarding denitrification versus respiration to ammonium, thermodynamic calculations

TABLE 2. CALCULATED FREE ENERGY (ΔG) YIELD FOR DISSIMILATORY NITRATE REDUCTION TO AMMONIUM VERSUS DENITRIFICATION DRIVEN BY SULFIDE OXIDATION AT 4°C (OR 25°C) AND 1 BAR PRESSURE

					$HS^- + NO_3^- + H^+ + H_2O \rightarrow SO_4^= + NH_4^+$	$HS^- + 1.6\,NO_3^- + 0.6\,H^+ \rightarrow SO_4^= + 0.8\,N_2 + 0.8\,H_2O$	
Entry #	pH	$[NO_3^-]$ (mM)	$[NH_4^+]$ (mM)	$[N_{2(aq)}]$ (mM)	ΔG per NO_3^- or HS^- (kJ/mol)	ΔG per NO_3^- (kJ/mol)	ΔG per HS^- (kJ/mol)
1	7.5	500	1.0	0.8	–436.2 (–428.9)	–446.7 (–442.6)	–714.7 (–708.2)
2	7.5	150	1.0	0.8	–433.4 (–426.0)	–443.9 (–439.6)	–710.3 (–703.4)
3	7.5	0.025	1.0	0.8	–413.4 (–404.4)	–423.9 (–418.1)	–678.2 (–668.9)
4	7.2	150	1.0	0.8	–435.0 (–427.7)	–444.5 (–440.3)	–711.2 (–704.4)
5	7.2	150	0.1	0.8	–440.3 (–433.4)	–444.5 (–440.3)	–711.2 (–704.4)
6	7.2	150	2.5	0.8	–432.9 (–425.4)	–444.5 (–440.3)	–711.2 (–704.4)
7	6.5	150	0.1	1.0	–444.0 (–437.4)	–445.6 (–441.5)	–713.0 (–706.4)

Notes: Equilibrium constants were calculated according to Johnson et al. (1992) using above-tabulated estimates of appropriate intracellular or pore-water solute concentrations, with $[HS^-]$ = 0.1 µM and $[SO_4^=]$ = 28 mM. Calculations assumed activity coefficients of 1.000, but substitution of coefficients appropriate to natural seawater (Millero, 2001) yielded free-energy values that differed by less than 0.2% from corresponding tabulated ones. Similarly insignificant changes resulted from the use of equilibrium constants calculated for pressure of 300 bars, which corresponds to an approximate water depth of 3km.

These ΔG-calculations ignore the energetic cost of concentrating nitrate into the vacuole against an environmental gradient. The minimum energy needed to concentrate a solute X-fold up such a gradient is: $\log_{10}(X) \cdot 5.7$ kJ mol^{-1}, which, for a typical 10,000-fold nitrate gradient, yields 22.8 kJ per mol. If we assume a one-third efficiency in capture of a proton gradient to perform useful work (Harold, 1986), roughly 15% of the energy liberated from respiratory reduction of a mole of nitrate driven by sulfide oxidation is needed to accumulate the nitrate. This conclusion is independent of whether nitrate is reduced through dissimilation or denitrification; therefore, neglect of the cost of assimilation does not change the implications of these tabulated data.

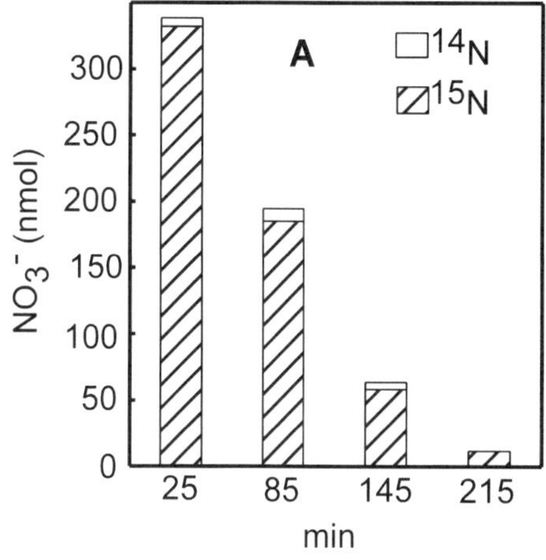

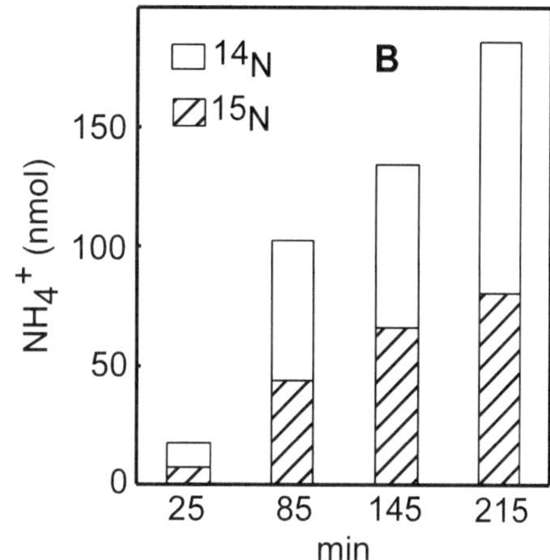

Figure 10. Nitrate reduction to ammonium by cleaned bundles of *Thioploca* trichomes collected fresh from sediments on the Chilean shelf. The bacteria were incubated with ^{15}N-labeled nitrate in the medium under a helium headspace and the production of ^{15}N-ammonium was monitored. (A) The nitrate was consumed over a 3–4 h period without a drop in the high specific label of ^{15}N, which shows that it was taken up without exchange with the large intracellular ^{14}N-nitrate pool. (B) The produced ammonium had only half the specific label of ^{15}N, which shows that half of the dissimilatory nitrate reduction was based on extracellular NO_3^- and half on intracellular NO_3^-. (Redrawn from Otte et al., 1999.)

were made taking into account reasonable environmental concentrations of key metabolites and waste products. The main conclusions of this exercise (Table 2) were as follows: Denitrifying bacteria should gain an ~60% advantage per mole of sulfide oxidized and thus would enjoy a large theoretical advantage if sulfide were the limiting nutrient. However, the free energy yields are quite similar for both processes when expressed in per mole of nitrate. The question is, therefore, whether nitrate or sulfide is the limiting substrate for the two types of nitrate reducers. Within the thin nitrate zone at the sediment surface (Fig. 8), the electron donor is likely to be limiting for denitrification. In contrast, nitrate is likely to be limiting for bacteria that carry it deep down into the sediment, where sulfate reduction rates are high. For the vacuolated sulfur bacteria, the ΔG per mol nitrate may therefore be the more relevant.

Until more is learned about the cellular location of these processes (i.e., whether they take place in the vacuole, cytoplasm, or periplasm) these sorts of calculations cannot be meaningfully refined. For example, the nitrate concentrations explored in Table 2 range over more than five orders of magnitude, which is the range between probable intracellular and external concentrations. Furthermore, ^{15}N tracer studies (Fig. 10) indicated that the nitrate reduced to ammonium originated both from the vacuolar and the external nitrate pools. Finally, it is important to point out that the limitation of certain key biochemical steps, rather than overall free energy gain, may dictate the metabolic waste products of these bacteria. For example, the biochemistry of denitrification is well studied in a variety of non-vacuolate bacteria, and several of the intermediate steps are characterized by strongly positive midpoint potentials and the need for metal cofactors such as copper (Zumft, 1992). Hence, in situations where highly sulfidic environments are inhabited by nitrate-accumulating bacteria, copper is strongly bound by sulfide and it may be difficult for denitrifying bacteria to accumulate sufficient free copper or to generate conditions that are sufficiently oxidizing.

COUPLING OF NITRATE REDUCTION AND SULFIDE OXIDATION

The nitrate accumulation by large sulfur bacteria and their dissimilatory nitrate reduction to ammonium in eutrophic shelf sediments may have a significant impact on both the nitrogen and the sulfur cycles in coastal marine ecosystems. We suggest that the occurrence of such bacteria is more widespread than generally appreciated because they may not be easily detected when they live buried within the mud. We also suggest that their nitrate reduction may be underestimated because the turnover of the large intracellular nitrate pool is sluggish and may not be detected during short-term experiments. Some examples will illustrate these points.

The most widespread of the large sulfur bacteria are *Beggiatoa* spp., which are known, in particular, from their formation of mats on sulfidic sediment surfaces. These mats are easily detected by observations along the coast, by diving, by underwater video on remotely operated vehicles, or from submersibles. Less known is their subsurface occurrence as scattered individual filaments in coastal sediments. Jørgensen (1977b) quantified such populations in sediments of Limfjorden, Denmark, and found high biomasses of 5–20 g fresh weight per m^2. There were no visible mats on the sediment surface but maximum numbers of filaments were in the suboxic zone at a few centimeters depth. This distribution remained puzzling for several decades, since the marine *Beggiatoa* were considered to be highly efficient gradient bacteria living at the O_2-H_2S interface as aerobic, lithoautotrophic sulfide oxidizers (Jørgensen and Revsbech, 1983; Nelson et al., 1986b). Their preferred environment in the Limfjorden sediment appeared, however, to be the anoxic but oxidized zone where neither O_2 nor H_2S were detectable (Jørgensen, 1982b).

Following these studies, few systematic surveys of "free-living" *Beggiatoa* spp. seem to have been undertaken, although identification of these bacteria is straightforward and the counts are not hampered by the shortcomings of viable counting or by the requirements of molecular techniques. The quantification requires microscopic technique and magnification rather similar to that used for meiofauna. The *Beggiatoa* filaments may be overlooked under the dissection microscope, and they are too few to show up in normal bacterial counts using, for example, DAPI staining. Yet, the *Beggiatoa* may occur in significant biomasses due to the large size of individual filaments. Recent studies from different North Sea and Baltic Sea sites indicate that non–mat-forming *Beggiatoa* of 10–20 μm diameter occur commonly in the upper 2–4 cm of the sediments (Mußmann et al., 2003; A. Preisler and L.P. Nielsen, 2003, personal comm.).

If nitrate storing sulfur bacteria such as *Beggiatoa* occur unnoticed in sediments, how might their nitrate pool and nitrate reduction be detected? Numerous studies have been made of nitrate distributions in the pore water of coastal marine sediments. It is commonly reported that nitrate penetrates one or several centimeters into the sediment and also that the nitrate decreases with depth to a background value of a few micromolar, rather than to zero. With the recent introduction of nitrate biomicrosensors, the first high resolution nitrate microprofiles in marine sediments were obtained (Larsen et al., 1997). The nitrate penetration into coastal sediments was shown to be in the range of several millimeters rather than several centimeters, and no background concentration was detected (Fig. 8 and 11; Kjær, 2000). The nitrate microprofiles either follow the oxygen microgradients with a slightly deeper penetration or they show a subsurface maximum due to nitrification of ammonium diffusing up into the lower part of the oxic zone.

Could the deeper nitrate penetration measured in extracted pore water samples be due to intracellular nitrate of *Beggiatoa*? Striking observations in this direction were made by Sayama (2001) when studying the nitrate uptake and denitrification in sediments of Tokyo Bay. Throughout his seasonal study, the NO_3^- flux was always directed from the water column into the sediment and showed a net uptake of nitrate. Yet, a very large peak of NO_3^- was detected just below the sediment surface in pore water samples extracted from frozen sediment sections. The uptake of nitrate against such a peak remained enigmatic (and unpublished) until the first reports of nitrate-storing sulfur bacteria led the author to check the pore water gradients without initial freezing. The "unfrozen" gradients were quite different and typical for coastal sediments. The results showed that freezing caused release of large amounts of intracellular nitrate (Fig. 12), due to the rupture of cells and vacuoles in *Beggiatoa* spp., that were visibly present in the Tokyo Bay sediments. A systematic study of pore water nitrate concentrations resulting from different sediment treatments showed that cell lysis by osmotic shock or freezing released nitrate that was not detected in carefully treated sediment samples.

It should be noted that microalgae also may accumulate nitrate to a lesser extent. This has been observed to provide an intracellular nitrate reservoir at the sediment surface just after the sedimentation of a phytoplankton spring bloom (Lomstein et al., 1990).

The discovery of nitrate vacuoles in the large marine *Beggiatoa* spp. inspired a revisit to the old Limfjorden stations of Jør-

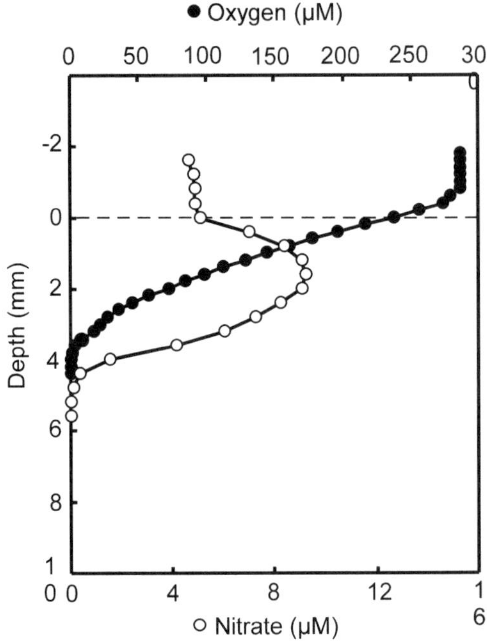

Figure 11. Oxygen and nitrate profiles in a coastal marine sediment from Aarhus Bay, Denmark, measured with O_2 and NO_3^- microsensors. The nitrate concentration of the overlying sea water was 4 μM and the subsurface nitrate peak was due to nitrification in the lower part of the oxic zone. Each curve shows the mean of three profiles. (Redrawn from Kjær, 2000.)

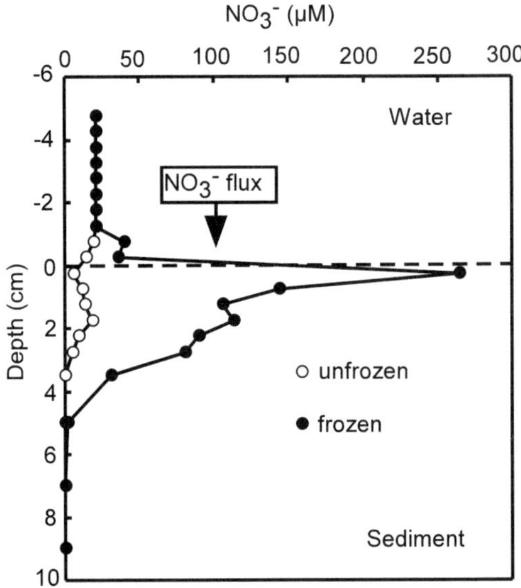

Figure 12. Nitrate concentration profiles measured in sediment cores from Tokyo Bay at 10 m water depth. The "unfrozen" samples were centrifuged immediately after collection for nitrate analysis in the pore water. The other series of samples was stored frozen until thawing and centrifugation. The large pool of nitrate obtained by freezing was released from intracellular nitrate in *Beggiatoa* vacuoles. The nitrate flux in intact sediment cores was always from the water into the sediment, which demonstrates the efficiency of nitrate uptake in *Beggiatoa*. (Redrawn from Sayama, 2001.)

gensen (1977b) to check whether these populations also contained vacuoles and whether this could explain the subsurface distribution. The check was indeed positive (Fig. 13; Mußmann et al., 2003). The *Beggiatoa* were distributed in the upper 30 mm with peaks near the surface and at 20 mm depth. Their diameters ranged from 5 to 30 µm, with the larger filaments occurring deeper in the sediment. Oxygen penetrated 2 mm into the sediment and nitrate 4 mm while free H_2S was detectable only below 25 mm. Most of the *Beggiatoa* population thus occurred in a sediment zone apparently devoid of their electron acceptors or donor. Analyses of individual filaments showed, however, that these had vacuoles with 100–200 mM NO_3^- and that their volumetric content of elemental sulfur was 300–500 mM. It is thus assumed that the *Beggiatoa* migrate freely up and down through the anoxic, oxidized sediment and efficiently take up nitrate when near the surface. This is a similar behavior as observed in *Thioploca*, although the vertical commute of *Thioploca* appears to be more efficiently directed by their oriented sheaths. Throughout the oxidized Limfjorden sediment, sulfate reduction produces H_2S at high rates (Fig. 13). There is apparently an ample supply of energy substrate for the *Beggiatoa* even though the H_2S is consumed rapidly and does not accumulate to a detectable level where they occur.

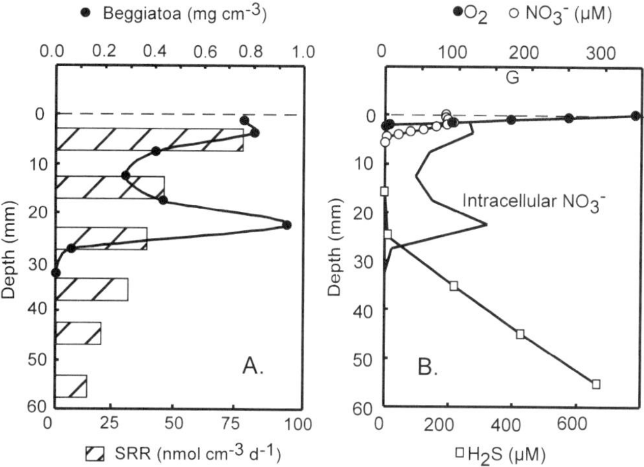

Figure 13. Microbiological and chemical profiles in sediment cores from Limfjorden, Denmark, taken in November 1997 at 10 m water depth. (A) Distribution of filamentous sulfur bacteria, *Beggiatoa* spp., and of sulfate reduction rates measured by ^{35}S-technique. (B) Pore water concentration of oxygen, nitrate, and hydrogen sulfide, and distribution of intracellular nitrate in vacuoles of *Beggiatoa*. (Data from Mußmann et al., 2003.)

Integrated studies of nitrogen and sulfur transformations in coastal marine sediments have indicated that nitrate reduction plays a minor quantitative role for the oxidation of sulfide (e.g., Sørensen and Jørgensen, 1987). Although there is a great potential for sulfide oxidation with nitrate in sediments (Elsgaard and Jørgensen 1992), the natural rates of sulfate reduction and sulfide oxidation are simply much higher than measured fluxes or reduction rates of nitrate. This does not, however, exclude that the reverse may be the case (i.e., that sulfide may provide an important substrate for dissimilatory nitrate reduction). The process may be catalyzed either through the nitrate reduction to N_2 by *Thiobacillus denitrificans*, *Thiomicrospira denitrificans*, and similar denitrifying bacteria, or through the dissimilatory nitrate reduction to ammonium (DNRA) by *Beggiatoa* and their relatives. It is important to understand under which conditions and to what extent DNRA competes with denitrification, since only the latter causes a loss of combined nitrogen in the marine ecosystem, whereas the DNRA retains the nitrogen as ammonium. Thus, *Beggiatoa* may prevent the removal of nitrogen by denitrification and thus enhance the effect of coastal eutrophication.

Early experimental studies of denitrification in marine sediments were done by the acetylene blockage technique by which the last step in the enzymatic process, the reduction of N_2O to N_2, was inhibited by addition of acetylene (Sørensen, 1978b). The transient accumulation of N_2O during incubation could thus be readily monitored and used as a measure of the uninhibited denitrification rate. The DNRA was found to be significant in highly reduced, sulfidic sediments (Sørensen, 1978a; Koike and Hattori, 1978). The results were later questioned, however, after it was found that sulfide, even in µM concentrations, may alleviate the acetylene blockage of N_2O reduction in denitrifying cultures of *Pseudomonas fluorescens* (Sørensen et al., 1987) and that *Thiobacillus denitrificans* may even oxidize sulfide with N_2O in the presence of acetylene (Dalsgaard and Bak, 1992). Thus, the experimental technique appeared to fail in exactly those sediments where nitrate storing sulfur bacteria could potentially play the greatest role.

Since the DNRA was initially calculated by the difference between the total nitrate reduction and that accounted for by denitrification, an underestimation of the latter due to inefficient acetylene blockage would lead to an overestimation of DNRA. It is striking that increasingly detailed studies of denitrification and DNRA at the same site in Norsminde Fjord, Denmark, over a 10 yr period concluded ever-decreasing contributions of DNRA, starting with up to 88% of total nitrate reduction and ending with 0% (Table 3). Since then, however, many independent studies have been made with partly improved techniques in sediments of eutrophic marine environments. In particular, the combined use of $^{15}NO_3^-$ and acetylene inhibition has helped to clear the picture. The present conclusion is that DNRA does indeed play a significant role in sulfidic sediments with high organic load but little or no role in more oxidized sediments (Table 3).

A clear example of this was provided by Christensen et al. (2000) who studied the two pathways of dissimilatory nitrate reduction in sediments of Horsens Fjord, Denmark (Fig. 14). In a transect of sediment stations from directly underneath the net cages of a fish farm and out to the undisturbed sea bed, they found a complete shift from dominance of DNRA to denitrification. The sediment underlying the cage received a high load of organic debris, and 86% of the nitrate reduction went to ammonium. In the unaffected sediment, all the nitrate reduction went by denitrification to N_2.

TABLE 3. DISSIMILATORY NITRATE REDUCTION TO AMMONIUM (DNRA) VERSUS DENITRIFICATION IN COASTAL MARINE SEDIMENTS

Site	Method*	Description of sediment	Denitrification†	DNRA†	% DNRA§	Reference
Limfjorden, Denmark	Acetylene	10 m, 6 °C, 0–3 cm	*870*	*750*	46%	Sørensen (1978a)
Norsminde Fjord, Denmark	Acetylene	Inner: 1 m, 10.2% C_{org}	4.3	31.2	88%	Jørgensen and Sørensen (1985)
		Outer: 1 m, 1.2% C_{org}	1.5	3.2	68%	
Norsminde Fjord, Denmark	$^{15}NO_3^-$ + acetylene	Inner: 1 m, June	120	23	16%	Jørgensen (1989)
		Inner: 1 m, December	17	33	66%	
Norsminde Fjord, Denmark	$^{15}NO_3^-$	Inner: 1 m, Oct. + March	3.2–34.9	0	0%	Binnerup et al. (1992)
Knebel Vig, Denmark	$^{15}NO_3^-$	14 m, *Beggiatoa* mat	1.0	3.0	75%	Risgaard-Petersen (1995)
Etang du Prévost, France	$^{15}NO_3^-$	1 m, 27 °C, dark	0.44	0.31	41%	Rysgaard et al. (1996)
Bassin d'Arcachon, France		Tidal flat, 25 °C, dark	0.009	0.13	94%	
Thau Lagoon, France	$^{15}NO_3^-$ + acetylene	Under shellfish farm, 8 m	*0.8*	*66*	99%	Gilbert et al. (1997)
		Reference site, 8 m	*0.4*	*11*	96%	
Gulf of Fos, France	$^{15}NO_3^-$ + acetylene	Eutrophic lagoon, 2.5%–7.3% C_{org}, 2–10 m	0–19.8	2.3–83.2	18%–100%	Bonin et al. (1998)
Horsens Fjord, Denmark	$^{15}NO_3^-$	Fish farm, 3 m, *Beggiatoa*	1.1	6.7	86%	Christensen et al. (2000)
		Reference site, 4 m	0.7	0.0	0%	
Concepción Bay, Chile	$^{15}NO_3^-$	24 m, *Beggiatoa* mat	0.18–1.30	~2.7–5	>50%	Graco et al. (2001)
Tokyo Bay, Japan	Acetylene	10 m, *Beggiatoa* mat	0.26	0.99	80%	Sayama (2001)

Note: A few characteristics of the sites are included in the table when available, such as water depth, temperature, or organic content in % dry weight.
*Methods: Acetylene—acetylene blockage technique; $^{15}NO_3^-$—experimental tracer incubation using ^{15}N-labeled nitrate.
†DNRA—Dissimilatory nitrate reduction to ammonium. Units: areal rates in mmol m^{-2} d^{-1}; italics: volume rates in nmol N cm^{-3} d^{-1}.
§DNRA in % of total dissimilatory nitrate reduction

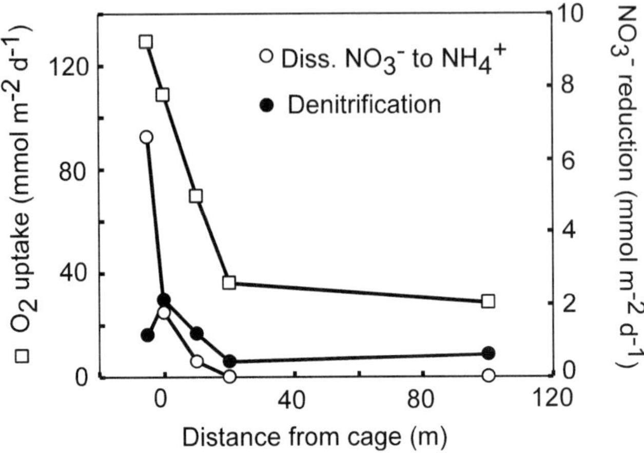

Figure 14. Sediment oxygen uptake and nitrate reduction via denitrification or dissimilatory nitrate reduction to ammonium in sediments along a transect in Horsens Fjord, Denmark. The transect in the shallow fjord extended from below suspended nets of a trout fish farm with very high organic deposition on the sea bed to 100 m distance where the fjord sediments were unaffected by fish farming. The oxygen uptake shows how the load of organic material increased under the nets. The pathway of nitrate reduction changed completely over the transect, from predominant DNRA under the net to exclusively denitrification in the unaffected sea bed. (Redrawn from Christensen et al., 2000.)

Sediments with relatively high DNRA are all characterized by very high organic load and, presumably, high H_2S production rates even near the sediment-water interface (Table 3). In several of the sediments, the respective authors also noted the presence of *Beggiatoa* on the sediment surface. Based on these data, we suggest that dissimilatory nitrate reduction to ammonium does indeed play an important role for the nitrogen cycling in high-organic sulfidic sediments. To what extent it plays a quantitative role for the sulfide oxidation is still unclear. Experiments with $^{15}NO_3^-$ must be adapted to also reveal the slow turnover of intracellular nitrate, which appears to escape detection during short-term incubations.

ACKNOWLEDGMENTS

We thank Jakob Zopfi and Mandy Joye for permission to present their unpublished data. We also thank Heide Schulz, Lars Peter Nielsen, André Preisler, Gaute Lavik, Elze Wieringa, and Dirk de Beer for stimulating discussions. We thank two anonymous reviewers for helpful comments that improved the paper. This study was supported by the Max Planck Society, the Fonds der Chemischen Industrie, the National Science Foundation, and the U.S. Department of Energy.

REFERENCES CITED

Aller, R.C., and Rude, P.D., 1988, Complete oxidation of solid phase sulfides by manganese and bacteria in anoxic marine sediments: Geochimica et Cosmochimica Acta, v. 52, p. 751–765, doi: 10.1016/0016-7037(88)90335-3.

Bak, F., and Pfennig, N., 1987, Chemolithotrophic growth of *Desulfovibrio sulfodismutans* sp. nov. by disproportionation of inorganic sulfur compounds: Archives of Microbiology, v. 147, p. 184–189.

Ballbjerg, V., Mouritsen, K.N., and Finster, K., 1998, Diel cycles of sulphate reduction rates in sediments of a *Zostera marina* bed (Denmark): Aquatic Microbial Ecology, v. 15, p. 97–102.

Benz, M., Brune, A., and Schink, B., 1998, Anaerobic and aerobic oxidation of ferrous iron at neutral pH by chemolithotrophic nitrate-reducing bacteria: Archives of Microbiology, v. 169, p. 159–165, doi: 10.1007/S002030050555.

Binnerup, S.J., Jensen, K., Revsbech, N.P., Jensen, M.H., and Sørensen, J., 1992, Denitrification, dissimilatory reduction of nitrate to ammonium, and nitrification in a bioturbated estuarine sediment as measured with ^{15}N and microsensor techniques: Applied and Environmental Microbiology, v. 58, p. 303–313.

Bonin, P., Omnes, P., and Chalamet, A., 1998, Simultaneous occurrence of denitrification and nitrate ammonification in sediments of the French Mediterranean Coast: Hydrobiologia, v. 389, p. 169–182, doi: 10.1023/A:1003585115481.

Brinkhoff, T., and Muyzer, G., 1997, Increased species diversity and extended habitat range of sulfur-oxidizing *Thiomicrospira* spp: Applied and Environmental Microbiology, v. 63, p. 3789–3796.

Brinkhoff, T., Santegoeds, C.M., Sahm, K., Kuever, J., and Muyzer, G., 1998, A polyphasic approach to study the diversity and vertical distribution of sulfur-oxidizing *Thiomicrospira* species in coastal sediments of the German Wadden Sea: Applied and Environmental Microbiology, v. 64, p. 4650–4657.

Burdige, D.J., 1993, The biogeochemistry of manganese and iron reduction in marine sediments: Earth Science Reviews v. 35, p. 249–284, doi: 10.1016/0012-8252(93)90040-E.

Canfield, D.E., 2001, Isotope fractionation by natural populations of sulfate-reducing bacteria: Geochimica et Cosmochimica Acta, v. 65, p. 1117–1124, doi: 10.1016/S0016-7037(00)00584-6.

Canfield, D.E., and Raiswell, R., 1992, The reactivity of sedimentary iron minerals toward sulfide: American Journal of Science, v. 292, p. 659–683.

Canfield, D.E., and Teske, A., 1996, Late Proterozoic rise in atmospheric oxygen concentration inferred from phylogenetic and sulphur-isotope studies: Nature, v. 382, p. 127–132, doi: 10.1038/382127A0.

Canfield, D.E., and Thamdrup, B., 1994, The production of ^{34}S-depleted sulfide during bacterial disproportionation of elemental sulfur. Science, v. 266, p. 1973–1975.

Canfield, D.E., Thamdrup, B., and Fleischer, S., 1998, Isotope fractionation and sulfur metabolism by pure and enrichment cultures of elemental sulfur disproportionating bacteria: Limnology and Oceanography, v. 43, p. 253–264.

Canfield, D.E., Thamdrup, B., and Hansen, J.W., 1993, The anaerobic degradation of organic matter in Danish coastal sediments: Iron reduction, manganese reduction, and sulfate reduction: Geochimica et Cosmochimica Acta, v. 57, p. 3867–3883, doi: 10.1016/0016-7037(93)90340-3.

Christensen, P.B., Rysgaard, S., Sloth, N.P., Dalsgaard, T., and Schwärter, S., 2000, Sediment mineralization, nutrient fluxes, denitrification and dissimilatory nitrate reduction to ammonium in an estuarine fjord with sea cage trout farms: Aquatic Microbial Ecology, v. 21, p. 73–84.

Dalsgaard, T., and Bak, F., 1992, Effect of acetylene on nitrous oxide reduction and sulfide oxidation in batch and gradient cultures of *Thiobacillus denitrificans*: Applied and Environmental Microbiology, v. 58, p. 1601–1608.

Eisenmann, E., Beuerle, J., Sulger, K., Kroneck, P.M.H., and Schumacher, W., 1995, Lithotrophic growth of *Sulfospirillum deleyianum* with sulfide as electron donor coupled to respiratory reduction of nitrate to ammonia: Archives of Microbiology, v. 164, p. 180–185, doi: 10.1007/S002030050252.

Elsgaard, L., and Jørgensen, B.B., 1992, Anoxic transformations of radiolabeled hydrogen sulfide in marine and freshwater sediments: Geochimica et Cosmochimica Acta, v. 56, p. 2425–2435, doi: 10.1016/0016-7037(92)90199-S.

Ferdelman, T.G., Lee, C., Pantoia, S., Harder, J., Bebout, B.M., and Fossing, H., 1997, Sulfate reduction and methanogenesis in a *Thioploca*-dominated sediment off the coast of Chile: Geochimica et Cosmochimica Acta, v. 61, p. 3065–3079, doi: 10.1016/S0016-7037(97)00158-0.

Finster, K., Liesack, W., and Thamdrup, B., 1998, Elemental sulfur and thiosulfate disproportionation by *Desulfocapsa sulfoexigens* sp. nov., a new anaerobic bacterium isolated from marine surface sediment: Applied and Environmental Microbiology, v. 64, p. 119–125.

Fossing, H., and Jørgensen, B.B., 1990, Oxidation and reduction of radiolabeled inorganic sulfur compounds in an estuarine sediment, Kysing Fjord, Denmark: Geochimica et Cosmochimica Acta, v. 54, p. 2731–2742, doi: 10.1016/0016-7037(90)90008-9.

Fossing, H., Thode-Andersen, S., and Jørgensen, B.B., 1992, Sulfur isotope exchange between ^{35}S-labeled inorganic sulfur compounds in anoxic marine sediments: Marine Chemistry, v. 38, p. 117–132, doi: 10.1016/0304-4203(92)90071-H.

Fossing, H., Gallardo, V.A., Jørgensen, B.B., Hüttel, M., Nielsen, L.P., Schulz, H., Canfield, D.E., Forster, S., Glud, R.N., Gundersen, J.K., Küver, J., Ramsing, N.B., Teske, A., Thamdrup, B., and Ollua, O., 1995, Concentration and transport of nitrate by the mat-forming sulphur bacterium *Thioploca*: Nature, v. 374, p. 713–715, doi: 10.1038/374713A0.

Gallardo, V.A., 1977, Large benthic microbial communities in sulphide biota under Peru-Chile Subsurface Countercurrent: Nature, v. 268, p. 331–332.

Garcia-Gil, L.J., and Golterman, H.L., 1993, Kinetics of FeS-mediated denitrification in sediments from the Camargue (Rhone delta, southern France): FEMS Microbiology Ecology, v. 13, p. 85–92, doi: 10.1016/0168-6496(93)90026-4.

Gilbert, F., Souchu, P., Bianchi, M., and Bonin, P., 1997, Influence of shellfish farming activities on nitrification, nitrate reduction to ammonium and denitrification at the water-sediment interface of the Thau lagoon, France: Marine Ecology Progress Series, v. 151, p. 143–153.

González, J.M., Kiene, R.P., and Moran, M.A., 1999, Transformation of sulfur compounds by n abundant lineage of marine bacteria in the a-subclass of the class Proteobacteria: Applied and Environmental Microbiology, v. 65, p. 3810–3819.

Graco, M., Farias, L., Molina, V., Gutiérrez, D., and Nielsen, L.P., 2001, Massive developments of microbial mats following phytoplankton blooms in a naturally eutrophic bay: implications for nitrogen cycling: Limnology and Oceanography, v. 46, p. 821–832.

Habicht, K.S., and Canfield, D.E., 1997, Sulfur isotope fraction during bacterial sulfate reduction in organic-rich sediments: Geochimica et Cosmochimica Acta, v. 61, p. 5351–5361, doi: 10.1016/S0016-7037(97)00311-6.

Habicht, K.S., and Canfield, D.E., 2001, Isotope fractionation by sulfate-reducing natural populations and the isotopic composition of sulfide in marine sediments: Geology, v. 29, p. 555–558, doi: 10.1130/0091-7613(2001)0292.0.CO;2.

Habicht, K.S., Canfield, D.E., and Rethmeier, J., 1998, Sulfur isotope fractionation during bacterial reduction and disproportionation of thiosulfate and sulfite: Geochimica et Cosmochimica Acta, v. 62, p. 2585–2595, doi: 10.1016/S0016-7037(98)00167-7.

Harold, F.M., 1986, The vital force: A study of bioenergetics: New York, W.H. Freeman and Co., 577 p.

Huettel, M., Ziebis, W., Forster, S., and Luther, G.W., III, 1998, Advective transport affecting metal and nutrient distributions and interfacial fluxes in permeable sediments: Geochimica et Cosmochimica Acta, v. 62, p. 613–631, doi: 10.1016/S0016-7037(97)00371-2.

Huettel, M., Forster, S., Klöser, S., and Fossing, H., 1996, Vertical migration in the sediment-dwelling sulfur bacteria *Thioploca* spp. in overcoming diffusion limitations. Applied and Environmental Microbiology, v. 62, p. 1863–1872.

Jannasch, H.W., Nelson, D.C., and Wirsen, C.O., 1989, Massive natural occurrence of unusually large bacteria (*Beggiatoa* sp.) at a hydrothermal deep-sea vent site: Nature, v. 342, p. 834–836, doi: 10.1038/342834A0.

Janssen, P.H., Schuhmann, A., Bak, F., and Liesack, W., 1996, Fermentation of inorganic sulfur compounds by the sulfate-reducing bacterium *Desulfocapsa thiozymogenes* gen. nov., sp. nov: Archives of Microbiology, v. 166, p. 184–192, doi: 10.1007/S002030050374.

Johnson, J.W., Oelkers, E.H., and Helgeson, H.C., 1992, SUPCRT 92, A software package for calculating the standard molal thermodynamic properties of minerals, gases, aqueous species and reactions: Computers & Geosciences, v. 18, p. 899–947, doi: 10.1016/0098-3004(92)90029-Q.

Jørgensen, B.B., 1977a, The sulfur cycle of a coastal marine sediment (Limfjorden, Denmark): Limnology and Oceanography, v. 22, p. 814–832.

Jørgensen, B.B., 1977b, Distribution of colorless sulfur bacteria (*Beggiatoa* spp.) in a coastal marine sediment: Marine Biology, v. 41, p. 19–28.

Jørgensen, B.B., 1982a, Mineralization of organic matter in the sea bed—The role of sulphate reduction: Nature, v. 296, p. 643–645.

Jørgensen, B.B., 1982b, Ecology of the bacteria of the sulphur cycle with special reference to anoxic-oxic interface environments: Philosophical Transactions of the Royal Society of London, B, v. 298, p. 543–561.

Jørgensen, B.B., 1990, A thiosulfate shunt in the sulfur cycle of marine sediments: Science, v. 249, p. 152–154.

Jørgensen, B.B., and Bak, F., 1991, Pathways and microbiology of thiosulfate transformations and sulfate reduction in a marine sediment (Kattegat, Denmark): Applied and Environmental Microbiology, v. 57, p. 847–856.

Jørgensen, B.B., and Gallardo, V.A., 1999, *Thioploca* spp.: filamentous sulfur bacteria with nitrate vacuoles: FEMS Microbiology Ecology, v. 28, p. 301–313, doi: 10.1016/S0168-6496(98)00122-6.

Jørgensen, B.B., and Revsbech, N.P., 1983, Colorless sulfur bacteria, *Beggiatoa* spp. and *Thiovulum* spp., in O_2 and H_2S microgradients: Applied and Environmental Microbiology, v. 45, p. 1261–1270.

Jørgensen, B.B., and Sørensen, J., 1985, Seasonal cycles of O_2, NO_3^- and SO_4^{2-} reduction in estuarine sediments: The significance of a NO_3^- reduction maximum in the spring: Marine Ecology Progress Series, v. 24, p. 65–74.

Jørgensen, B.B., Bang, M., and Blackburn, T.H., 1990, Anaerobic mineralization in marine sediments from the Baltic Sea–North Sea transition: Marine Ecology Progress Series, v. 59, p. 39–54.

Jørgensen, B.B., Böttcher, M.E., Lüschen, H., Neretin, L.N., and Volkov, I.I., 2004a, Anaerobic methane oxidation and a deep H_2S sink generate isotopically heavy sulfides in Black Sea sediments: Geochimica et Cosmochimica Acta, v. 68, p. 2095–2118.

Jørgensen, B.B., Teske, A., and Ahmad, A., 2004b, *Thioploca* Lauterborn 1907, 242[AL], *in* Krieg, N.R., Staley, J.T., and Brenner, D.J., eds., Bergey's Manual of Determinative Bacteriology, Vol. 2: Baltimore, Williams & Wilkins, (in press).

Jørgensen, K.S., 1989, Annual pattern of denitrification and nitrate ammonification in estuarine sediment: Applied and Environmental Microbiology, v. 55, p. 1841–1847.

Kelly, D.P., 1982, Biochemistry of the chemolithotrophic oxidation of inorganic sulphur: Philosophical Transactions of the Royal Society of London, Series B, Biological Sciences, v. 298, p. 499–528.

Kjær, T., 2000, Development and application of new biosensors for microbial ecology [Ph.D. Thesis]: Denmark, University of Aarhus, 324 p.

Koike, I., and Hattori, A., 1978, Denitrification and ammonia formation in anaerobic coastal sediments: Applied and Environmental Microbiology, v. 35, p. 853–857.

Krämer, M., and Cypionka, H., 1989, Sulfate formation via ATP sulfurylase in thiosulfate- and sulfite-disproportionating bacteria: Archives of Microbiology, v. 151, p. 232–237.

Kuenen, J.G., 1979, Growth yields and "maintenance energy requirement" in *Thiobacillus* species under energy limitation: Archives of Microbiology, v. 122, p. 183–188.

Kuenen, J.G., and Veldkamp, H., 1972, *Thiomicrospira pelophila*, gen. n., sp. n., a new obligately chemolithotrophic colourless sulphur bacterium: Antonie van Leeuwenhoek, v. 38, p. 241–256.

Kuenen, J.G., Robertson, L.A., and van Gemerden, H., 1985, Microbial interactions among aerobic and anaerobic sulfur-oxidizing bacteria, *in* Marshall, K.C., ed., Advances in microbial ecology, vol. 8: Plenum Press, p. 1–59.

Kuenen, J.G., Robertson, L.A., and Tuovinen, O.H., 1992, The genera *Thiobacillus, Thiomicrospira*, and *Thiosphera, in* Balows, A., Trüper, H.G., Dworkin, M., Harder, W., and Schleifer, K.-H., eds., The Prokaryotes, Vol. 1: Springer Verlag, p. 2638–2658.

Larsen, L.H., Kjær, T., and Revsbech, N.P., 1997, A microscale NO_3^- biosensor for environmental applications: Analytical Chemistry, v. 69, p. 3527–3531, doi: 10.1021/AC9700890.

Llobet-Brossa, E., Roselló-Mora, R., and Amann, R., 1998, Microbial community composition of Wadden Sea sediments as revealed by fluorescence in situ hybridization: Applied and Environmental Microbiology, v. 64, p. 2691–2696.

Lomstein, E., Jensen, M.H., and Sørensen, J., 1990, Intracellular NH^+_4 and NO^-_3 pools associated with deposited phytoplankton in a marine sediment (Aarhus Bight, Denmark): Marine Ecology Progress Series, v. 61, p. 97–105.

Lowson, R.T., 1982, Aqueous oxidation of pyrite by molecular oxygen: Chemical Reviews, v. 82, p. 461–497.

Luther, G.W., III, 1987, Pyrite oxidation and reduction: molecular orbital theory considerations: Geochimica et Cosmochimica Acta, v. 51, p. 3193–3199, doi: 10.1016/0016-7037(87)90127-X.

Luther, G.W., III, 1991, Pyrite synthesis via polysulphide compounds: Geochimica et Cosmochimica Acta, v. 55, p. 2839–2849, doi: 10.1016/0016-7037(91)90449-F.

Maier, S., and Gallardo, V.A., 1984, Nutritional characteristics of two marine Thioplocas determined by autoradiography: Archives of Microbiology, v. 139, p. 218–220.

Mason, J., and Kelly, D.P., 1988, Thiosulfate oxidation by obligately heterotrophic bacteria: Microbial Ecology, v. 15, p. 123–134.

McHatton, S.C., 1998, Ecology and physiology of autotrophic sulfur bacteria from sulfide-rich seeps and marine sediments [Ph.D. dissertation]: Davis, California, University of California, 168 p.

McHatton, S.C., Barry, J.P., Jannasch, H.W., and Nelson, D.C., 1996, High nitrate concentrations in vacuolate, autotrophic marine *Beggiatoa* spp: Applied and Environmental Microbiology, v. 62, p. 954–958.

Millero, F.J., 2001, Physical chemistry of natural waters: New York, Wiley-Interscience, 654 p.

Morse, J.W., 1991, Oxidation kinetics of sedimentary pyrite in seawater: Geochimica et Cosmochimica Acta, v. 55, p. 3665–3667, doi: 10.1016/0016-7037(91)90064-C.

Moses, C.O., and Herman, J.S., 1991, Pyrite oxidation at circumneutral pH: Geochimica et Cosmochimica Acta, v. 55, p. 471–482, doi: 10.1016/0016-7037(91)90005-P.

Mußmann, M., Schulz, H.N., Strotmann, B., Kjaer, T., Nielsen, L.P., Roselló-Mora, R., Amann, R., and Jørgensen, B.B., 2003, Nitrate-storing *Beggiatoa* spp. under anoxic conditions in Baltic and North Sea sediments: Environmental Microbiology, v. 5, p. 523–533.

Nelson, D.C., Jørgensen, B.B., and Revsbech, N.P., 1986a, Growth pattern and yield of a chemoautotrophic *Beggiatoa* sp. in oxygen-sulfide microgradients: Applied and Environmental Microbiology, v. 52, p. 225–233.

Nelson, D.C., Revsbech, N.P., and Jørgensen, B.B., 1986b, Microoxic-anoxic niche of *Beggiatoa* spp.: Microelectrode survey of marine and freshwater strains: Applied and Environmental Microbiology, v. 52, p. 161–168.

Otte, S., Kuenen, J.G., Nielsen, L.P., Paerl, H.W., Zopfi, J., Schulz, H.N., Teske, A., Strotmann, B., Gallardo, V.A., and Jørgensen, B.B., 1999, Nitrogen, carbon, and sulfur metabolism in natural *Thioploca* samples: Applied and Environmental Microbiology, v. 65, p. 3148–3157.

Podgorsek, L., and Imhoff, J.F., 1999, Tetrathionate production by sulfur oxidizing bacteria and the role of tetrathionate in the sulfur cycle of Baltic Sea sediments: Aquatic Microbial Ecology, v. 17, p. 255–265.

Pruysers, P.A., 1998, Early diagenetic processes in sediments of the Angola Basin, eastern South Atlantic [Ph.D. thesis]: The Netherlands, University of Utrecht, 135 p.

Risgaard-Petersen, N., 1995, Denitrification and dissimilative nitrate reduction to ammonium in mats of *Beggiatoa* spp. on marine sediments [Ph.D. thesis]: Denmark, University of Aarhus.

Robertson, L.A., and Kuenen, J.G., 1992, The colorless sulfur bacteria, *in* Balows, A., Trüper, H.G., Dworkin, M. Harder, W. and Schleifer, K.H., eds., The Prokaryotes, vol. 1: Springer Verlag, p. 385–413.

Rusch, A., Töpken, H., Böttcher, M.E., and Höpner, T., 1998, Recovery from black spots: results of a loading experiment in the Wadden Sea: Journal of Sea Research, v. 40, p. 205–219, doi: 10.1016/S1385-1101(98)00030-6.

Rysgaard, S., Risgaard-Petersen, N., and Sloth, N.P., 1996, Nitrification, denitrification, and nitrate ammonification in sediments of two coastal lagoons in Southern France: Hydrobiologia, v. 329, p. 133–141.

Sayama, M., 2001, Presence of nitrate-accumulating sulfur bacteria and their influence on nitrogen cycling in a shallow coastal marine sediment: Applied and Environmental Microbiology, v. 67, p. 3481–3487, doi: 10.1128/AEM.67.8.3481-3487.2001.

Schippers, A., and Jørgensen, B.B., 2001, Oxidation of pyrite and iron sulfide by manganese in marine sediments: Geochimica et Cosmochimica Acta, v. 65, p. 915–922, doi: 10.1016/S0016-7037(00)00589-5.

Schippers, A., and Jørgensen, B.B., 2002, Biogeochemistry of pyrite and iron sulfide oxidation in marine sediments: Geochimica et Cosmochimica Acta, v. 66, p. 85–92.

Schmaljohann, R., Drews, M., Walter, S., Linke, P., von Rad, U., and Imhoff, J.F., 2001, Oxygen-minimum zone sediments in the northeastern Arabian Sea off Pakistan: a habitat for the bacterium *Thioploca*: Marine Ecology Progress Series, v. 211, p. 27–42.

Schulz, H.N., and Jørgensen, B.B., 2001, Big bacteria: Annual Review of Microbiology, v. 55, p. 105–137, doi: 10.1146/ANNUREV.MICRO.55.1.105.

Schulz, H.N., Jørgensen, B.B., Fossing, H.A., and Ramsing, N.B., 1996, Community structure of filamentous, sheath-building sulfur bacteria, *Thioploca* spp., off the coast of Chile: Applied and Environmental Microbiology, v. 62, p. 1855–1862.

Schulz, H.N., Strotmann, B., Gallardo, V.A., and Jørgensen, B.B., 2000, Population study of the filamentous sulfur bacteria *Thioploca* spp. off the Bay of Concepción, Chile: Marine Ecology Progress Series, v. 200, p. 117–126.

Schulz, H.N., Brinkhoff, T., Ferdelman, T.G., Hernandez Marine, M., Teske, A., and Jørgensen, B.B., 1999, Dense populations of a giant sulfur bacterium in Namibian shelf sediments: Science, v. 284, p. 493–495, doi: 10.1126/SCIENCE.284.5413.493.

Sievert, S.M., Kuever, J., and Muyzer, G., 2000, Identification of 16S ribosomal DNA-defined bacterial populations at a shallow submarine hydrothermal vent near Milos Island (Greece): Applied and Environmental Microbiology, v. 66, p. 3102–3109, doi: 10.1128/AEM.66.7.3102-3109.2000.

Sørensen, J., 1978a, Capacity for denitrification and reduction of nitrate to ammonia in a coastal marine sediment: Applied and Environmental Microbiology, v. 35, p. 301–305.

Sørensen, J., 1978b, Denitrification rates in a marine sediment as measured by the acetylene inhibition technique: Applied and Environmental Microbiology, v. 36, p. 139–143.

Sørensen, J., and Jørgensen, B.B., 1987, Early diagenesis in sediments from Danish coastal waters: Microbial activity and Mn-Fe-S geochemistry: Geochimica et Cosmochimica Acta, v. 51, p. 1583–1590, doi: 10.1016/0016-7037(87)90339-5.

Sørensen, J., Rasmussen, L.K., and Koike, I., 1987, Micromolar sulfide concentrations alleviate acetylene blockage of nitrous oxide reduction by denitrifying *Pseudomonas fluorescens*: Canadian Journal of Microbiology, v. 33, p. 1001–1005.

Sorokin, D.Y., 1996, Oxidation of sulfide and elemental sulfur to tetrathionate by chemoorganoheterotrophic bacteria: Microbiology, v. 65, p. 5–9.

Straub, K.L., Benz, M., Schink, B., and Widdel, F., 1996, Anaerobic, nitrate-dependent microbial oxidation of ferrous iron: Applied and Environmental Microbiology, v. 62, p. 1458–1460.

Sweerts, J.-P.R.A., de Beer, D., Nielsen, L.P., Verdouw, H., van den Heuvel, J.C., Cohen, Y., and Cappenberg, T.E., 1990, Denitrification by sulphur oxidizing *Beggiatoa* spp. mats on freshwater sediments: Nature, v. 344, p. 762–763, doi: 10.1038/344762A0.

Teske, A., Ramsing, N.B., Küver, J., and Fossing, H., 1995, Phylogeny of *Thioploca* and related filamentous sulfide-oxidizing bacteria: Systematic and Applied Microbiology, v. 18, p. 517–526.

Teske, A., Brinkhoff, T., Muyzer, G., Moser, D.P., Rethmeier, J., and Jannasch, H.W., 2000, Diversity of thiosulfate-oxidizing bacteria from marine sediments and hydrothermal vents: Applied and Environmental Microbiology, v. 66, p. 3125–3133, doi: 10.1128/AEM.66.8.3125-3133.2000.

Thamdrup, B., and Canfield, D.E., 1996, Pathways of carbon oxidation in continental margin sediments off central Chile: Limnology and Oceanography, v. 41, p. 1629–1650.

Thamdrup, B., Finster, K., Hansen, J.W., and Bak, F., 1993, Bacterial disproportionation of elemental sulfur coupled to chemical reduction of iron or manganese: Applied and Environmental Microbiology, v. 59, p. 101–108.

Thamdrup, B., Fossing, H., and Jørgensen, B.B., 1994a, Manganese, iron, and sulfur cycling in a coastal marine sediment, Aarhus Bay, Denmark: Geochimica et Cosmochimica Acta, v. 58, p. 5115–5129, doi: 10.1016/0016-7037(94)90298-4.

Thamdrup, B., Finster, K., Fossing, H., Hansen, J.W., and Jørgensen, B.B., 1994b, Thiosulfate and sulfite distributions in porewater of marine sediments related to manganese, iron and sulfur geochemistry: Geochimica et Cosmochimica Acta, v. 58, p. 67–73, doi: 10.1016/0016-7037(94)90446-4.

Troelsen, H., and Jørgensen, B.B., 1982, Seasonal dynamics of elemental sulfur in two coastal sediments: Estuarine Coastal Shelf Science, v. 15, p. 255–266.

Tuttle, J.H., and Jannasch, H.W., 1972, Occurrence and types of thiobacillus-like bacteria in the sea: Limnology and Oceanography, v. 17, p. 532–543.

van den Ende, F.P., and van Gemerden, H., 1993, Sulfide oxidation under oxygen limitation by a *Thiobacillus thioparus* isolated from a marine microbial mat: FEMS Microbiology Ecology, v. 13, p. 69–78, doi: 10.1016/0168-6496(93)90042-6.

Vargas, A., and Strohl, W.R., 1985, Utilization of nitrate by *Beggiatoa alba*: Archives of Microbiology, v. 142, p. 279–284.

Zopfi, J., Kjær, T., Nielsen, L.P., and Jørgensen, B.B., 2001, Ecology of *Thioploca* spp.: nitrate and sulfur storage in relation to chemical microgradients and influence of *Thioploca* spp. on the sedimentary nitrogen cycle: Applied and Environmental Microbiology, v. 67, p. 5530–5537, doi: 10.1128/AEM.67.12.5530-5537.2001.

Zumft, W.G., 1992, The denitrifying prokaryotes, *in* Balows, A., Truper, H.G., Dworkin, M., Harder, W., and Schleifer, K.H., eds., The prokaryotes, second edition: Berlin, Springer-Verlag, p. 554–582.

Manuscript Accepted by the Society March 28, 2004

Geological Society of America
Special Paper 379
2004

Formation and degradation of seafloor hydrothermal sulfide deposits

Katrina J. Edwards*
Geomicrobiology Group, Department of Marine Chemistry & Geochemistry, Woods Hole Oceanographic Institution, Woods Hole, Massachusetts 02536, USA

ABSTRACT

Sulfide weathering in the environment plays a critical role in balancing the global biogeochemical sulfur cycle. The rates, pathways, and role(s) of microorganisms in the oxidative transformation of sulfide minerals have been studied in detail for over a century. However, nearly all studies to date have focused on terrestrial environments, specifically on regionally restricted massive sulfide deposits, and on corresponding weathering at low pH. This attention has been warranted because the low-pH conditions caused by weathering in terrestrial environments often result in serious environmental damage due to acid mine drainage. However, from a global perspective, the weathering of massive sulfides in well-buffered, near-neutral pH conditions below Earth's ocean likely plays a larger role in controlling the oxidative portion of the biogeochemical sulfur cycle. Only recently have studies begun to elucidate these processes and describe some of the microbiological communities that mediate them. In the past decade, novel, diverse groups of sulfur- and iron-oxidizing microorganisms have been cultured in the laboratory and studied in the field at deep-sea hydrothermal habitats. Their role in forming a trophic base for a deep-sea food web has been suggested and may rival what is supported by reduced chemical species within hydrothermal fluids. Furthermore, in the carbon-poor deep ocean, where hydrothermal activity is geologically ephemeral, it is becoming recognized that sulfide minerals may serve as a long-term, stable source of electrons to support chemosynthesis. This chapter briefly reviews formation of massive sulfide at the seafloor, synthesizes and reviews recent studies concerning the biological degradation of these deposits, and presents new data concerning the effect neutrophilic iron-oxidizing bacteria have on the kinetics of metal sulfide dissolution.

Keywords: iron oxidation, bacteria, chemosynthesis, weathering, mid-ocean ridge.

INTRODUCTION

The formation and degradation of sulfide minerals exert fundamental control on the biogeochemical sulfur cycle. At present, most sulfides on Earth are present in the form of disseminated sedimentary sulfides, either in consolidated rocks such as shales in the terrestrial environment or in coastal marine sediments. Modern massive sulfide deposits are also prevalent at mid-ocean ridge (MOR) axes, and ancient deposits are recognized on all continents. The mechanisms for sulfide formation in the marine environment, globally the most significant source of metal sulfides, have been the subject of considerable study within coastal and hydrothermal systems. In contrast, sulfide weathering, although ubiquitous in both terrestrial and marine environments, has been studied to a far

*katrina@whoi.edu.

Edwards, K.J., 2004, Formation and degradation of seafloor hydrothermal sulfide deposits, 2004, *in* Amend, J.P., Edwards, K.J., and Lyons, T.W., eds., Sulfur Biogeochemistry—Past and Present: Geological Society of America Special Paper 379, p. 83–96. For permission to copy, contact editing@geosociety.org.

greater extent in terrestrial settings, mainly within the context of the environmental problem known as acid mine drainage (AMD).

Sulfide weathering and AMD formation in terrestrial rivers and streams have been recognized for centuries. Weathering of massive sulfide deposits does occur and always has occurred without human intervention. In fact, acidic runoff was used as a diagnostic indicator of metal-rich sulfide deposits by humans probably as far back as the Bronze Age (Jenkins, 1995). Oftentimes, however, a significant amount of weathering occurs prior to terrestrial deposition, resulting in a massive iron oxide "cap" (discussed further, below). These observations, along with numerous similar observations made on modern seafloor deposits, confirm that as with terrestrial sulfides, seafloor sulfide deposits undergo oxidative alteration resulting in the formation of iron-oxyhydroxide minerals that are often preserved in the rock record. However, the nature of seafloor weathering of sulfide deposits has not been studied in as great detail as has terrestrial sulfide weathering, in large part due to the relatively recent discovery of seafloor hydrothermal systems, and because these systems are poorly accessible by comparison to continental sulfides.

For more than half a century, it has been recognized that sulfide weathering in terrestrial habitats supports unique chemosynthetic microorganisms that harness the energy from various redox reactions involved with the oxidation of reduced chemical species in sulfides. Principally, these include Fe^{2+} and various reduced S compounds (thiosulfate, elemental sulfur, etc.), coupled with O_2 reduction and autotrophic CO_2 reduction for growth and cellular biomass production. Some of these organisms and these processes are discussed in detail by Schippers (this volume). Indeed, chemosynthetic Fe- and S-oxidizing chemoautotrophs fundamentally control the kinetics, pathways, and end products of sulfide oxidation in terrestrial habitats. Similarly, submarine hydrothermal environments, where sulfide deposits are formed, also support unique biological communities based largely on the primary biomass generated in situ by chemosynthetic microorganisms (Jannasch, 1985; Karl, 1995). By and large, however, scientists have principally focused on the contributions of syntrophic microbes and other free-living autotrophs that obtain their energy from dissolved inorganic compounds in vent fluids, such as H_2S, CH_4, and H_2 (e.g., de Angelis et al., 1993; Karl, 1995; Jannasch, 1985; Winn et al., 1995). Aqueous reduced chemical species in vent fluids are the most readily harnessed source of energy for chemosynthetic microorganisms. However, similar to the utilization of solid sulfides by microorganisms on continents, minerals in the deep sea may also be considered potential energy reservoirs that can be harnessed for growth. Microbial weathering of, and growth from, seafloor sulfides has recently been the subject of several independent studies. The principal purpose of this chapter is to review and synthesize recent studies conducted in my laboratory and others concerning microbial utilization and weathering of seafloor sulfides. Particular attention is given to a group of microorganisms long hypothesized, but only recently demonstrated, to be involved in seafloor weathering processes: the neutrophilic iron-oxidizing bacteria.

MASSIVE SULFIDE FORMATION AT SEAFLOOR HYDROTHERMAL VENTS

The physical, chemical, and rheological controls on massive sulfide deposits have been the subjects of a number of recent reviews, including an entire *Reviews in Economic Geology* volume (Barrie and Hannington, 1999a) and several chapters in an American Geophysical Union Monograph (Humphris et al., 1995). Most of what is briefly summarized here draws from these sources.

Volcanic-hosted massive sulfide (VMS) deposits vary widely, both in terms of tectonic setting and chemical composition. The unifying feature is that most modern and ancient massive sulfide deposits form as the result of venting of hot, metal-containing solutions at the seafloor. Additionally, most VMS deposits have at least some component of mafic volcanic rock within the host stratigraphic succession, implying that heat from the upper mantle is fundamentally linked with mineralization (Hannington et al., 1995). VMS deposits are stratiform sulfide mineral accumulations that precipitate at or below the seafloor. Often, they occur within volcanic and sedimentary stratigraphic successions and are generally coeval with the volcanic rocks (Hannington et al., 1995). Variations in VMS deposit chemistry can be attributed in part to variations in the host-rock or sub-adjacent igneous rock. Hence, classification schemes have been devised principally based on host-rock composition, such as described by Barrie and Hannington (1999b). This type of scheme classifies VMS deposits based on the mafic, felsic, siliciclastic, or bi-modal composition of the host rock.

Though classification of VMS deposits usually emphasizes pre-alteration composition of host-rocks, other factors can contribute to both the chemical characteristics and physical growth of sulfides at the seafloor. For example, it is recognized that seawater chemistry, in particular the presence or absence of oxygen, plays a role in some of the observed chemical variation in VMS deposits and certainly plays a role in the degree of postdepositional alteration/preservation (Eastoe and Gustin, 1996). Additionally, it is becoming increasingly apparent that biological activity plays a role—in both a passive and active manner—in mineralization processes. Passive mineralization generally occurs when biological materials simply serve as nucleation sites for precipitation and do not directly exert influence on the process. Active mineralization has been suggested when mineral nucleation serves as a biological detoxification mechanism, or when bacteria appear to specifically concentrate and nucleate metals in a manner that is inconsistent with what would be expected abiotically (Juniper et al., 1992; Zierenberg and Schiffman, 1990). The importance of biological contributions to sulfide mineralization has not yet been fully evaluated, but it appears as though colonization of newly formed chimney structures may be an important factor in early growth stages of sulfide structures at the seafloor (Juniper et al., 1992; Tunnicliffe and Juniper, 1990).

The factors that contribute to large variations in sulfide deposit size, both in modern and ancient seafloor environments, also exert influence on postdepositional processes. In some cases, the size of a particular deposit reflects the length of time active venting occurred at a particular locale, for example, at the Trans-Atlantic GeoTraverse (TAG), where hydrothermal activity has occurred intermittently over approximately the past 50,000 yr (Lalou et al., 1993; Lalou et al., 1990). This has resulted in a very large body of sulfide (~4 million tonnes; Hannington et al., 1998) compared to most modern deposits (Hannington et al., 1995). However, most modern marine deposits, which are on the order of tens of tonnes, are significantly smaller than the land-based deposits that represent ancient hydrothermal systems. It is difficult, however, to directly compare the factors that contribute to the size variation among modern versus ancient deposits, because most ancient hydrothermal deposits were not formed at mid-ocean ridges such as they are today, but rather formed in settings such as back-arc basins or in rifted continental margins. Hannington et al. (1995) suggests that the significance of the many large sulfide deposits in the geologic past may relate to fundamental differences in the magnitude of hydrothermal systems in the past compared to those operating at mid-ocean ridges today.

SULFIDE WEATHERING AT THE SEAFLOOR

The interaction between oxygenated seawater and hydrothermally deposited sulfide minerals on the seafloor results in oxidative weathering. These reaction pathways and products are discussed in detail elsewhere in this volume (Schippers). The terminal end product of oxidative weathering of the sulfur moiety (S) in sulfide minerals is sulfate, which is largely soluble in seawater at low temperature. In contrast, the end product of oxidative weathering of metals in sulfide minerals, most notably iron (Fe), commonly produces insoluble oxy-hydroxide minerals such as ferrihydrite, goethite, and hematite. These accumulate as crusts or caps on sulfide deposits at the seafloor that are sometimes referred to as "gossans" (Herzig and Hannington, 1995, and references therein).

Although microbiologists have been studying the microbial ecology of hydrothermal vent habitats since their discovery more than of quarter of a century ago, it has only been in recent years that attention has been paid to microbiological communities that may participate in weathering. Despite this only very recent interest, significant findings have been made concerning sulfide weathering at the seafloor by S- and Fe-oxidizing microorganisms based on theoretical, laboratory, and field studies.

Theoretical Considerations: Energetics of Microbial Sulfide Oxidation

The oxidation of Fe and S in sulfide minerals during weathering involves a large change in free energy, which, if harnessed by microorganisms, could be used for metabolic growth. This potential has recently been explored from a theoretical perspective by McCollom (2000). McCollom (2000) calculated the amount of energy that might be available from the oxidation of metallic sulfide minerals that could be produced in a hypothetical seafloor hydrothermal plume, as compared with the energy that would be available from oxidation reduced chemical species that would be predicted in coeval hydrothermal vent fluids (calculations used measured vent fluid chemistry from the East Pacific Rise, 21°N; Table 1). While this is rather different from oxidation of massive sulfide deposits at or below the seafloor, the reaction energetics, given similar conditions (particularly the availability of oxygen), should apply in principal. The results of this study indicate that a substantial amount of energy is available in sulfide minerals to support chemolithoautotrophic microbial growth, if it can be harnessed. Furthermore, these calculations indicate that sulfide minerals represent a far larger reservoir of potential energy than co-occurring aqueous chemical species in plumes (Table 1). Most of this energy is available from the oxidation of either elemental sulfur or the sulfide moiety within sulfide minerals (>98%; Table 1). This study offers the first support of the theory that, on a global basis, sulfide minerals at the seafloor could represent a vast potential "food source" (electrons) to support chemosynthetic microbial growth.

Laboratory and Field Studies

Microbial Sulfur Oxidation

The use of solid seafloor minerals by S-oxidizing microorganisms was first reported in 1993 by researchers studying indigenous sulfide-colonizing microbial populations from Mid-Atlantic Ridge hydrothermal vent sites (Wirsen et al., 1993). Wirsen et al. (1993) sought to explain the microbial processes responsible for the occurrence of vast populations of eyeless shrimp unique to Mid-Atlantic Ridge sulfide deposits. Mid-Atlantic Ridge shrimp are grazers that feed in part by scraping surfaces of sulfide minerals (Polz et al., 1998; Van Dover et al., 1988). A series of experiments were performed that included the examination of $^{14}CO_2$ fixation and enzymatic (RuBisCo) activity, both in the lab and field, and enrichment and isolations of various S-oxidizing strains capable of growth in the presence of elemental sulfur and sulfide minerals. Results of these experiments led Wirsen et al. (1993) to conclude that the transformation of sulfide minerals to microbial

TABLE 1. ESTIMATES OF THE METABOLIC ENERGY AVAILABLE FROM VARIOUS CHEMOLITHOAUTOTROPHIC REACTIONS IN A SUBMARINE HYDROTHERMAL PLUME

Chemolithoautotrophic energy source	Available energy (cal/kg vent fluid)†
Dissolved substrates	
Methanotrophy	13.
Sulfide oxidation	<0.0001
Fe oxidation	0.001
Mn oxidation	0.6
Hydrogen oxidation (Knallgas reaction)	161.
Solid substrates	
Sulfur (S^0) oxidation	610.
Sulfide mineral oxidation†	557.

Note: (Data fom McCollom, 2000).
†Total of pyrite, pyrrhotite, chalcopyrite, and sphalerite. >98% of energy is from oxidation of the sulfur moiety, with remainder from Fe^{2+} oxidation.

biomass at the Mid-Atlantic Ridge was based on lithoautotrophic oxidation processes. Further studies of sulfide supported growth of S-oxidizing bacteria conducted by Eberhard et al. (1995) found that rates of chemosynthesis were dependent on both the type of sulfide mineral available and the S-oxidizing strain. In this study, they found that mixed polymetal sulfides, such as those rich in chalcopyrite ($CuFeS_2$), supported higher activities than sphalerite (ZnS), galena (PbS), or chalcocite (CuS_2) supported alone (Eberhard et al., 1995). These studies provided the first definitive laboratory and field evidence to support the hypothesis that massive sulfide deposits at seafloor hydrothermal vents could be a long-term source of electrons for chemosynthetic production of biomass in the deep sea. They also underscored the importance of biological activity for deep-sea weathering of sulfides.

It is important to recognize that in the case above, with the S-oxidizing autotrophs and shrimp, it is not only the S-oxidizing prokaryotes that play a role in sulfide weathering via chemical transformations, but the grazing shrimp also play an important role in weathering. The physical scraping that occurs during grazing removes both surface oxides and primary sulfide material. Surface oxide removal is a mechanism that may continually provide fresh, unweathered surfaces that can be acted on chemically and biologically. Oxides may otherwise accumulate on the surface until it is impervious to oxidants. Primary sulfide materials have also been observed to occur in high abundance in shrimp guts (Van Dover et al., 1988), and it has been suggested that chemosynthetic sulfide mineral oxidation by gut-hosted microbial communities could significantly contribute to the nutritional support of the shrimp (Polz et al., 1998). From a geochemical standpoint, this chemical and physical processing and mixing ultimately results in enhanced exposure of the sulfide to oxidants, which influences the overall rate and mechanism by which these minerals degrade in the deep sea.

Microbial Iron Oxidation

It has long been speculated that in addition to microbial S-oxidizing bacteria, Fe-oxidizing bacteria such as *Gallionella ferruginea* and *Acidithiobacillus ferrooxidans* (formerly *Thiobacillus ferrooxidans*; Kelly and Wood, 2000) may play a role in the formation of Fe oxide deposits at the seafloor. This is largely due to the frequent observation of Fe oxides associated with deep-sea weathering deposits consistent with the morphologies of biogenic Fe oxides (e.g., Alt, 1988; Juniper and Fouquet, 1988; Juniper and Tebo, 1995; Tunnicliffe and Fontaine, 1987; Wirsen et al., 1993). Most of these morphologically distinct oxides are filamentous, and they often coincide with amorphous silica deposition. Fe-silica deposits containing filamentous forms, of putative microbial origin, are also recognized in ancient hydrothermal deposits at the seafloor or in terrestrial deposits of ancient marine origin (e.g., Duhig et al., 1992; Hofmann and Farmer, 2000; Juniper and Fouquet, 1988; Juniper and Tebo, 1995; Reysenbach and Cady, 2001). These distinctive Fe oxide particles often closely resemble biogenic Fe oxides produced by the neutrophilic Fe-oxidizing bacteria *G. ferruginea* ("stalks"), *Leptothrix discophora* ("sheaths"), and the recently cultured strain PV-1 ("branching filaments") (Emerson, 2000).

In addition to these Fe oxide particles, specific pitting patterns on sulfides are thought to be the products of biocorrosion and have been used to infer the activity of Fe-oxidizing species at the seafloor (Verati et al., 1999). In this case, the actions of acidophilic Fe-oxidizing bacteria such as *A. ferrooxidans* are implied. Though the bulk seawater in the vicinity of weathering sulfide minerals in the deep sea is well buffered and generally close to neutral, it is possible that in restricted microenvironments acidic conditions prevail and could support the activity of acidophiles. See Schippers (this volume), for discussion of the appropriate pH range for growth of common acidophiles.

Historically, definitively implicating biological involvement in a process such as mineral dissolution or precipitation based solely on morphology of either degradation "footprints" or of extracellular minerals has been problematic. Even if an organism has been cultured from the environment and demonstrated to produce similar mineral forms or biocorrosion pits, simple environmental associations should not be considered acceptable forms of proof of process (Juniper and Tebo, 1995). As one example, for many years it was inferred that the presence of rod-shaped corrosion pits on sulfide minerals that were reacted in the presence of *A. ferrooxidans* implied a direct contact reaction between the mineral and cell surface (e.g., Bennett and Tributsch, 1978; Berry and Murr, 1978; Rodriguez-Leiva and Tributsch, 1988). It was recently shown, however, that the general size and shape of pits that can be produced during reaction with *A. ferrooxidans* can also be produced abiotically (Edwards et al., 2001) (Figure 1). Hence, mineral pitting does not require biological involvement. Similar caution must be used when interpreting putative biominerals in the environment (Juniper and Tebo, 1995).

Recent field and laboratory experiments may offer the first definitive evidence of what has long been suggested by morphological observations: that Fe-oxidizing bacteria actively participate in sulfide mineral weathering at seafloor hydrothermal vent sites. In July of 2000, Edwards et al. (2003a) conducted seafloor incubation studies with a variety of naturally occurring sulfide and sulfur minerals. After two months of reaction at ambient seafloor conditions, all surfaces were observed to be colonized by bacteria, but to very different degrees. A piece of natural chimney sulfide (mainly pyritic with some Cu) was extremely heavily colonized, particularly within pits and pores that occurred on the surface. Other minerals, such as chalcopyrite and sphalerite were not as heavily colonized (Edwards et al., 2003a). This result is consistent with the large variation in colonization of seafloor sulfides at vents; reports range from the observation of dense microbial mats (Wirsen et al., 1993) to the virtual absence of any surface community (Gebruk et al., 1993). Colonization densities on the sulfide minerals, which ranged from $\sim 7–50 \times 10^4$ cells/mm^2 was found to correlate with reactivity of the starting material to oxidizing chemical species, with the most "reactive" being the most heavily colonized (Edwards et al., 2003a). It was also observed that colonization densities were quite variable across

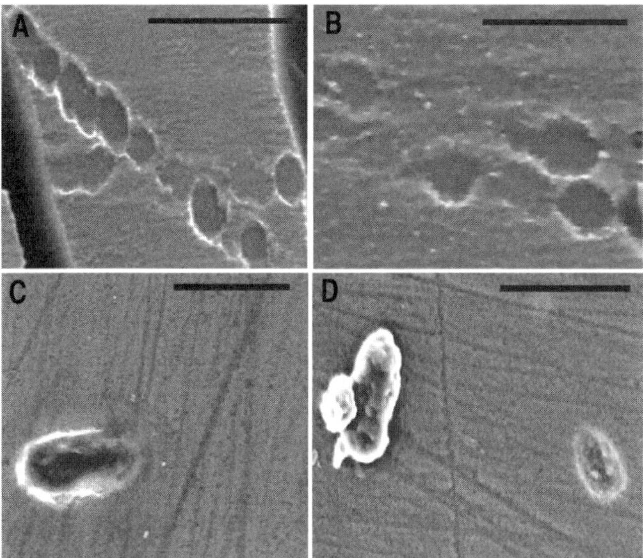

Figure 1. Examples of corrosion pits produced on pyrite surfaces during reaction with ferric chloride (A, B) and *Acidothiobacillus ferrooxidans* (C, D) (modified after Edwards et al., 2001). Scale bars in A, C, and D are 5 µm; scale bar in B is 1 µm. Small, elongated, bacillus-shaped pits developed on biologically and abiotically reacted surfaces.

surfaces of the minerals that were very heavily colonized, most notably, the chimney sulfide. Bacterial cells occurred in very dense colonies within pits and pores on the surface; these pits and pores occurred on the surface due to the nature of the starting material. Naturally occurring hydrothermal chimney sulfide is often characterized by very high porosity. This material was prepared for seafloor reaction by cutting and polishing, exposing many pores to the surface.

A second correlation of interest in this study was between these local sites of heavy colonization on the sulfide chimney and the sites that had the densest accumulations of Fe oxides (Edwards et al., 2003a). Not only did bacterial cells occur in highest density within pits, but Fe oxide accumulations were enormously abundant in the vicinity of pits as well (Fig. 2). Because the surfaces were free of alteration products and debris prior to seafloor reaction, the oxides must have formed in situ during seafloor reaction. The co-occurrence of bacterial cells and pits provides some evidence of biological activity in their formation. The oxides also have a characteristic appearance that is consistent with the morphologies of Fe oxides produced by known Fe-oxidizing bacteria (Fig. 2). Interestingly, the localization of Fe oxides within pits on the surface provides perhaps the strongest support for the involvement of neutrophilic Fe-oxidizing bacteria in their formation. Neutrophilic Fe-oxidizing bacteria must compete with the very rapid abiotic reaction kinetics between ferrous Fe (Fe^{2+}) and oxygen. They are best able to successfully accomplish this via residing at a physical-chemical redox gradient where some oxygen is present, but at a low enough level that they are able to effectively compete with abiotic Fe^{2+} oxidation. This is best visualized with laboratory gradient culture growth. The gradient culture growth method for the neutrophilic Fe-oxidizing bacteria was first developed by Kucera and Wolfe (1957) for the bacterium *G. ferruginea.* An example is shown in Figure 3. Reflecting on the seafloor reactions conducted by Edwards et al. (2003a), it is important to consider what type of microenvironments may have developed on the surface over a two-month period. When the surfaces were first submerged, they would have been flushed with large amounts of well-oxygenated, buffered seawater. Following initial colonization, it is likely that oxygen-utilizing microorganisms would modulate surface oxygen levels. All surface sites may not be depleted at equal rates or to equal levels; rather, it is likely that sites restricted from free advective and diffusive exchange with bulk seawater would most rapidly become, and remain, low in oxygen. Therefore, it is likely that pits and pores on the surface represent ideal sites for colonization by oxygen-sensitive physiological groups such as the Fe-oxidizing bacteria.

Further support for the presence and activity of neutrophilic Fe-oxidizing bacteria comes from the results of culture studies using deep-sea hydrothermal weathered minerals. Edwards et al. (2003b) initiated enrichment cultures using some of the incubation surfaces described above (Edwards et al., 2003a) as well as a variety of weathered materials collected from Middle Valley and the main Endeavour segment of Juan de Fuca, such as brecciated rubble and metalliferous sediments. Packed sediment columns (Fig. 4) were inoculated with these materials and incubated with artificial seawater medium (ASW) that was devoid of supplemental organic carbon for a period of about six months (Edwards et al., 2003b). Following this enrichment, gradient tubes were used to obtain axenic cultures by performing successive dilution series to extinction. Fe-oxidizing bacteria grow approximately one cm from the surface of gradient tubes in narrow bands that occur at the oxic-anoxic interface, defined using an oxygen microelectrode, picoammeter, and micromanipulator (Edwards et al., 2003b). These characteristics are consistent with what would be expected for microaerophilic growth at neutral-pH growth by lithoautotrophic Fe-oxidizers, and this suggestion was ultimately confirmed by measuring the rate of $H^{13}CO_2^-$ incorporation during culture growth (Edwards et al., 2003b). Surprisingly, two additional findings were also made that might not be considered "normal" and expected for neutrophilic Fe-oxidizing bacteria: all strains are psychrophilic, with optimal growth at 3–10 °C, and all strains are capable of growth anaerobically with nitrate as the terminal electron acceptor (Edwards et al., 2003b). These findings extend the likely range of habitats in which we may predict autotrophic Fe-oxidizing bacteria to occur at the seafloor. It is important to recognize that low-temperature weathering habitats prevail over the transient moderate- to high-temperature conditions characteristic of ephemeral hydrothermal vents. Long after active venting has stopped, the activities of microorganisms that are both capable of harnessing the potential energy within these sulfide minerals and are optimized to grow at very low, ambient seafloor temperature conditions that hover around 0 °C should prevail and perhaps dominate these formerly high-temperature systems.

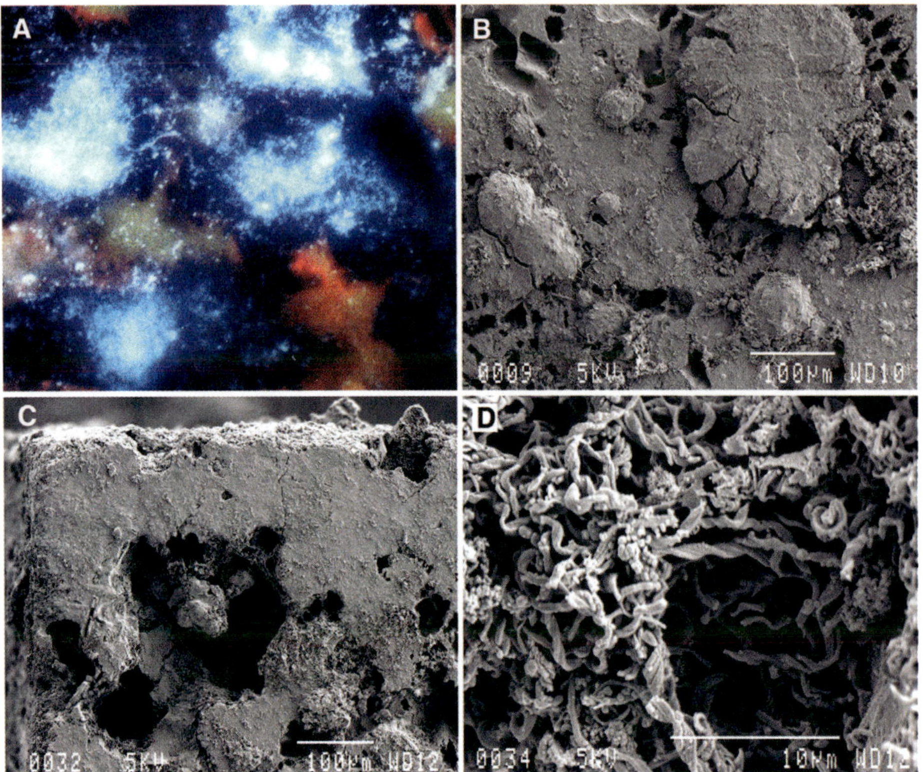

Figure 2. Colonization patterns and Fe oxide development on chimney sulfide surface during seafloor reaction (modified after Edwards et al., 2003a). (A) (~100 μm across): DAPI stained cells colonizing pits on chimney sulfide surfaces. Cells are bright blue dots and masses; edges of pits are deep blue and surround the cell clusters. (B) Large masses of Fe oxides (~200 μm across) on chimney sulfide surface, covering pits and pores. (C) Higher resolution image of Fe oxide development within pits on chimney sulfide surface. (D) Higher resolution of the particle morphology of Fe oxides within pits in C.

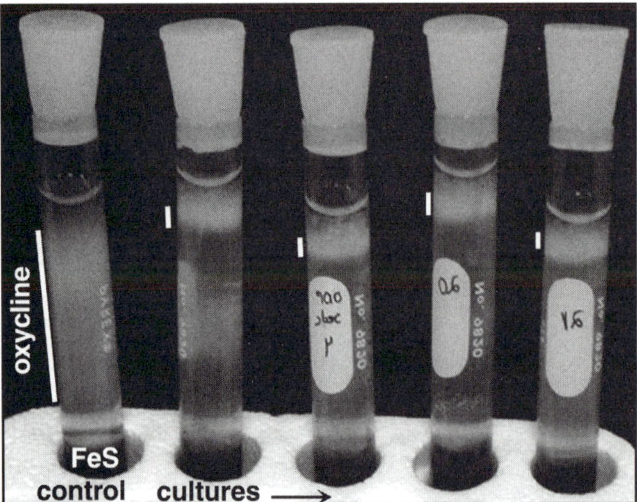

Figure 3. Image of gradient culture for the enrichment and growth of Fe-oxidizing bacteria after Kucera and Wolfe (1957). Far left control: bottom of tube contains synthetic FeS that is overlain by slush-agarose; hazy appearance indicates that oxygen is present through most of the tube. The distribution of the oxycline is shown with the white bar (left of tube). Remaining four tubes have been inoculated with Fe-oxidizing bacteria. Their presence has modulated the oxygenation of the tubes; Fe-oxidation occurs only near top of tubes in a discreet band that coincides with bacterial growth.

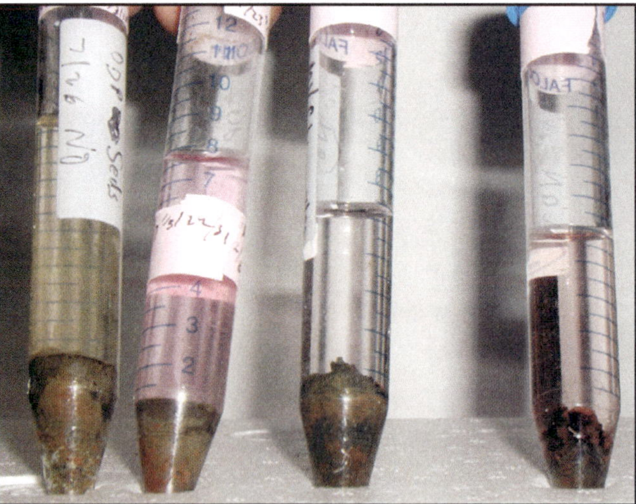

Figure 4. Packed sediment columns used for the enrichment of Fe-oxidizing bacteria from the Juan de Fuca Ridge axis. Red color is due to accumulation of oxy-hydroxide alteration minerals.

Figure 5 shows the phylogenetic relationships between some representatives of the Fe-oxidizing isolates reported by Edwards et al. (2003b) based on comparisons of 16S rRNA sequences. Many of the Fe-oxidizing isolates fall within the gamma-subdivision of the Proteobacterial lineage. Interestingly, this group of isolates bears no close phylogenetic relationship to any previously known Fe-oxidizing bacteria, such as the first described neutrophilic Fe-oxidizers, *G. ferruginea* and *L. discophora*, or to more recently cultured strains (PV-1, ES-1) (Fig. 5). Furthermore, they bear no close relationship to any previously known autotrophs (Fig. 5). Rather, they are phylogenetically most closely related to a group of widely recognized, successful marine heterotrophs: the *Marinobacter* and *Halomonas*. These bacteria are ubiquitous in the world's oceans, with known habitats ranging from the coastal ocean to the deep-sea. They are recognized for their ability to degrade hydrocarbons (Cohen, 2002), for halotolerance (Kaye and Baross, 2000), siderophore production (Martinez et al., 2000), and other physiological capacities, but not, as yet, for lithoautotrophic Fe-oxidation.

While the findings from the above physiological laboratory and field experimental studies suggest a role for neutrophilic Fe-oxidizing bacteria in sulfide weathering at the seafloor, a direct link between what is observed in the environment and studied in the laboratory is required. This is in part because it is commonly found

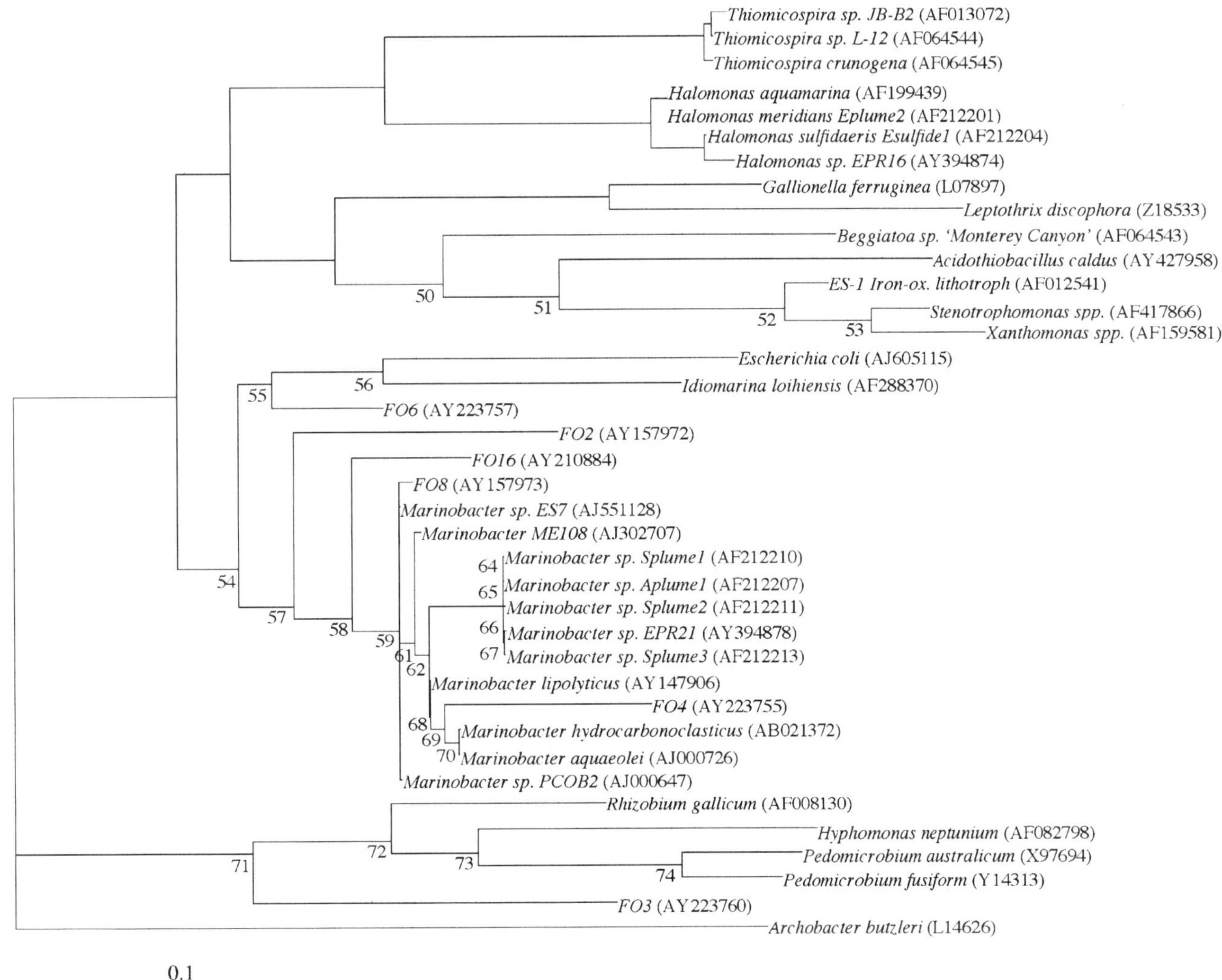

Figure 5. Phylogenetic relationships among Fe-oxidizing bacteria from Edwards et al. (2003b) (FO numbers), uncultured deep-sea strains from Rogers et al. (2003) (operational taxonomic unit numbers), and other relevant autotrophs/Fe-oxidizers. Tree was constructed using maximum likelihood (DAMBE software package, DNAml). Sequences were aligned using ClustalW and cropped to ca. 800bp before constructing topology. Scale bar represents percent change per nucleotide position. Bootstraps are percent values out of 1000 iterations; values 50 or greater are shown. Sequence accession numbers are shown in parentheses. More complete methods for tree construction are as described by Rogers et al. (2003).

that microorganisms that can be cultured in the laboratory may not be active in the environment. The standard methods for evaluating microbial communities without cultivation involve molecular techniques that allow genetic comparisons between different microorganisms; for example, by comparing 16S rRNA gene sequences (Olsen et al., 1986; Pace et al., 1986). A few studies have used molecular methods to examine the composition of microbial communities associated with deep-sea sulfides, but rarely for the purpose of examining populations associated with weathering. The purpose of most molecular studies on sulfide-associated microbial communities has been to explore the diversity of thermophilic chimney-hosted microorganisms and to address issues related to the upper temperature limits at which life can be supported (i.e., within high-temperature chimney walls). Hence, the information we can glean from these studies with respect to weathering is limited, though it may provide some context.

In one of the earliest molecular studies on sulfide-hosted microbial communities in the deep sea, Harmsen et al. (1997) used fluorescent in situ hybridizations (FISH) to examine the distribution of thermophilic subpopulations within the walls of a diffuse venting "beehive" sulfide structure. This study found roughly equal proportions of bacteria and archaea within the structure, and found that populations increased toward the exterior and top of the beehive (Harmsen et al., 1997). In a study by Takai et al. (2001), the archaeal community associated with a high-temperature (250 °C maximum T at venting orifice) chimney sulfide structure was examined using a 16S-based sequencing approach. These authors also found that the total population size (inferred based on the quantity of DNA) associated with the outer, weathered portion of the chimney was significantly higher than the interior (Takai et al., 2001). Most recently, a study by Schrenk et al. (2003) that combined a 16S-based community survey, FISH, and lipid analysis on a high-temperature (302 °C maximum T at venting orifice) sulfide chimney, showed that the highest cell density occurred just inside of the outer chimney wall. This study also found that archaea dominated the microbial population, and that bacterial numbers were higher toward the outside of the chimney. In sum, though a causal relationship between any of the organisms detected and weathering processes cannot be directly inferred from these studies, the following two points may be relevant: (1) both bacteria and archaea are present and apparently active on the exterior of diffuse and actively venting seafloor sulfide structures (i.e., where weathering reactions are occurring), and (2) the population of microorganisms appears to be highest where weathering reactions occur, on or near the exterior of sulfide structures.

In light of the above findings, one recent study sought to compare the community structure of surface-associated microbial communities on sulfides as a function of both temperature and degree of weathering. Rogers et al. (2003) performed restriction fragment length polymorphism (RFLP), 16S rDNA sequencing, and mineralogical analyses (X-ray diffraction) on the surface communities of five deep-sea sulfide samples that ranged from intact, venting chimneys to rubbly debris and sulfide sediments weathering at ambient seafloor conditions (~4 °C) (Fig. 6). For all materials except sediments, the outer surfaces of the sulfides were scraped and the resulting scrapings used for analyses. Interestingly, only bacteria were detected in this study and no archaea. It is possible that archaeal DNA was simply not successfully extracted. Alternatively, these results may reflect the fact that, in contrast to the above-described studies in which interior populations, and specifically archaea, were the principal targets for analyses (Harmsen et al., 1997; Schrenk et al., 2003; Takai et al., 2001), only surface-associated populations were ana-

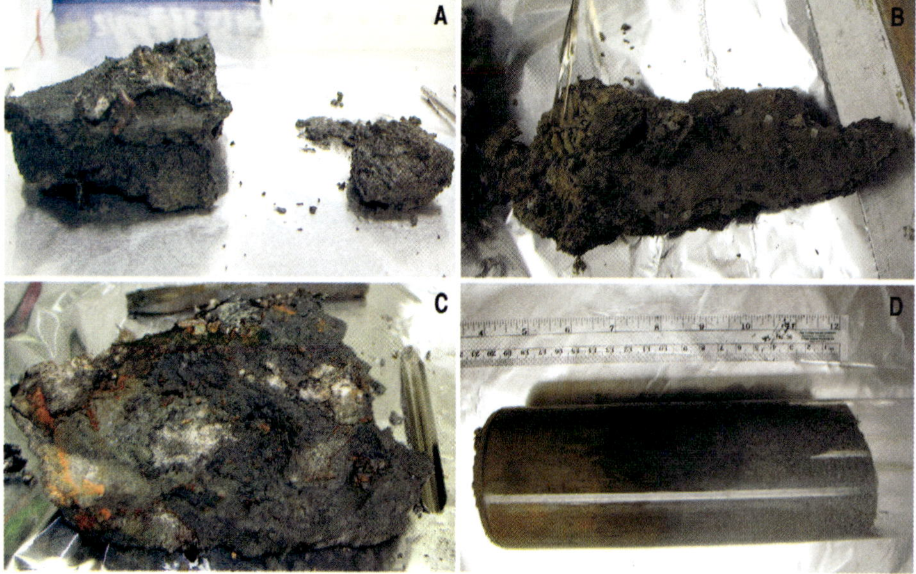

Figure 6. Examples of sulfide materials used for studies comparing the diversity of surface-colonizing microorganisms in deep-sea weathering deposits (modified after Rogers et al., 2003). (A) Highest temperature, least weathered material (~250 °C maximum T at venting orifice). (B) Intermediate temperature/weathered (diffuse venting; ~80 °C maximum T at venting orifice) sample. (C) Low-temperature (no venting; ~4 °C), moderately weathered sulfide. (D) Low-temperature (no venting; 4 °C), heavily weathered sulfide sediment.

lyzed. The apparent absence of archaea is also consistent with the results of FISH analyses conducted as part of the seafloor incubation studies discussed above, where the surfaces were exclusively colonized by bacteria (Edwards et al., 2003a). The RFLP analyses performed by Rogers et al. (2003) allowed assignment of phylogenetic groupings, or "operational taxonomic units" (OTUs), based on the banding patterns observed. Representatives of different OTUs were the targets for sequencing. This study revealed two trends. First, the highest temperature samples contained the highest numbers of OTUs (maximum diversity), while the lower temperature samples contained less diversity. The second trend involved the degree of weathering; samples with more weathering products present (oxides, clay minerals) harbored less diversity than fresher, unweathered samples. In fact, the most weathered, lowest temperature sample (weathered sulfide sediments; Fig. 6) harbored the least diversity within this sample set, remarkably containing only a single OTU. These data could suggest that the communities associated with deep-sea sulfide weathering are dependent on the source (niche) and the amount of energy available. In the higher temperature samples, energy is available from both the solid rock and the hydrothermal fluids. Because of the abundance and diversity of energy sources at this site, the diversity of the community supported is higher than at the colder, more oxidized sites. In low-temperature, highly oxidized samples, less biologically available energy sources would result in fewer niches and hence, lower diversity. Fewer niches result in more strenuous competition for resources, leading to the dominance of populations of microorganisms best suited to exploit the niche.

The phylogenetic relationships between the OTUs determined in Rogers et al. (2003) are shown along with the Fe-oxidizing isolates from Edwards et al. (2003b) in Figure 5. The OTUs shown include a broad diversity of Proteobacteria. The Mound sample, which was an actively venting spire, contained all OTUs (Fig. 5). The low-temperature sulfide sediment sample, which contained abundant oxides and clay minerals, contained only OTU1 (Fig. 5). OTU1 was present in all samples, and dominated several, indicating that it is likely an important constituent of surface communities at low to moderate temperatures. Interestingly, OTU1 falls within the *Marinobacter*/*Halomonas* grouping and is very closely related to some of the deep-sea Fe-oxidizing strains from Edwards et al. (2003b). This is perhaps surprising because it is often thought that the dominant or physiologically and/or ecologically important microorganisms can only rarely be cultured from the environment. However, the remarkable physiological versatility and success of the *Marinobacter*/*Halomonas* that has previously been recognized (Kaye and Baross, 2000) and that has now been augmented with the recognition of chemoautotrophic activity among some members (above) may account for this occurrence. Regardless, it seems highly likely that the taxonomic group represented by OTU1 includes a collection of previously unrecognized Fe-oxidizing bacteria. Moreover, these findings suggest that Fe-oxidizing bacteria are quite prevalent and active in the environment. And finally, an important role for Fe-oxidizing bacteria in sulfide mineral weathering—in the modern, and by analogy, in the past—has been established.

Remaining Questions and Future Directions

The studies discussed herein have only served to lay the groundwork for future studies in this field. Though deep-sea hydrothermal vents and sulfide deposits were discovered more than 25 years ago, the gap between what is known concerning sulfide weathering in terrestrial versus marine systems remains wide. Questions remaining include:

- How abundant are S- and Fe-oxidizing microorganisms associated with deep-sea sulfides?
- What is the ecological interaction between S- and Fe-oxidizers? Are there similarities between the ecological interactions between S- and Fe-oxidizers in the terrestrial and marine environments?
- What is the phylogenetic diversity among deep-sea, sulfide weathering S- and Fe-oxidizing microorganisms? For example, a thermophilic, anaerobic Fe-oxidizing archaeon has been cultured from Fe-bearing fluids at a hydrothermal seamount (Hafenbrandl et al., 1996). Could this or other archaea be involved in sulfide weathering in the deep sea?
- What is the functional diversity among deep-sea, sulfide weathering S- and Fe-oxidizing microorganisms? Very little is known about the biomolecules and pathways of S- and Fe-oxidation, but it is well recognized that these capabilities are distributed among many diverse lineages (Lane et al., 1992), suggesting either multiple evolutionary origins or remarkable plasticity and mobility among the Fe- and S-oxidase genetic elements. Deep-sea sulfide deposits may be the ideal natural laboratory for examining these relationships.
- Can we substantiate the theoretical work of McCollom (2000), which suggests that chemolithautotrophic microbial activity associated with sulfide weathering could play an important tropic role as a source of new carbon to the seafloor?
- Do sulfide weathering microbial communities persist in the sub-seafloor? What bioalteration signatures or biomarkers for S- and/or Fe-oxidizers might be developed and applied to ancient hydrothermal samples to determine when, if, and/or how microbial sulfide weathering occurred in the past?
- Finally, how do neutrophilic S- and/or Fe-oxidizing bacteria affect the kinetics and pathway of sulfide mineral weathering?

It is well established that in the terrestrial environment Fe-oxidizing bacteria fundamentally control the rates and mechanisms of sulfide weathering (see Schippers, this volume). It is this interaction that is thought to shape the ecological interactions between Fe- and S-oxidizing microorganisms. The following section of this chapter offers a preliminary and tantalizing glimpse at part of the answer to the last question posed above.

Preliminary Studies: Kinetics of Sulfide Weathering by Neutrophilic Fe-Oxidizing Bacteria

Although neutrophilic Fe- (and the better studied Mn-) oxidizing bacteria have been recognized in the environment for nearly two centuries (Ehrenberg, 1836, 1838) and have been recognized as playing a role in biocorrosion, particularly in stainless steel ennoblement (Dickinson et al., 1996; Dickinson and Lewandowski, 1996), the specific role that they play in the kinetics of solid material transformations has not been elucidated. This contrasts with their terrestrial counterparts, the acidophilic Fe-oxidizers, who have long been recognized as ultimately controlling the rate of sulfide mineral dissolution by controlling the rate of Fe^{2+} oxidation (Singer and Stumm, 1970). Under acidic conditions, however, Fe^{2+} oxidation is kinetically slow by comparison to neutral-pH oxidation (see Schippers, this volume). Even so, recent studies have shown that microbial Fe^{2+} oxidation at neutral pH is ~18% higher than abiotic oxidation (Neubauer et al., 2002). Furthermore, this study also showed that in a bioreactor system, microbial Fe-oxidation accounted for 62% of the total Fe oxidized in the system, demonstrating that Fe-oxidizers effectively outcompete abiotic processes for electrons (Neubauer et al., 2002).

In order to understand the effect that the activities of neutrophilic Fe-oxidizing bacteria have on sulfide mineral dissolution, dissolution experiments were performed using diagenetic and hydrothermal pyrite and a deep-sea Fe-oxidizing strain cultured from the Juan de Fuca Ridge (Edwards et al., 2003b).

Methods

Fe-oxidizing strain FO10 was grown anaerobically in a bicarbonate-buffered (pH 7.5, 2mM $NaHCO_3$) artificial seawater medium (ASW modified from Jannasch, 1985) lacking thiosulfate and tris buffer (Trizma base; see Edwards et al., 2003b). Anaerobic conditions were achieved through gently boiling and cooling the ASW under a nitrogen atmosphere. Resazurin was used as an oxygen indicator. The medium was then supplemented with 1mM $NaNO_3$ to serve as a terminal electron acceptor. The ASW medium was dispensed into 37 mL serum bottles containing 130 mg of one of the following Fe-bearing substrates: natural hydrothermal massive sulfide ("chimney"; mainly pyritic, with some Cu; as described in Edwards et al., 2003a) and whole diagenetic pyrite cubes ("pyrite"; Ward's Natural Science, New York). Minerals were ground and sieved, and the 150–300 µm size fraction was autoclaved, etched with 1N HCl and washed in ethanol. Microbiological experiments were inoculated by syringe, the headspace purged with nitrogen. Corresponding controls were identically prepared but without microbial inoculum. Serum bottles were incubated in the dark at room temperature (~25 °C), unshaken.

One mL of sample was taken by syringe from the biological and abiological vials at the start of the experiment and thereafter on days 3, 7, 14, 18, 21, 25, 29, 32, 44, and 57. Iron was determined using the FerroZine method (Stookey, 1970) as modified by Viollier et al. (2000). Briefly, Fe^{2+} was determined using 1 mL of sample added directly to 100 µl of 10 mM FerroZine (100 mM ammonium acetate buffer). Absorbance was measured at 562 nm. Fe^{3+} was measured by adding 0.5 mL of the Fe^{2+}-FerroZine solution to 150 µl of 1.4 M hydroxylamine hydrochloride ($H_2NOH*HCl$) in 2 M analytical grade HCl. This solution was incubated at room temperature for 10 min before adding 50 µl of 10 mM ammonium acetate buffer (pH 9.5) and measuring the absorbance at 562 nm. Standards were made from an anaerobic $FeCl_2$ solution, a $FeCl_3$ solution, and a ferrous iron standard (200 ppm) in sulfuric acid.

Results

Figure 7 and Table 2 show the concentration of Fe (µM) detected in solution over the course of the 57-day experiment, for the two microbial experiments and their corresponding abiotic controls. Only Fe^{2+} was detected in solution. Data for one analysis is not shown (day 7, pyrite control) because the measurement was five times higher than any other Fe measurement in the series; contamination of the syringe by Fe is suspected. All data, including the errant Fe measurement, are shown in Table 2. For both the pyrite and chimney experiments, approximately five times more Fe was released to solution in the microbial experiments than in the corresponding controls. Fe levels reached maximal levels in less than 10 days for controls and at ~15 days in the presence of strain FO10.

Discussion and Conclusions

The data shown in Figure 7 and Table 2 demonstrate that the presence of Fe-oxidizing strain FO10 resulted in the release of more Fe to solution compared to controls. These data

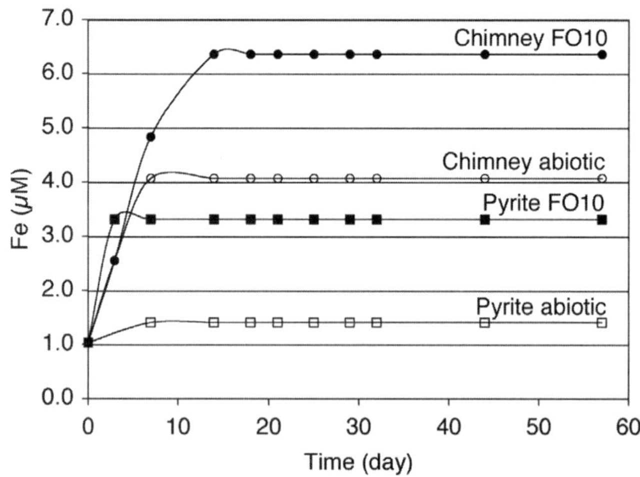

Figure 7. Plot of the Fe concentration in solution over time for abiotic and microbial sulfide dissolution experiments. Microbial experiments utilized Fe-oxidizing strain FO10. "Fe (µM)" represents the concentration of Fe^{2+} detected in solution; Fe^{3+} was below detection for all time points. Data for the microbiological experiments is plotted with filled symbols; data for abiotic runs is plotted with open symbols. Results for the "chimney" pyrite experiments are shown with circles and the diagenetic "pyrite" data shown with squares.

TABLE 2. DATA FROM SULFIDE DISSOLUTION EXPERIMENTS

	Chimney				Pyrite			
	FO10		Control		FO10		Control	
Day	mol ($x10^{-9}$)	μM	mol ($x10^{-9}$)	μM	mol $x10^{-9}$)	μM	mol ($x10^{-9}$)	μM
0	2.36	1.03	2.36	1.03	2.36	1.02	2.36	1.02
3	5.87	2.55	5.87	2.55	7.62	3.31	8.49	3.69
7	11.1	4.83	9.37	4.07			3.24	1.04
14	14.6	6.36						

Note: Data is not shown for measurements that are the same as previous analyses within our detection limits (when concentration of Fe plateaus). All data correspond to $Fe^{2+}_{(aq)}$ measurements (see text).

should be considered preliminary because the use of Fe as a proxy for weathering at neutral pH is problematic because of the insolubility of Fe in ferric oxidation state. These experiments were conducted anaerobically because experiments have shown that metal sulfides cannot be oxidized with nitrate at neutral pH (Bonnissel-Gissinger et al., 1998; Schippers and Jorgensen, 2001, 2002). Hence, the low levels of Fe^{2+} detected in the controls should reflect simple leaching and not oxidative dissolution. In the case of the microbial experiments, however, oxidation is taking place, resulting in the precipitation of ferric oxy-hydroxide minerals ("2-line ferrihydrite"; Edwards et al., 2003a). The Fe^{2+} measured in solution therefore must underestimate the total Fe mobilized from the sulfide. It is difficult to accurately assess the amount of Fe oxide produced, but Fe minerals appear abundant and are readily observed via light or electron microscopy (e.g., Fig. 8).

The fact that Fe^{2+} is detected in solution may suggest that either the rate of Fe release from the sulfide is faster than the oxidation rate, or that there is simply more Fe available than these cultures are able to process. Their growth rates are slow and growth yields are low compared to many other microorganisms (Edwards et al., 2003b), which could suggest that the latter interpretation is most likely the case here. An interesting observation, however, is that while their cell numbers grow only slowly and never reach very high concentrations, the bacteria do produce prolific quantities of exopolymeric material (EPS). This can be seen in Figure 8, where the EPS appears significantly more abundant than the cells themselves. EPS production is an energetically expensive proposition for microorganisms; chemoautotrophic Fe-oxidizing bacteria, particularly when growing anaerobically, would have to expend a significant proportion of their total energy on polymer production to achieve the volume observed (Fig. 8). The emphasis on polymer production over cell production may account for the slow growth and low cellular yields. It seems reasonable to presume that there is some underlying physiological advantage conferred on these bacteria as a consequence of producing this abundant EPS. One might speculate that the increase in release of Fe from the sulfides, the presence of Fe in the ferrous oxidation state, and EPS production by these bacteria are linked in some yet undescribed manner. It would be of significant interest to investigate the EPS for the presence of specific biomolecules that act on these minerals to promote dissolution.

Though the mechanism remains unknown, these data provide first evidence that neutrophilic Fe-oxidizing bacteria actively promote the dissolution of sulfide minerals. This may have significant ramifications for the rate and pathway of sulfide dissolution in the world's oceans and clearly needs further focused study, both in the laboratory and in the field, for a more accurate, quantitative understanding of the process.

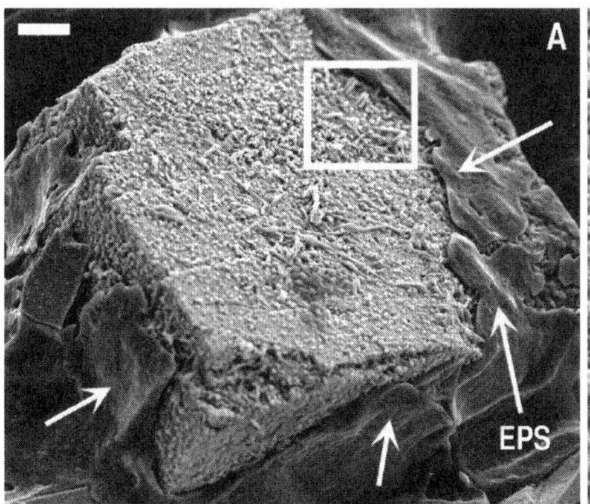

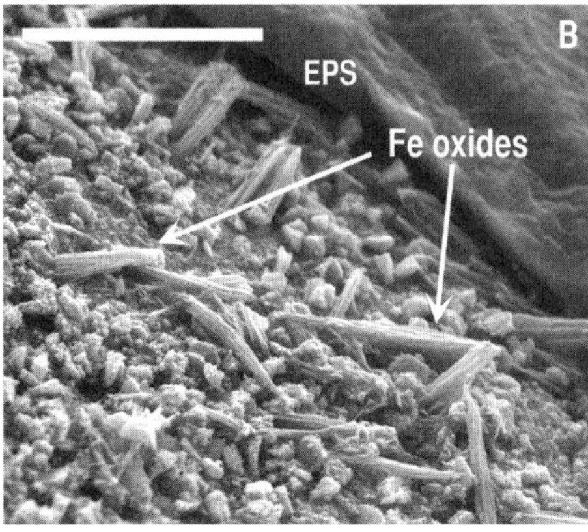

Figure 8. Scanning electron micrograph of pyrite surface reacted with FO10. Scale bar is 10 μm in both images. B is higher magnification of the region outlined with a box in A. A shows prolific exopolymeric material draping the corners of the pyrite crystal following reaction with strain FO10. B shows higher magnification of this region and the occurrence of Fe-oxides in proximity to the surface of the pyrite crystal.

SEAFLOOR WEATHERING, PAST AND PRESENT

The degree to which hydrothermally precipitated sulfides are weathered is fundamentally linked to depositional characteristics. Size, depositional setting (e.g., MOR versus back-arc, etc.), degree of above-ground versus subsurface deposition, sedimentation rate, and many other characteristics ultimately control the degree to which fluids, oxidants, and microorganisms can access and act on seafloor massive sulfide deposits. In the terrestrial environment, the weathering of sulfide is greatly enhanced by human mining activities, because the process of blasting, drilling, and grinding sulfide greatly increases the exposure of surfaces to the necessary fluids, oxidants, and microorganisms. At the seafloor, hydrothermal circulation and the activities of macro benthic communities, such as the grazing shrimp discussed above, can facilitate weathering both above and below the seafloor. Erosion, mass wasting, and resedimentation of weathered sulfide are abundantly evident at the seafloor. For example, at the TAG hydrothermal field, the largest modern seafloor deposit, bright red metalliferous sediment, derived from mass wasting of the mound, flanks the deposit and forms an apron that extends up to 50 m (Hannington et al., 1995).

The studies discussed above on chemosynthetic microbial growth supported by seafloor sulfides establish important, quantitative linkages between oxidative weathering and biomass production. In the modern seafloor environment, it is likely that at any accessible location where conditions are suitable, microorganisms harness the electrons from weathering reactions for growth—at and below the seafloor. McCollom (2000) estimated that the oxidative weathering of plume particles alone, if harnessed for growth by chemosynthetic microorganisms, could produce ~10^{12} g dry wt biomass carbon per year. While this is a small fraction of global carbon production, locally it may be an important source of new carbon to the oligiotrophic seafloor environment. Production of biomass carbon and poorly crystallized oxy-hydroxide minerals ("2-line ferrihydrite") by Fe-oxidizers, for example, coupled with the local depletion of oxygen could enrich Fe-reducing heterotrophic bacteria, at or below the seafloor. Local production of biomass via chemosynthesis may serve as an important trophic food-web base in sub-seafloor bare-rock systems.

In the geologic past, it is likely that microbial populations participated in sulfide weathering processes for as long as S and Fe metabolisms have evolved and the conditions have been suitable. The question of when these metabolisms arose is an open question. Most of the debate concerning early Earth microbial metabolisms has revolved around S and Fe-reduction pathways and the question of which is more primordial (e.g., Pace, 1991; Vargas et al., 1998), while the antiquity of microbial S- and Fe-oxidation has received little attention. At the seafloor (i.e., below the euphotic zone), however, the evidence discussed above indicates that these metabolisms are dependent on the presence of either oxygen or nitrate. The exact timing of delivery of these oxidants to the ocean floor remains uncertain. A growing body of evidence suggests that the oceans became stratified with respect to oxygen and sulfate in the mid-Proterozoic (~1.5 Ga) and were probably not fully oxidized until about the Neoproterozoic (~1 Ga) (Canfield, 1998; Shen et al., 2003). During this time, the capacity for microbial S-oxidation is believed to have evolved (Canfield and Teske, 1996), so as soon as oxygen was present in the lower ocean, microbiological participation in weathering could have begun. It is likely that nitrate was available much earlier in the oceans, deriving from an atmospheric source (Yung and McElroy, 1979; Kasting and Walker, 1981; Kasting, 1990). Hence, it is possible that microbial Fe-oxidation could have evolved before the ocean floors became oxygenated and participated in seafloor weathering.

Our ability to comprehensively understand the history, timing, extent, and evolution of microbial weathering of seafloor sulfides is hampered by a lack of laboratory and field studies of the process. The likely antiquity of microbial seafloor weathering, however, indicates that a better mechanistic and ecologic understanding of this process should provide insight to biogeochemical processes on the ancient seafloor, and the co-evolution of Earth, life, and the global weathering cycles.

ACKNOWLEDGMENTS

I thank the present and former members of the Geomicrobiology Group at the Woods Hole Oceanographic Institution. In particular, I appreciate the work done by Dan Rogers and the helpful discussions I have had with him and W. Bach regarding these studies. G. Druschel and an anonymous reviewer provided comments that greatly improved this manuscript, and I thank Sheila Clifford for her careful editing. This work has been supported by grants from the National Science Foundation Division of Ocean Sciences (OCE-0096992 and OCE-0241791). This is WHOI contribution number 11058.

REFERENCES CITED

Alt, J.C., 1988, Hydrothermal oxide and nontronite deposits on seamounts in the eastern Pacific: Marine Geology, v. 81, p. 227–239, doi: 10.1016/0025-3227(88)90029-1.

Barrie, C.T., and Hannington, M.D., editors, 1999a, Volcanic-associated massive sulfide deposits: Processes and examples in modern and ancient settings: Littleton, Colorado, Society of Economic Geologists, Inc., Reviews in Economic Geology, v. 8, 408 p.

Barrie, C.T., and Hannington, M.D., 1999b, Classification of volcanic-assisted massive sulfide deposits based on host-rock composition, *in* Barrie, C.T., and Hannington, M.D., eds., Volcanic-associated massive sulfide deposits: Processes and examples in modern and ancient settings: Littleton, Colorado, Society of Economic Geologists, Inc., Reviews in Economic Geology, v. 8, p. 1–11.

Bennett, J.C., and Tributsch, H., 1978, Bacterial leaching patterns on pyrite crystal surfaces: Journal of Bacteriology, v. 134, p. 310–326.

Berry, V.K., and Murr, L.E., 1978, Direct observations of bacteria and quantitative studies of their catalytic role in the leaching of low-grade copper bearing waste, *in* Murr, A.E., and Torma, A.E., eds., Metallurgical applications of biological leaching and related phenomena: New York, Academic Press, p. 526.

Bonnissel-Gissinger, P., Alnot, M., Ehrhardt, J.-J., and Behra, P., 1998, Surface oxidation of pyrite as a function of pH: Environmental Science & Technology, v. 32, p. 2839–2845, doi: 10.1021/ES980213C.

Canfield, D., 1998, A new model for Proterozoic ocean chemistry: Nature, v. 396, p. 450–453, doi: 10.1038/24839.

Canfield, D.E., and Teske, A., 1996, Late Proterozoic rise in atmospheric oxygen concentration inferred from phylogenetic and sulphur-isotope studies: Nature, v. 382, p. 127–132, doi: 10.1038/382127A0.

Cohen, Y., 2002, Bioremediation of oil by marine microbial mats: International Microbiology, v. 5, p. 189–193, doi: 10.1007/S10123-002-0089-5.

de Angelis, M.A., Lilley, M.D., Olson, E.J., and Baross, J.A., 1993, Methane oxidation in deep-sea hydrothermal plumes of the Endeavour Segment of the Juan de Fuca Ridge: Deep-Sea Research I, v. 40, p. 1169–1186.

Dickinson, W.H., and Lewandowski, Z., 1996, Manganese biofouling and the corrosion behavior of stainless steel: Biofouling, v. 10, p. 79–93.

Dickinson, W.H., Caccovo, F.J., and Lewandowski, Z., 1996, The ennoblement of stainless steel by manganic oxide biofouling: Corrosion Science, v. 38, p. 1407–1422, doi: 10.1016/0010-938X(96)00031-5.

Duhig, N., Davidson, G., and Stolz, J., 1992, Microbial involvement in the formation of Cambian sea-floor silica-iron oxide deposits, Australia: Geology, v. 20, p. 511–514, doi: 10.1130/0091-7613(1992)0202.3.CO;2.

Eastoe, C.J., and Gustin, M.M., 1996, Volcanogenic massive sulfide deposits and anoxia in the Phanerozoic oceans: Ore Geology Reviews, v. 10, p. 179–197, doi: 10.1016/0169-1368(95)00022-4.

Eberhard, C., Wirsen, C.O., and Jannasch, H.W., 1995, Oxidation of polymetal sulfides by chemolithoautotrophic bacteria from deep-sea hydrothermal vents: Geomicrobiology Journal, v. 13, p. 145–164.

Edwards, K.J., Hu, B., Hamers, R.J., and Banfield, J.F., 2001, A new look at microbial leaching patterns on sulfide minerals: FEMS Microbiology Ecology, v. 34, p. 197–206, doi: 10.1016/S0168-6496(00)00094-5.

Edwards, K.J., McCollom, T.M., Konishi, H., and Buseck, P.R., 2003a, Seafloor bio-alteration of sulfide minerals: Results from *in-situ* incubation studies: Geochimica et Cosmochimica Acta, v. 67, p. 2843–2856.

Edwards, K.J., Rogers, D.R., Wirsen, C.O., and McCollom, T.M., 2003b, Isolation and characterization of novel psychrophilic, neutrophilic, Fe-oxidizing chemolithoautotrophic α- and γ-Proteobacteria from the deep sea: Applied and Environmental Microbiology, p. 2906–2913.

Ehrenberg, C.G., 1836, Vorlage mettheilugen ueber das wirklige vorkommen fossiler infusorien und ihre grosse verbreitung: Poggendorff's Annalen der Physik und Chemie, v. 38, p. 213–227.

Ehrenberg, C.G., 1838, *Gallionella ferruginea*: Taylor's Scientific Memoirs, v. 1, p. 402.

Emerson, D., 2000, Microbial oxidation of Fe(II) and Mn(II) at circumneutral pH, *in* Lovley, D.R., ed., Environmental microbe-metal interactions: Washington, D.C., ASM Press, p. 31–52.

Gebruk, A.V., Pimenov, N.V., and Savvichev, A.S., 1993, Feeding specialization of bresiliid shrimps in the TAG site hydrothermal community: Marine Ecology Progress Series, v. 98, p. 247–253.

Hafenbrandl, D., Keller, M., Dirmeier, R., Rachel, R., Ronagel, P., Burggraf, S., Huber, H., and Stetter, K.O., 1996, *Ferroglobus placidus* gen. new., sp. nov., a novel hyperthemophilic archaeum that oxidizes Fe^{2+} at neutral pH under anoxic conditions: Archives of Microbiology, v. 166, p. 308–314, doi: 10.1007/S002030050388.

Hannington, M.D., Jonasson, I.R., Herzig, P.M., Petersen, S., and Herzig, P.M., 1995, Physical and chemical processes of seafloor mineralization at mid-ocean ridges, *in* Humphris, S.E., Zierenberg, R.A., Mullineaux, L.S., and Thomson, R.E., eds., Seafloor hydrothermal systems: Washington, D.C., American Geophysical Union Geophysical Monograph 91, p. 115–157.

Hannington, M.D., Galley, A.G., Herzig, P.M., and Petersen, S., 1998, Comparison of the TAG mound and stockwork complex with Cyprus-type massive sulfide deposits, *in* Herzig, P.M., Humphris, S.E., Miller, D.J., and Zierenberg, R.A., eds., Proceedings of the Ocean Drilling Program, Scientific Results, Volume 158: College Station, Texas, Ocean Drilling Program, p. 389–416.

Harmsen, H.J., Prieur, D., and Jeanthon, C., 1997, Distribution of microorganisms in deep-sea hydrothermal vent chimneys investigated by whole-cell hybridization and enrichment culture of thermophilic subpopulations: Applied and Environmental Microbiology, v. 63, p. 2876–2883.

Herzig, P.M., and Hannington, M.D., 1995, Polymetallic massive sulfides at the modern seafloor: A review: Ore Geology Reviews, v. 10, p. 95–115, doi: 10.1016/0169-1368(95)00009-7.

Hofmann, B.A., and Farmer, J.D., 2000, Filamentous fabrics in low-temperature mineral assemblages: Are they fossil biomarkers? Implications for the search for a subsurface fossil record on the early Earth and Mars: Planetary and Space Science, v. 48, p. 1077–1086, doi: 10.1016/S0032-0633(00)00081-7.

Humphris, S.E., Zierenberg, R.A., Mullineaux, L.S., and Thomson, R.E., editors, 1995, Seafloor hydrothermal systems: Washington, D.C., American Geophysical Union Geophysical Monograph 91, p. 466.

Jannasch, H.W., 1985, The chemosynthetic support of life and microbial diversity at deep-sea hydrothermal vents: Proceedings of the Royal Society of London, Series B, Biological Sciences, v. 225, p. 277–297.

Jenkins, D.A., 1995, Parys copper mines, Anglesey: Archaeology in Wales, v. 35, p. 35–37.

Juniper, S.K., and Fouquet, Y., 1988, Filamentous iron-silica deposits from modern and ancient hydrothermal sites: Canadian Mineralogist, v. 26, p. 859–869.

Juniper, S.K., Jonasson, I.R., Tunnicliffe, V., and Southward, A.J., 1992, Influence of a tube-building polychaete on hydrothermal chimney mineralization: Geology, v. 20, p. 895–898, doi: 10.1130/0091-7613(1992)0202.3.CO;2.

Juniper, S.K., and Tebo, B.M., 1995, Microbe-metal interactions and mineral deposition at hydrothermal vents, *in* Karl, D.M., ed., The microbiology of deep-sea hydrothermal vents: Boca Raton, American Chemical Society, p. 219–253.

Karl, D., 1995, Ecology of free-living, hydrothermal vent microbial communities, *in* Karl, D., ed., The microbiology of deep-sea hydrothermal vents: Boca Raton, CRC Press, p. 35–124.

Kasting, J.F., 1990, Bolide impacts and the oxidation state of carbon in the Earth's early atmosphere: Origins of life and evolution of the biosphere, v. 20, p. 199–231.

Kasting, J.F., and Walker, J.C., 1981, Limits on oxygen concentrations in the prebiological atmosphere and the rate of abiotic fixation of nitrogen: Journal of Geophysical Research, v. 86, p. 1147–1158.

Kaye, J.Z., and Baross, J.A., 2000, High incidence of halotolerant bacteria in Pacific hydrothermal-vent and pelagic environments: FEMS Microbiology Ecology, v. 32, p. 249–260, doi: 10.1016/S0168-6496(00)00035-0.

Kelly, D.P., and Wood, A.P., 2000, Reclassification of some species of *Thiobacillus* to the newly designated genera *Acidithiobacillus* gen. nov., *Halothiobacillus* gen. nov., and *Thermithiobacillus* gen. nov: International Journal of Systematic Bacteriology, v. 50, p. 511–516.

Kucera, S., and Wolfe, R.S., 1957, A selective enrichment method for *Gallionella ferruginea*: Journal of Bacteriology, v. 74, p. 344–349.

Lalou, C., Tompson, M., Arnold, M., Brichet, M., Druffel, E.R., and Rona, P.A., 1990, Geochronology of TAG and Snakepit hydrothermal fields, Mid-Atlantic Ridge: Witness to a long and complex hydrothermal history: Earth and Planetary Science Letters, v. 97, p. 113–128, doi: 10.1016/0012-821X(90)90103-5.

Lalou, C., Reyss, J.-L., Brichet, M., Arnold, G., Tompson, G., Fouquet, Y., and Rona, P.A., 1993, New age data for Mid-Atlantic Ridge hydrothermal sites: TAG and Snakepit chronology revisited: Journal of Geophysical Research, v. 98, p. 9705–9713.

Lane, D.J., Harrison, A.P., Jr, Stahl, D., Pace, B., Giovannoni, S.J., Olsen, G.J., and Pace, N.R., 1992, Evolutionary relationships among sulfur- and iron-oxidizing eubacteria: Journal of Bacteriology, v. 174, p. 269–278.

Martinez, J.S., Zhang, G.P., Holt, P.D., Jung, H.-T., Carrano, C.J., Haygood, M.G., and Butler, A., 2000, Self-assembling amphiphilic siderophores from marine bacteria: Science, v. 287, p. 1245–1247, doi: 10.1126/SCIENCE.287.5456.1245.

McCollom, T.M., 2000, Geochemical constraints on primary productivity in submarine hydrothermal vent plumes: Deep-sea Research I, v. 47, p. 85–101.

Neubauer, S.C., Emerson, D., and Megonigal, J.P., 2002, Life at the energetic edge: Kinetics of circumneutral iron oxidation by lithotrophic iron-oxidizing bacteria isolated from the wetland-plant rhizosphere: Applied and Environmental Microbiology, v. 68, p. 3988–3995, doi: 10.1128/AEM.68.8.3988-3995.2002.

Olsen, G.J., Lane, D.J., Giovannoni, S.J., Pace, N.R., and Stahl, D.A., 1986, Microbial ecology and evolution: a ribosomal RNA approach: Annual Review of Microbiology, v. 40, p. 337–365, doi: 10.1146/ANNUREV.MI.40.100186.002005.

Pace, R.N., Stahl, D.A., Lane, D.J., and Olsen, G.J., 1986, The analysis of natural microbial population by ribosomal RNA sequences: Advances in Microbial Ecology, v. 9, p. 1–55.

Pace, R.N., 1991, The origin of life—facing up to the physical setting: Cell, v. 65, p. 531–533.

Polz, M.F., Robinson, J.J., Cavanaugh, C.M., and Van Dover, C.L., 1998, Trophic ecology of massive shrimp aggregations at a Mid-Atlantic Ridge hydrothermal vent site: Limnology and Oceanography, v. 43, p. 1631–1638.

Reysenbach, A.-L., and Cady, S.L., 2001, Microbiology of ancient and modern hydrothermal systems: Trends in Microbiology, v. 9, p. 79–86, doi: 10.1016/S0966-842X(00)01921-1.

Rodriguez-Leiva, L.M., and Tributsch, H., 1988, Morphology of bacterial leaching patterns by *Thiobacillus ferrooxidans* on synthetic pyrite: Archives of Microbiology, v. 149, p. 401–405.

Rogers, D.R., Santelli, C.M., and Edwards, K.J., 2003, Geomicrobiology of deep-sea deposits: Estimating community diversity from low-temperature seafloor rocks and minerals: Geobiology Journal, v. 1 p. 109–117.
Schippers, A., and Jorgensen, B.B., 2001, Oxidation of pyrite and iron sulfide by manganese dioxide in marine sediments: Geochimica et Cosmochimica Acta, v. 65, p. 915–922, doi: 10.1016/S0016-7037(00)00589-5.
Schippers, A., and Jorgensen, B.B., 2002, Biogeochemistry of pyrite and iron sulfide oxidation in marine sediments: Geochimica et Cosmochimica Acta, v. 66, p. 85–92, doi: 10.1016/S0016-7037(01)00745-1.
Schrenk, M.O., Kelly, D.S., Delaney, J.R., and Baross, J.A., 2003, Incidence and diversity of microorganisms within the walls of an active deep-sea sulfide chimney: Applied and Environmental Microbiology, v. 69, p. 3580–3592, doi: 10.1128/AEM.69.6.3580-3592.2003.
Shen, Y., Knoll, A.H., and Walter, M.R., 2003, Evidence for low sulphate and anoxia in a mid-Proterozoic marine basin: Nature, v. 423, p. 632–635, doi: 10.1038/NATURE01651.
Singer, P.C., and Stumm, W., 1970, Acidic mine drainage: The rate-determining step: Science, v. 167, p. 1121–1123.
Stookey, L.L., 1970, Ferrozine—A new spectrophotometric reagent for iron: Analytical Chemistry, v. 42, p. 779–781.
Takai, K., Komatsu, T., Inagaki, F., and Horikoshi, K., 2001, Distribution of archaea in a black smoker chimney structure: Applied and Environmental Microbiology, v. 67, p. 3618–3629, doi: 10.1128/AEM.67.8.3618-3629.2001.
Tunnicliffe, V., and Fontaine, A.R., 1987, Faunal composition and organic surface encrustations at hydrothermal vents on the southern Juan de Fuca Ridge: Journal of Geophysical Research, v. 92, p. 303–311.
Tunnicliffe, V., and Juniper, S.K., 1990, Dynamic character of the hydrothermal vent habitat and the nature of sulfide chimney fauna: Progress in Oceanography, v. 24, p. 1–13, doi: 10.1016/0079-6611(90)90015-T.
Van Dover, C.L., Fry, B., Grassle, J.F., Humphris, S., and Rona, P.A., 1988, Feeding biology of the shrimp Rimicaris exoculata at hydrothermal vents on the Mid-Atlantic Ridge: Marine Biology, v. 98, p. 209–216.
Verati, C., de Donato, P., Prieur, D., and Lancelot, J., 1999, Evidence of bacterial activity from micrometer-scale layer analyses of black-smoker sulfide structures (Pito Seamount Site, Easter microplate): Chemical Geology, v. 158, p. 257–269, doi: 10.1016/S0009-2541(99)00054-6.
Viollier, E., Inglett, P.W., Hunter, K., Roychoudhury, A.N., and Van Cappellen, P., 2000, The ferrozing method revisited: Fe(II)/Fe(III) determination in natural waters: Applied Geochemistry, v. 15, p. 785–790, doi: 10.1016/S0883-2927(99)00097-9.
Winn, C.D., Cowen, J.P., and Karl, D.M., 1995, Microbes in deep-sea hydrothermal vent plumes, *in* Karl, D.M., ed., The microbiology of deep-sea hydrothermal vents: Boca Raton, CRC Press, p. 255–273.
Wirsen, C.O., Jannasch, H.W., and Molyneaux, S.J., 1993, Chemosynthetic microbial activity at Mid-Atlantic Ridge hydrothermal vent sites: Journal of Geophysical Research, v. 98, p. 9693–9703.
Vargas, M., Kashefi, K., Blunt-Harris, E.L., and Lovley, D.R., 1998, Microbiological evidence for Fe(III) reduction on early Earth: Nature, v. 395, p. 65–67, doi: 10.1038/25720.
Yung, Y.L., and McElroy, M.B., 1979, Fixation of nitrogen in the prebiotic atmosphere: Science, v. 203, p. 1002–1004.
Zierenberg, R.A., and Schiffman, P., 1990, Microbial control of silver mineralization at a sea-floor hydrothermal site on the northern Gorda Ridge: Nature, v. 348, p. 155–157, doi: 10.1038/348155A0.

Manuscript Accepted by the Society March 28, 2004

Printed in the USA

Geological Society of America
Special Paper 379
2004

Distribution and fate of sulfur intermediates—sulfite, tetrathionate, thiosulfate, and elemental sulfur—in marine sediments

J. Zopfi*
T.G. Ferdelman
H. Fossing*
Max Planck Institute for Marine Microbiology, Biogeochemistry Department, Celsiusstrasse 1, D-28359 Bremen, Germany

ABSTRACT

Most of the sulfide produced in surface marine sediments is eventually oxidized back to sulfate via sulfur compounds of intermediate oxidation state in a complex web of competing chemical and biological reactions. Improved handling, derivatization, and chromatographic techniques allowed us to more closely examine the occurrence and fate of the sulfur intermediates elemental sulfur (S^0), thiosulfate ($S_2O_3^{2-}$), tetrathionate ($S_4O_6^{2-}$), and sulfite (SO_3^{2-}) in Black Sea and North Sea sediments. Elemental sulfur was the most abundant sulfur intermediate with concentrations ~3 orders of magnitude higher than the dissolved species, which were typically in the low micromolar range or below. Turnover times of the intermediate sulfur compounds were inversely correlated with concentration and followed the order: $SO_3^{2-} \approx S_4O_6^{2-} > S_2O_3^{2-} > S^0$. Experiments with anoxic but non-sulfidic surface sediments from the Black Sea revealed that added sulfide and sulfite disappeared most rapidly, followed by thiosulfate. Competing chemical reactions, including the reaction of sulfite with sedimentary S^0 that led to temporarily increased thiosulfate concentrations, resulted in the rapid disappearance of SO_3^{2-}. Conversely, low thiosulfate concentrations in the Black Sea sediments (<3μM) were attributed to the activity of thiosulfate-consuming bacteria. Experiments with anoxic but non-sulfidic sediments revealed that 1 mol of tetrathionate was rapidly converted to 2 moles of thiosulfate. This tetrathionate reduction was bacterially mediated and occurred generally much faster than thiosulfate consumption. The rapid reduction of tetrathionate back to thiosulfate creates a cul-de-sac in the sulfur cycle, with thiosulfate acting as a bottleneck for the oxidation pathways between sulfide and sulfate.

Keywords: sulfide oxidation, sulfur cycle, diagenesis, tetrathionate, thiosulfate, sulfite.

INTRODUCTION

Sulfur exists in the marine environment predominately in its most oxidized state as sulfate (oxidation state of +VI), and in the reduced form as sulfide and pyrite (oxidation states of −II and −I respectively). In between the oxidized and reduced states, a wide

*Zopfi also affiliated with: Danish Center for Earth System Science and Institute of Biology, University of Southern Denmark, Campusvej 55, DK-5230 Odense M, Denmark. Current addresses, Zopfi: Laboratoire de Microbiologie, Institut de botanique, Université de Neuchâtel, Emile Argand 9, CH-2007 Neuchâtel, Switzerland; Fossing: Department of Lake and Estuarine Ecology, National Environmental Research Institute, Vejlsøvej 25, DK-8600 Silkeborg, Denmark.

Zopfi, J., Ferdelman, T.G., and Fossing, H., 2004, Distribution and fate of sulfur intermediates—sulfite, tetrathionate, thiosulfate, and elemental sulfur—in marine sediments, *in* Amend, J.P., Edwards, K.J., and Lyons, T.W., eds., Sulfur biogeochemistry—Past and present: Boulder, Colorado, Geological Society of America Special Paper 379, p. 97–116. For permission to copy, contact editing@geosociety.org.

variety of sulfur compounds of intermediate oxidation states have been identified. Although they do not form an appreciable quantity of the overall sulfur mass in marine environments, their low concentrations belie their role in a number of biogeochemical reactions and processes within the sulfur cycle. For instance, sulfur intermediates have been shown to influence trace metal solubility and mobility by complexation with polysulfides and thiosulfate (Jacobs and Emerson, 1982; Morse et al., 1987). Polysulfides are suspected to be involved in the formation of pyrite (Luther, 1991), thiols, and organic polysulfides (Vairavamurthy and Mopper, 1989; Kohnen et al., 1989). Sulfonates have been proposed to be formed by the reaction of sulfite or thiosulfate with reactive organic matter (Vairavamurthy et al., 1994). The bacterial disproportionation reactions of sulfite, thiosulfate, and elemental sulfur have been shown to have a strong impact on the fractionation of stable sulfur isotopes (Canfield and Thamdrup, 1994; Cypionka et al., 1998; Habicht et al., 1998) and the interpretation of the sulfur isotope record (Jørgensen, 1990a; Canfield and Teske, 1996).

The formation of sulfur intermediates in marine sediments principally occurs through the oxidation of sulfide produced during bacterial sulfate reduction (Fig. 1, Table 1). Although bacterial sulfate reduction is usually the second most important terminal electron acceptor process for the degradation of organic matter after aerobic respiration in most continental margin sediments, mass balance considerations show that only 10–20% of the produced sulfide is buried in the sediment in its reduced form, principally as pyrite sulfur (Jørgensen, 1982; Ferdelman et al.,

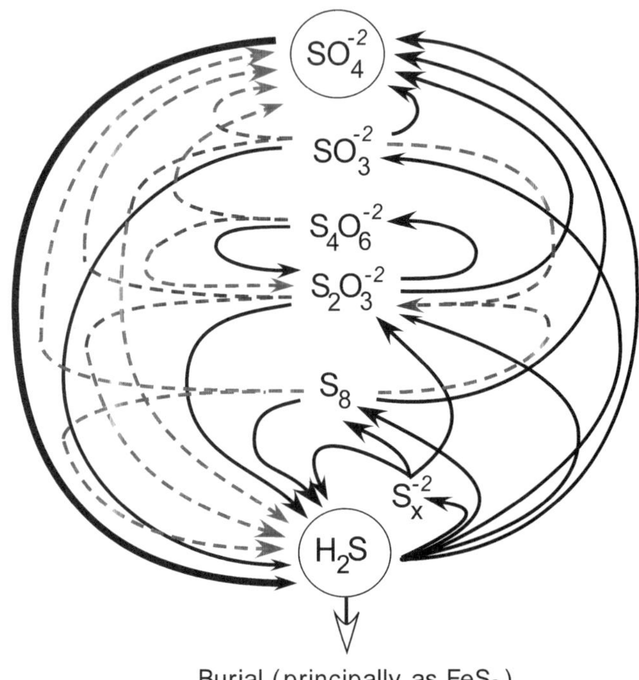

Figure 1. Schematic figure of the sedimentary sulfur cycle where important reductive (left-side, downward arrows) and oxidative (right-side, upward arrows) pathways are shown. Broken lines on the left signify bacterial disproportionation reactions. The cycle is driven by the degradation of organic matter through sulfate-reducing bacteria (thick arrow on the left). Burial of iron-sulfur minerals, mostly FeS_2, represents the dominant sink for reduced sulfur in marine sediments.

TABLE 1. PRODUCTS OF CHEMICAL OR BIOLOGICAL OXIDATION OF MAJOR REDUCED SULFUR COMPOUNDS IN MARINE SEDIMENTS

S-species	Oxidant	Products	Comments§	Reference
H_2S	O_2	SO_4^{2-}, $S_2O_3^{2-}$, $\mathit{SO_3^{2-}}$	C	Zhang and Millero, 1993
	O_2	SO_4^{2-} $S_2O_3^{2-}$, $\mathit{S_n^{2-}}$, S^0	C	Chen and Morris, 1972
	O_2	SO_4^{2-}, $\mathit{S_2O_3^{2-}}$, $\mathit{SO_3^{2-}}$	M	Kelly, 1989
	O_2	S^0, $S_2O_3^{2-}$, SO_4^{2-}, $\mathit{S_nO_6^{2-}}$	M	van den Ende and van Gemerden, 1993
	NO_3^-	S^0, SO_4^{2-}	S	Elsgaard and Jørgensen, 1992
	NO_3^-	S^0, SO_4^{2-}	M	Otte et al., 1999
	Mn_{IV}	S^0, $S_2O_3^{2-}$, SO_4^{2-}, $\mathit{SO_3^{2-}}$	C	Yao and Millero, 1996; Burdige and Nealson, 1986
	Fe_{III}	S^0, $S_2O_3^{2-}$, $\mathit{SO_3^{2-}}$	C	Pyzik and Sommer, 1981
S_n^{2-}	O_2	$S_2O_3^{2-}$, S^0	C	Steudel et al., 1986; Chen and Morris, 1972
FeS	O_2	S^0, $S_nO_6^{2-}$, $S_2O_3^{2-}$, SO_4^{2-}	C	von Rège, 1999
	NO_3^-	SO_4^{2-}	M	Straub et al., 1996
	Mn_{IV}	S^0, SO_4^{2-}	C, S	Schippers and Jørgensen, 2001
	Fe_{III}	$\mathit{SO_4^{2-}}$*†	S	Aller and Rude, 1988
FeS_2	O_2	SO_4^{2-}, $S_nO_6^{2-}$, $S_2O_3^{2-}$	C	Moses et al., 1987
	Mn_{IV}	SO_4^{2-}, $S_nO_6^{2-}$, $S_2O_3^{2-}$	C	Schippers and Jørgensen, 2001

Note: The order of products from the left to the right signifies their quantitative importance. Only results from studies conducted at circumneutral pH are included. Intermediates, which are unstable under the experimental conditions or which are only observed in trace quantities are given in italics. For experimental details, we refer to the original literature.

*No sulfur intermediates determined.

†Only weak sulfate production. See also Schippers and Jørgensen (2001) for additional comments.

§Type of study: C—chemical, M—microbiological, S—sediment incubation.

1999). The remaining 80–90% is eventually recycled back to sulfate through sulfur compounds of intermediate oxidation state in a complex web of competing chemical and biological reactions (Fig. 1) (Jørgensen, 1987; Fossing and Jørgensen, 1990; Jørgensen and Bak, 1991). A brief review of some of the important reactions leading to the formation of sulfur intermediates follows.

Review of Sulfide Oxidation Pathways

Oxic Sulfide Oxidation

Where dissolved sulfide (H_2S and HS^-) comes in contact with oxygen, sulfide may be chemically oxidized by dissolved oxygen according to the overall reaction

$$HS^- + 2O_2 \Rightarrow SO_4^{2-} + H^+ \quad (1)$$

However, the chemistry of the reaction is not as simple as the stoichiometry implies, and the exact reaction mechanism still remains to be elucidated (Zhang and Millero, 1993). A number of studies have shown that the oxidation of sulfide does not directly lead to sulfate but passes through several intermediates of different oxidation states (e.g., Avrahami and Golding, 1968; Cline and Richards, 1969; Chen and Morris, 1972; Zhang and Millero, 1993). Among them, sulfite is usually the first product formed (Equation 2).

$$HS^- + 1.5O_2 \Rightarrow HSO_3^- \quad (2)$$

The rapid oxidation of sulfite with oxygen explains the sulfate formation that is commonly observed during sulfide oxidation experiments (Equation 3). Sulfite can also react with HS^- to form thiosulfate ($S_2O_3^{2-}$) (Equation 4).

$$SO_3^{2-} + 0.5O_2 \Rightarrow SO_4^{2-} \quad (3)$$

$$HS^- + SO_3^{2-} + 0.5O_2 \Rightarrow S_2O_3^{2-} + OH^- \quad (4)$$

In most chemical studies, thiosulfate and sulfate were the only stable oxidation products that accumulated during the course of the experiments.

Tetrathionate, $S_4O_6^{2-}$, has been proposed as an intermediate in the incomplete oxidation of thiosulfate to sulfate (Jørgensen, 1990a; Schippers, this volume, Chapter 4). Based on thermodynamic considerations alone, thiosulfate will be oxidized to tetrathionate in the presence of various oxidants, such as O_2, Fe(III), Mn(IV), and I_2. (For instance, the conversion of thiosulfate to tetrathionate in the presence of iodine forms the basis of classic iodometric methods). The reaction between O_2 and $S_2O_3^{2-}$ is kinetically inert, although Xu and Schoonen (1995) have demonstrated that pyrite catalyzes this reaction at pH values of up to 8.6. Thiosulfate, which is the first intermediate product during pyrite oxidation (Moses et al., 1987; Luther 1987), is oxidized by Fe(III) to tetrathionate and eventually through to sulfate in the "thiosulfate-mechanism" of pyrite oxidation (Schippers et al., 1996; Schippers, this volume). MnO_2 will also oxidize thiosulfate to tetrathionate (Schippers and Jørgensen, 2001).

In the presence of trace metals, as is typical for natural environments, the formation of elemental sulfur in the initial step of sulfide oxidation is also possible (Equation 5) (Steudel, 1996; Zhang and Millero, 1993).

$$2HS^- + O_2 \Rightarrow 2S^0 + 2OH^- \quad (5)$$

Elemental sulfur can react with sulfite and sulfide to form thiosulfate (Equation 6) and polysulfides (Equation 7), respectively.

$$S^0 + SO_3^{2-} \Rightarrow S_2O_3^{2-} \quad (6)$$

$$(n-1)S^0 + HS^- \Rightarrow HS_n^- \quad (7)$$

Polysulfides are not stable under oxic conditions and rapidly decompose to thiosulfate and elemental sulfur (Steudel et al., 1986).

Although sulfide is basically a waste product of sulfate-reducing bacteria, it still contains a considerable amount of the energy originally stored in the biomass of primary producers. Aerobic lithotrophic bacteria can thrive on the oxidation of sulfide or sulfur intermediates with oxygen. The main product of biological sulfide oxidation is sulfate. Sulfur intermediates are mostly formed transiently under changing environmental conditions and severe oxygen limitation (van den Ende and van Gemerden, 1993). Because chemical sulfide oxidation can be very rapid in the environment, bacteria have had to develop strategies to successfully compete for sulfide. The most important adaptations are high enzyme affinities toward O_2 and sulfide and motility. Motility enables the organisms to position themselves in the oxic/anoxic interface where both oxygen and sulfide are present in low concentrations and are only supplied by diffusion (Jørgensen, 1987). Under such low reactant conditions, chemical sulfide oxidation becomes much slower due to the second order kinetics of the reaction (Zhang and Millero, 1994). Because of the Michaelis-Menthen kinetics of biological oxidation and the very low saturation constants for oxygen and sulfide of 1 μM or below in chemolithotrophic sulfur bacteria (Kuenen and Bos, 1989; van den Ende and van Gemerden, 1993), these organisms can still metabolize at maximal rates and may out-compete the chemical sulfide oxidation (Zopfi et al., 2001a).

Anoxic Sulfide Oxidation

In most marine sediments, sulfide does not diffuse to the sediment surface, but is removed from the pore water below the oxidized surface layer, in the suboxic zone, by oxidation and precipitation. The suboxic zone is characterized by the absence of oxygen and sulfide and increased concentrations of dissolved reduced iron and manganese. For the chemical oxidation of sulfide in marine sediments, only Mn(IV)oxides (Equation 8) and Fe(III)oxides (Equation 9) are of importance, because the reaction with nitrate is kinetically unfavorable. Similar to the oxic pathways of sulfide oxidation, sulfur intermediates are also formed during anoxic oxidation of sulfide.

$$\delta MnO_2 + HS^- + 3H^+ \Rightarrow Mn^{2+} + S^0 + 2H_2O \quad (8)$$

For instance, elemental sulfur is a main product of the sulfide oxidation with Mn(IV) (Burdige and Nealson, 1986), but with increasing MnO_2/H_2S ratios, thiosulfate and especially sulfate become more important as products (Yao and Millero, 1996). The stoichiometry in Equation 8 is thus an oversimplification and describes only approximately the situation for a 1:1 ratio between sulfide and manganese. Manganese is a powerful oxidant and reacts also with solid phases such as FeS and FeS_2. Tetrathionate and thiosulfate have been reported as intermediates during the oxidation of pyrite with Mn(IV) oxide (Schippers and Jørgensen, 2001).

In most marine sediments, iron is much more abundant than manganese and is responsible for the efficient removal of dissolved sulfide from the interstitial water (Canfield, 1989). Unlike manganese, Fe(III) oxide is a rather poor oxidant for the complete oxidation of sulfide to sulfate (Aller and Rude, 1988; King, 1990; Elsgaard and Jørgensen, 1992). During the reaction of sulfide with Fe(III)oxides, dissolved ferrous iron and elemental sulfur are produced (Equation 9).

$$2FeOOH + HS^- + 5H^+ \Rightarrow 2Fe^{2+} + S^0 + 4H_2O \quad (9)$$

Furthermore, if sulfide is present in excess, dissolved ferrous iron will be precipitated as FeS. However, the formation of polysulfides and small amounts of thiosulfate and sulfite has also been reported (Peiffer et al., 1992; Pyzik and Sommer, 1981; dos Santos and Stumm, 1992).

The sulfur intermediates that are formed during sulfide oxidation may be further transformed by microorganisms. In the presence of an electron donor (i.e., organic matter, hydrogen), all of the sulfur intermediates can be reduced back to sulfide by sulfate-reducing bacteria and others (e.g., *Shewanella* sp., *Dethiosulfovibrio* sp., *Desulfitobacterium* sp., *Clostridium sp.*). Sulfur intermediates are also further oxidized to sulfate when a suitable electron acceptor becomes available. Under anoxic conditions, nitrate and possibly Mn(IV)oxides have been shown to be used by microorganisms as electron acceptors for complete sulfide oxidation (Elsgaard and Jørgensen, 1992; Lovley and Phillips, 1994).

The third type of metabolism responsible for the anaerobic transformation of sulfur intermediates is the so-called disproportionation (Bak and Cypionka, 1987; Thamdrup et al., 1993; Wentzien and Sand, 1999), which is described as a type of inorganic fermentation, where the substrate serves as electron donor as well as electron acceptor (Equations 10–13).

$$4SO_3^{2-} + H^+ \Rightarrow 3SO_4^{2-} + HS^- \quad (10)$$

$$S_2O_3^{2-} + H_2O \Rightarrow SO_4^{2-} + HS^- + H^+ \quad (11)$$

$$4S^0 + 4H_2O \Rightarrow SO_4^{2-} + 3HS^- + 5H^+ \quad (12)$$

$$4S_4O_6^{2-} + 4H_2O \Rightarrow 6S_2O_3^{2-} + S_3O_6^{2-} + SO_4^{2-} + 8H^+ \quad (13)$$

By using radiotracers, it was shown that the disproportionation of thiosulfate is a key process in the sedimentary sulfur cycle (Jørgensen, 1990a).

Scope of this Study

Despite the importance of sulfur intermediates for the biogeochemical cycling of carbon, manganese, iron, and trace metals, comparatively little is known about their occurrence in nature. However, improvements in sample handling and analytical methods now allow us to take another look at the distribution and cycling of sulfur intermediates in marine systems. This study represents a composite of a number of field investigations and experiments made over the past decade using these methods. We provide detailed descriptions of the applied analytical methods and sample processing where necessary, because proper handling and analysis is critical to the determination of these often ephemeral and redox-sensitive compounds. In this report, we present new data on the distribution of sulfur intermediates (mostly S^0, $S_2O_3^{2-}$, and SO_3^{2-}) along a transect extending from the oxygenated shelf to the permanently anoxic waters of the Black Sea. Through a series of amendment experiments, we explore the fate of sulfur intermediate compounds in marine sediments and the extent to which they are regulated by microbial or inorganic reactions. These experiments were performed using Black Sea, estuarine (Weser Estuary, Germany), and continental slope (Skagerrak, Denmark) sediments. Although certainly not all-inclusive, these sites are typical of continental margin sediments where the sulfur cycle plays an important role in the overall cycling of carbon and other elements.

METHODS

Study Sites and Sampling

Black Sea

Sediment for pore-water and solid phase sulfur speciation was collected during a cruise along a transect from the Romanian shelf to the abyssal plain with R/V *Petr Kotsov* in 1997. The sediment surface at Station 2 (77 m deep, 7.2 °C, 213 µM O_2) was covered with a layer of *Modiolus phaesolinus* shells (Wenzhoefer et al., 2002). The underlying muddy sediment was carbonate-rich and light gray. The total mineralization rate was 1110 nmol C cm^{-2} d^{-1}, and about half of the organic matter in the top centimeter was degraded via Mn reduction (Thamdrup et al., 2000). Sulfate reduction accounted for ~15% of the total mineralization rate (Weber et al., 2001). Station 4 at the shelf break was located at the upper boundary of the chemocline (130 m, 7.8 °C, <5 µM O_2). The sediment surface was covered with a 1.5 cm thick layer of dead mussel shells followed by homogenous gray sediment beneath. Between 8 and 17 cm a second, a very porous band of buried mussel shells was observed. Organic matter mineralization was dominated by sulfate reduction (60–80%) and proceeded at a rate of 50–122 nmol C cm^{-2} d^{-1} (Weber et al., 2001). Station 6 was located in the permanently anoxic part of the Black Sea at a depth of 396 m. Sulfide concentration in the bottom water was 75 μM. The sediment was finely laminated, and organic matter was degraded solely by sulfate reduction at a rate of 112 nmol C cm^{-2} d^{-1} (Weber et al., 2001).

Skagerrak

Sediments were obtained from two sites in the Skagerrak basin of the North Sea using a multi-corer from on board the F/S *Victor Hensen*. Station S4 at 190 m was a sandy silt with total carbon oxidation rates of 200–300 nmol cm^{-3} d^{-1} in the upper 5 cm of sediment, with sulfate reduction accounting for ~60% of the total organic carbon degradation (Canfield et al., 1993). Station S9 at 695 m was a clayey-silt with a high concentration of manganese oxide (3–4% by weight). Organic carbon degradation (50–200 nmol cm^{-3} d^{-1}) was dominated by dissimilatory manganese oxide reduction in the upper 5 cm, and sulfate reduction was virtually absent at the same depths (Canfield et al., 1993).

Weser Estuary

The upper 5 cm of sediment from an intertidal mud flat located on the lower Weser Estuary (Weddewarden, 5 km north of Bremerhaven, Germany) was sampled during low tide and stored in buckets with 2–3 cm of overlying water at 4 °C until use in incubation experiments. Due to the relatively high iron contents of the predominately fine-grained silts, free dissolved sulfide is rarely ever present in the uppermost 5 cm of this sediment (Sagemann et al., 1996).

Pore-Water Sampling

Pore water from sediment cores was extracted by pressure filtration (0.45 μm Millipore PTFE filters) at 8 °C in a N_2-filled glove bag. The pore water was directly led into 1.5 mL reaction tubes containing either a 0.3 mL 20% Zn-acetate dihydrate solution for sulfate and sulfide measurements or the derivatization-mixture (see monobromobimane [MBB] method) for thiosulfate and sulfite determination. Unless the fixed samples were not analyzed within 24 h, they were frozen and stored at −20 °C.

Sediment and Slurry Incubation Experiments

Time-course studies on the fate of sulfide, thiosulfate, tetrathionate, or sulfite-amended sediments were performed on sediments obtained from the upper three (Black Sea) or upper five (Weser Estuary and Skagerrak) centimeters of sediment. The Black Sea sediment was—after removing mussel shells—homogenized under a N_2 atmosphere and directly poured into gas-tight plastic bags (Canfield et al., 1993). Sediments from the Weser Estuary and Skagerrak were diluted with water (1vol/1vol) from the corresponding site before being poured into the bags. The bags were equipped with glass outlets that were closed with rubber stoppers (sediment incubations) or connected to a three-way Luer stopcock (slurry experiments) to allow for the hermetic removal of sample into a syringe.

Sulfide, thiosulfate, and sulfite amendments were performed with Black Sea sediments. All manipulations of the Black Sea sediments were done in a N_2-filled glove bag at 8 °C. Amendments of sulfide, thiosulfate, and sulfite were made to a final concentration of ~20–40 μM. The μM concentrations added were not expected to affect the pH of the well-buffered (mM range) marine sediments. At specific times sediment was withdrawn with truncated 1 ml plastic syringes and transferred into 1.5 mL centrifuge tubes for monobromobimane derivatization of sulfide, thiosulfate, and sulfite.

Tetrathionate experiments were performed on Skagerrak and Weser Estuary slurries, which were incubated, unless otherwise indicated, in the dark for 24 h (Skagerrak at 6–7 °C; Weser Estuary at room temperature). After a zero time-point sample was taken, 3–5 mL of 20 mM tetrathionate, freshly prepared in deoxygenated water, was injected into the bag (250–300 cm^{-3}) and mixed thoroughly. Subsamples were taken with 20 mL plastic syringes through the stopcock. Typically, 10 mL of slurry was removed, placed into a centrifuge tube, and spun down. The supernatant was then filtered through 0.4 μm Gelman syringe filters and analyzed by anion-exchange HPLC (high performance liquid chromatography) within one day. Thiosulfate and tetrathionate concentrations in darkened, refrigerated samples were determined to be stable for at least seven days (three days at room temperature). Various pre-treatments or amendments were performed on the Weser Estuary slurries to elucidate the role of bacterial versus inorganic reactions with tetrathionate, and these are described later in this paper. In some experiments, this included the addition of 20 MBq of $^{35}SO_4^{2-}$ (Amersham) in order to follow rates of sulfate reduction in the slurries.

Analytical Methods

Tetrathionate and Thiosulfate (Ion Chromatography [IC] method)

Initially, tetrathionate and thiosulfate were determined using the anion-exchange HPLC method described by Bak et al. (1993), using a Sykam S2100 pump, with an all–polyether-ether-ketone (PEEK) pumphead, a Rheodyne 9175 PEEK injector (50 or 20 μL sample loop), PEEK tubing, a LCA08 anion-exchange column (a silicon-based, polymer-coated material from Sykam), and a Linear Instruments UV/VIS (Ultraviolet/Visible) detector set for measurement at 216 nm. The eluent consisted of 11.7 g L^{-1} NaCl (Alfa, ultra-pure) dissolved in 64% acetonitrile and 10% methanol. The column was thermostated at 30 °C. With a flow rate of 1 mL min^{-1}, tetrathionate and thiosulfate eluted at 9.1 and 13.6 min, respectively. Due to the relative long-term degradation of the LCA08 column, we switched to a LCA09 (polymer-based, Sykam) anion column part-way through the experiments. Although tetrathionate and thiosulfate could not be measured on the same isocratic run, retention time stability and peak resolution improved greatly. Tetrathionate was determined using an eluent described above and eluted at 5.81 min. Thiosulfate was determined using an eluent mix of 5.84 g NaCl in 10% methanol (100 mM NaCl) and eluted at 4.82 min. Standard solutions of thiosulfate (from sodium thiosulfate pentahydrate, Merck) and tetrathionate (sodium tetrathionate, 99% pure, Aldrich) were prepared freshly each day of analysis.

Thiosulfate and Sulfite (MBB Method)

Samples for thiosulfate ($S_2O_3^{2-}$) and sulfite (SO_3^{2-}), typically 500 µL, were derivatized at room temperature in the dark with a mixture of 50 µL monobromobimane (Sigma; 45 mM in acetonitrile) and 50 µL HEPES-EDTA buffer (pH 8, 500 mM, 50 mM) (Fahey and Newton, 1987; Vetter et al., 1989). The derivatization reaction was stopped after 30 min by adding 50 µL methanesulfonic acid (324 mM). Samples were frozen at −20 °C until analysis within the next few days. In order to ensure a rapid and complete derivatization reaction, the amount of bimane in the assay was set to be at least twice as high as the total reduced sulfur content (Vetter et al., 1989).

A Sykam gradient controller S2000 (low pressure mixing system) combined with a LiChrosphere 60RP select B column (125 × 4 mm, 5 µm; Merck) and a Waters 470 scanning fluorescence detector (excitation at 380 nm; detection at 480 nm) were used for analysis. Eluent A was 0.25% (v/v) acetic acid pH 3.5 (adjusted with 5N NaOH), eluent B was 100% HPLC-grade methanol, and the flow rate was 1 mL min^{-1}. A modification of the gradient conditions described by Rethmeier et al. (1997) was used: start, 10% B; 7 min, 12% B; 15–19 min, 30% B; 23 min, 50% B; 30 min, 100% B; 33 min, 100% B; 34 min, 10% B; 39 min, 10% B; injection of the next sample. Separate standards for sulfite, thiosulfate, and sulfide were prepared in anoxic Milli-Q water in a N_2-filled glove bag. No difference was observed between calibration curves with standards prepared in seawater or in Milli-Q water. With an injection volume of 100 µL, the detection limits for thiosulfate and sulfite were ~0.05 µM, and the precision for measurements of 10 µM standards was better than ±3% standard deviation. Although some authors reported that MBB derivates were stable at room temperature (Fahey and Newton, 1987), we observed that (for example) thiosulfate values changed with time. We suggest, therefore, using a cooled autosampler (4 °C) and to keep derivatized samples at −70 °C for long-term storage.

Elemental Sulfur

Sediment samples for elemental sulfur (S^0) were sliced, fixed in zinc acetate dihydrate (20% w/v) solution and stored in 50 mL polyethylene centrifuge tubes at −20 °C. Elemental sulfur in this study is defined as the sulfur extracted with methanol from sediment samples and measured as cyclo-S_8 by Reversed-Phase HPLC. Methanol is as effective as or better than other commonly employed extraction solvents for elemental sulfur, such as acetone or toluene/methanol mixtures or non-polar solvents such as cyclohexane, toluene, and carbon disulfide (Ferdelman, 1994; Ferdelman and Fossing, unpublished data). Elemental sulfur was extracted from a subsample (~0.3 g wet weight) of the fixed sediment for 12–16 h on a rotary shaker with pure methanol. The sample-to-extractant ratio was ~1/10–1/30 (wet weight/vol), depending on the sulfur content. Elemental sulfur in the extracts was determined by reversed-phase chromatography as originally described by Möckel (1984a, 1984b). A Sykam pump (S1100), a UV-VIS Detector (Sykam S3200), a Zorbax octadecylsilane (ODS) column (125 × 4 mm, 5 µm; Knauer, Germany), and 100% methanol (HPLC grade) at a flow rate of 1 mL min^{-1} were employed. S_8 eluted after 3.5 min and was detected at 265 nm; the detection limit was <0.5 µM, and the analytical precision of the method was ±0.5% relative standard deviation. A 2 mM stock solution of S^0 was made by dissolving 16 mg S^0 in 25 mL dichloromethane. After S^0 was completely dissolved, methanol (HPLC-grade) was added to a final volume of 250 mL. Dilutions for secondary standards (1–1000 µM) were prepared in methanol. The stock solution and standards of higher concentrations were stable at 4 °C for >6 months.

Sulfide

Dissolved sulfide was either determined on Zn-preserved pore-water samples by the colorimetric methylene blue method of Cline (1969) or by using the MBB method. In highly sulfidic sediments, however, the quantification of sulfide with the MBB method was often impaired by neighboring peaks of polysulfide- and thiol-derivates; thus, the Cline (1969) method was used instead

Sulfate Reduction Measurements

Sulfate reduction was determined on the $^{35}SO_4^{2-}$ labeled slurry experiments. At each time point, 10 mL of slurry sample would be injected into 10 mL of 20% (wt/vol) zinc acetate dihydrate solution and frozen. The recovery of radiolabeled reduced sulfur compounds followed the two-step acidic-chromium reduction procedure as described by Fossing and Jørgensen (1989). ^{35}S-radioactivity was determined using a Canberra-Packard Tri-Carb 2400 TR liquid scintillation detector (scintillation fluid: Packard Ultima Gold). Sulfate was determined by non-suppressed ion chromatography and conductivity detection (Ferdelman et al., 1997).

RESULTS AND DISCUSSION

Distribution of the Sulfur Intermediates Sulfite, Thiosulfate, and Elemental Sulfur

Pore-water distributions of sulfur intermediates were determined on both Black Sea and Weser Estuary sediments. No SO_3^{2-} was detected in Weser Estuary sediment and only a few samples showed a small $S_2O_3^{2-}$ peak (data not shown). Since the detection limit was only 0.5 µM at that time, no further conclusion can be made other than thiosulfate was generally ≤0.5 µM. Attempts to measure tetrathionate ($S_4O_6^{2-}$) at the same site with anion exchange HPLC showed that ambient tetrathionate concentrations were also below the detection limit of 0.5 µM (data not shown). Therefore, further discussion will focus on sulfur distributions in Black Sea sediments.

Black Sea Pore-Water Characteristics

Depth profiles of dissolved and solid phase sulfur species at three stations in the Black Sea are shown in Figure 2. The Black Sea stations selected for study represent sediment sites underlying oxic (Station 2), dysoxic (<5 µM O_2, Station 4), and anoxic, sulfidic (Station 6) waters. The overlying water conditions are partially reflected in the sedimentary sulfide distributions. At the oxic shelf Station 2, sulfide in the pore water was not detected down to 6 cm, and never exceeded 0.7 µM down to 20 cm depth. Despite oxygen concentrations of less than 5 µM (Weber et al., 2001) in the bottom water at Station 4, sulfide concentrations in the top 10 cm were below 0.2 µM. Maximum sulfide concentrations in this core reached ~3 µM and were detected at intermediate depth between 10 and 20 cm. At Station 6, pore-water sulfide concentrations increased steadily with depth

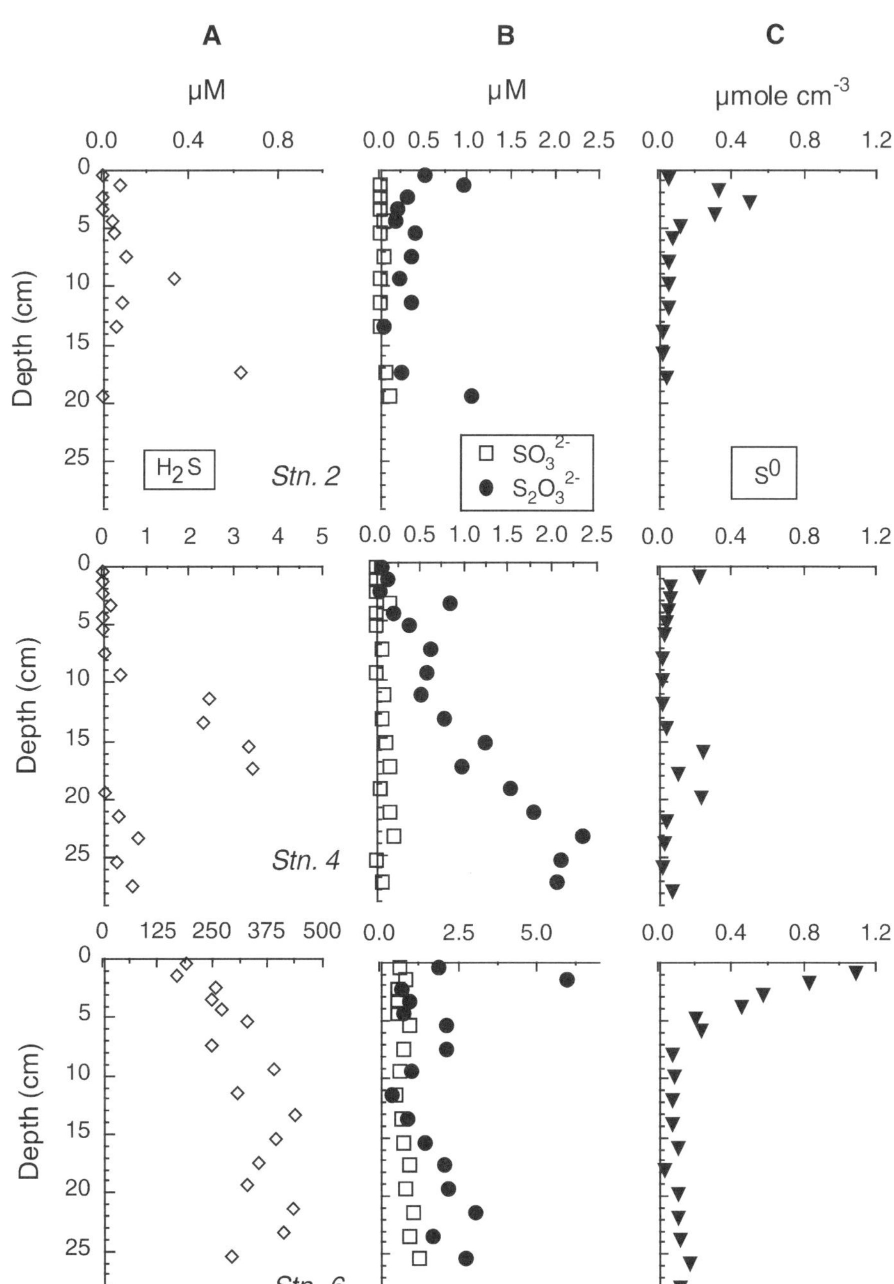

Figure 2. Depth profiles of (A) pore-water sulfide, (B) thiosulfate and sulfite, and (C) solid phase elemental sulfur in Black Sea sediments. Stn—Station. Stn. 2: Oxic bottom water. Stn. 4: Redox transition zone. Stn. 6: anoxic bottom water.

and reached maximum concentrations of 435 μM at 19 cm. A sulfide efflux from the sediment of 27 nmol cm^{-2} d^{-1} was calculated from the concentration profile; however, this value is only half of the sulfide production that was determined by in situ ^{35}S radiotracer incubations at the same station (Weber et al., 2001). In the following, we discuss the distribution of each of the sulfur intermediates (S^0, $S_2O_3^{2-}$, and SO_3^{2-}) in these three distinct Black Sea environments.

Distribution of Elemental Sulfur (S^0)

Elemental sulfur is the main reaction product of sulfide oxidation by Mn(IV)oxides and Fe(III)oxides (e.g., Yao and Millero, 1993, 1996; Pyzik and Sommer, 1981). Sulfur is also formed during oxic and anoxic FeS oxidation (Moses et al., 1987; Schippers and Jørgensen, 2001), and microorganisms produce S^0 as an intermediate or final product during bacterial oxidation of sulfide and thiosulfate (Taylor and Wirsen, 1997; Kelly, 1989). In contrast to sulfide, polysulfides, and sulfite, cyclic elemental sulfur is almost insoluble and can best be described as a Lewis acid. It is much less reactive and accumulates in the sediment to higher concentrations (Table 2) than other sulfur intermediates (Table 3). This greatly facilitates quantification, which is either done by cyanolysis and subsequent spectrophotometry (Bartlett and Skoog, 1954; Troelsen and Jørgensen, 1982), sulfitolysis and subsequent thiosulfate measurement (Luther et al., 1985; Ferdelman et al., 1991), or by reversed phase liquid chromatography and UV-detection (Möckel, 1984a, 1984b). During the last few years, the HPLC method has been applied to a variety of samples and has proved to be very sensitive and robust (e.g., Ramsing et al., 1996; Ferdelman et al., 1997; Henneke et al., 1997; Zopfi et al., 2001a, 2001b). The ease by which elemental sulfur is extracted by a relatively polar organic solvent such as methanol suggests that elemental sulfur in marine sediment (extracellular and intracellular) exists principally in the form of colloidal sols (Steudel, 1989), rather than as highly insoluble, crystalline elemental sulfur.

Peak concentrations of S^0 in the three Black Sea stations were between 0.22 and 1.08 μmol cm^{-3}. This is at the lower end of what has been reported previously (Table 2), but in the same range that Wijsman et al. (2001) found along the northwestern margin of the Black Sea. Although there are some exceptions, it appears that S^0 concentrations are higher in environments with increased sulfate reduction rates. The S^0 content in the three Black Sea stations fits this hypothesis because the sulfate reduction rates (0.5–0.8 mmol m^{-2} d^{-1}) are comparatively low (Skyring, 1987). Similarly, Moeslund et al. (1994) found during a seasonal study of bioturbated coastal sediment that S^0 concentrations increased from spring to late fall as sulfate reduction rates and bioturbation activities increased. In wintertime, S^0-consuming processes outweigh S^0 production until settling detritus from the spring bloom refuels higher benthic sulfate reduction rates (Moeslund et al., 1994). Schimmelmann and Kastner (1993) observed in the Santa Barbara Basin that sediments deposited during periods of decreased productivity and more oxygenated conditions in the water column were depleted in total organic carbon and S^0. Exceptionally high concentrations (>10 μmol cm^{-3}) are only found in very active and dynamic environments such as sulfureta, salt marshes, and organic-rich sediments from upwelling areas (see Table 2).

Although the concentrations are fairly comparable between the three Black Sea stations, the distribution of S^0 is different. Station 2, for example, exhibits a subsurface maximum of S^0 as is frequently found in bioturbated coastal marine sediments (e.g., Troelsen and Jørgensen, 1982; Sørensen and Jørgensen, 1987; Thode-Andersen and Jørgensen, 1989; Moeslund et al., 1994; Thamdrup et al., 1994a, 1994b; Zopfi, 2000). The balance between producing and consuming processes determines the concentration of S^0 in the sediment. Assuming that all pore-water sulfide is first oxidized to S^0 and after that to sulfate, the turnover time for S^0 can be calculated by dividing the S^0 pool (μmol cm^{-3}) by the sulfate reduction rate (μmol cm^{-3} d^{-1}) in the same depth interval. The average turnover time of S^0 in the top 2 cm at Station 2 is only 10 days, but rapidly increases to 66 days (3–4 cm) and falls again to ~27 d below 5 cm depth. Thus, the S^0 peak at 3 cm rather represents a turnover minimum than a production maximum. Above the S^0 peak, S^0 is rapidly produced, but also rapidly oxidized further to sulfate. The required oxidants, O_2, NO_3^- and Mn(IV), may be supplied by bioturbation (Aller and Rude, 1988) or advection (Huettel et al., 1998). At 3–4 cm depth, the supply of oxidants may be sufficient to remove sulfide from the pore water, but not for the complete oxidation of the produced S^0 to sulfate. Below that depth, S^0-consuming processes, such as dissimilative S^0 reduction, S^0 disproportionation, and pyrite formation dominate and lead to decreasing concentrations with depth. Whether a subsurface S^0 peak indeed indicates bioturbation activity and whether the location of the maximum may be a measure for the average bioturbation depth needs to be established by more detailed studies that should include combined S^0 and ^{234}Th and ^{210}Pb measurements.

At Stations 4 and 6, maximum S^0 concentrations were determined at the sediment-water interface. Similar distributions have been observed in sulfidic sediments and sediments overlain by anoxic bottom water (Thode-Andersen and Jørgensen, 1989; Troelsen and Jørgensen, 1982; Zopfi, 2000). Since elemental sulfur is only produced during oxidative pathways in the sulfur cycle (Fig. 1), the distribution of S^0 at Station 6 suggests that a part of the pore-water sulfide in the uppermost centimeters of the core is oxidized to S^0. At this depth, oxygen and nitrate can be excluded as oxidants. Although in the sulfidic water column of the Black Sea most settling iron reaches the sediment surface as FeS or FeS_2, some Fe(III)oxides or Mn(IV)oxides with a lower reactivity toward sulfide must become deposited and buried as well. They will finally react with pore-water sulfide. The produced S^0 then reacts further with sulfide and forms a range of polysulfides, depending on the pH in the sediment (Jacobs and Emerson, 1982; Morse et al., 1987). Polysulfides are more reactive nucleophiles than sulfide and are expected to play an important role in formation of organosulfur compounds and pyrite (Vairavamurthy and Mopper, 1989; Luther, 1991)

TABLE 2. SOLID PHASE AND PORE-WATER CONCENTRATIONS OF ZEROVALENT SULFUR (S^0) IN BRACKISH AND MARINE SEDIMENTS

Ecosystem/Site	Concentration S^0		Method	Reference
	Solid phase $\mu mol\ cm^{-3}$	Pore water μM		
Marshes				
Spartina-Salt Marshes	up to 500	up to 555	RP-HPLC/Sulfitosis	Ferdelman, 1994
Brackish Water Marshes	0.22–95		RP-HPLC	Ferdelman, 1994
Sulfureta				
Texel, Netherlands	6.2		UV-spectroscopy	Visscher and van Gemerden, 1993
Kalø Lagoon, Denmark	6.8		Cyanolysis	Thode-Andersen and Jørgensen, 1989
Aarhus Bay, Denmark	10–17		Cyanolysis	Troelsen and Jørgensen, 1982
Marine sediments				
Weser Estuary, Germany	0.8		RP-HPLC	Canfield and Thamdrup, 1996
Aarhus Bay, Denmark	<1–4.6		Cyanolysis	Thode-Andersen and Jørgensen, 1989; Thamdrup et al., 1994b; Moeslund et al.,
North Sea, Denmark	2.2		Cyanolysis	Sørensen and Jørgensen, 1987
Kattegat, Denmark	0.5		Cyanolysis	Sørensen and Jørgensen, 1987
Skagerrak, Denmark	0.2		Cyanolysis	Sørensen and Jørgensen, 1987
Saguenay Fjord, Canada		5.5–21	UV-spectroscopy	Gagnon et al, 1996
Central Peru, upwelling area	0–1.6		Cyanolysis	Fossing, 1990
Central Chile, upwelling area	2–15 (76)	4–59	RP-HPLC	Ferdelman et al. 1997; Thamdrup and Canfield, 1996; Zopfi, 2000
Black Sea Shelf, Romania	0.22–0.67		RP-HPLC	This study
Mid-Atlantic Bight, USA	0.04–13.7	<1–19.0	RP-HPLC/Sulfitosis	Ferdelman, 1994
Euxinic sediment				
Gotland Deep, Baltic	6.4		Cyanolysis	Podgorsek, 1998
Arkona Deep, Baltic	0.34		Cyanolysis	Podgorsek, 1998
Black Sea	1.1		RP-HPLC	This study
Tyro Basin, Mediterranean	2.5*		RP-HPLC	Hennecke et al., 1997
Bannock Basin, Mediterranean	4.8*		RP-HPLC	Hennecke et al., 1997

*Recalculated from $\mu mol\ g^{-1}$ dry weight (d.w.) by using a water content of 75% and a sediment density of 1.1 g cm^{-3}. (Original data: 11 $\mu mol\ g^{-1}$ d.w. Tyro and 21 $\mu mol\ g^{-1}$ d.w. Bannock). UV—ultraviolet. RP-HPLC—reverse phase high performance liquid chromatography.

TABLE 3. CONCENTRATION RANGES OF SULFITE (SO_3^{2-}) AND THIOSULFATE ($S_2O_3^{2-}$) DETERMINED IN MARINE SEDIMENTS

Ecosystem/Site	Concentration (μM)		Method	Reference
	$S_2O_3^{2-}$	SO_3^{2-}		
Spartina-Salt Marshes				
Great Marsh, Delaware, USA	130–530	<0.1–177	Hg^{2+}-titration	Boulègue et al., 1982
Sippewissett, Massachusetts, USA	<15–340	n.a.	Cyanolysis	Howarth et al., 1983
Sippewissett, Massachusetts, USA; Great Marsh, Delaware, USA	<0.2–1000	0–7.3	Voltametry	Luther et al., 1985, 1986, 1991; Luther and Church, 1988
Mission Bay, California, USA	0–45	0–6	Bimane HPLC	Vetter et al., 1989
Unvegetated Marshes				
Mission Bay, California, USA	2–84	3–56	Bimane HPLC	Vetter et al., 1989
New York, New York, USA	3–50	n.a.	Cyanolysis	Swider and Mackin, 1989
Skallingen, Denmark	<0.05–0.6	0.1–1.1	DTNP HPLC	Thamdrup et al., 1994b
Sulfureta				
Orkneys, UK	0–2500	n.a.	Cyanolysis	van Gemerden et al., 1989
Texel, Netherlands	0–35	n.a.	Cyanolysis	Visscher et al. 1992
Marine Sediments				
Walvis Bay, Namibia	15–145	n.a.	Hg^{2+}-titration	Boulègue and Denis, 1983
Kysing Fjord, Denmark	<1–10	n.a.	Cyanolysis	Troelsen and Jørgensen, 1982
Gulf of California	0-500	n.a.	?	Lein, 1984
Orleans, Massachusetts, USA	400–1300	6.8–7.9	Voltametry	Luther et al., 1985
Mid-Atlantic Bight, USA	<1–89	n.a.	Voltametry	Ferdelman, 1994
Aarhus; Skagerrak, Denmark	<0.05–0.45	0.1–0.8	DTNP HPLC	Thamdrup et al., 1994b
Chesapeake Bay, Maryland, USA	0.1–7.1	n.a.	IP RP-HPLC	MacCrehan and Shea, 1995
Saguenay Fjord, Canada	0–15		Hg^{2+}-titration	Gagnon et al., 1996
Venice Lagoon, Italy	0–35	n.a.	Voltametry	Bertolin et al., 1997

Note: Table modified and extended from Thamdrup et al., 1993. n.a.—not analyzed. DTNP—2,2′-dithiobis(5-nitropyridine). IP-RP-HPLC—ion pair reverse phase high performance liquid chromatography.

Polysulfides are not easy to quantify in environmental samples since they decompose to ZnS and S^0 as soon as the sediment is fixed with Zn-acetate. Thus, S^0 concentration determined in sulfidic sediments always includes the sulfane sulfur from polysulfides. Under the simplified assumption that all S^0 is transformed into polysulfides if sulfide is present in excess, S^0 concentrations can be used as an upper estimate for the total polysulfide concentration. For Station 6 at 7 cm and below, a polysulfide concentration of 115 µM is calculated by using the average porosity and S^0 values from the same depths (0.1 µmol S^0 cm^{-3}/0.87 ml cm^{-3} = 0.115 µmol mL^{-1} = 115 µM).

Distribution of Thiosulfate ($S_2O_3^{2-}$) and Sulfite (SO_3^{2-})

Table 3 summarizes the results from previous determinations of thiosulfate and SO_3^{2-} in marine sediments and illustrates the large variability in the measured concentrations, ranging from low nM to mM. As already pointed out by Thamdrup et al. (1994b), a variety of different methods have been used for quantification, and it is thus unclear to what extent the variability in the data is due to environmental conditions, sample treatment, or method applied. Since thiosulfate and SO_3^{2-} concentrations in the Black Sea sediments (Fig. 2), an intertidal mud flat in the Weser Estuary, eutrophic sediments off the coast of Central Chile, and a hypersaline cyanobacterial mat (Table 4) were all determined by the MBB derivatization method, a comparison between different systems is now possible. Together with earlier MBB data from salt marsh sediments (Table 3; Vetter et al., 1989), it appears that thiosulfate and SO_3^{2-} concentrations in normal marine sediments are typically in the low micromolar range or below. The low concentrations indicate a high turnover and suggest a tight coupling between sulfur intermediate producing and consuming processes. As for S^0, increased concentrations were mostly observed in highly active and/or dynamic environments, where non–steady-state conditions lead to transient accumulation of sulfur intermediates. For instance, high thiosulfate concentrations in salt marsh sediment are likely caused by intense pyrite oxidation (Luther et al., 1991). In microbial mats, thiosulfate and SO_3^{2-} may be produced in large amounts during the incomplete oxidation of sulfide by cyanobacteria or anoxygenic phototrophic microorganisms (Rabenstein et al., 1995, Wieland et al., 2004).

The values for thiosulfate and SO_3^{2-} presented in this study are in the same range as Thamdrup et al. (1994b) found by 2,2′-dithiobis(5-nitropyridine) (DTNP) derivatization. Despite the report by Witter and Jones (1998) that derivatization with DTNP perturbs coupled equilibria between reactive sulfur species and may lead to a 33% overestimation of thiosulfate, the derivatization methods tend to result in lower concentrations than other methods (Table 3). This suggests that the history of a sample (e.g., exposure to O_2, manipulations and additions, temperature and pH changes) can affect the sulfur speciation even more significantly. Also, the time between sampling and analysis is critical because sulfur speciation can change within minutes if the conditions are unfavorable. The advantage of derivatization methods is therefore that labile sulfur species like sulfite, sulfide, and thiols are rapidly fixed, and reactions between the compounds or with oxygen are excluded. The risk of typical oxidation artifacts, such as the loss of sulfite and increased thiosulfate concentrations, is thereby minimized.

Whereas in some environments maximum thiosulfate concentrations were detected close to the sediment-water interface (Station 2, Fig. 2; Zopfi, 2000; Troelsen and Jørgensen, 1982) where sulfide oxidation is most intense, a similar distribution was not observed at Station 4. There, thiosulfate concentrations increased steadily with depth but did not correlate with pore-water sulfide, thus making an oxidation artifact unlikely. In contrast to S^0, thiosulfate can also be a product of reductive processes (Fitz and Cypionka, 1990). The formation of extracellular thiosulfate has been observed in sulfate-reducing cultures growing under substrate limiting conditions (Vainshtein et al., 1980; Sass et al., 1992). The mineralization rates at Station 4 were very low, and the quality of organic matter decreases typically with sediment depth. Thus, the distribution of thiosulfate could be explained by the incomplete reduction of sulfate under starvation conditions.

TABLE 4. SUMMARY OF SO_3^{2-} AND $S_2O_3^{2-}$ MEASUREMENTS WITH THE MONOBROMOBIMANE DERIVATIZATION METHOD

Site	Concentration (µM)		Comments
	$S_2O_3^{2-}$	SO_3^{2-}	
Weser Estuary, Germany	≤0.5	n.d.	Mud flat, no free sulfide
Black Sea Stn. 2	0.1–1.0	0–0.1	77 m depth, 213 µM O_2*
Black Sea Stn. 4	0.07–2.3	0–0.15	130 m depth, <5 µM O_2*, RTZ†
Black Sea Stn. 6	0.7–3.0	0.5–1.2	396 m depth, 75 µM H_2S*
Chile (Concepción Bay)	1.3–5.7	0.4–2.6	Sulfidic sediment, >1mM H_2S§
Chile (shelf)	0.2–2.2	0.1–0.5	H_2S in sediment <5µM, *Thioploca*
Cyanobacterial mat (Camargue, France)	360	45	Hypersaline, *Microcoleus* sp. dominated

Note: Data from this study, Zopfi (2000), and Wieland et al. (2003).
*Bottom water concentration.
†Redox transition zone.
§Pore-water concentration

This hypothesis could be tested by stimulating sulfate reduction through the addition of organic substrates to intact sediment cores and monitoring changes in thiosulfate concentrations.

Pore-water sulfite concentrations at the three Black Sea stations were typically lower than 1.2 µM. Although SO_3^{2-} is observed in many sulfide oxidation reactions (Table 1), it does not reach high concentrations in the environment, most likely due to its high chemical reactivity.

Sulfide, Thiosulfate, and Sulfite Transformations

Surface sediment (0–3 cm) from Station 2 in the Black Sea was amended with sulfide, thiosulfate, and sulfite in incubation experiments designed to provide insight into the observed thiosulfate and sulfite pore-water distributions. The experiments were performed in duplicates, but as all of them showed qualitatively identical results, only data from one bag of each amendment experiment is shown in Figure 3.

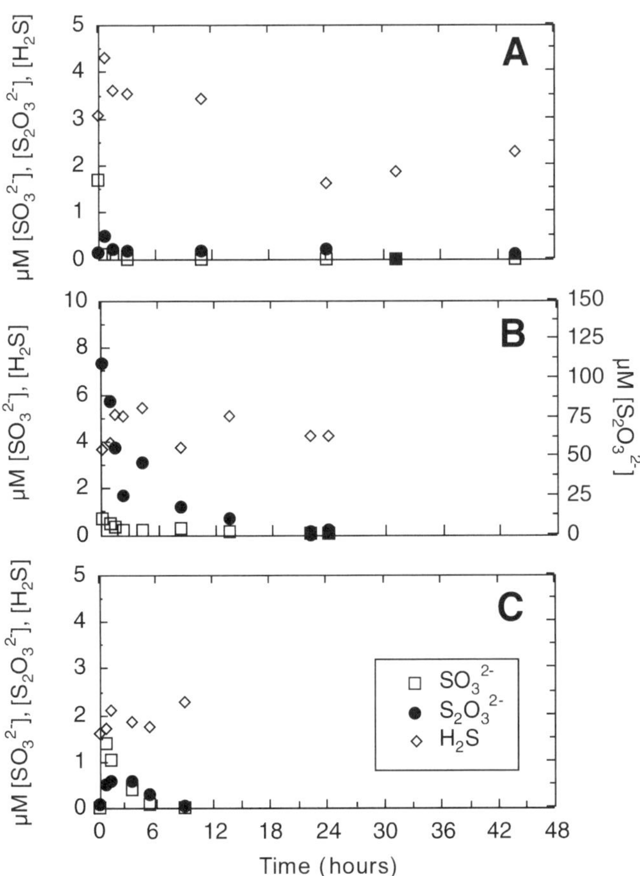

Figure 3. Sulfide, thiosulfate, and sulfite concentrations during a time series experiment with surface sediment from Station 2 in the Black Sea and different amendments: (A) sulfide, (B) thiosulfate, and (C) sulfite addition.

Sulfide Amendment

Sulfide was added to the bag from a freshly prepared stock (2 mM) to obtain a final concentration of ~30–40 µM. The sulfide concentration in the bag was initially 3 µM, but was only slightly higher (4.3 µM) 40 min after the addition. Sulfide then slowly decreased to a minimum concentration of 1.6 µM at 24 h, but increased again toward the end of the experiment, probably due to bacterial sulfate reduction. The sediment in the first 1.5 cm was particularly rich in particulate manganese (125 µmol cm^{-3}) and contained up to 45 µmol cm^{-3} Fe(III)oxides (Thamdrup et al., 2000). Most likely, sulfide was rapidly removed from the pore water by oxidation and precipitation by reactive metal oxides. The concentration of thiosulfate before the addition was 0.14 µM, slightly lower than observed in the pore-water depth profiles, but reached a transient maximum of 0.5 µM immediately after the amendment. Thereafter, the concentrations fell to a rather constant value of 0.2 µM, which is comparable to the pore-water concentration. Sulfite was only measurable immediately after the addition, and concentrations did not exceed 0.08 µM.

Thiosulfate Amendment

By mistake, thiosulfate was added to a much higher concentration than in the other incubations. However, this allowed us to observe the strong rate dependence of the thiosulfate concentration. The disappearance rate was 42 µM h^{-1} at 82 µM $S_2O_3^{2-}$, 8.5 µM h^{-1} at 21 µM $S_2O_3^{2-}$, and only 1.1 µM h^{-1} at a concentration of 6 µM. Despite the addition of 120 µM thiosulfate, the sulfide concentration increased only transiently from 3.6 µM to 5.4 µM. Sulfite immediately rose to 0.7 µM and then fell rapidly to 0.18 µM after 2 h. (In the duplicate bag where the thiosulfate concentration reached only 40 µM, sulfide production was also stimulated, but no dynamics in pore-water sulfite were observed.)

Interestingly, a transient sulfite accumulation accompanied the addition of relatively high concentrations of thiosulfate. This demonstrates a tight coupling between the two species, although the reason for sulfite formation is not yet clear. Sulfite may be produced from thiosulfate by enzymatic reduction according to Equation 14:

$$S_2O_3^{2-} + 2\,[H] \rightarrow HSO_3^- + HS^- \quad (14)$$

where [H] represents a reducing equivalent delivered by the thiosulfate reductase (Barrett and Clark, 1987). The ability to reduce thiosulfate (and tetrathionate; see below) is widely spread in the domains of Bacteria and Archaea. Most sulfate-reducing bacteria reduce thiosulfate to sulfide by soluble enzymes located within the cytoplasm. In contrast, other microorganisms reduce thiosulfate by a periplasm facing membrane-enzyme. Since many of them are unable to use the formed sulfite as an additional electron acceptor (Barrett and Clark, 1987), it is released to the environment. The increase in extracellular sulfite during the incubation experiment is therefore consistent with a partial reduction of thiosulfate by non–sulfate-reducing bacteria. The sulfite released may then react further with extracellular S_8 to form more thiosulfate. Such a "sulfur clearing" mechanism has been proposed for the growth of *Salmonella enterica* (Hinsley and Berks, 2002). Since sulfite is also an intermediate of the bacterial thiosulfate disproportionation (Cypionka et al., 1998), a contribution by this process cannot be excluded; however, thiosulfate disproportionation is a cytoplasmatic process and the appearance of extracellular sulfite is probably less likely.

Sulfite Amendment

Added SO_3^{2-} disappeared very rapidly and reached similar concentrations as found in the pore water of an undisturbed core. Sulfite was not detected in the bag pore water before the amendment and the concentration only increased to 1.4 µM 40 min after the addition. A fraction of the sulfite was transformed into thiosulfate, which rapidly built up to 0.6 µM and decreased again to the same concentration as at the beginning of the experiment (0.07 µM). This may reflect a reaction with S^0 or sulfide to form thiosulfate as observed in laboratory experiments (Atterer, 1960; Chen and Morris, 1972; Heunisch, 1977). As in the thiosulfate experiment, sulfite led to increased sulfide concentrations in the bag. A sample taken after 21 h in the duplicate bag indicated that this sulfide increase was only transient and concentrations decreased again later. Whether this sulfide production was due to disproportionation or dissimilatory reduction of sulfite by sulfate-reducing bacteria cannot be deduced from this experiment. Pure culture studies with sulfate-reducing bacteria, however, showed that sulfite (and thiosulfate) is preferred over sulfate as an electron acceptor, because sulfite reduction precludes the highly energy demanding step of sulfate activation (Widdel, 1988). In recent years, an increasing number of non–sulfate-reducing bacteria have been found to use SO_3^{2-} as an electron acceptor, including members of the genera *Desulfitobacter* sp. (Lie et al., 1999) and *Shewanella* sp. (Perry et al., 1993).

Most of the SO_3^{2-} added to the surface sediment was not recovered in any measured sulfur pool. It is possible that SO_3^{2-} was oxidized to sulfate by reacting with Fe(III)oxides or Mn(IV)oxides. Because sulfite is a strong nucleophile, it could also have reacted with organic molecules to form sulfonates (R-SO_3^-), which have been recognized as a major class of organic sulfur compounds in marine sediments (Vairavamurthy et al., 1994; Vairavamurthy et al., 1995). A reactant half-life of ~5 min has been reported, indicating that the reaction between SO_3^{2-} and organic molecules can be very fast (Vairavamurthy et al., 1994).

Thamdrup et al. (1994b) observed similar variations of SO_3^{2-} and thiosulfate with sediment depth, which was explained either by an oxidative production at a fixed ratio or by coupled transformations as described in Equation 6. In the Black Sea sediments, a covariation of the two sulfur intermediates was not observed, and thiosulfate concentrations were, as is also found in other environments (Tables 3 and 4), typically higher than SO_3^{2-}. Although both compounds can be oxidized, reduced, or disproportionated by bacteria, there are clear differences in terms of their chemical reactivity. Thiosulfate is chemically stable in absence of microorganisms under pH neutral conditions (Millero, 1991) and is less reactive toward organic compounds (Vairavamurthy et al., 1994). Thus, while competing chemical reactions contribute to the rapid disappearance of SO_3^{2-}, the low thiosulfate concentrations in the Black Sea sediments (<3 µM) are mostly due to the activity of thiosulfate-consuming bacteria.

Measurements of Tetrathionate in Natural Environments

Polythionates such as tetrathionate appear as products of the chemical oxidation of H_2S, FeS, and FeS_2 (Table 1). Tetrathionate also forms as an intermediate during the aerobic microbial oxidation of sulfide or thiosulfate to sulfate (e.g., Kelly, 1989; Kelly et al., 1997; van den Ende and van Gemerden, 1993; Podgorsek and Imhoff, 1999). Chemoorganoheterotrophic bacteria oxidizing sulfide and S^0 to tetrathionate as the sole product have been described recently by Sorokin (1996). Under anoxic conditions, tetrathionate is abiotically formed from thiosulfate by oxidation with Mn(IV)oxide (Schippers and Jørgensen, 2001). The anaerobic formation of tetrathionate from thiosulfate with NO_3^- as oxidant, however, is bacterially mediated (Sorokin et al., 1999).

In contrast to the results from laboratory experiments, measurements of tetrathionate in natural environments are few. This is partially due to the lack of simple and sensitive analytical methods, but probably more importantly to the fact that tetrathionate is not a major constituent of dissolved sulfur pools in marine sediment pore waters. It is presently also not possible to directly fix and store tetrathionate with compounds such as monobromobimane or other additives. With a few exceptions, such as salt marsh sediments (300 µM, Luther et al., 1986), concentrations fall below detection limits of ~0.01 µM in Kysing Fjord, Denmark (Bak et al., 1993); 0.5 µM in sediments of intertidal Weser Estuary and Chilean continental shelf (Ferdelman and Fossing, unpublished); and 1 µM in the chemocline of Mariager Fjord (Ramsing et al., 1996). Podgorsek and Imhoff (1999) report finding detectable concentrations of tetrathionate (up to 21.6 µM) in Baltic Sea sediments that were anoxic and contained relatively high concentrations of dissolved hydrogen sulfide. As sulfide readily reacts with tetrathionate to form elemental sulfur and thiosulfate (Atterer, 1960; Steudel, 1989), according to Equation 15

$$S_4O_6^{2-} + H_2S \rightarrow 2\ S_2O_3^{2-} + 2\ H^+ + S^0 \qquad (15)$$

they suggested that the rate of tetrathionate formation must therefore be exceeding its consumption. They proposed a model of sulfide oxidation whereby sulfide is oxidized to zero-valent sulfur in the presence of catalytic amounts of tetrathionate, which in turn is regenerated through the oxidation of thiosulfate (Podgorsek and Imhoff, 1999); however, no possible oxidants for thiosulfate under such reducing conditions were named. Conversely, tetrathionate was not detected in sediment depths that contained low concentrations of hydrogen sulfide (Podgorsek and Imhoff, 1999).

Transformations of Tetrathionate Added to Marine Sediments

Oxidized versus Reduced Sediment

Any tetrathionate that may be formed through either biological or chemical reactions is readily removed from pore-water solution to concentrations below 1 µM. Figure 4 shows the typical course of tetrathionate addition to both oxidized and reduced (but not sulfidic) sediment slurries. In this particular experiment, the effects of sediment reduced substances and oxidation state of the sediment on tetrathionate dynamics were examined by comparing an artificially oxidized sediment with a minimally altered sediment (i.e., reduced). Two slurries were prepared. One of the slurries was vigorously bubbled with air until the normally black sediment had taken on a browner, oxidized appearance. After two hours had elapsed, tetrathionate was added to both slurries, and the tetrathionate and thiosulfate concentrations were measured over time. Additionally, 20 MBq of carrier-free $^{35}SO_4^{2-}$ (Amersham) was added to the anoxic bag (giving an approximate activity of 80 kBq cm^{-3}) in order to track sulfate reduction.

In the reduced slurry (Fig. 4A), tetrathionate disappeared within several hours, at a rate of 31.8 µM h^{-1}, and thiosulfate concentrations increased with a 2:1 $S_2O_3^{2-}$:$S_4O_6^{2-}$ ratio at a rate of 64.7 µM h^{-1}. After the tetrathionate sank to concentrations below 10 µM, the thiosulfate concentrations peaked and began decreasing, albeit at a substantially slower rate (5.9 µM h^{-1}). The oxidized sediments (Fig. 4B) exhibited a somewhat decreased rate of tetrathionate consumption by 25%. Correspondingly, the rate of thiosulfate increase in the oxidized sediment slurry was also slightly lower than in the untreated, reduced slurry, hence the 2:1 stoichiometry between tetrathionate consumption and thiosulfate remained constant. In contrast, the rate of thiosulfate concentration decrease, after the build-up of thiosulfate, was similar for both the reduced and oxidized slurries (5.9 and 6.4 µM h^{-1}, respectively). In neither slurry was dissolved sulfide measurable at any time point. Interestingly, the oxidized sediment exhibited a small lag of one hour before the onset of tetrathionate consumption in the oxidized slurry, and repeated additions of tetrathionate had the effect of increasing tetrathionate consumption (data not shown). These and numerous following incubation experiments confirm the initial observations of Bak et al. (1993) that demonstrate a complete consumption of tetrathionate in anoxic sediments with a concomitant and stoichiometric release of thiosulfate

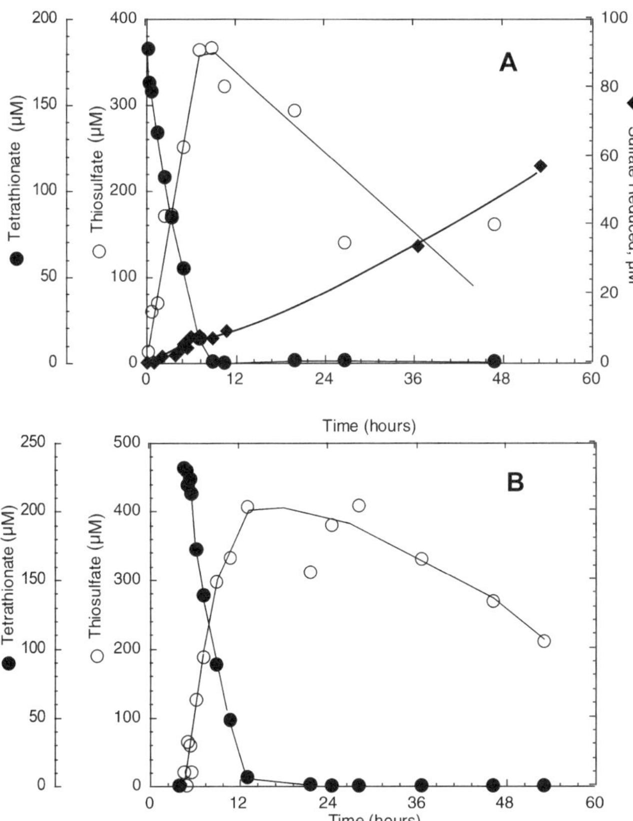

Figure 4. Tetrathionate and thiosulfate concentrations during a time series experiment with (A) reduced and (B) oxidized Weser Estuary sediments. The amount of sulfate reduced in the reduced slurry as measured by ^{35}S-sulfate labeling is also depicted in A.

Inhibition of Microbial Activity

Bak et al. (1993) suggested that the reduction of tetrathionate to thiosulfate is a microbially mediated process. Our experiments with Weser Estuary sediment also show that this conversion is principally a microbial process. We inhibited microbial activity in the sediments either by formaldehyde poisoning (final concentration of 0.1%; Tuominen et al., 1994) or heat sterilization (tyndallization). Formaldehyde treatment and heat sterilization strongly inhibited the rate of tetrathionate reduction relative to the control experiment (85% and 94% inhibition, respectively; data not shown). These inhibition experiments and the temperature response (see below) of tetrathionate consumption clearly indicate a role for bacteria in the reduction of tetrathionate to thiosulfate.

Role of Temperature

Figure 5 shows the rate of tetrathionate degradation in seawater and in Weser Estuary sediment slurries as a function of temperature. Five mL of slurry was added to each of 148 10 mL glass test tubes, fitted with rubber stoppers. The overlying headspace was purged with N_2 and stored at 11 °C overnight (in situ temperature). The filled test tubes were placed in ~2 °C intervals between 10–60 °C in a temperature-gradient block. After the slurry samples were allowed to equilibrate within the temperature gradient block (~1 hr), an exact amount of tetrathionate (170 µM) was then injected into each of the test tubes through the stopper. The test tubes were briefly shaken to equally distribute sediment and tetrathionate and placed back into the temperature gradient block. For each temperature, incubations were stopped at four time points, generally between 10 and 150 min. The incubations were stopped by immediately plunging the test tube into an ice bath until the slurry could be filtered through a 0.4 µm cellulose acetate (Millipore) filter using a pneumatic pore-water squeezer.

In a separate experiment, a series of test tubes containing tetrathionate-amended seawater (no sediment) were run to examine the inorganic decomposition of tetrathionate between 11 and 78 °C. In tetrathionate-amended slurries, tetrathionate consumption increased with rising temperature and peaked at temperatures between 35 °C and 41 °C before decreasing. Without sediment, tetrathionate exhibited only very low rates of chemical degradation at temperatures below 50 °C in seawater. Only at temperatures >50 °C did the rates increase considerably. The peak in tetrathionate reduction at temperatures between 30 and 40 °C (Fig. 5) suggests the role of an enzymatic or biologically catalyzed reaction typical of a mesophilic bacterial population.

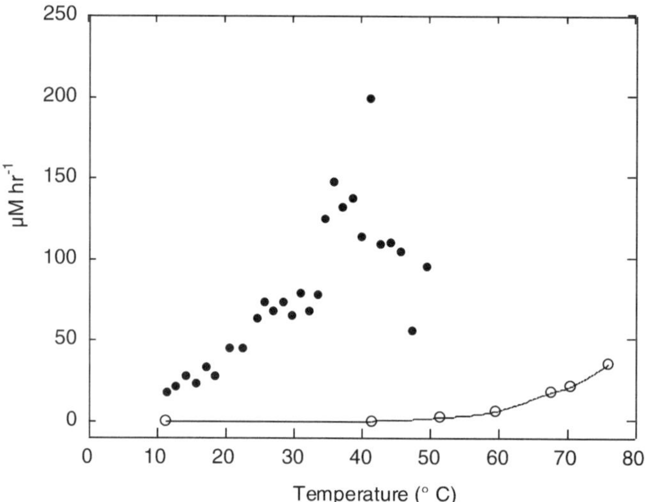

Figure 5. Response of the rate of tetrathionate reduction in Weser Estuary sediments (February 1994) to temperature (closed circles). Open circles indicate the disappearance rate of tetrathionate dissolved in seawater.

Role of Reduced Inorganic Compounds

These experiments do not provide conclusive proof that bacteria directly participate in tetrathionate reduction in these sediments. As shown in Equation 15, dissolved sulfide readily reduces tetrathionate to form thiosulfate and zero-valent sulfur. However, sulfide or other reduced substances do not appear to be chemically reducing tetrathionate in these experiments. In both the Weser Estuary and Skagerrak sediments, dissolved sulfide was not detectable (<1 μM). Oxidizing the sediments to remove sulfides, either free in solution, adsorbed to surfaces, or present as iron sulfides, had little impact on the rate of tetrathionate consumption (Fig. 4). The addition of another reduced compound, Fe(II), to a concentration of 500 μM increased the rate of tetrathionate consumption only slightly over that of the control (16% increase), and concentrations of dissolved iron remained constant throughout the experiment as measured using the Ferrozine method (Stookey, 1970).

Another source of sulfide for the reduction of the tetrathionate could be the continuous production of hydrogen sulfide due to sulfate reduction. Sorokin et al. (1996) propose such a mechanism as a means of regenerating thiosulfate from tetrathionate for further oxidation of thiosulfate and subsequent energy gain in *Catenococcus thiocycli*. Podgorsek and Imhoff (1999) propose a similar mechanism to explain observed tetrathionate concentrations in sulfidic Baltic Sea sediments. We measured the production of sulfide via the turnover of ^{35}S-labeled sulfate in the experiment with the reduced slurry. Sulfide was continually produced from sulfate reduction in the reduced sediment slurry (Fig. 4A); however, the rate of sulfate reduction was much lower than the disappearance rate of tetrathionate.

We sought to exclude sulfide reduction of tetrathionate by blocking sulfate reduction with the addition of molybdate, which is a well-known inhibitor of sulfate reduction. Sodium molybdate was added to slurry to give a final concentration of 20 mM MoO_4^{2-} (approximately equivalent to the sulfate concentration). A second slurry was not treated with molybdate. Within 30 min, tetrathionate was added to both slurries and sampling commenced for the determination of thiosulfate and tetrathionate concentrations. Sulfate reduction was also measured in these slurries. Twenty hours prior to molybdate addition, $^{35}SO_4^{2-}$ was added to both bags, and samples were taken for sulfate reduction rate measurements during, before, and after the molybdate-tetrathionate additions.

In the molybdate-untreated slurry, sulfate reduction proceeded in the first 20 h before addition of tetrathionate at a rate of 4.5 μM h^{-1} (Fig. 6A). Addition of tetrathionate to a concentration of 180 μM had no immediate effect on the sulfate reduction rate. The tetrathionate concentration decreased at a rate of 36.6 μM h^{-1} with a concurrent rise in thiosulfate concentration of 87.2 μM h^{-1}. At maximum thiosulfate concentration and when tetrathionate was fully consumed, a break in the rate of sulfate reduction was observed and the sulfate reduction rate decreased to 2.0 μM h^{-1}, until thiosulfate concentrations fell below 50 μM, at which point sulfate reduction rates increased to 3.3 μM h^{-1}. Thiosulfate decreased in the untreated slurry at a rate of 13.5 μM h^{-1}.

In the slurry that had been treated with molybdate, sulfate reduction initially proceeded at a rate of 3.6 μM h^{-1} until molybdate was added, at which point sulfate reduction ceased for the remainder of the experiment (Fig. 6B). Tetrathionate, added after the molybdate addition, decreased in concentration at a rate of 26.4 μM h^{-1} (72.1% of the rate in the untreated slurry). As with the molybdate-free slurry, stoichiometric increases in thiosulfate matching the decrease in tetrathionate were observed (at a rate of 62.6 μM h^{-1}). Thiosulfate consumption, however, was significantly lower than the molybdate-free slurry (at 1.0 μM h^{-1} or 7.5% of the rate of thiosulfate consumption in the untreated slurry). The experiments demonstrate that although sulfate reduction was fully inhibited by molybdate (and thiosulfate reduction was significantly inhibited), tetrathionate reduction was only partially affected (by ~25–26%). Moreover, rates of tetrathionate reduction significantly exceeded those for sulfate reduction (between 7.5- and 27-fold higher). Thus, sulfide from sulfate reduction could not be titrating the tetrathionate added to the slurries. We therefore conclude that a direct microbial reduction must be responsible for the rapid rates of tetrathionate reduction that were observed.

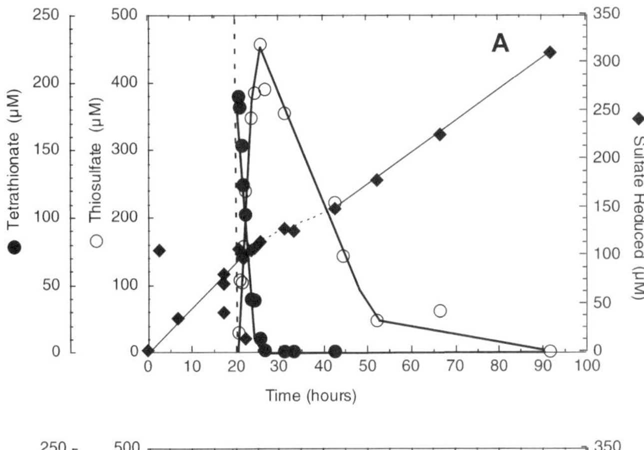

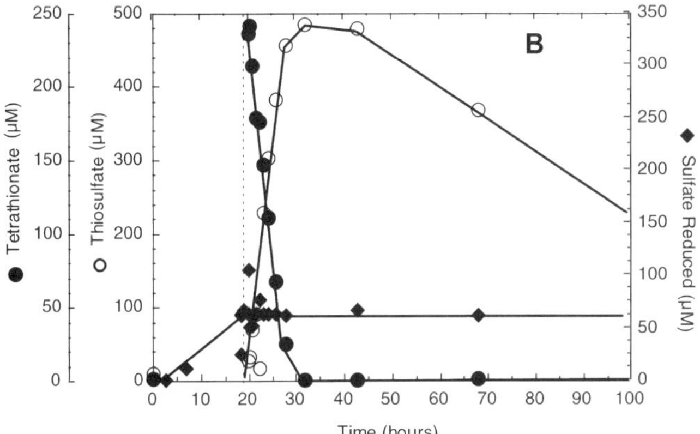

Figure 6. Tetrathionate and thiosulfate concentrations during a time series experiment with (A) untreated and (B) molybdate treated Weser Estuary sediments. Sulfate reduction was also measured in both experiments (^{35}S-sulfate labeling). The vertical dashed line indicates the time the tetrathionate was added to the slurry.

Possible Ecological Role of Tetrathionate Reduction in Marine Sediment

In a review of tetrathionate reduction by non–sulfate-reducing bacteria, Barrett and Clark (1987) suggested that the ability to reduce tetrathionate using the enzyme tetrathionate reductase is more common among anaerobes than the ability to reduce sulfite, the latter being a distinguishing feature of sulfate-reducing bacteria. Tetrathionate reductase catalyzes the following reaction:

$$S_4O_6^{2-} + 2\,[H] \rightarrow 2\,S_2O_3^{2-} + 2\,H^+, \qquad (16)$$

where [H] represents tetrathionate reductase containing reducing equivalents. Tetrathionate reductase is membrane bound, functions best at a pH >7, is regulated by the presence of oxygen and nitrate, and may be part of a reversible enzyme system that catalyzes both the oxidation of thiosulfate and the reduction of tetrathionate (Tuttle and Jannasch, 1973; Tuttle, 1980; Barrett and Clark, 1987). The redox couple of $S_4O_6^{2-}/S_2O_3^{2-}$ lies at a relatively high potential of +170 mV (Barrett and Clark, 1987). The free energies of reaction for the oxidation of organic matter (CH_2O) under standard biochemical conditions (pH = 7.0), via sulfate and tetrathionate reduction, respectively, are shown below (as calculated from compiled $\Delta G'_0$values in Thauer, 1989).

$$SO_4^{2-} + 2\,CH_2O \rightarrow 2\,HCO_3^- + HS^- + H^+ \quad -195.5 \text{ kJ/reaction} \qquad (17)$$

$$2\,S_4O_6^{2-} + CH_2O + 2\,H_2O \rightarrow HCO_3^- + 5\,H^+ + 4\,S_2O_3^{2-} \quad -190.8 \text{ kJ/reaction} \qquad (18)$$

Per mole of reduced carbon or H_2 tetrathionate reduction is more energetically favorable than sulfate reduction (−190.8 kJ/mol versus −97.8 kJ/mol, respectively). Thus, tetrathionate reduction may become favorable when the electron donating substrate is limiting, which is the typical situation in most sediments.

Substrate Amendment

Our experiments indicate that tetrathionate reduction, unlike dissimilatory sulfate or thiosulfate reduction, is not directly coupled as a terminal electron acceptor to the oxidation of organic matter. We base this conclusion on the observation that tetrathionate reduction takes place at substantially higher rates than observed for either sulfate reduction or even thiosulfate consumption. Assuming that the slurries are substrate (organic carbon) limited, the rate of tetrathionate reduction should be only fourfold that of sulfate reduction, based on the stoichiometries in Equations 17 and 18; however, they fell between 7.5 and 27 times the sulfate reduction rate in all experiments where both sulfate reduction and tetrathionate reduction were measured.

The effect of organic matter availability on tetrathionate reduction was studied in a substrate addition experiment (data not shown). Four different slurries were prepared: (a) no substrate, no molybdate, (b) no substrate plus molybdate (ca. 20 mM), (c) substrate, no molybdate, and (d) substrate plus molybdate. The substrate additions consisted of a cocktail containing formate, acetate, propionate, butyrate, and lactate that yielded a 1 mM concentration of each fatty acid in the slurry. These fermentation products are typical substrates for sulfate-reducing bacteria. Molybdate was added to block indirect tetrathionate reduction via sulfide production from dissimilatory sulfate reduction. Addition of substrate yielded only a slight increase in the rate of tetrathionate reduction (221 and 168 $\mu M\ h^{-1}$ with and without substrate, respectively). The slurries where sulfate reduction was inhibited showed a similar pattern, albeit at slightly lower rates (142 and 124 $\mu M\ h^{-1}$ with and without substrate, respectively). These results suggest that tetrathionate reduction is not necessarily linked to the terminal oxidation of substrate to CO_2 and that, more specifically, sulfate reducing bacteria are only minimally involved in tetrathionate reduction in marine sediments.

Moreover, tetrathionate had no effect on the sulfate reduction rate, unlike the subsequent appearance of thiosulfate, which significantly depressed the sulfate reduction rate. Thiosulfate consumption also exhibits an immediate and strong response to the addition of molybdate, whereas tetrathionate reduction decreases by less than one-fourth (see Figs. 4 and 6). This effect of thiosulfate on the sulfate reduction rate has been attributed to the greater energy gain due to thiosulfate reduction over sulfate reduction (Widdel, 1988; Jørgensen, 1990b). In pure cultures of some fermenting heterotrophs (e.g., *Salmonella enterica* [Hinsley and Berks, 2002] and *S. typhimurium* [Hensel et al., 1999]), tetrathionate is also the preferred electron acceptor over thiosulfate. In marine sediments, however, tetrathionate apparently plays no such similar role as preferred electron acceptor, because the concentration of tetrathionate appears to have no direct impact on either the rate of sulfate or thiosulfate reduction.

Alternatives to Dissimilatory Tetrathionate Reduction

If it is not being used as a terminal electron acceptor for sulfate-reducing bacteria, what possible role could tetrathionate reduction have in the microbial community? Anaerobic disproportionation of 4 moles of tetrathionate (Equation 13) to form 6 moles of thiosulfate, 1 mol of trithionate, and 1 mol of sulfate (1.5:1 $S_2O_3^{2-}$:$S_4O_6^{2-}$ ratio) has been shown for the facultative heterotroph *Thiomonas intermedia* K12 (Wentzien and Sand, 1999) at circumneutral pH. Disproportionation of other intermediate sulfur compounds in marine sediments has been demonstrated (Jørgensen, 1990a; Jørgensen and Bak, 1991; Canfield and Thamdrup, 1994, 1996), and there is no reason to think that tetrathionate disproportionation may not occur as well. The major argument, however, that tetrathionate disproportionation is not the principal pathway of tetrathionate consumption, is that the stoichiometry of thiosulfate formation to tetrathionate disappearance is closer to the 2:1 stoichiometry of tetrathionate reduction (Equation 16) than to that of disproportionation (Equation 13). Furthermore, we observed no trithionate formation, which should have appeared during the chromatographic runs.

Tetrathionate reduction as expressed in Equation 16 may also be linked to fermentation, which conforms well to our earlier observation that sulfate- and tetrathionate-reducing bacteria do not have the same substrate spectrum. Fermenting bacteria have a problem getting rid of excess reducing power they generate in form of NADH or NADPH in the oxidative branches of fermentation pathways. Many of them have developed means of releasing electrons to syntrophic partner organisms or external electron acceptors. Such an external electron sink allows fermenters to regenerate NAD(P), and thus to oxidize organic matter further, which results in more ATP production per substrate. Moreover, Barrett and Clark (1987) suggested that tetrathionate reduction may even be coupled with the production of ATP through oxidative phosphorylation. Fermentative bacteria have been shown to dump electrons onto, for example, elemental sulfur, humic substances, and iron oxide and other metal oxides (e.g., Jones et al., 1984; Stal and Moezelaar, 1997; Benz et al., 1998). We speculate that, in sediment where the sulfur cycle is active and tetrathionate may arise through sudden oxidation events, the ability to channel electrons through a membrane-bound tetrathionate reductase may be widespread among facultative and strictly anaerobic bacteria and not just among those involved in sulfate reduction or thiosulfate consumption (reduction or disproportionation).

Tetrathionate Dynamics in the Presence of Oxidants

Although this study has focused principally on the fate of tetrathionate added to sediment slurries under anaerobic conditions, there are indications that the thiosulfate-tetrathionate system is altered in the presence of oxidants such as oxygen, nitrate, and manganese oxides. Where air was continually bubbled through the slurry, tetrathionate consumption decreased to 41.8% of the untreated control (data not shown). In the two experiments where nitrate was added to a final concentration of 200 μM, the rates of tetrathionate consumption decreased to 89% and 55% of the unamended rates. Nitrate addition tended to flatten out the thiosulfate response (Fig. 7). The initial increase in thiosulfate was only 36.4% of the unamended rate, and the decrease was also lower (27.9%). Both of these experiments conform to the observation from pure culture studies that tetrathionate reductase is repressed by higher redox potential electron acceptors such as oxygen and nitrate (Barrett and Clark, 1987).

Manganese oxides may also inhibit tetrathionate reduction, as shown by the results from the two Skagerrak sites (Fig. 8). At Station S4, where sulfate reduction rates vary between 8 and 12 μM h^{-1} (Canfield et al., 1993), tetrathionate disappeared at a rate of 35.7 μM h^{-1} and exhibited a nearly stoichiometric increase in thiosulfate concentration (60.9 μM h^{-1}). At this typical continental margin site, tetrathionate decreased to below detection limits within 8 h, and thiosulfate, after its initial build-up, decreased to near 10 μM within 32 h. In contrast, the behavior of tetrathionate and thiosulfate in the manganese oxide-rich sediments of Station S9 was strikingly different. A lag time of 8 h was required before any tetrathionate reduction occurred. At

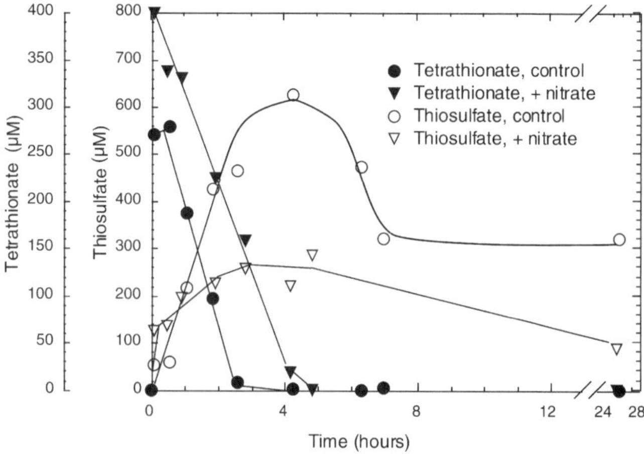

Figure 7. Tetrathionate and thiosulfate concentrations during a time series experiment with untreated and nitrate amended Weser Estuary sediments.

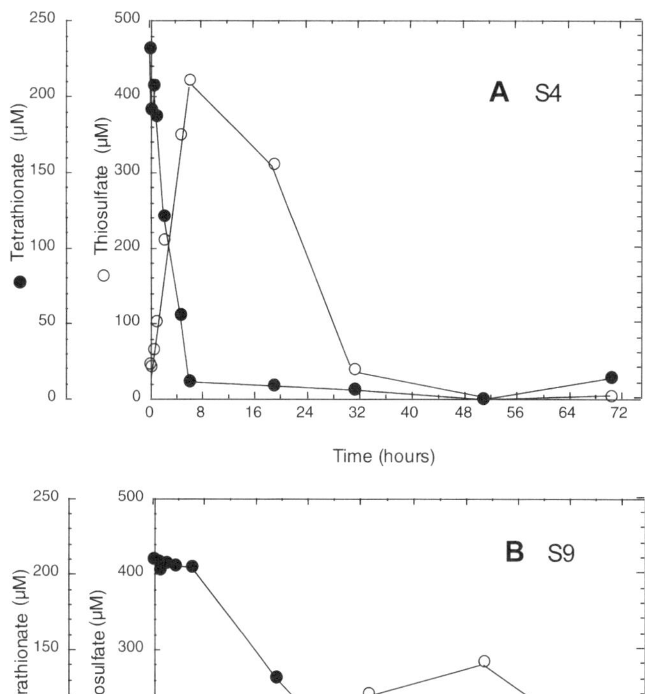

Figure 8. Tetrathionate and thiosulfate concentrations during time series experiments with sediment from (A) Station S4 (190 m water depth) and (B) Station S9 (695 m water depth) from the continental slope of the Skagerrak region of the North Sea.

this point, tetrathionate consumption commenced, but at a much lower rate of 5.1 μM h^{-1}, with a corresponding increase in thiosulfate of 8.9 μM h^{-1}. Furthermore, tetrathionate concentrations never went to zero. Rather, they remained constant at near 60 μM or even slightly increased over the remaining 36 h of the experiment, which may reflect the concurrent reoxidation of thiosulfate to tetrathionate by MnO_2 (Schippers and Jørgensen, 2001). The increase in thiosulfate also exhibited the characteristically flat response, as seen in the experiments with aerated and nitrate amended sediments.

In oxidized sediments, tetrathionate typically disappeared only after a time lag of up to several hours, which suggests that the capacity to reduce tetrathionate must first be induced. However, in most marine coastal sediments, the response to tetrathionate additions is immediate, suggesting that the bacteria are primed and waiting for tetrathionate arising from various sulfide oxidation events.

CONCLUSIONS

This work demonstrates that in most marine sediments the concentrations of SO_3^{2-}, and $S_2O_3^{2-}$, and $S_4O_6^{2-}$ are in the sub-micromolar range with maximum values not exceeding a few micromoles per liter. Elemental sulfur is the most abundant sulfur intermediate in coastal marine sediments. In sediments deposited under oxic conditions, a distinct subsurface maximum of S^0 is often observed, possibly associated with the depth of the bioturbation zone, whereas in anoxic environments (e.g., in the Black Sea), the highest values of S^0 are found at the sediment-water interface.

The low concentrations of the dissolved intermediates reflect equilibrium conditions where the rates of production and consumption are tightly coupled. Disequilibrium conditions due to bioturbation events or rapid temperature changes, for example, may lead to sudden and high concentration excursions in one or more of the intermediate sulfur compounds, but they will rapidly return to low equilibrium concentrations.

Both chemical and biochemical pathways are operating to maintain such low concentrations. Sulfite disappeared rapidly and was, most likely, chemically oxidized to sulfate or reacted with other sulfur compounds, such as elemental sulfur or sulfide. Tetrathionate is readily reduced in the presence of excess sulfide to give thiosulfate and polysulfides. However, in non-sulfidic sediments, which comprise the majority of surface marine sediments, tetrathionate and thiosulfate are chemically stable. Under such conditions, both tetrathionate and thiosulfate are consumed directly in bacterially mediated processes that drive the concentrations of both tetrathionate and thiosulfate to low equilibrium concentrations.

The rates at which the concentrations of sulfur intermediates return to equilibrium decrease in the order: $SO_3^{2-} \approx S_4O_6^{2-} > S_2O_3^{2-} > S^0$. Elemental sulfur and thiosulfate are the key intermediates in sulfide oxidation, based both on their concentration and on their lower rates of turnover. For example, thiosulfate is consumed much more slowly than tetrathionate is reduced to thiosulfate. If tetrathionate is formed during any of the various sulfide oxidation pathways, it will primarily be reduced back to thiosulfate, and thus, sulfur cycling through tetrathionate acts mostly as a closed-loop under anoxic conditions. Therefore, the processes regulating thiosulfate consumption are rate-determining steps, or bottlenecks, in the oxidative half of the sulfur cycle.

ACKNOWLEDGMENTS

We thank the crew of the R/V *Petr Kottsov* and B.B. Jørgensen and A. Weber for leading and organizing the Black Sea Cruise. We also thank the crew of the F/S *Victor Hensen* and Chief Scientist S. Forster for their assistance under less than ideal weather conditions, and K. Neumann and D. Ganzhorn for assistance in the laboratory. We thank A. Schippers for his detailed and helpful review, J. Amend for his patience and helpful editorial comments, and finally, one anonymous reviewer, who pointed out the possibility of tetrathionate as an electron sink for fermenting bacteria. This research was sponsored by the Max-Planck Society; J.Z. was additionally supported in the writing process by the Swiss National Science Foundation (Grant No. 83 EU-062451).

REFERENCES CITED

Aller, R.C., and Rude, P.D., 1988, Complete oxidation of solid phase sulfides by manganese and bacteria in anoxic marine sediments: Geochimica et Cosmochimica Acta, v. 52, p. 751–765, doi: 10.1016/0016-7037(88)90335-3.

Atterer, M., 1960, Gmelins Handbuch der Anorganischen Chemie, Element S: Schwefel, "Teilband B.": Weinheim, Germany, Verlag Chemie, 853–953.

Avrahami, M., and Golding, R.M., 1968, The oxidation of the sulfide ion at very low concentrations in aqueous solutions: Journal of the Chemical Society, v. A, p. 647–651.

Bak, F., and Cypionka, H., 1987, A novel type of energy metabolism involving fermentation of inorganic sulphur compounds: Nature, v. 326, p. 891–892, doi: 10.1038/326891A0.

Bak, F., Schuhmann, A., and Jansen, K.-H., 1993, Determination of tetrathionate and thiosulfate in natural samples and microbial cultures by a new, fast and sensitive ion chromatographic technique: FEMS Microbiology Ecology, v. 12, p. 257–264, doi: 10.1016/0168-6496(93)90049-D.

Barrett, E.L., and Clark, M.A., 1987, Tetrathionate reduction and production of hydrogen sulfide from thiosulfate: Microbiological Reviews, v. 51, p. 192–205.

Bartlett, J.K., and Skoog, D.A., 1954, Colorimetric determination of elemental sulfur in hydrocarbons: Analytical Chemistry, v. 26, p. 1008–1011.

Benz, M., Schink, B., and Brune, A., 1998, Humic acid reduction by *Propionibacterium freudenreichii* and other fermenting bacteria: Applied and Environmental Microbiology, v. 64, p. 4507–4512.

Bertolin, A., Mazzocchin, G.A., Rudello, D., and Ugo, P., 1997, Seasonal and depth variability of reduced sulfur species and metal ions in mud-flat pore-waters of the Venice lagoon: Marine Chemistry, v. 59, p. 127–140, doi: 10.1016/S0304-4203(97)00075-3.

Boulègue, J., Lord, C.J., III, and Church, T.M., 1982, Sulfur speciation and associated trace metals (Fe, Cu) in the pore waters of Great Marsh, Delaware: Geochimica et Cosmochimica Acta, v. 46, p. 453–464, doi: 10.1016/0016-7037(82)90236-8.

Boulègue, J., and Denis, J., 1983, Sulfide speciations in upwelling areas. In: Coastal upwelling. Its sedimentary record. Part A (ed. J. Thiede and E. Suess) New York, Plenum Press, p. 439–454.

Burdige, D.J., and Nealson, K.H., 1986, Chemical and microbiological studies of sulfide-mediated manganese reduction: Geomicrobiology Journal, v. 4, p. 361–387.

Canfield, D.E., 1989, Reactive iron in marine sediments: Geochimica et Cosmochimica Acta, v. 53, p. 619–632, doi: 10.1016/0016-7037(89)90005-7.

Canfield, D.E., and Teske, A., 1996, Late Proterozoic rise in atmospheric oxygen concentration inferred from phylogenetic and sulfur-isotope studies: Nature, v. 382, p. 127–132, doi: 10.1038/382127A0.

Canfield, D.E., and Thamdrup, B., 1994, The production of ^{34}S-depleted sulfide during bacterial disproportionation of elemental sulfur: Science, v. 266, p. 1973–1975.

Canfield, D.E., and Thamdrup, B., 1996, Fate of elemental sulfur in an intertidal sediment: FEMS Microbiology Ecology, v. 19, p. 95–103, doi: 10.1016/0168-6496(95)00083-6.

Canfield, D.E., Thamdrup, B., and Hansen, J.W., 1993, The anaerobic degradation of organic matter in Danish coastal sediments: Iron reduction, manganese reduction, and sulfate reduction: Geochimica et Cosmochimica Acta, v. 57, p. 3867–3883, doi: 10.1016/0016-7037(93)90340-3.

Chen, K.Y., and Morris, J.C., 1972, Kinetics of oxidation of aqueous sulfide by O_2: Environmental Science & Technology, v. 6, p. 529–537.

Cline, J.D., 1969, Spectrophotometric determination of hydrogen sulfide in natural waters: Limnology and Oceanography, v. 14, p. 454–458.

Cline, J.D., and Richards, F.A., 1969, Oxygenation of hydrogen sulfide in seawater of constant salinity, temperature, and pH: Environmental Science & Technology, v. 3, p. 838–843.

Cypionka, H., Smock, A.M., and Böttcher, M.E., 1998, A combined pathway of sulfur compound disproportionation in *Desulfovibrio desulfuricans*: FEMS Microbiology Letters, v. 166, p. 181–186, doi: 10.1016/S0378-1097(98)00330-9.

dos Santos, A.M., and Stumm, W., 1992, Reductive dissolution of iron(III)hydroxides by hydrogen sulfide: Langmuir, v. 8, p. 1671–1675.

Elsgaard, L., and Jørgensen, B.B., 1992, Anoxic transformation of radiolabeled hydrogen sulfide in marine and freshwater sediments: Geochimica et Cosmochimica Acta, v. 56, p. 2425–2435, doi: 10.1016/0016-7037(92)90199-S.

Fahey, R.C., and Newton, G.L., 1987, Determination of low-weight thiols using monobromobimane fluorescent labeling and high-performance liquid chromatography: Methods in Enzymology, v. 143, p. 85–96.

Ferdelman, T.G., 1994. Oceanographic and geochemical controls on sulfur diagenesis in coastal sediments [Ph.D. Dissertation]: Newark, Delaware, University of Delaware, 140 p.

Ferdelman, T.G., Church, T.M., and Luther, G.W., III, 1991, Sulfur enrichment of humic substances in a Delaware salt marsh sediment core: Geochimica et Cosmochimica Acta, v. 55, p. 979–988, doi: 10.1016/0016-7037(91)90156-Y.

Ferdelman, T.G., Lee, C., Pantoja, S., Harder, J., Bebout, B., and Fossing, H., 1997, Sulfate reduction and methanogenesis in a *Thioploca*-dominated sediment off the coast of Chile: Geochimica et Cosmochimica Acta, v. 61, p. 3065–3079, doi: 10.1016/S0016-7037(97)00158-0.

Ferdelman, T.G., Fossing, H., Neumann, K., and Schulz, H.D., 1999, Sulfate reduction in surface sediments of southeast Atlantic continental margin sediments between 15°38′ S and 27°57′ S (Angola and Namibia): Limnology and Oceanography, v. 44, p. 650–661.

Fitz, R.M., and Cypionka, H., 1990, Formation of thiosulfate and trithionate during sulfite reduction by washed cells of *Desulfovibrio desulfuricans*: Archives of Microbiology, v. 154, p. 400–406.

Fossing, H., 1990, Sulfate reduction in shelf sediments in the upwelling region off Central Peru: Continental Shelf Research, v. 10, p. 355–367, doi: 10.1016/0278-4343(90)90056-R.

Fossing, H., and Jørgensen, B.B., 1989, Measurement of bacterial sulfate reduction in sediments: Evaluation of a single-step chromium reduction method: Biogeochemistry, v. 8, p. 205–222.

Fossing, H., and Jørgensen, B.B., 1990, Oxidation and reduction of radiolabeled inorganic sulfur compounds in an estuarine sediment, Kysing Fjord, Denmark: Geochimica et Cosmochimica Acta, v. 54, p. 2731–2742, doi: 10.1016/0016-7037(90)90008-9.

Gagnon, C., Mucci, A., and Pelletier, E., 1996, Vertical distribution of dissolved sulphur species in coastal marine sediments: Marine Chemistry, v. 52, p. 195–209, doi: 10.1016/0304-4203(95)00099-2.

Habicht, K., Canfield, D.E., and Rethmeier, J., 1998, Sulfur isotope fractionation during bacterial reduction and disproportionation of thiosulfate and sulfite: Geochimica et Cosmochimica Acta, v. 62, p. 2585–2595, doi: 10.1016/S0016-7037(98)00167-7.

Henneke, E., Luther, G.W., III, de Lange, G.J., and Hoefs, J., 1997, Sulphur speciation in anoxic hypersaline sediments from the eastern Mediterranean Sea: Geochimica et Cosmochimica Acta, v. 61, p. 307–321, doi: 10.1016/S0016-7037(96)00355-9.

Hensel, M., Hinsley, A.M., Nikolaus, T., Sawers, G., and Berks, B.C., 1999, The genetic basis of tetrathionate respiration in *Salmonella typhimurium*: Molecular Microbiology, v. 32, p. 275–288, doi: 10.1046/J.1365-2958.1999.01345.X.

Heunisch, G.W., 1977, Stoichiometry of the reaction of sulfites with hydrogen sulfide ion: Inorganic Chemistry, v. 16, p. 1411–1413.

Hinsley, A.P., and Berks, B.C., 2002, Specificity of respiratory pathways involved in the reduction of sulfur compounds by *Salmonella enterica*: Microbiology, v. 148, p. 3631–3638.

Howarth, R.W., Giblin, A.E., Gale, J., Peterson, B.J., and Luther, G.W., III, 1983, Reduced sulfur compounds in porewaters of a New England salt marsh, *in* Hallberg, R.O., ed., Environmental Biogeochemistry, Ecological Bulletin (Stockholm), v. 35, p.135–152.

Huettel, M., Ziebis, W., Forster, S., and Luther, G.W., III, 1998, Advective transport affecting metal and nutrient distribution and interfacial fluxes in permeable sediments: Geochimica et Cosmochimica Acta, v. 62, p. 613–631, doi: 10.1016/S0016-7037(97)00371-2.

Jacobs, L., and Emerson, S., 1982, Trace metal solubility in an anoxic fjord: Earth and Planetary Science Letters, v. 60, p. 237–252, doi: 10.1016/0012-821X(82)90006-1.

Jones, J.G., Gardener, S., and Simon, B.M., 1984, Reduction of ferric iron by heterotrophic bacteria in lake sediments: Journal of General Microbiology, v. 130, p. 45–51.

Jørgensen, B.B., 1982, Mineralization of organic matter in the sea bed—the role of sulphate reduction: Nature, v. 296, p. 643–645.

Jørgensen, B.B., 1987, Ecology of the sulphur cycle: oxidative pathways in sediments. *in* Cole, J.A., and Ferguson, S., eds., The nitrogen and sulphur cycles: Society for General Microbiology Symposium 42, Cambridge University Press, p. 31–36.

Jørgensen, B.B., 1990a, A thiosulfate shunt in the sulfur cycle of marine sediments: Science, v. 249, p. 152–154.

Jørgensen, B.B., 1990b, The sulfur cycle of freshwater sediments: role of thiosulfate: Limnology and Oceanography, v. 35, p. 1329–1342.

Jørgensen, B.B., and Bak, F., 1991, Pathways and microbiology of thiosulfate transformations and sulfate reduction in a marine sediment (Kattegat, Denmark): Applied and Environmental Microbiology, v. 57, p. 847–856.

Kelly, D.P., 1989, Oxidation of sulphur compounds, *in* Cole, J.A., and Ferguson, S., eds., The nitrogen and sulphur cycles: Society for General Microbiology Symposium 42, Cambridge University Press, p. 65–98.

Kelly, D.P., Shergill, J.K., Lu, W.-P., and Wood, A.P., 1997, Oxidative metabolism of inorganic sulfur compounds by bacteria: Antonie van Leeuwenhoek, v. 71, p. 95–107, doi: 10.1023/A:1000135707181.

King, G.M., 1990, Effects of added manganic and ferric oxides on sulfate reduction and sulfide oxidation in intertidal sediments: FEMS Microbiology Ecology, v. 73, p. 131–138, doi: 10.1016/0378-1097(90)90659-E.

Kohnen, M.E.L., Sinninghe Damsté, J.S., ten Haven, H.L., and de Leeuw, J.W., 1989, Early incorporation of polysulfides in sedimentary organic matter: Nature, v. 341, p. 640–641, doi: 10.1038/341640A0.

Kuenen, J.G., and Bos, P., 1989, Habitats and ecological niches of chemolitho(auto)trophic bacteria, *in* Schlegel, H.G., and Bowien, B., eds., Biology of autotrophic bacteria: Berlin, Springer-Verlag, p. 52–80.

Lein, Y.A., 1984, Anaerobic consumption of organic matter in modern marine sediments: Nature, v. 312, p. 148–150.

Lie, T.J., Godchaux, W., and Leadbetter, E.R., 1999, Sulfonates as terminal electron acceptors for growth of sulfite-reducing bacteria (*Desulfitobacterium* sp.) and sulfate-reducing bacteria: Effects of inhibitors of sulfidogenesis: Applied and Environmental Microbiology, v. 65, p. 4611–4617.

Lovley, D.R., and Phillips, E.J.P., 1994, Novel processes for anaerobic sulfate production from elemental sulfur by sulfate-reducing bacteria: Applied and Environmental Microbiology, v. 60, p. 2394–2399.

Luther, G.W., III, 1987, Pyrite oxidation and reduction: molecular orbital considerations: Geochimica et Cosmochimica Acta, v. 51, p. 3193–3199, doi: 10.1016/0016-7037(87)90127-X.

Luther, G.W., III, 1991, Pyrite formation via polysulfide compounds: Geochimica et Cosmochimica Acta, v. 55, p. 2839–2849.

Luther, G.W., III, Giblin A.E., and Vorsolona, R., 1985, Polarographic analysis of sulfur species in marine porewaters: Limnology and Oceanography, v. 30, p. 727–736.

Luther, G.W., III, Church, T.M., Scudlark, J.R., and Cosman, M., 1986, Inorganic and organic sulfur cycling in salt-marsh pore waters: Science, v. 232, p. 746–749.

Luther, G.W., III, and Church, T.M., 1988, Seasonal cycling of sulfur and iron in porewaters of a Delaware salt marsh: Marine Chemistry, v. 23, p. 295–309, doi: 10.1016/0304-4203(88)90100-4

Luther, G.W., III, Ferdelman, T.G., Kostka, J.E., Tsmaskis, E.J., and Church, T.M., 1991, Temporal and spatial variability of reduced sulfur species (FeS_2, $S_2O_3^{2-}$) and pore water parameters in salt marsh sediments: Biogeochemistry, v. 14, p. 57–88.

MacCrehan, W., and Shea, D., 1995, Temporal relationship of thiols to inorganic sulfur compounds in anoxic Chesapeake Bay sediment porewater, *in* Vairavamurthy, M.A., and Schoonen, M.A.A., eds., Geochemical transformations of sedimentary sulfur: Washington, D.C., American Chemical Society Symposium Series 612, p. 295–310.

Millero, F., 1991, The oxidation of H_2S in Black Sea waters: Deep-Sea Research II, v. 38, Supplement 2, p. S1139–S1150.

Möckel, H.J., 1984a, Retention of sulphur and sulphur organics in reversed-phase liquid chromatography: Journal of Chromatography, v. 317, p. 589–614, doi: 10.1016/S0021-9673(01)91699-1.

Möckel, H.J., 1984b, The retention of sulphur homocycles in reversed-phase HPLC: Analytical Chemistry, v. 318, p. 327–334.

Moeslund, L., Thamdrup, B., and Jørgensen, B.B., 1994, Sulfur and iron cycling in a coastal sediment: Biogeochemistry, v. 27, p. 129–152.

Morse, J.W., Millero, F.J., Cornwell, J.C., and Rickard, D., 1987, The chemistry of the hydrogen sulfide and iron sulfide systems in natural waters: Earth Science Reviews, v. 24, p. 1–42, doi: 10.1016/0012-8252(87)90046-8

Moses, C.O., Nordstrom, D.K., Herman, J.D., and Mills, A.L., 1987, Aqueous pyrite oxidation by dissolved oxygen and by ferric iron: Geochimica et Cosmochimica Acta, v. 51, p. 1561–1571, doi: 10.1016/0016-7037(87)90337-1.

Otte, S., Kuenen, J.G., Nielsen, L.P., Pearl, H.W., Zopfi, J., Schulz, H.N., Teske, A., Strotmann, B., Gallardo, V.A., and Jørgensen, B.B., 1999, Nitrogen, carbon and sulfur metabolism in natural *Thioploca* samples: Applied and Environmental Microbiology, v. 65, p. 3148–3157.

Perry, K.A., Kostka, J.A., Luther, G.W., III, and Nealson, K.H., 1993, Mediation of sulfur speciation by a Black Sea facultative anaerobe: Science, v. 259, p. 801–803.

Peiffer, S., dos Santos Alfonso, M., Wehrli, B., and Gächter, R., 1992, Kinetics and mechanisms of the reaction of H_2S with lepidocrocite: Environmental Science & Technology, v. 26, p. 2408–2413.

Podgorsek, L., 1998, Oxidative Prozesse des Schwefelzyklus in den Sedimenten der Ostsee: Aerobe, bakterielle Umsetzung von Thiosulfat [Ph.D. Dissertation]: Kiel, Germany, Christian-Albrechts Universität Kiel, 105 p. (in German)

Podgorsek, L., and Imhoff, J.F., 1999, Tetrathionate production by sulfur oxidizing bacteria and the role of tetrathionate in the sulfur cycle of Baltic Sea sediments: Aquatic Microbial Ecology, v. 17, p. 255–265.

Pyzik, A.J., and Sommer, S.E., 1981, Sedimentary iron monosulfides: Kinetics and mechanism of formation: Geochimica et Cosmochimica Acta, v. 45, p. 687–698, doi: 10.1016/0016-7037(81)90042-9.

Rabenstein, A., Rethmeier J., and Fischer, U., 1995, Sulphite as intermediate sulphur compound in anaerobic sulphide oxidation to thiosulfate by marine cyanobacteria: Zeitschrift für Naturforschung, v. 50c, p. 769–774.

Ramsing, N.B., Fossing, H., Ferdelman, T.G., Andersen, F., and Thamdrup, B., 1996, Distribution of bacterial populations in a stratified fjord (Mariager Fjord, Denmark) quantified by in situ hybridization and related to chemical gradients in the water column: Applied and Environmental Microbiology, v. 62, p. 1391–1404.

Rethmeier, J., Rabenstein, A., Langer, M., and Fischer, U., 1997, Detection of traces of oxidized and reduced sulfur compounds in small samples by combination of different high-performance liquid chromatography methods: Journal of Chromatography A, v. 760, p. 295–302, doi: 10.1016/S0021-9673(96)00809-6.

Sagemann, J., Skowronek, F., Dahmke, A., and Schulz, H.D., 1996, Pore-water response on seasonal environmental changes in intertidal sediments of the Weser Estuary, Germany: Environmental Geology, v. 27, p. 362–369, doi: 10.1007/S002540050070.

Sass, H., Steuber, J., Kroder, M., Kroneck, P.M.H., and Cypionka, H., 1992, Formation of thionates by freshwater and marine strains of sulfate-reducing bacteria: Archives of Microbiology, v. 158, p. 418–421.

Schimmelmann, A., and Kastner, M., 1993, Evolutionary changes over the last 1000 years of reduced sulfur phases and organic carbon in varved sediments of the Santa Barbara Basin, California: Geochimica et Cosmochimica Acta, v. 57, p. 67–78, doi: 10.1016/0016-7037(93)90469-D.

Schippers, A., Josza, P.G., and Sand, W., 1996, Sulfur chemistry in bacterial leaching of pyrite: Applied and Environmental Microbiology, v. 62, p. 3424–3431.

Schippers, A., and Jørgensen, B.B., 2001, Oxidation of pyrite and iron sulfide by manganese dioxide in marine sediments: Geochimica et Cosmochimica Acta, v. 65, p. 915–922, doi: 10.1016/S0016-7037(00)00589-5.

Skyring, G.W., 1987, Sulfate reduction in coastal ecosystems: Geomicrobiology Journal, v. 5, p. 295–374.

Sørensen, J., and Jørgensen, B.B., 1987, Early diagenesis in sediments from Danish coastal waters: Microbial activity and Mn-Fe-S geochemistry: Geochimica et Cosmochimica Acta, v. 51, p. 1583–1590, doi: 10.1016/0016-7037(87)90339-5.

Sorokin, D.Y., 1996, Oxidation of sulfide and elemental sulfur to tetrathionate by chemoorganoheterotrophic bacteria: Microbiology, v. 65, p. 1–5. (English translation)

Sorokin, D.Y., Robertson, L.A., and Kuenen, J.G., 1996, Sulfur cycling in *Catenococcus thiocyclus*: FEMS Microbiology Ecology, v. 19, p. 117–125, doi: 10.1016/0168-6496(95)00085-2.

Sorokin, D.Y., Teske, A., Robertson, L.A., and Kuenen, J.G., 1999, Anaerobic oxidation of thiosulfate to tetrathionate by obligately heterotrophic bacteria, belonging to the *Pseudomonas stutzeri* group: FEMS Microbiology Ecology, v. 30, p. 113–123, doi: 10.1016/S0168-6496(99)00045-8.

Stal, L.J., and Moezelaar, R., 1997, Fermentation in cyanobacteria: FEMS Microbiology Reviews, v. 21, p. 179–211, doi: 10.1016/S0168-6445(97)00056-9.

Steudel, R., 1989, On the nature of "elemental sulfur" (S0) produced by sulfur oxidizing bacteria–a model for S^0 globules, *in* Schlegel, H.G., and Bowien, B., eds., Biology of Autotrophic Bacteria: Berlin, Springer-Verlag, p. 289–303.

Steudel, R., 1996, Mechanisms of the formation of elemental sulfur from aqueous sulfide in chemical and microbiological desulfurization processes: Industrial Engineering Chemical Research, v. 35, p. 1417–1423.

Steudel, R., Holdt, G., and Nagorka, R., 1986, On the autoxidation of aqueous sodium polysulfide: Zeitschrift Naturforschung, v. 41b, p. 1519–1522.

Stookey, L., 1970, Ferrozine—a new spectrophotometric reagent for iron: Analytical Chemistry, v. 42, p. 779–781.

Straub, K.L., Benz, M., Schink, B., and Widdel, F., 1996, Anaerobic, nitrate-dependent microbial oxidation of ferrous iron: Applied and Environmental Microbiology, v. 62, p. 1458–1460.

Swider, K.T., and Mackin, J.E., 1989, Transformations of sulfur compounds in marsh-flat sediments: Geochimica et Cosmochimica Acta, v. 53, p. 2311–2323, doi: 10.1016/0016-7037(89)90353-0.

Taylor, C.D., and Wirsen, C.O., 1997, Microbiology and ecology of filamentous sulfur formation: Science, v. 277, p. 1483–1485, doi: 10.1126/SCIENCE.277.5331.1483.

Thamdrup, B., and Canfield, D.E., 1996, Pathways of carbon oxidation in continental margin sediments off central Chile: Limnology and Oceanography, v. 41, p. 1629–1650.

Thamdrup, B., Finster, K., Hansen, J.W., and Bak, F., 1993, Bacterial disproportionation of elemental sulfur coupled to chemical reduction of iron or manganese: Applied and Environmental Microbiology, v. 59, p. 101–108.

Thamdrup, B., Fossing, H., and Jørgensen, B.B., 1994a, Manganese, iron, and sulfur cycling in a coastal marine sediment, Aarhus bay, Denmark: Geochimica et Cosmochimica Acta, v. 58, p. 5115–5129, doi: 10.1016/0016-7037(94)90298-4.

Thamdrup, B., Finster, K., Fossing, H., Hansen, J.W., and Jørgensen, B.B., 1994b, Thiosulfate and sulfite distributions in pore water of marine sediments related to manganese, iron, and sulfur geochemistry: Geochimica et Cosmochimica Acta, v. 58, p. 67–73, doi: 10.1016/0016-7037(94)90446-4.

Thamdrup, B., Roselló-Mora, R., and Amann, R., 2000, Microbial manganese and sulfate reduction in Black Sea shelf sediments: Applied and Environmental Microbiology, v. 66, p. 2888–2897, doi: 10.1128/AEM.66.7.2888-2897.2000.

Thauer, R.K., 1989, Energy metabolism of sulfate-reducing bacteria, in Schlegel, H.G., and Bowien, B., eds., Autotrophic bacteria. E.: Berlin, Springer-Verlag, p. 397–413.

Thode-Andersen, S., and Jørgensen, B.B., 1989, Sulfate reduction and the formation of ^{35}S-labeled FeS, FeS_2, and S^0 in coastal marine sediments: Limnology and Oceanography, v. 34, p. 793–806.

Troelsen, H., and Jørgensen, B.B., 1982, Seasonal dynamics of elemental sulfur in two coastal sediments: Estuarine, Coastal and Shelf Science, v. 15, p. 255–266.

Tuominen, L., Kairesalo, T., and Hartikainen, H., 1994, Comparison of methods for inhibiting bacterial activity in sediment: Applied and Environmental Microbiology, v. 60, p. 3454–3457.

Tuttle, J.H., 1980, Thiosulfate oxidation and tetrathionate reduction by a marine pseudomonad 16B: Applied and Environmental Microbiology, v. 39, p. 1159–1166.

Tuttle, J.H., and Jannasch, H.W., 1973, Dissimilatory reduction of inorganic sulfur by facultatively anaerobic marine bacteria: Journal of Bacteriology, v. 115, p. 732–737.

Vainshtein, M.B., Matrosov, A.G., Baskunov, V.P., Zyakun, A.M., Ivanov, M.V., 1980, Thiosulfate as an intermediate product of bacterial sulfate reduction: Mikrobiologiya, v. 49, p. 855–858. (English translation)

Vairavamurthy, A., and Mopper, K., 1989, Mechanistic studies of organosulfur (thiol) formation in coastal marine sediments, *in* Saltzman, E.S., and Cooper, W.J., eds., Biogenic sulfur in the environment: Washington, D.C., American Chemical Society, p. 231–242.

Vairavamurthy, A., Zhou, W., Eglington, T., and Manowitz, B., 1994, Sulfonates: A novel class of organic sulfur compounds in marine sediments: Geochimica et Cosmochimica Acta, v. 58, p. 4681–4687, doi: 10.1016/0016-7037(94)90200-3.

Vairavamurthy, M.A., Orr, W.L., and Manowitz, B., 1995, Geochemical transformations of sedimentary sulfur: an introduction, *in* Vairavamurthy, M.A., and Schoonen, M.A.A., Geochemical transformations of sedimentary sulfur: Washington, D.C., American Chemical Society Symposium Series 612, p. 1–14.

van den Ende, F.P., and van Gemerden, H., 1993, Sulfide oxidation under oxygen limitation by a *Thiobacillus thioparus* isolated from a marine microbial mat: FEMS Microbiology Ecology, v. 13, p. 69–78, doi: 10.1016/0168-6496(93)90042-6.

van Gemerden, H., Tughan, S.S., de Wit, R., and Herbert, R.A., 1989, Laminated microbial ecosystems on sheltered beaches in Sapa Flow, Orkney Islands: FEMS Microbiology Ecology, v. 62, p. 87–102, doi: 10.1016/0378-1097(89)90018-9.

Vetter, R.D., Matrai, P.A., Javor, B., and O'Brien, J., 1989, Reduced sulfur compounds in the marine environment, *in* Saltzman, E.S., and Cooper, W.J., eds., Biogenic sulfur in the environment: Washington, D.C., American Chemical Society, p. 243–261.

Visscher, P.T., and van Gemerden, H., 1993, Sulfur cycling in laminated marine microbial ecosystems, *in* Oremland, R.S., ed., Biogeochemistry of global change: Radiatively active trace gases: London, Chapman and Hall, p. 672–690.

Visscher, P.T., Prins, R.A., and van Gemerden, H., 1992, Rates of sulfate reduction and thiosulfate consumption in a marine microbial mat: FEMS Microbiology Ecology, v. 86, p. 283–294, doi: 10.1016/0378-1097(92)90792-M.

von Rège, H., 1999, Bedeutung von Mikroorganismen des Schwefelkreislaufes für die Korrosion von Metallen [Ph.D. Dissertation]: Hamburg, Germany, Universität Hamburg, 140 p. (in German).

Weber, A., Riess, W., Wenzhoefer, F., and Jørgensen, B.B., 2001, Sulfate reduction in Black Sea sediments: in situ and laboratory radiotracer measurements from the shelf to 2000m depth: Deep-Sea Research I, v. 48, p. 2073–2096.

Wentzien, S., and Sand, W., 1999, Polythionate metabolism in *Thiomonas intermedia* K12, *in* Amils, R., and Ballester, A., eds., Biohdyrometallurgy and the environment toward the Mining of the 21st Century: Amsterdam, Elsevier, p 787–797.

Wenzhoefer, F., Riess, W., and Luth, U., 2002, In situ macrofaunal respiration rates and their importance for benthic carbon mineralization on the northwestern Black Sea shelf: Ophelia, v. 56, p. 87–100.

Widdel, F., 1988, Microbiology and ecology of sulfate- and sulfur-reducing bacteria, *in* Zehnder, A.J.B., ed., Biology of anaerobic microorganisms: New York, John Wiley and Sons, p. 469–585.

Wieland, A., Zopfi, J., Benthien, M., and Kühl, M., 2004, Biogeochemistry of an iron-rich hypersaline microbial mat (Camargue, France): Microbial Ecology (in press).

Wijsman, J.W.M., Middelburg, J.J., Herman, P.M.J., Böttcher, M.E., and Heip, C.H.R., 2001, Sulfur and iron speciation in surface sediments along the northwestern margin of the Black Sea: Marine Chemistry, v. 74, p. 261–278, doi: 10.1016/S0304-4203(01)00019-6.

Witter, A.E., and Jones, A.D., 1998, Comparison of methods for inorganic sulfur speciation in a petroleum production effluent: Environmental Toxicology and Chemistry, v. 17, p. 2176–2184.

Xu, Y., and Schoonen, M.A.A., 1995, The stability of thiosulfate in the presence of pyrite in low-temperature aqueous solutions: Geochimica et Cosmochimica Acta, v. 59, p. 4605–4622, doi: 10.1016/0016-7037(95)00331-2.

Yao, W., and Millero, F.J., 1993, The rate of sulfide oxidation by δMnO_2: Geochimica et Cosmochimica Acta, v. 57, p. 3359–3365, doi: 10.1016/0016-7037(93)90544-7.

Yao, W., and Millero, F.J., 1996, Oxidation of hydrogen sulfide by hydrous Fe(III)oxides in seawater: Marine Chemistry, v. 52, p. 1–16, doi: 10.1016/0304-4203(95)00072-0.

Zhang, J.-Z., and Millero, F.J., 1993, The products from the oxidation of H_2S in seawater: Geochimica et Cosmochimica Acta, v. 57, p. 1705–1718, doi: 10.1016/0016-7037(93)90108-9.

Zhang, J.-Z., and Millero, F.J., 1994, Kinetics of oxidation of hydrogen sulfide in natural waters, *in* Alpers, C.N. and Blowes, D.W., eds., Environmental geochemistry of sulfide oxidation: Washington, D.C., American Chemical Society Symposium Series 550, p. 393–409.

Zopfi, J., 2000, Sulfide oxidation and speciation of sulfur intermediates in marine environments [Ph.D. Dissertation]: Bremen, Germany, University of Bremen, 132 p.

Zopfi, J., Ferdelman, T., Jørgensen, B.B., Teske, A., and Thamdrup, B., 2001a, Influence of water column dynamics on sulfide oxidation and other major biogeochemical processes in the chemocline of the stratified Mariager Fjord (Denmark): Marine Chemistry, v. 74, p. 29–51, doi: 10.1016/S0304-4203(00)00091-8.

Zopfi, J., Kjær, T., Nielsen, L.P., and Jørgensen, B.B., 2001b, Ecology of *Thioploca* spp: NO_3^- and S^0 storage in relation to chemical microgradients and influence on the sedimentary nitrogen cycle: Applied and Environmental Microbiology, v. 67, p. 5530–5537, doi: 10.1128/AEM.67.12.5530-5537.2001.

Manuscript Accepted by the Society March 28, 2004

Printed in the USA

Geological Society of America
Special Paper 379
2004

Mechanisms of sedimentary pyrite formation

Martin A.A. Schoonen*
Center for Environmental Molecular Science, Department of Geosciences, Stony Brook University, Stony Brook, New York 11794-2100, USA

ABSTRACT

The mechanisms of pyrite formation are reviewed. Advances since 1994 in our understanding of the mechanisms and rate of pyrite formation, the role of bacteria in the formation of pyrite, framboid formation, and incorporation of impurities into pyrite are emphasized. Both field studies as well as laboratory studies designed to better represent natural environments have provided significant new insights. Field studies suggest that hydrogen sulfide can sulfidize amorphous FeS and form pyrite. The reaction rate as determined in the field is orders of magnitude slower than in laboratory experiments, and questions remain about the role of the FeS surface and the electron acceptor involved in the conversion. It is also becoming increasingly clear that sulfate-reducing bacteria play a more important role than simply providing hydrogen sulfide for the reaction. Experiments with in vitro cultures demonstrate the role of cell walls in directing and promoting the precipitation process. Synthesis of nanoscale pyrite, trace element incorporation, and formation of defects in pyrite are new research directions that are examined.

Keywords: pyrite, marine sediments, acid-volatile-sulfides, trace metals, kinetics, electron transfer reactions, redox reactions.

INTRODUCTION

Pyrite formation in sediments is an important process in the global cycles of iron, sulfur, atmospheric oxygen, and carbon (Canfield et al., 2000; Berner, 2001; Holland, 2002). For example, approximately half of the sedimentary organic matter in marine coastal environments is metabolized with sulfate as the terminal electron acceptor (Jørgenson, 1982). The electron acceptor sulfate is reduced to sulfide in the process, and some of this sulfide becomes sequestered as pyrite. Sedimentary pyrite formation is by no means restricted to marine coastal environments, although that is where most sedimentary pyrite is formed. Other sedimentary environments where pyrite and/or its dimorph, marcasite, are formed include aquifers (Kimblin and Johnson, 1992; Bottrell et al., 1995; Brown et al., 1999a; Brown et al., 1999b), lakes (Marnette et al., 1993; Suits and Wilkin, 1998), swamps (Dellwig et al., 2002), soils (Brennan and Lindsay, 1996), and waste ponds (Fortin and Beveridge, 1997; Fortin et al., 2000a; Gammons and Frandsen, 2000; Goulet and Pick, 2001; Paktunc and Dave, 2002). Given its geological importance, it is appropriate that pyrite formation, or more broadly, iron disulfide formation, has been studied for more than a century with a wide range of research strategies and tools. Despite the extensive research to understand the mechanism of pyrite formation and specifically to derive reaction rates, there are still some significant gaps in our knowledge.

While pyrite formation remains relevant to geologists, it has also become of importance to environmental scientists. The interest in pyrite formation by environmental scientists stems largely from the role it plays as a control on metal and metalloid contaminants in anoxic coastal sediments. There is also an interest in

*martin.schoonen@sunysb.edu

Schoonen, M.A.A., 2004, Mechanisms of sedimentary pyrite formation, *in* Amend, J.P., Edwards, K.J., and Lyons, T.W., eds., Sulfur biogeochemistry—Past and present: Boulder, Colorado, Geological Society of America Special Paper 379, p. 117–134. For permission to copy, contact editing@geosociety.org.

understanding how the formation process dictates the incorporation of defects and impurities in pyrite. Defects and impurities in pyrite, a semiconductor, affect its reactivity, its electronic properties, and its surface chemistry. Hence, there is a growing interest in the fundamental molecular steps involved in the formation of pyrite, how these steps may be altered by the presence of impurities, how impurities are incorporated in the pyrite structure, and how electronic defects arise in the pyrite structure. Finally, the formation of pyrite in the form of nanoscale crystallites has received attention because the electronic properties of pure pyrite change significantly when crystallites are confined to a size below a few nanometers (Alivisatos, 1996; Wilcoxon et al., 1996). Understanding the formation process in detail may open up new avenues for synthesis of nanoscale pyrite, which may one day be deployed as a reactant or catalyst in environmental remediation technologies.

Sedimentary iron disulfide formation has been reviewed several times (Berner, 1970; Berner, 1984; Morse et al., 1987; Rickard et al., 1995). The last comprehensive review of the subject was published in 1995 as part of an American Chemical Society symposium devoted to the geochemical transformations of sedimentary sulfur (Rickard et al., 1995). That last review covered the literature up to 1994 and is still a very useful resource. This contribution differs in scope from the last review. In this contribution, in addition to reviewing and updating the state of knowledge on the mechanism of iron disulfide formation as it relates to sedimentary iron disulfide formation, there will also be a brief review and discussion of the incorporation of impurities and the formation of electronic defects during the formation process.

IRON DISULFIDE FORMATION: AN UPDATED REVIEW OF THE PROCESS

Before reviewing the mechanism of pyrite or marcasite formation, it is useful to briefly review the thermodynamic constraints on the occurrences of the iron sulfide mineral phases and their distributions in sedimentary environments.

Thermodynamic Constraints

Equilibrium thermodynamic calculations unequivocally show that pyrite is the stable iron sulfide phase in anoxic low-temperature environments over much of the pe-pH parameter space (Fig. 1A). Iron monosulfides are predicted to be stable under a very narrow set of pe-pH conditions (e.g., mackinawite in Fig. 1A). Marcasite is metastable with respect to pyrite under all *P* and *T* conditions (Gronvold and Westrum, 1976); hence, it never shows up in equilibrium pe-pH diagrams, unless pyrite is excluded from the calculations (see Fig. 1B; also see Anderko and Shuler, 1997). A comparison of Figure 1A and B shows that exclusion of pyrite does not change the topology of the diagram. This is due to the fact that the difference in free energy between marcasite and pyrite is only ~2 kJ/mole (Table 1 summarizes the standard free energies of formation for common iron sulfide phases; the values in this table are the same as those used in the pe-pH diagrams). In Figure 1A and B, generated using Geochemist's Workbench™ (Bethke, 2002), the iron disulfides (pyrite or marcasite) dominate the diagram. The stability of iron disulfides derives from the fact that the presence of the S_2^{2-} moiety in pyrite and marcasite stabilizes the iron in a low-spin configuration; by contrast, iron in the monosulfides is in a high-spin configuration. The 1995 review, as well as more recent publications (Theberge and Luther 1997), present an in-depth discussion of the molecular orbital structure of Fe-S phases. Troilite, the thermodynamically stable FeS phase, and pyrrhotite do not normally occur in sedimentary environments and have been excluded from these calculations. Exclusion of the iron disulfides as well as troilite and pyrrhotite produces pe-pH diagrams that show the metastability fields for sedimentary iron monosulfides (amorphous FeS, mackinawite, greigite; Fig. 1C and D).

It should be noted that the shape and size of the fields occupied by iron sulfides in pe-pH diagrams depends on the solution composition. This is illustrated by two examples. In the first example (Fig. 1E), the calculations were conducted with world average river water rather than seawater. As in Figure 1C, greigite and mackinawite are allowed to form (i.e., troilite, pyrrhotite, marcasite, and pyrite are excluded), but they occupy a smaller area on the diagram (Fig. 1E) than in the equivalent diagram computed with seawater (Fig. 1D). This is directly related to the smaller amount of sulfur in average world river water than in seawater. Many published pe-pH diagrams or Eh-pH diagrams show a stability field for siderite, $FeCO_3$. The presence or absence of siderite depends on the activity of C(IV) species in solution as well as the iron activity. Increasing the iron activity from 10^{-6} to 10^{-3} produces a field for siderite (Fig. 1F).

Distribution of Iron Sulfides in Sedimentary Environments

On the basis of the thermodynamic stability relationships discussed above, only pyrite would be expected to occur in low-temperature sedimentary environments if these systems attained equilibrium. By contrast, there are a large number of studies that show the presence of iron monosulfides. In some environments, marcasite is present. For the purpose of this review, which is centered on the mechanisms of pyrite formation, distributions of iron sulfides in sedimentary environments are only very generally presented. With most pyrite forming in marine sediments, it is appropriate to focus this section on the distribution of iron sulfides in marine sediments, complemented by a few comments on the distribution in other sedimentary environments.

The distribution of iron sulfide phases in marine environments is routinely determined. Typically, a sequential acid digestion protocol is used to determine the distribution of iron sulfides in sediment cores (Allen and Parkes, 1995; Cutter and Kluckhohn, 1999; Billon et al., 2001). On the basis of differential

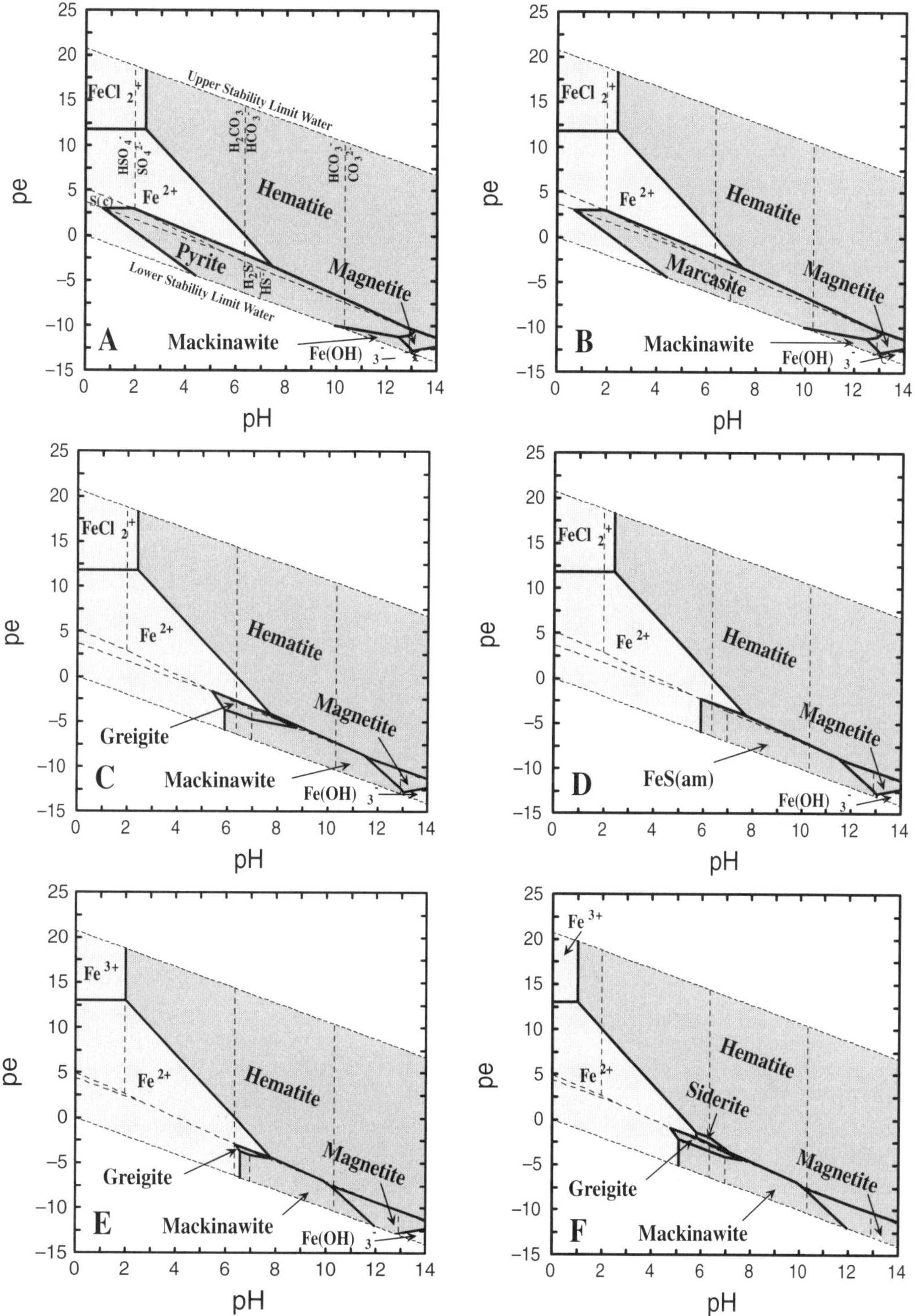

Figure 1. pe-pH diagrams to illustrate the thermodynamic limits on sedimentary iron sulfides (25 °C, 1bar). A. Equilibrium diagram for iron sulfides in seawater (see water composition from Parkhurst [Parkhurst and Appelo, 1999]), iron activity 10^{-6}, sulfur activity $10^{-2.551}$, C(IV) activity $10^{-3.001}$, troilite and pyrrhotite suppressed. B. Same as previous panel, but pyrite, troilite, and pyrrhotite suppressed. C. Same as previous panel, but marcasite suppressed in addition. D. Same as previous panel, but greigite and mackinawite suppressed in addition. E. Same as C but solution composition changed from seawater to world average river water (Berner and Berner, 1996), with iron activity 10^{-6}, sulfur activity $10^{-3.902}$, C(IV) activity $10^{-3.06}$. F. Same as C but iron activity 10^{-6} increased to 10^{-3} and C(IV) increased to $10^{-2.5}$. All diagrams calculated using Geochemist's Work Bench™ (Bethke, 2002). In all diagrams, the thin, dashed lines define the speciation of the sulfur and C(VI) species (i.e., redox equilibration for carbon suppressed). For clarity, these lines are only labeled in panel A.

acid dissolution kinetics, the distribution is often reported using two parameters: the concentration of acid volatile sulfides (AVS) and total reduced sulfur (TRS). The difference between AVS and TRS values provides an estimate of the concentration of iron disulfide in the sediment. (A protocol developed by Cutter and Kluckhohn [1999] allows for the discrimination of three different pools of iron sulfides: AVS, greigite, and pyrite). The mineralogical composition of the AVS fraction is difficult to establish in a given sample; small particle size (perhaps nanoscale particles) and rapid oxidation make it challenging to determine the mineralogy of the AVS fraction with techniques that require extensive sample handling and preparation. It is thought that amorphous FeS and poorly ordered mackinawite contribute significantly to the AVS fraction, while greigite may be relatively rare (Morse and Cornwell, 1987; Morse et al., 1987). Although perhaps rare, greigite has been found in sediments of Cretaceous age (Reynolds et al., 1994). Given the fact that greigite is a ferromagnetic mineral, its formation and persistence in sediments is of interest in the context of paleomagnetic studies (Roberts and Turner, 1993; Reynolds et al., 1994; Dekkers and Schoonen, 1996; Passier et al., 1998; Dekkers et al., 2000; Jiang et al., 2001; Strechie et al., 2002).

What is clear from mineralogical studies of marine sediments is that pyrite is the only iron disulfide present (i.e., marcasite is absent from marine sediments). Pyrite may be present as euhedral crystals or spherical aggregates of very small crystallites. The spherical aggregates, referred to as pyrite framboids, are typically 1–10 µ in diameter (see the extensive study by Wilkin and Barnes, 1996, for more details). The crystallites within the aggregate show a remarkably narrow size distribution. Among framboids, the average size of the crystallites ranges from 0.2 to 2 µ. A single framboid can contain as many as 10^2 to 10^5 crystallites (Wilkin et al., 1996).

The picture that emerges from studies of marine sediments is that in most sediments, iron monosulfide phases and pyrite coexist. Cores retrieved from fine-grained terrigenous sediments overlain by oxic seawater show the presence of iron monosulfide at the onset of sulfate reduction. Deeper in the core, the amount of iron monosulfides diminishes while the amount of pyrite increases. While variable, the AVS fraction is typically smaller than the pyrite pool. Seasonal fluctuations in the redox conditions within the sediment column can lead to oxidation of a significant fraction of the AVS and pyrite formed. For example, a detailed study in the East China Sea showed that 96% of the sedimentary iron sulfide formed is reoxidized (Lin et al., 2000). It should be noted that in some marine and estuarine sediments, the AVS fraction remains the dominant form of iron sulfide. For example, a study of the iron sulfide distribution in the Saguenay Fjord, Canada, shows that 50% or more of the iron sulfides are in the form of AVS phases (i.e., the ratio of AVS to pyrite is >1) (Gagnon et al., 1995). Hurtgen et al. (1999) reported anomalous enrichment of AVS in euxinic marine sediments in Effingham Inlet, Orca Basin, Canada, and the Black Sea. Anomalously high AVS/pyrite ratios have also been reported for shallow, heavily-vegetated, sandy marine sediments (Morse, 1999).

TABLE 1. SELECTED THERMODYNAMIC DATA FOR SEDIMENTARY IRON SULFIDES*

Mineral	Stoichiometry	ΔG_f° (kJ/mol)
Am. FeS[†]	FeS_{1-x} 0.13 >x >0.08	–83.7
Mackinawite[§]	FeS_{1-x} 0.07> x >0.04	–87.7
Greigite	Fe_3S_4	–273.8
Marcasite	FeS_2	–158.42
Pyrite	FeS_2	–160.23

*Thermodynamic data based on compilation and work by Benning et al (2000), complemented with data for marcasite (Robie et al., 1978).

[†]Stoichiometry based on work by Sweeney and Kaplan (1973). However, in calculations and reactions, amorphous FeS is assumed to have a 1:1 stoichiometry.

[§]Slightly iron rich in stochiometry, but in calculations and reactions, assumed to be FeS (i.e., x = 0).

The initial AVS/pyrite ratio in a sediment and the rate of AVS-to-pyrite conversion depend on the interplay of a number of factors that vary among sedimentary environments. The rate of hydrogen sulfide formation in relation to the availability of reactive iron exerts an important control on the initial AVS/pyrite ratio. The availability of metabolizable organic matter constrains the rate of hydrogen sulfide formation via sulfate reduction. The availability of reactive iron and metabolizable organic matter can vary laterally as a result of sedimentation rates. Upon burial below the zone of bioturbation, the amount of reactive iron available for sulfidation decreases along with the pool of metabolizable organic matter. It is beyond the scope of this review to discuss these relationships in detail; however, these relationships have been extensively studied in recent years (for examples, see Raiswell [1998]; Raiswell et al. [1994]; Raiswell and Canfield [1996]; Suits and Arthur [2000]; Lin [2000]; Roychoudhury [2003]; Hurtgen [1999]; Wijsman [2001]). The rate of AVS to pyrite conversion depends on the mechanism by which the conversion takes place. This in turn is dictated by the availability of reactants (H_2S, other sulfur sources, and oxidants) and will be discussed in a later section.

Where sediments are overlain by anoxic waters, iron sulfides form in the water column as well as in the sediment. Recent detailed work on two of these euxinic environments, Framvaren Fjord and the Black Sea, showed that amorphous FeS or mackinawite, greigite, and pyrite were formed in the water column (Wilkin et al., 1997; Cutter and Kluckhohn, 1999; Wilkin and Arthur, 2001). The pyrite is exclusively present as framboidal pyrite. Wilkin and coworkers (1996) have shown that framboids formed in the water column tend to be smaller than framboids formed within the sediment. Hence, this difference in size distribution provides a new tool to constrain paleoredox environments in marine sedimentary environments (Passier et al., 1997; Wilkin et al., 1997; Böttcher and Lepland, 2000; Wilkin and Arthur, 2001) as well as in lake sediments (Suits and Wilkin, 1998). However, in a recent study in which pyrite morphology and size distribution was determined in two areas within the same salt marsh, the universal applicability of these parameters as paleoenvironmental indicators was questioned (Roychoudhury et al., 2003). Sulfur isotopic signatures of pyrite have also been used to reconstruct paleoenvironments (Dellwig et al., 1999; Böttcher and Lepland, 2000; Lyons et al., 2000; Luepke and Lyons, 2001; Dellwig et al., 2001; Werne et al., 2002). Pyrite formed in anoxic water will closely match the sulfur isotopic value of dissolved hydrogen sulfide and show little variation (Calvert et al., 1996; Lyons, 1997; Wijsman et al., 2001). By contrast, the S-isotopic value of pyrite formed within sediments often shows a wide distribution of values. This distribution is dictated by the relative rate of sulfate replenishment and with respect to sulfate reduction. If the reduction rate exceeds the rate of replenishment, the resulting pyrite will vary in S-isotopic composition from depleted in ^{34}S to enriched in ^{34}S (Lyons, 1997).

Recent studies in non-marine environments have yielded new insights on the distribution of iron sulfides. A detailed study of the distribution of heavy minerals within the Long Island aquifer system showed that both pyrite and marcasite are present in the Magothy Formation (Brown et al., 1999b). The Cretaceous Magothy unit consists of sands with intercalated clay and lignite lenses. The Magothy, recognized as a major sedimentary unit along the entire North American Atlantic seaboard (Trapp and Meisler, 1992), was deposited in a shallow marine environment. Pyrite, often as framboids, is always associated with the lignite and inferred to have formed shortly after deposition. Marcasite formed later and occurs as cements in clayey zones. On the basis of pore water composition and determinations of reactive iron abundances, Brown et al. (1999a) have argued that marcasite forms in clayey microenvironments within the aquifer where sulfate reducing bacteria (SRB) have converted trapped seawater sulfate to hydrogen sulfide. Although not necessarily representative of the microenvironments present in the Magothy aquifer, the pH of water in the Magothy aquifer can be as low as 5.5. The presence of marcasite rather than pyrite is probably a function of pH. As pointed out in a seminal study by Murowchick and Barnes (1986) and confirmed by a number of subsequent studies, marcasite forms under acid conditions (pH <5), whereas pyrite forms in neutral to alkaline solutions. The formation of marcasite, rather than pyrite, is thought to reflect a difference in crystal growth kinetics that is controlled by the protonation of the S_2^{2-} moiety (see Murowchick and Barnes, 1986). Rakovan et al. (1995) studied the occurrence of iron disulfides recovered from the construction of the Tunnel and Reservoir project in Chicago, Illinois. These specimens, found in vugs in the Racine Dolomite (Silurian), show spectacular overgrowths of marcasite on pyrite. The unusual overgrowths are thought to be a result of a decrease in pH as the iron disulfides formed. While it is tempting to use the presence of marcasite versus pyrite as an indication of paleo-pH conditions, some caution is warranted. At higher temperatures, pyrite may dominate marcasite. For example, observations by my group have found that pyrite is present in several acid hot springs in Lassen Volcanic National Park, California. In addition, marcasite will convert to pyrite over time (Murowchick, 1992).

Proposed Mechanisms for the Formation of Pyrite

Pyrite formation has been the subject of many laboratory and field studies. The experimental designs in laboratory studies have varied widely, with some types of studies more relevant to sedimentary environments than others. Increasingly, field studies are carried out to evaluate mechanisms and reaction pathways proposed on the basis of laboratory studies. The ultimate goal of such approaches is to derive a quantitative model for the rate of pyrite formation under geologically relevant conditions along with other diagenetic reactions (Soetaert et al., 1996; Meysman et al., 2003). This goal is still elusive, but significant progress has been made over the last decades and in particular over the last eight years.

Any model describing pyrite formation must explain the following key observations: (1) metastable iron monosulfides coexist with pyrite and appear to form first in anoxic sediments, (2) upon burial, the AVS to pyrite ratio typically decreases, (3) pyrite is often present in the framboidal texture, and (4) marcasite is the dominant form of FeS_2 below a pH of ~5.

Formation of FeS Precursors

The formation of metastable iron sulfides as precursors to pyrite has been the subject of several studies. Schoonen and Barnes (1991a) showed that pyrite does not form directly from a homogeneous solution. Pyrite does form, however, from solutions initially supersaturated with respect to amorphous FeS. Under those conditions, pyrite forms at the expense of the iron monosulfide precursors. The formation of metastable precursors can be explained on the basis of classical nucleation theory (Schoonen and Barnes, 1991a). Nucleation is the first step in the precipitation of a mineral in the absence of crystal seeds. Classical nucleation theory predicts that the rate of nucleation depends strongly on the surface tension of the phase formed and the degree of supersaturation. Although based on an empirical relationship, surface free energy decreases sharply with increased solubility (Sohnel, 1982). Hence, if we consider two separate solutions, each supersaturated to the same degree with respect to a mineral phase, nucleation is predicted to be faster for the more soluble mineral. Schoonen and Barnes (1991a) argued that for a solution supersaturated with respect to amorphous FeS and pyrite, the nucleation rate for amorphous FeS is so much higher than for pyrite that the rapid nucleation of FeS(am) prevents the nucleation of pyrite. In other words, rapid FeS precipitation precludes building up supersaturation with respect to pyrite to a degree that is sufficiently high to initiate pyrite nucleation. The presence of pyrite seed crystals lifts the nucleation barrier. Hence, it is possible to grow pyrite from solutions undersaturated with respect to amorphous FeS or any of the other monosulfide phases, but saturated with respect to pyrite. The work by Schoonen and Barnes (1991a) has since been corroborated with studies conducted over a wide range of experimental conditions (Wang and Morse, 1995; Wang and Morse, 1996; Harmandas et al., 1998; Benning et al., 2000). Hence, the high nucleation barrier of pyrite is responsible for the formation of AVS phases in sediments.

The structure and composition of the initial precipitate or precursor amorphous FeS phase, as well as its mechanism of formation, has been the subject of several recent studies. Results of a low-angle X-ray diffraction study suggest that amorphous FeS consists of nanophase crystallites with an average size of ~4 nm (Wolthers, 2003). Particle size measurements using dynamic light scattering of freshly precipitated FeS indicate a size ranging from 45 nm to 300 nm (F.M. Michel, 2003, personal commun.). Most likely, aggregation in solution is responsible for the large size obtained in the dynamic light scattering measurements. The structure of amorphous FeS is thought to be a poorly ordered mackinawite, an iron-rich monosulfide phase (Lennie et al., 1995; Wolthers, 2003). In addition to distinct particles, it has been shown that FeS clusters exist in sulfide-rich waters (Theberge and Luther 1997; Rickard et al., 1999). These clusters or complexes develop rapidly in solution where amorphous FeS is precipitated. These clusters may be an important reactant in anoxic waters and sediments (Theberge and Luther 1997; Rickard et al., 1999; Grimes et al., 2001).

Conversion of Precursors to Pyrite

The conversion of the AVS fraction to pyrite has been the subject of a large number of studies, and considerable progress been made since 1995 on this topic. It is important to point out that after some small fraction of AVS is converted to FeS_2 nuclei, FeS_2 may grow from solution via a different mechanism. In this section, the focus is on the conversion of AVS to pyrite, which evidently circumvents the high kinetic barrier to the direct nucleation of pyrite.

The conversion of amorphous FeS or poorly ordered mackinawite to pyrite requires an electron acceptor and a change in the molar Fe/S ratio from close to 1:1 to 1:2. An electron acceptor is required to oxidize the S(-II) component in FeS to an oxidation state of −I (i.e., oxidation state of S in the S_2^{2-} moiety). Concomitant with this oxidation, the Fe/S ratio has to decrease either via the addition of sulfur or the loss of iron. There are three general reaction pathways by which these two requirements can be met. These three pathways are

1. FeS conversion via sulfur addition, with the incorporated sulfur species as electron acceptor;
2. FeS conversion via sulfur addition with a non-sulfur electron acceptor;
3. FeS conversion via iron loss, combined with an electron acceptor.

The first conversion pathway (sulfur addition with the sulfur species acting as electron acceptor) has received considerable attention. Traditionally, this mechanism has been represented by the following reaction (Berner, 1970; Berner, 1984):

$$FeS + S(0) \rightarrow FeS_2 \quad (1).$$

The stoichiometry of this reaction is rooted in a common experimental technique for the synthesis of pyrite. Precipitation of amorphous FeS in the presence of elemental sulfur yields pyrite. Elemental sulfur is a reagent but most likely not the true reactant in this process. The hydrolysis of the sulfur, as well as reactions of elemental sulfur with H_2S, creates polysulfide species, which are more likely to be reactants (Luther, 1991; Schoonen and Barnes, 1991b; Wilkin and Barnes, 1996). In addition, metastable sulfur oxyanions have been suggested as possible reactants in this conversion (Schoonen and Barnes, 1991b). Wilkin and Barnes (1996) conducted an extensive study in which they evaluated the conversion of freshly precipitated amorphous FeS in the presence of metastable sulfur oxyanions, polysulfides, colloidal elemental sulfur, as well as several organic S-bearing compounds. They found that only solutions that contained polysulfide species or colloidal elemental sulfur yielded pyrite. Furthermore, on the basis of several experiments in which the S-isotope signatures of the FeS, the sulfur source, and the resulting pyrite were measured, it was concluded that there was no evidence that any of the sulfur atoms from the sulfur source were incorporated into the pyrite. The sulfur species appear to merely act as electron acceptors, while the Fe/S ratio is converted via iron loss. The results by Wilkin and Barnes (1996) call into question the longstanding belief that pyrite formation can proceed via sulfur addition by a "zerovalent" sulfur species.

One of the hottest controversies in this field has centered on the second conversion mechanism: sulfur addition combined with a non-sulfur electron acceptor. In most anoxic sediments, hydrogen sulfide is by far the most abundant dissolved sulfur source available. Hence, it is logical to explore the possibility of reacting FeS with H_2S or HS^- to pyrite, a reaction that is thermodynamically favorable (Rickard, 1997; Rickard and Luther, 1997; Theberge and Luther 1997).

$$FeS(am) + H_2S(aq) \rightarrow FeS_2(pyrite) + H_2(g) \quad (2)$$

If the reaction proceeds as written, a significant amount of hydrogen gas is expected to form. However, early attempts by Berner (1970) and many other researchers since 1970 to form pyrite by precipitating FeS in a solution with excess H_2S under strictly anoxic conditions failed to produce pyrite. Only mackinawite is formed in these types of precipitation/aging experiments. Hence, reaction 2 appears to be kinetically inhibited. By contrast, Rickard (1997) found that reacting freeze-dried FeS with H_2S-containing solutions yields pyrite. In fact, the reaction is fast, and rate equations and activation energies for the conversion reaction have been obtained (Rickard, 1997). The following four-step mechanism has been proposed for the conversion (Rickard, 1997):

$$FeS(am) \rightarrow FeS(aq) \quad (3a)$$

$$FeS(aq) + H_2S \rightarrow [FeS\text{-}SH_2] \quad (3b)$$

$$[FeS\text{-}SH_2] \rightarrow [FeS_2.H_2] \quad (3c)$$

$$[FeS_2.H_2] \rightarrow FeS_2(pyrite) + H_2(g) \quad (3d)$$

The controversy alluded to above centers on two issues: (1) although some hydrogen was found at the conclusion of the experiment, the yield is much lower than expected, and (2) the effects of freeze-drying the starting FeS are unclear. The unexpectedly low yields of hydrogen could be the result of analytical problems or point to a fundamental deviation from the reaction as proposed. Rickard (1997) initially proposed that the low yields could be accounted for by incorporation of hydrogen, either molecular or atomic, in the pyrite structure. In essence, the conversion proceeds through reaction 3c, but the last step (3d) is largely incomplete. This notion was based in part on the result of a digestion of the product FeS_2 using reduced Cr(II) as reductant, which produced significant amounts of hydrogen. Rickard (1997) attributed this release to hydrogen incorporated in pyrite; however, Cr(II) is unstable in water and can reduce water to hydrogen. Hence, the results of the Cr(II) reductive digestion are inconclusive. The presence of hydrogen in pyrite can be tested using neutron diffraction experiments. The hypothesis would be that pyrite formed via reactions 3a–d would contain structural hydrogen, while pyrite formed via high temperature vapor deposition processes would be devoid of hydrogen. With the increased availability of neutron scattering facilities, it is becoming more realistic to conduct this test.

It is also of interest to consider the possibility that electron acceptors other than protons, as suggested by Rickard and Luther (1997), are involved in reaction 3. This notion is supported by a number of experimental studies that were conducted with FeS and H_2S as reagents, as well as observations in natural systems. Work by Heinen and Lauwers (1996, 1997) has shown that CO_2 can be reduced to thiols in a system with FeS and H_2S as reactants. Schoonen and Xu (2001), as well as Dorr et al. (2003), have shown that some N_2 is reduced to ammonia in solutions containing FeS and H_2S. These studies illustrate that other reagents in the experiments by Rickard (1997) could have served as electron acceptors, which could explain some of the unaccounted hydrogen. Finally, it is important to realize that in natural environments there are a number of possible electron acceptors present that are absent in lab experiments. Besides carbonic acid, organic compounds, Fe(III)-O-H phases, and Mn(V)-O-H phases serve as electron acceptors in natural systems (Aller and Rude, 1988). It is generally assumed that reactive ferric iron phases and manganese compounds with oxidations states in excess of +II would have been exhausted as electron acceptors before sulfate reduction, and, by extension, pyrite formation becomes important. However, this may not always be the case. Bioturbation and seasonal fluctuations in redox fronts within shallow sediments could supply electron acceptors to zones where pyrite forms with H_2S as reactant.

In a study spurred by the work of Rickard (1997) and Rickard and Luther (1997), Benning et al. (2000) conducted a series of experiments to carefully evaluate the importance of the freeze-drying step. This work shows unequivocally that freeze-drying activates the conversion process. Freshly precipitated amorphous FeS kept in a hydrogen sulfide solution does become more crystalline over time and ultimately stabilizing as mackinawite, but it does not convert to pyrite on the time scale of months. The conversion of amorphous FeS to mackinawite has been studied with synchrotron-based, in situ time-resolved X-ray diffraction (Cahill et al., 2000). The issue then is what freeze-drying does to the FeS used as starting material in the experiments conducted by Rickard (1997). Benning et al. (2000) argue that freeze-drying partially oxidizes the material, a notion supported by recent work on extraction techniques (Brumbaugh and Arms, 1996; Zhang et al., 2001). Even a small degree of oxidation of amorphous FeS may lead to facile formation of some pyrite nuclei as the FeS is exposed to H_2S. Once the pyrite nuclei form, growth can commence. In experiments in which the FeS precipitate is never removed from solutions (Benning et al., 2000), pyrite nucleation is evidently much slower or effectively inhibited.

The coexistence of appreciable amounts of H_2S and AVS overlain by a deep anoxic-sulfidic water column supports the notion that the conversion of FeS to pyrite via reaction 2 is slow. Sediments overlain by a deep anoxic-sulfidic (euxinic) water column provide a unique opportunity to study pyrite formation kinetics, as outlined by Hurtgen et al. (1999). Formation of intermediate sulfur species, such as polysulfides, and elemental sulfur is restricted to a narrow zone just below the chemocline, the transition from oxic to anoxic waters. Conversion of FeS via reactions involving intermediate sulfur species and elemental sulfur leads to pyrite formation in the water column and prevents these sulfur sources from reaching the sediment. Under these conditions, sulfidic sulfur remains as the only sulfur source in the sediment. In sediments, such as those in the permanently euxinic basins of the Black Sea, the supply of oxidants, such as iron oxides and manganese oxides, down into the sediments is very limited. Despite the lack of oxidants, the conversion of AVS to pyrite formation does proceed within sediment, albeit slowly (Hurtgen et al., 1999). A comparative study of pyrite formation in three anoxic basins (Effingham Inlet, an anoxic fjord on the west coast of Vancouver Island with high H_2S concentrations in sediment pore waters; Orca Basin, a highly saline basin within the Gulf of Mexico with negligible H_2S pore water concentrations; and the Black Sea with high H_2S pore water concentrations) showed that pyrite forms faster in those basins with appreciable H_2S pore water concentrations. However, the estimated rate of pyrite formation in the presence of H_2S is on the order of decades to centuries, rather than hours as predicted on the basis of the rate laws derived by Rickard (1997) and Rickard and Luther (1997) (Hurtgen et al., 1999). It is possible that the conversion proceeds via a direct reaction of AVS with H_2S producing H_2 (i.e., reaction 2), but it is also possible that the conversion proceeds with H_2S as the sulfur source and a non-sulfur electron acceptor (e.g., bicarbonate or dinitrogen).

The third possible pathway, iron loss combined with a non-sulfur electron acceptor, has received less attention. Work by Wilkin and Barnes (1996) indicates that, at least under certain conditions, iron loss may dominate over sulfur addition. Furukawa and Barnes (1995) have argued on the basis of calculations of volumes of reactions that iron loss should be the dominant process responsible for the change in S/Fe ratio. The calculations are based on the partial molar volumes for crystalline, macroscopic Fe-S phases. In natural environments, some of the precursors, such as amorphous FeS and mackinawite, are poorly crystalline and present as nanophase materials (Wolthers, 2003). It is not clear whether the partial molar volume relationships as calculated by Furukawa and Barnes (1995) still hold if relaxation effects and disorder in nanophase solids are taken into account (Alivisatos, 1996). Experimental work by Lennie et al. (1997) shows that mackinawite can readily transform to greigite via iron loss, but the next step to pyrite has not been demonstrated thus far.

The Role of Microorganisms and Organic Matter in AVS Conversion to Pyrite

Thus far, this review of AVS conversion studies has been focused on abiotic studies. This reflects the tendency, including in my own work (Schoonen, 1989; Schoonen and Barnes, 1991a, 1991b, 1991c), to study pyrite formation in abiotic experiments designed to avoid the complexity of the natural environment. For example, it is tacitly assumed in abiotic experiments that the role of dissimilatory sulfate-reducing bacteria is restricted to the formation of hydrogen sulfide. Following that logic, the role of these bacteria, which are ubiquitous in sediment (Konhauser, 1998), can be replaced by adding H_2S reagent to the system. This simplifies the experimental design, and the concentration of H_2S can either be kept constant or easily determined. However, the underlying assumption that the bacteria's role is limited to the formation of hydrogen sulfide is incorrect, as shown in a number of recent studies.

Experiments with in vitro cultures of a species of sulfate-reducing bacteria as well as bacteria capable of disproportionating elemental sulfur indicate that the role of bacteria goes beyond supplying hydrogen sulfide. For example, work with sulfate reducing bacteria by Donald and Southam (1999) showed rapid pyrite formation and incorporation of sulfur initially added to the system as a radiolabeled sulfur-bearing amino acid (cysteine). The cysteine was initially entirely incorporated into the organic cell matter, but over the course of the experiment some of the cysteine was transformed and incorporated into the pyrite formed on the outside of the cell. Donald and Southam (1999) proposed that the cysteine was converted to a labile form of sulfur, possibly H_2S, before it became incorporated. This finding is important because it indicates addition of sulfur to FeS. These results contradict the conclusions by Wilkin and Barnes (1996). They evaluated cysteine as a sulfur source in a strictly abiotic system and found no evidence for incorporation of cysteine-bound sulfur in pyrite. The work by Donald and Southam (1999) points to the role bacteria can play in converting organic sulfur compounds to reactants that can take part in pyrite formation. The high rates of pyrite formation in the cultures were attributed by the authors to the formation of the FeS precursor as a thin film on the cell wall. As pointed out by Donald and Southam (1999), as well as by Konhauser (1998), cell walls of microorganisms contain anionic sites capable of binding ferrous iron. It is postulated that these sites induce the nucleation of the FeS phase in the form of a thin film on the outside of the cell wall. Conversion of the FeS leads to the formation of pyrite nuclei, which can grow to become macroscopic crystals. In experiments with plant material in the presence of sulfate-reducing bacteria, Grimes et al. (2001) also showed that the organic matter exerts a control on the location of the formation of pyrite. In experiments with twigs and celery as organic substrates, the first precipitate on cell walls is FeS(am), followed with pyrite at the expense of the FeS(am).

Pyrite also formed very rapidly in experiments with a variety of sulfur-disproportionating bacteria (Canfield et al., 1998). On the basis of the sulfur isotope signature of the pyrite formed in these experiments, Canfield et al. (1998) argued that pyrite formed both by addition of zerovalent sulfur and by the reaction between H_2S and FeS at rates that exceeded those measured in abiotic experiments (Rickard, 1975; Rickard, 1997; Rickard and Luther, 1997) by at least four orders of magnitude. The extremely high reaction rates reported by Canfield et al. (1998) may not be directly relevant to natural systems. The rate of pyrite formation in these experiments was enhanced by the presence of ferrihydrite, the most reactive Fe(III)-hydroxide phase known (Canfield et al., 1992). In natural systems, the lack of reactive iron phases allows for H_2S to build up in sediments (Hurtgen et al., 1999). This H_2S can then react slowly with FeS to form pyrite, or it can form pyrite slowly via reactions with Fe(III)-bearing phases (Neal et al., 2001). The use of pure cultures under optimal growth conditions in the experimental work by Canfield et al. (1998) may have contributed to conversion rates that are extremely high and inconsistent with field observations (Hurtgen et al., 1999).

While the results by Donald and Southam (1999), Grimes et al. (2001), and Canfield et al. (1998) show that biomatter plays an important role in pyrite formation, these studies also corroborate the notion that pyrite formation proceeds via the conversion of a FeS precursor and not via direct nucleation of pyrite (Schoonen and Barnes, 1991a). Hence, with respect to the role of FeS precursors, the abiotic experiments and biotic experiments are in agreement; however, experiments with in vitro cultures may be the only type of laboratory experiment that incorporates some of the essential complexity to be meaningful to natural sedimentary systems.

Growth Mechanisms

The actual mechanism of iron disulfide crystal growth under sedimentary conditions is still largely unresolved. In most experimental studies there is little or no effort made to distinguish between the mechanism of FeS_2 nucleation via AVS conversion and the growth of FeS_2 nuclei to macroscopic particles. However, crystal growth kinetics and mechanisms are often very different from the nucleation kinetics and mechanisms.

One of the important results of the few studies in which solutions undersaturated with AVS, but supersaturated with respect to FeS_2, were seeded with pyrite was that pyrite growth can take place under those conditions (Schoonen and Barnes, 1991a;

Benning et al., 2000). This indicates that there is a mechanism by which pyrite, and presumably marcasite, can grow that does not involve an AVS precursor. Murowchick and Barnes (1986) proposed that growth takes place by incorporation of S_2^{2-} species. In fact, they postulated that the pH dependence of the protonation of the S_2^{2-} species dictates whether marcasite or pyrite form (see Murowchick and Barnes, 1986, for details). While the argument is compelling, the problem is that disulfide and polysulfide concentrations are generally in the low micromolar range (MacCrehan and Shea, 1995; Rozan et al., 2000). Furthermore, the interaction between polysulfides and Fe^{2+} leads to the decomposition of polysulfides into elemental sulfur and hydrogen sulfide (Luther, 1991; Schoonen and Barnes, 1991a; Wang and Morse, 1996).

It is possible that pyrite grows from FeS-undersaturated solutions via a mechanism in which the S-S bond is formed on the surface of a growing pyrite crystal. As illustrated schematically in Figure 2, two hydrogen sulfide molecules interacting with a growing pyrite surface may form an S-S group. Ab initio calculations of the interaction of H_2S with the (100) surface of pyrite (Stirling et al., 2003) suggest that this interaction will be restricted to defects, because the dissociation of H_2S on the perfect (100) surface is energetically very unfavorable. It has been estimated on the basis of XPS studies that ~10–20% of the (100) surface is comprised of an intrinsic, sulfur-deficient defect (Guevremont et al., 1998c). These defects are capable of dissociating H_2S among other compounds (Guevremont et al., 1997; Guevremont et al., 1998a; Guevremont et al., 1998c, 1998d; Elsetinow et al., 2000). Of course, an electron acceptor is required to complete the reaction depicted in Figure 2. It is commonly assumed that the electron transfer between an oxidant and a reductant (in this example, the $H_2S–SH_2$ surface complex) requires the formation of a reaction intermediate between oxidant and reductant. However, as a semiconductor, pyrite itself may act as a conduit for the transfer of electrons between an oxidant and a reductant adsorbed on the pyrite surface (Xu and Schoonen, 2000). A pyrite-mediated electron transfer circumvents the need for formation of a reaction intermediate and alleviates possible orbital mismatches between the reactants. This concept has been demonstrated in a series of experimental studies on the oxidation of thiosulfate by molecular oxygen, in which pyrite and ZnS catalyze the reaction by mediating the electron transfer between the two reactants (Xu and Schoonen, 1995; Xu et al., 1996). In natural environments and in experiments with microorganisms, the mechanism illustrated in Figure 2 is plausible because organic matter, aqueous electron acceptors, and mineral electron acceptors (e.g., Fe(III)-O-H phases) are in direct contact with the pyrite surface.

Under conditions in which the system is saturated with respect to amorphous FeS, aqueous Fe-S-H complexes (Davison et al., 1999) and/or a FeS(aq) cluster (Theberge and Luther 1997) may be important reactants in FeS_2 growth. Laboratory experiments and measurements in pore waters indicate that the presence of particulate amorphous FeS is often accompanied by the presence of either Fe-HS complexes or FeS(aq) clusters. There is some debate about the stoichiometry of the complexes (Theberge and Luther 1997; Davison et al., 1999), and the existence of FeS clusters has been called into question (Davison et al., 1999). Despite the uncertainty in the exact nature of the mobile FeS fraction, it has been suggested that this pool of FeS is important in advecting Fe and S to a growing FeS_2 crystal in sediments and in and around biomatter (Grimes et al., 2001). Similar to the mechanism illustrated in Figure 2, it is possible that pyrite itself could play a role as a conduit between a $FeS–SH_2$ surface complex and an adsorbed electron acceptor (see Fig. 3). Although speculative at this point, the mechanism illustrated in Figure 3 allows species other than protons to function as an electron acceptor.

A number of studies have deployed silica gel techniques to better represent the transport conditions and saturation levels in sedimentary environments (Wang and Morse, 1995; Wang and Morse, 1996; Morse and Wang, 1997; Harmandas et al., 1998; Allen, 2002). In this type of experiment, the iron source is dispersed and immobilized in a silica gel. The sulfur source is loaded on top of the gel and allowed to diffuse into the gel. Wang and Morse (1995; 1996) exploited this technique extensively. Their work shows that the conversion of amorphous FeS or mackinawite to pyrite proceeds very slowly. Experiments with greigite as the starting AVS phase showed much faster conversions to pyrite. These findings are consistent with earlier work using batch experiments (Schoonen and Barnes, 1991b). The experiments with the silica gel technique by Wang and Morse (1996) also yielded interesting results regarding the morphology of the products. In addition to the formation of aggregates with

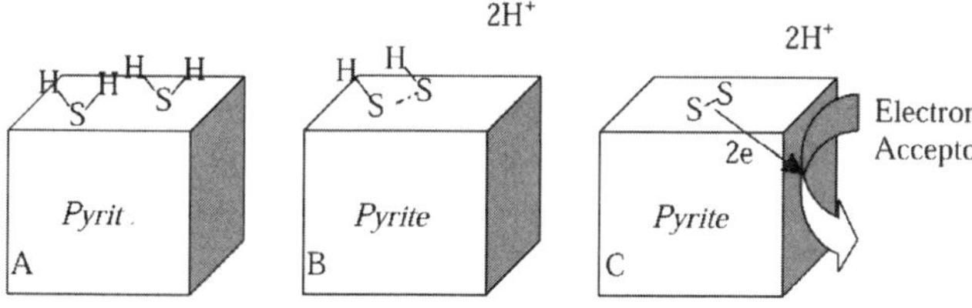

Figure 2. Conceptual diagram illustrating the role pyrite may play as a conduit for the formation of the S_2^{2-} moiety on its surface. A. Two adsorbed hydrogen sulfide molecules. B. Formation of a H-S–S-H surface complex. C. Formation of S_2^{2-} complex via electron transfer through pyrite to adsorbed electron acceptor.

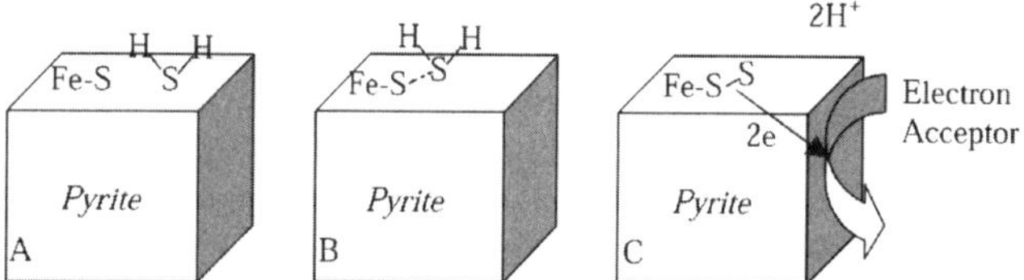

Figure 3. Conceptual diagram illustrating the role pyrite may play in facilitating growth with H_2S and FeS as reactants. A. adsorbed hydrogen sulfide molecule and FeS complex or cluster.; B. Formation of a Fe-S-S-H surface complex. C. Formation of FeS_2 monomer via electron transfer through pyrite to adsorbed electron acceptor.

a framboid-like texture, Wang and Morse (1996) demonstrated that the degree of supersaturation controls the form of individual crystals. The dominant crystal form changes from cubic to cubo-octahedral to octahedral to spherulitic with increasing supersaturation. In a follow-up study, Morse and Wang (1997) found that the addition of organic matter decreased the rate of pyrite formation significantly. Harmandas et al. (1998), also using the silica gel method, found that the presence of organophosphate compounds inhibited pyrite growth.

The experiments by Allen (2002) using the silica gel technique explored the effect of iron availability on the distribution of Fe-S phases around decaying organic matter inoculated with an active culture of sulfate-reducing bacteria. In experiments simulating an iron-rich environment, the formation of pyrite was confined to a zone directly adjacent to the organic matter. This distribution is caused by the fact that any H_2S produced within the organic matter is sequestered as an iron monosulfide precursor as it diffuses out of the organic matter. The FeS precursor is converted to pyrite over time. In experiments simulating an iron-limited environment, the iron sulfide precipitation occurred in bands, a distribution caused by the same process that causes Liesegang banding in sediments. As iron sulfide is formed around the organic matter, iron diffuses into this zone, which leads to a zone depleted in reactive iron. As sulfide continues to be formed, it will at some point in time exhaust the amount of available iron immediately adjacent to the organic matter and diffuse outward. As it diffuses outward, it first passes through an iron-depleted zone where the solution remains undersaturated with respect to the FeS precursor. Once the hydrogen sulfide encounters higher dissolved iron concentrations, it causes FeS to precipitate, which sets up a new iron hydrogen sulfide and iron diffusion pattern.

Formation of Framboidal Pyrite

Since the last major review of the formation of sedimentary iron sulfides (Rickard et al., 1995) there have been several experimental studies focused on resolving the mechanism by which pyrite precipitates in the framboid texture. Before the mid-1990s, there were only a handful of experimental studies in which pyrite precipitated in the framboidal texture (see Wilkin and Barnes, 1997, for a review). The most influential of these studies was by Sweeney and Kaplan (1973). The conclusion of that study was that the formation of greigite was a necessary step for the formation of framboids. The problem has been that many experimental studies before and since the work of Sweeney and Kaplan (1973) showed the formation of greigite without the formation of framboidal pyrite (i.e., pyrite precipitated as euhedral crystals). Given that a very significant fraction of sedimentary pyrite, if not most of it, is present as framboidal pyrite, it seemed logical to conclude that nearly all laboratory experiments were conducted under conditions that were not entirely relevant to the conditions in sediments. However, subsequent experimental studies (Wang and Morse, 1995; Wang and Morse, 1996; Wilkin and Barnes, 1996; Butler and Rickard, 2000) have shown that framboidal pyrite can be formed readily in laboratory experiments in which the FeS-to-FeS_2 conversion rate is fast. There is, however, some controversy regarding the role of greigite. Wilkin and Barnes (1997) have argued that greigite plays a critical role because its magnetic properties drive the aggregation that leads to the framboidal texture. They presented a detailed model that takes into account the magnetic moment between greigite particles as well as other factors. By contrast, Butler and Rickard (2000) have argued that greigite formation is not a necessary step in framboid formation. Their argument is based on experiments in which previously freeze-dried mackinawite was exposed to H_2S. It is possible that the use of freeze-dried mackinawite influences the results.

NEW RESEARCH DIRECTIONS

The purpose of this section is to highlight new directions research on pyrite formation is taking. It is beyond the scope of this contribution to present in depth discussions of each of the directions; instead, the emphasis is on the major research questions that have emerged.

Incorporation of Impurities

Elemental analyses of natural pyrite or marcasite often show that iron disulfides contain significant amounts of heavy metals (e.g., Ni, Cu, Co, Hg) and metalloids (see Table 2). For example, the pyrite muds that flooded the northern section of the Doñana National Park in southern Spain in 1998, as a result of the collapse of a dam retaining a pond filled with pyritic mine waste, contained an array of toxic elements (Table 2; Alastuey et al., 1999). Pyrite formed in coastal sediments can also contain significant amounts of metals and metalloids (Huerta-Diaz and Morse, 1990; see Table 2). The association of toxic elements with iron sulfide phases in coastal sediments has received considerable attention because it is recognized that benthic invertebrates may mobilize the toxic metals through ingestion and incorporation into their tissue (Mayer et al., 1996; Kaag et al., 1998; Chen and Mayer, 1999; Wang and Chapman, 1999; Wang et al., 1999; Griscom et al., 2000; Lee et al., 2000; Griscom et al., 2002). As a result of this process, toxic elements enter into the food chain and may adversely affect public health. AVS phases and pyrite may also be sinks for metal contaminants in fresh water sediments (Huerta-Diaz et al., 1993; van den Berg et al., 1998; Mikac et al., 2000; Grabowski et al., 2001; van den Berg et al., 2001; Yu et al., 2001; van Griethuysen et al., 2003). Anoxic conditions in tailings ponds and wetlands associated with mine operations can induce the formation of pyrite or marcasite (Fortin and Beveridge, 1997; Fortin et al., 2000a; Fortin et al., 2000b; Gammons and Frandsen, 2000; Paktunc and Dave, 2002). FeS_2 in these artificial environments may sequester significant amounts of toxic elements and contribute to their retention within the wetland.

The trace-element composition of ancient sedimentary pyrite and coal pyrite has been extensively studied. The trace element composition of ancient pyrites has been used to reconstruct paleodepositional environments (Dill and Kemper, 1990;

TABLE 2. MINOR ELEMENT AND TRACE ELEMENT COMPOSITION OF PYRITES FROM DIFFERENT ORIGINS*

Environment	Marine (Recent)	Marine (Jurrasic)	Brackish (Oligocene)	Lacustrine (Triassic)	Lacustrine (Recent)	Coal	Coal (cell fill)	Coal (cell overgrowth)	Coal (vein)	Ore
Origin	Gulf of Mexico	Germany	Germany	Germany	Canada	China	Warrior Basin, AL	Warrior Basin, AL	Warrior Basin, AL	Spain
Reference[†]	1	2	2	2	3	4	5	5	5	6
no. analyses	42	4	3	1		2	4	6	35	2
Mn	315–23466 (4867)	504–730	2–48	24	4	N.A.	N.A.	N.A.	N.A.	N.A.
Cu	683–89890 (21214)	63–132	3–5	14	741	200–200	0–3 (1)	<DL	N.A.	1800–2200
Pb	8–82213 (2574)	N.A.	N.A.	N.A.	5	N.A.	<DL	0–0.3	0–18	8000–10000
Zn	349–33740 (11065)	N.A.	N.A.	N.A.	1035	N.A.	<DL	<DL	0–80 (10)	8000–11000
As	24–5776 (2048)	32–290	66–90	19	N.A.	10000–27500	0–90 (38)	300–17350 (8810)	80–27400 (4130)	5000–6000
Sb	N.A.	N.A.	N.A.	N.A.	N.A.	400–900	<DL	<DL	0–2 (0.4)	500–800
Se	N.A.	N.A.	N.A.	N.A.	N.A.	N.A.	<DL	<DL	<DL	N.A.
Co	15–18848 (1380)	35–142	12–35	14	93	N.A.	N.A.	N.A.	N.A.	N.A.
Cd	0.6–97 (12)	N.A.	N.A.	N.A.	N.A.	N.A.	N.A.	N.A.	N.A.	30–40
Ag	N.A.	N.A.	N.A.	N.A.	N.A.	N.A.	N.A.	N.A.	N.A.	30–35
Bi	N.A.	N.A.	N.A.	N.A.	N.A.	N.A.	N.A.	N.A.	N.A.	50–70
Hg	0–1271 (351)	N.A.	N.A.	N.A.	N.A.	1400	<DL	<DL	7–39 (17)	25–35
Cr	74–4101 (1777)	N.A.	N.A.	N.A.	N.A.	N.A.	N.A.	N.A.	N.A.	50–60
Ni	29–32973 (8464)	138–694	172–310	25	N.A.	N.A.	N.A.	N.A.	N.A.	15–20
Tl	N.A.	N.A.	N.A.	N.A.	N.A.	N.A.	0.01–0.9 (0.3)	1–10 (6)	1–13 (7)	40–60

*All concentrations in μg/g; for some entries the range is given as well as an average in parentheses.

[†]Sources: 1—Huerta-Diaz and Morse, 1992; 2—Dill et al., 1997; 3—Huerta-Diaz et al., 1993; 4—Ding et al., 2001; 5—Diehl et al., 2002; 6—Alastuey et al., 1999.

Dill et al., 1997; Dellwig et al., 2002). The study of trace element compositions of coal pyrite has been driven by health and environmental concerns (Finkelman, 1995; Finkelman, 1999; Finkelman and Gross, 1999). Iron disulfides in coal often contain high arsenic concentrations (Kolker et al., 2000; Ding et al., 2001; Zhang et al., 2002) and are the primary host for antimony, cobalt, mercury, selenium, and thallium (Finkelman, 1995; Bool et al., 1997; Senior et al., 1997; Kolker and Finkelman, 1998; Goodarzi, 2002). The liberation of toxic elements during the combustion of coal (Zeng et al., 2001) and leaching of toxic elements from fly ash are major health problems in some countries (Finkelman, 1999; Finkelman and Gross, 1999). An example of an As-rich coal pyrite is given in Table 2. The distribution of As in pyrite in U.S. coal has been studied extensively due to the growing concern regarding As in drinking water and in acid mine drainage (Kolker and Koeppen, 1998; Kolker et al., 2000; Kolker and Nordstrom, 2001; Goldhaber et al., 2002; Kolker et al., 2003).

Given the complexities in the formation of FeS_2, it is not immediately clear how impurities are incorporated in pyrite or marcasite. A key question is what the fate is of impurities initially associated with AVS phases as these are oxidized or converted to FeS_2. Oxidation of the AVS phase leads to a redistribution of the associated trace elements between the pore water and other mineral phases. As reported in work on the distribution of Hg and Se in deep sea sediments, the position of the redox boundary is a primary control on the distribution of these elements (Mercone et al., 1999). For the fraction of AVS that escapes oxidation, the mode of conversion to pyrite—dissolution-reprecipitation versus a solid-state conversion—may exert an important control on the fate of the impurities. In a dissolution-reprecipitation mechanism, the impurities may be released to solution as the initial AVS phase dissolves. Incorporation into the FeS_2 phase will then be controlled by the interaction of the dissolved species with growth faces. As demonstrated by Reeder and Rakovan (1998) for the incorporation of metals and various anions into calcite, precipitation rate, surface complexation, surface structure of the host, and coordination and size of the surface complex may all be important factors (Reeder and Rakovan, 1998). Morse and Luther (1999) argued that the rate of ligand exchange and the rates of redox reactions involving the trace elements are important factors in determining whether a given element is incorporated into sedimentary iron sulfides. The presence of cobaltite, (Co,Fe)AsS, in association with pyrite framboids in the Kupferschiefer in Germany is thought to be the result of the liberation of Co and/or As from an AVS precursor phase followed by precipitation of cobaltite (Large et al., 1999). Evidently, the incorporation of Co and As into pyrite was either kinetically or thermodynamically hindered compared to the formation of the cobaltite.

Post-depositional processes can also influence the trace element composition of iron disulfides. Sediments often have multiple stages of pyrite formation, commonly in the form of overgrowths. For example, detailed studies of the occurrence of arsenic in pyrite found in coal from the Warrior Basin, Alabama, USA, show that the introduction of hydrothermal solutions led to the formation of arsenic-rich pyrite overgrowths on syngenetic or early-diagenetic framboids (Goldhaber et al., 2002). Similarly, several stages of overgrowths are found on massive pyrite crystals that fill in plant cells. Two analyses from the same pyrite grain, as well as an analysis of pyrite formed in a late vein in the same coal sample from the Warrior Basin, Alabama, illustrate the variability in trace-element concentration within a pyrite grain and among pyrites in the same sample (see Table 2). While not all coal samples or sediment samples will show such high degrees of variability in pyrite-hosted trace elements, it does serve as a reminder that reconstruction of ancient depositional environments on the basis of the trace element composition of ancient pyrite is wrought with problems, as the initial composition may be greatly overprinted by multiple stages of later pyrite precipitation. Advances in analytic tools, such as micro-particle-induced X-ray emission (PIXE) (Graham and Robertson, 1995), micro-synchrotron X-ray fluorescence (SXRF), and laser ablation inductively coupled plasma–mass spectrometry (LA-ICP-MS) (Kolker et al., 2002), in combination with automated element mapping, make it possible to determine the spatial distribution of trace elements within single pyrite grains or framboids on the scale of a few microns. Using these advanced techniques is probably the only way to reconstruct ancient depositional environments and epigenetic pyrite formation events.

Development of experimental protocols to determine the bioavailability of metals and metalloids associated with iron monosulfides in modern anoxic sediments is an active area of research (van den Berg et al., 1998; Cooper and Morse, 1999; van den Berg et al., 1999; Wilkin and Ford, 2002; van Griethuysen et al., 2003). However, as illustrated by a recent study of As extraction from sulfide-rich sediments (Wilkin and Ford, 2002), the results of extraction methods to estimate bioavailability have to be used with caution. Diffuse gradient in thin film (DGTF) probe techniques are emerging as useful new in situ tools to study the dynamics and spatial distribution of metal and sulfide chemistry in anoxic sediments on a scale of millimeters (Motelica-Heino et al., 2003). This relatively new technology relies on the diffusion and subsequent sequestration of dissolved metals and/or dissolved sulfide in a layered gel probe (Motelica-Heino et al., 2003). The probes can be deployed in the water column (Hamilton-Taylor et al., 1999; Torre et al., 2000; Odzak et al., 2002), soil (Zhang et al., 1998; Ernstberger et al., 2002), or sediments (Zhang et al., 1995; Fones et al., 2001; Motelica-Heino et al., 2003).

While the use of more advanced analytical techniques yields new insight into the distribution of trace elements associated with iron sulfides, new or improved spectroscopic techniques make it possible to study the processes that control the fate of trace elements associated with iron sulfide phases at a molecular level. Recent advances in the sensitivity of X-ray detectors and brightness of synchrotrons now make it possible to study the local coordination of impurities present at low concentrations in a host mineral. Extended X-ray absorption fine structure (EXAFS) can be used to study the coordination of an element of interest, while X-ray absorption near-edge spectroscopy (XANES) can be used to study the valence state of a contaminant as it interacts or is incorporated with the iron sulfide (Brown et al., 1999c). Examples of these types of studies include work by Fuhrmann et al. (1998), Bostic et al. (2000), Wharton et al. (2000), Farquhar et al. (2002), and Bostick and Fendorf (2003). The combination of a better understanding of the surface chemistry of pyrite at the molecular level (Guevremont et al., 1998c, 1998d) with studies of the macroscopic surface properties of pyrite (e.g., the development of surface charge, Bebié et al., 1998) are also important in regard to the fate of trace elements.

Incorporation of Electronic Defects into the Pyrite Structure

The incorporation of electronic defects during pyrite formation is of importance in understanding its reactivity. Pyrite oxidation in mine waste leads to the formation of acid mine drainage, which affects streams in many mining districts around the world (Evangelou, 1995). Research in our group has shown that sulfur-deficient defects in pyrite are responsible for its initial reactivity as it is exposed to water (Guevremont et al., 1997; Guevremont et al., 1998a; Guevremont et al., 1998b; Elsetinow et al., 2000). Upon exposure to water, hydroxyl radicals are created at the pyrite surface (Borda et al., 2001; Borda et al., 2003). The hydroxyl radicals are thought to be a key intermediate in the pyrite oxidation process and may also degrade organic compounds added to waste piles to abate the formation of acid mine waters. Pyrite-induced formation of hydroxyl radicals may have also played an important role on the early Earth. It has recently been shown that RNA decomposes rapidly in the presence of pyrite (Cohn et al., 2003). The hypothesis is that hydroxyl radical formed in the pyrite-water reaction is oxidizing RNA. Pyrite-induced hydroxyl formation may have been a driving force for the evolution from anoxygenic to oxygenic photosynthesis (Borda et al., 2001).

Hydroxyl is formed as the result of the presence of sulfur-deficient defects in the pyrite structure. Studies of the pyrite surface using ultra-high vacuum techniques invariably show the presence of a small fraction of sulfur in the S(-II) oxidation state (Schaufuss et al., 1998; Elsetinow et al., 2000). To maintain electroneutrality, this fraction of sulfidic sulfur is compensated in charge by the presence of ferric iron. This ferric iron, surrounded by S_2^{2-} and some S^{2-}, has been shown to be very reactive. When exposed to water, it is capable of extracting an electron from the water and converting it to a hydroxyl radical ($OH^{\bullet}$) and a proton.

$$\text{Fe(III) (pyrite)} + H_2O \rightarrow \text{Fe(II)(pyrite)} + OH^{\bullet} + H^+ \quad (4)$$

The hydroxyl radical is one of the strongest oxidants known. The combination of two hydroxyl radicals leads to the formation of hydrogen peroxide:

$$2OH^{\bullet} = H_2O_2 \quad (5)$$

A central research question that emerges in the context of this mechanism is how the defects form. Research in our lab has just started to address this question. The working hypothesis is that during pyrite growth the formation of the S-S bond in the S_2^{2-} moiety takes place on the surface. It is proposed that a defect is created if this bond breaks and or if the process is incomplete and a S^- species, rather than a S_2^{2-} species, is incorporated into the structure. One of the related questions is whether the defect density can be influenced by the incorporation of impurities. Research has shown that NiS_2 can also form hydroxyl radicals upon exposure to water (Borda et al., 2001). In fact, the amount of hydrogen peroxide formed when pure NiS_2 is exposed to water is higher than when synthetic pyrite is exposed to water (Borda et al., 2001). We are currently synthesizing pyrites with variable amounts of Ni to evaluate whether incorporation of Ni increases the defect density. Pyrite with a higher defect density may produce more hydroxyl radicals and be better suited as a material for the decomposition of organic solvents.

Formation of Nanoscale Pyrite Particles

At the nanoscale, the electronic and optical properties of pyrite differ from bulk pyrite. The band gap of pyrite increases from 0.95 eV for bulk pyrite to 2.25eV for 2 nm scale pyrite (Wilcoxon et al., 1996). This change in band gap has major implications for the reactivity of pyrite, in particular in the context of photochemical reactions (Schoonen et al., 1998, provides a general introduction to this topic for geoscientists). Given the fact that pyrite is potentially of use as a (photo)catalyst in the decomposition of organic solvents (Kriegmann-King and Reinhard, 1994; Weerasooriya and Dharmasena, 2001), there is interest in synthesizing nanoscale pyrite. Currently, there are two synthetic routes that have been used to achieve this. One method is to synthesize pyrite and separate out the smallest size fraction (Liu and Bard, 1989). An ensemble of nanoscale pyrites can be formed if the synthesis is restricted to micelles. With this technique, pioneered by Wilcoxon et al. (1996), nearly monodisperse pyrite crystals can be formed with crystal sizes down to 2 nm. While the pyrite crystals synthesized by this technique are suitable for fundamental optical and electronic studies, these nanocrystals have limited use in geochemical studies because the surfaces are encapsulated with surfactants. Alternative methods that rely on the sulfidation of nanoscale FeS(am) are now being explored in our group. The notion is that nanoscale FeS_2 may be produced if the conversion of FeS to FeS_2 can be arrested before growth takes place. Another method is to sulfidize the iron hydroxide core of the protein ferritin. Ferritin consists of a protein cage surrounding a nanoscale iron hydroxide particle (Harrison and Arosio, 1996; Chasteen and Harrison, 1999). Our research group is currently involved in a project to synthesize pyrite with ferritin as a reagent. The objective is to convert the core without decomposing the protein cage.

CONCLUSIONS

Pyrite formation remains an active area of research. Significant progress has been made in understanding the mechanism of pyrite formation, the role of bacteria in the formation of pyrite, framboid formation, and incorporation of impurities into pyrite. Despite this progress, some aspects of pyrite formation are yet to be resolved. Perhaps the most important issue is the kinetics and mechanism of the reaction between H_2S and FeS_2. In many anoxic environments, H_2S is the most abundant source of sulfur for the conversion of FeS to FeS_2. Estimates of the rate of this reaction based on field studies (Hurtgen et al., 1999) indicate that the rate of conversion is too slow to be studied in the laboratory at temperatures relevant for sedimentary environments. Following the lead by Hurtgen et al. (1999), it may be possible to put better constraints on the rate of this process with additional field studies. The second important challenge is to understand the fate of metals and metalloids associated with AVS upon its conversion to pyrite. This has major environmental implications, and it is also important to understand this process for the use of trace-element composition in reconstructing the depositional environment of ancient pyrite. The crystal growth mechanism of pyrite from solutions that are essentially devoid of S_2^{2-} is an interesting problem that may also have some bearing on the formation of sulfur-deficient defects site, sites that render pyrite some unique reactivity. A better understanding of the pyrite growth mechanisms may also lead to new methods for the synthesis of nanoscale pyrite, which has potential in photochemical reaction processes.

ACKNOWLEDGMENTS

Some of the material in this contribution is directly related to ongoing or past research in our lab. This work is impossible without the continued support of our research group by a number of funding agencies (the Department of Energy, the National Aeronautics and Space Administration, the National Science Foundation, and the Environmental Protection Agency). Our current work on the fate of AVS-associated metalloids is conducted is collaboration with colleagues at the Center for Environmental Molecular Science at Stony Brook, which is funded by the National Science Foundation–Chemistry. Allan Kolker, of the U.S. Geological Survey, is thanked for his assistance and guidance on the topic of trace element composition in coal pyrite. Alex Smirnov, State University of New York at Stony Brook, is thanked for his assistance in conducting the calculations necessary for the production of the pe-pH diagrams. Mike Borda, University of Delaware, is thanked for valuable comments on an earlier version of this manuscript. Reviews by Martin Goldhaber, U.S. Geological Survey–Denver, Timothy Lyons, University of Missouri, and an anonymous reviewer greatly improved the manuscript. A special thanks to Timothy Lyons for editing and handling the manuscript.

REFERENCES CITED

Alastuey, A., Garcia-Sanchez, A., Lopez, F., and Querol, X., 1999, Evolution of pyrite mud weathering and mobility of heavy metals in the Guadiamar valley after the Aznalcollar spill, southwest Spain: The Science of the Total Environment, v. 242, p. 41–55, doi: 10.1016/S0048-9697(99)00375-7.

Alivisatos, A.P., 1996, Perspectives on the physical chemistry of semiconductor nanocrystals: Journal of Physical Chemistry, v. 100, p. 13226–13239, doi: 10.1021/JP9535506.

Allen, R.E., 2002, Role of diffusion-precipitation reactions in authigenic pyritization: Chemical Geology, v. 182, p. 461–472, doi: 10.1016/S0009-2541(01)00334-5.

Allen, R.E., and Parkes, R.J., 1995, Digestion procedures for determining reduced sulfur species in bacterial cultures and in ancient and recent sediments, *in* Vairavamurthy, M.A., and Schoonen, M.A.A., eds., Geochemical transformations of sedimentary sulfur: ACS Symposium Series: Washington, ACS, p. 243–259.

Aller, R.C., and Rude, P.D., 1988, Complete oxidation of solid phase sulfides by manganese and bacteria in anoxic sediments: Geochimica et Cosmochimica Acta, v. 52, p. 751–765, doi: 10.1016/0016-7037(88)90335-3.

Anderko, A., and Shuler, P.J., 1997, A computational approach to predicting the formation of iron sulfide species using stability diagrams: Computers & Geosciences, v. 23, p. 647–658, doi: 10.1016/S0098-3004(97)00038-1.

Bebié, J., Schoonen, M.A.A., Strongin, D.R., and Fuhrmann, M., 1998, Surface charge development on transition metal sulphides: an electrokinetic study: Geochimica et Cosmochimica Acta, v. 62, p. 633–642, doi: 10.1016/S0016-7037(98)00058-1.

Benning, L.G., Wilkin, R.T., and Barnes, H.L., 2000, Reaction pathways in the Fe-S system below 100°C: Chemical Geology, v. 167, p. 25–51, doi: 10.1016/S0009-2541(99)00198-9.

Berner, E.K., and Berner, R.A., 1996, Global environment: Upper Saddle River, New Jersey, Prentice-Hall, 376 p.

Berner, R.A., 1970, Sedimentary pyrite formation: American Journal of Science, v. 268, p. 1–23.

Berner, R.A., 1984, Sedimentary pyrite formation: An update: Geochimica et Cosmochimica Acta, v. 48, p. 605–615, doi: 10.1016/0016-7037(84)90089-9.

Berner, R.A., 2001, Modeling atmospheric O_2 over Phanerozoic time: Geochimica et Cosmochimica Acta, v. 65, p. 685–694, doi: 10.1016/S0016-7037(00)00572-X.

Bethke, C.M., 2002, The Geochemist's Workbench—release 4.0, A user's guide to Rxn, Act2, Tact, React and Gtplot: Champaign-Urbana, University of Illinois, 224 p.

Billon, G., Ouddane, B., Laureyns, J., and Boughriet, A., 2001, Chemistry of metal sulfides in anoxic sediments: Physical Chemistry Chemical Physics, v. 3, p. 3586–3592, doi: 10.1039/B102404N.

Bool, L.E., III, Senior, C.L., Huggins, F., Huffman, G.P., Shah, N., Wendt, J.O.L., Shadman, F., Peterson, T., Seames, W., Sarofim, A.F., Olmez, I., Zeng, T., Crowley, S., Finkelman, R.B., Helble, J.J., and Wornat, M.J., 1997, Toxic substances from coal combustion—A comprehensive assessment: Pittsburgh, U.S. Department of Energy, Pittsburgh Energy Technology Center.

Borda, M., Elsetinow, A., Schoonen, M., and Strongin, D., 2001, Pyrite-induced hydrogen peroxide formation as a driving force in the evolution of photosynthetic organisms on an early Earth: Astrobiology, v. 1, p. 283–288, doi: 10.1089/15311070152757474.

Borda, M.J., Elsetinow, A.R., Strongin, D.R., and Schoonen, M.A., 2003, A mechanism for the production of hydroxyl radical at surface defect sites on pyrite: Geochimica et Cosmochimica Acta, v. 67, p. 935–939, doi: 10.1016/S0016-7037(02)01222-X.

Bostic, B.C., Fendorf, S., and Fendorf, M., 2000, Disulfide disproportionation and CdS formation upon cadmium sorption on FeS_2: Geochimica et Cosmochimica Acta, v. 64, p. 247–255, doi: 10.1016/S0016-7037(99)00295-1.

Bostick, B.C., and Fendorf, S., 2003, Arsenite sorption on troilite (FeS) and pyrite (FeS_2): Geochimica et Cosmochimica Acta, v. 67, p. 909–921, doi: 10.1016/S0016-7037(02)01170-5.

Böttcher, M.E., and Lepland, A., 2000, Biogeochemistry of sulfur in a sediment core from the west-central Baltic Sea: Evidence from stable isotopes and pyrite textures: Journal of Marine Systems, v. 25, p. 299–312, doi: 10.1016/S0924-7963(00)00023-3.

Bottrell, S.H., Hayes, P.J., Bannon, M., and Williams, G.M., 1995, Bacterial sulfate reduction and pyrite formation in a polluted sand aquifer: Geomicrobiology Journal, v. 13, p. 75–90.

Brennan, E.W., and Lindsay, W.L., 1996, The role of pyrite in controlling metal ion activities in highly reduced soils: Geochimica et Cosmochimica Acta, v. 60, p. 3609–3618, doi: 10.1016/0016-7037(96)00162-7.

Brown, C.J., Coates, J.D., and Schoonen, M.A.A., 1999a, Localized sulfate-reducing zones in the Magothy Aquifer, Long Island, New York: Ground Water, v. 37, p. 505–516.

Brown, C.J., Rakovan, J., and Schoonen, M.A.A., 1999b, Heavy mineral and sedimentary organic matter in Pleistocene and Cretaceous sediments on Long Island, New York, with emphasis on pyrite and marcasite in the Magothy aquifer: Coram, New York, U.S. Geological Survey Water Resources Investigation Report 99-4216.

Brown, G.E., Henrich, V.E., Casey, W.H., Clark, D.L., Eggleston, C., Felmy, A., Goodman, D.W., Gratzel, M., Maciel, G., McCarthy, M.I., Nealson, K.H., Sverjensky, D.A., Toney, M.F., and Zachara, J.M., 1999c, Metal oxide surfaces and their interactions with aqueous solutions and microbial organisms: Chemical Reviews, v. 99, p. 77–174, doi: 10.1021/CR980011Z.

Brumbaugh, W.G., and Arms, J.W., 1996, Quality control considerations for the determination of acid-volatile sulfide and simultaneously extracted metals in sediments: Environmental Toxicology and Chemistry, v. 15, p. 282–285.

Butler, I.B., and Rickard, D., 2000, Framboidal pyrite formation via the oxidation of iron (II) monosulfide by hydrogen sulphide: Geochimica et Cosmochimica Acta, v. 64, p. 2665–2672, doi: 10.1016/S0016-7037(00)00387-2.

Cahill, C.L., Benning, L.G., Barnes, H.L., and Parise, J.B., 2000, In situ, time-resolved X-ray diffraction of iron sulfides during hydrothermal pyrite growth: Chemical Geology, v. 167, p. 53–63, doi: 10.1016/S0009-2541(99)00199-0.

Calvert, S.E., Thode, H.G., Yeung, D., and Karlin, R.E., 1996, A stable isotope study of pyrite formation in the Late Pleistocene and Holocene sediments of the Black Sea: Geochimica et Cosmochimica Acta, v. 60, p. 1261–1270, doi: 10.1016/0016-7037(96)00020-8.

Canfield, D.E., Habicht, K.S., and Thamdrup, B., 2000, The Archean sulfur cycle and the early history of atmospheric oxygen: Science, v. 288, p. 658–660, doi: 10.1126/SCIENCE.288.5466.658.

Canfield, D.E., Raiswell, R., and Bottrell, S., 1992, The reactivity of sedimentary iron minerals toward sulfide: American Journal of Science, v. 292, p. 659–683.

Canfield, D.E., Thamdrup, B., and Fleischer, S., 1998, Isotope fractionation and sulfur metabolism by pure and enrichment cultures of elemental sulfur-disproportionating bacteria: Limnology and Oceanography, v. 43, p. 253–264.

Chasteen, N.D., and Harrison, P.M., 1999, Mineralization in ferritin: An efficient means of iron storage: Journal of Structural Biology, v. 126, p. 182–194, doi: 10.1006/JSBI.1999.4118.

Chen, Z., and Mayer, L.M., 1999, Sedimentary metal bioavailability determined by the digestive constraints of marine deposit feeders: gut retention time and dissolved amino acids: Marine Ecology Progress Series, v. 176, p. 139–151.

Cohn, C.A., Borda, M.J., and Schoonen, M.A., 2003, Fate of RNA in the presence of pyrite: Relevance to the origin of life [abstract]: New Orleans, American Chemical Society 225th National Meeting.

Cooper, D.C., and Morse, J.W., 1999, Selective extraction chemistry of toxic metal sulfides from sediments: Aquatic Geochemistry, v. 5, p. 87–97, doi: 10.1023/A:1009672022351.

Cutter, G.A., and Kluckhohn, R.S., 1999, The cycling of particulate carbon, nitrogen, sulfur, and sulfur species (iron monosulfide, greigite, pyrite, and organic sulfur) in the water columns of Framvaren Fjord and the Black Sea: Marine Chemistry, v. 67, p. 149–160, doi: 10.1016/S0304-4203(99)00056-0.

Davison, W., Phillips, N., and Tabner, B.J., 1999, Soluble iron sulfide species in natural waters: reappraisal of their stoichiometry and stability constants: Aquatic Sciences, v. 61, p. 23–43, doi: 10.1007/S000270050050.

Dekkers, M.J., Passier, H.F., and Schoonen, M.A.A., 2000, Magnetic properties of hydrothermally synthesized griegite (Fe_3S_4)–II High- and low-temperature characteristics: Geophysical Journal International, v. 141, p. 809–819.

Dekkers, M.J., and Schoonen, M.A.A., 1996, Magnetic properties of hydrothermally synthesized greigite (Fe_3S_4). 1. Rock magnetic parameters at room temperature: Geophysical Journal International, v. 126, p. 360–368.

Dellwig, O., Böttcher, M.E., Lipinski, M., and Brumsack, H.-J., 2002, Trace metals in Holocene coastal peats and their relation to pyrite formation (NW Germany): Chemical Geology, v. 182, p. 423–442, doi: 10.1016/S0009-2541(01)00335-7.

Dellwig, O., Watermann, F., Brumsack, H.-J., and Gerdes, G., 1999, High-resolution reconstruction of a Holocene coastal sequence (NW Germany) using inorganic geochemical data and diatom inventories: Estuarine, Coastal and Shelf Science, v. 48, p. 617–633, doi: 10.1006/ECSS.1998.0462.

Dellwig, O., Watermann, F., Brumsack, H.-J., Gerdes, G., and Krumbein, W.E., 2001, Sulphur and iron geochemistry of Holocene coastal peats (NW Germany): A tool for paleoenvironmental reconstruction: Palaeogeography, Palaeoclimatology, Palaeoecology, v. 167, p. 359–379, doi: 10.1016/S0031-0182(00)00247-9.

Diehl, S.F., Goldhaber, M.B., Hatch, J.R., Kolker, A., Pashin, J.C., and Koenig, A.E.P., 2002, Mineralogic residence and timing of emplacement of trace elements in coals of the Warrior Basin, Alabama, *in* Proceedings of the Nineteenth Annual International Pittsburgh Coal Conference, p. 14 (CD-ROM).

Dill, H., and Kemper, E., 1990, Crystallographic and chemical variations during pyritization in the upper Barremian and lower Aptian dark claystone from the Lower Saxonian Basin (NW Germany): Sedimentology, v. 37, p. 427–443.

Dill, H.G., Eberhard, E., and Hartmann, B., 1997, Use of variations in unit cell length, reflectance and hardness for determining the origin of Fe disulphides in sedimentary rocks: Sedimentary Geology, v. 107, p. 281–301, doi: 10.1016/S0037-0738(96)00031-0.

Ding, Z., Zheng, B., Long, J., Belkin, H.E., Finkelman, R.B., Chen, C., Zhou, D., and Zhou, Y., 2001, Geological and geochemical characteristics of high arsenic coals from endemic arsenosis areas in southwestern Guizhou Province, China: Applied Geochemistry, v. 16, p. 1353–1360, doi: 10.1016/S0883-2927(01)00049-X.

Donald, R., and Southam, G., 1999, Low temperature anaerobic bacterial diagenesis of ferrous monosulfide to pyrite: Geochimica et Cosmochimica Acta, v. 63, p. 2019–2023, doi: 10.1016/S0016-7037(99)00140-4.

Dorr, M., Kasbohrer, J., Grunert, R., Kreisel, G., Brand, W.A., Werner, R.A., Geilmann, H., Apfel, C., Robl, C., and Weigand, W., 2003, A possible prebiotic formation of ammonia from dinitrogen on iron oxide surfaces: Angewandte Chemie International Edition, v. 42, p. 1540–1543 (in English), doi: 10.1002/ANIE.200250371.

Elsetinow, A.R., Guevremont, J.M., Strongin, D.R., and Schoonen, M.A.A., 2000, Oxidation of {111} and {100} planes of pyrite: effects of surface atomic structure and preparation method: American Mineralogist, v. 85, p. 623–626.

Ernstberger, H., Davison, W., Zhang, H., Tye, A., and Young, S., 2002, Measurement and dynamic modeling of trace metal mobilization in soils using DGT and DIFS: Environmental Science & Technology, v. 36, p. 349–355.

Evangelou, V.P., 1995, Pyrite oxidation and its control: New York, CRC Press, 293 p.

Farquhar, M.L., Charnock, J.M., Livens, F.R., and Vaughan, D.J., 2002, Mechanisms of arsenic uptake from aqueous solution by interaction with goethite, lepidochrocite, mackinawite, and pyrite: an X-ray absorption spectroscopy study: Environmental Science & Technology, v. 36, p. 1757–1762, doi: 10.1021/ES010216G.

Finkelman, R.B., 1995, Modes of occurrence of environmentally-sensitive trace elements in coal, *in* Swaine, D. J., and Goodarzi, F., eds., Environmental aspects of trace elements in coal: Dordrecht, Kluwer, p. 24–44.

Finkelman, R.B., 1999, Health impacts of domestic coal use in China: Proceedings of the National Academy of Sciences of the United States of America, v. 96, p. 3427–3431, doi: 10.1073/PNAS.96.7.3427.

Finkelman, R.B., and Gross, P.M.K., 1999, The types of data needed for assessing the environmental and human health impacts of coal: International Journal of Coal Geology, v. 40, p. 91–101, doi: 10.1016/S0166-5162(98)00061-5.

Fones, G.R., Davison, W., Holby, O., Jørgensen, B.B., and Thamdrup, B., 2001, Notes—High-resolution metal gradients measured by in situ DGT/DET deployment in Black Sea sediments using an autonomous benthic lander: Limnology and Oceanography, v. 46, p. 982–989.

Fortin, D., and Beveridge, T.J., 1997, Microbial sulfate reduction within sulfidic mine tailings: Formation of diagenetic Fe sulfides: Geomicrobiology Journal, v. 14, p. 1–21.

Fortin, D., Goulet, R., and Roy, M., 2000a, Seasonal cycling of Fe and S in a constructed wetland: The role of sulfate-reducing bacteria: Geomicrobiology Journal, v. 17, p. 221–235, doi: 10.1080/01490450050121189.

Fortin, D., Roy, M., Rioux, J.-P., and Thibault, P.-J., 2000b, Occurrence of sulfate-reducing bacteria under a wide range of physico-chemical conditions in Au and Cu-Zn mine tailings: FEMS Microbiology Ecology, v. 33, p. 197–208, doi: 10.1016/S0168-6496(00)00062-3.

Fuhrmann, M., Bajt, S., and Schoonen, M.A.A., 1998, Sorption processes of iodine on minerals: Applied Geochemistry, v. 13, p. 127–141, doi: 10.1016/S0883-2927(97)00068-1.

Furukawa, Y., and Barnes, H.L., 1995, Reactions forming pyrite from precipitated amorphous ferrous sulfide, *in* Vairavamurthy, M.A., and Schoonen, M.A. A., eds., Geochemical Transformations of Sedimentary Sulfur: Washington, D.C., American Chemical Society Symposium Series 612, p. 194–205.

Gagnon, C., Mucci, A., and Pelletier, E., 1995, Anomalous accumulation of acid volatile sulphides (AVS) in a coastal marine sediment, Saguenay Fjord, Canada: Geochimica et Cosmochimica Acta, v. 59, p. 2663–2675, doi: 10.1016/0016-7037(95)00163-T.

Gammons, C.H., and Frandsen, A.K., 2000, Fate and transport of metals in H_2S-rich waters at a treatment wetland: Geochemical Transactions, v. 1, p. 1–15 (online).

Goldhaber, M.B., Lee, R., Hatch, J., Pashin, J., and Treworgy, J., 2002, The role of large-scale-fluid flow in subsurface arsenic enrichment, *in* Welch, A., and Stollenwerk, K.G., eds., Occurrence and geochemistry of arsenic in ground water: New York, Kluwer Academic, p. 127–176.

Goodarzi, F., 2002, Mineralogy, elemental composition and modes of occurrence of elements in Canadian feed-coals: Fuel, v. 81, p. 1199–1213, doi: 10.1016/S0016-2361(02)00023-6.

Goulet, R.R., and Pick, F.R., 2001, The effects of cattails (*Typha latifolia L.*) on concentrations and partitioning of metals in surficial sediments of surface-flow constructed wetlands: Water, Air, and Soil Pollution, v. 132, p. 275–291, doi: 10.1023/A:1013246614159.

Grabowski, L.A., Houpis, J.L.J., Woods, W.I., and Johnson, K.A., 2001, Seasonal bioavailability of sediment-associated heavy metals along the Mississippi river floodplain: Chemosphere, v. 45, p. 643–651, doi: 10.1016/S0045-6535(01)00037-6.

Graham, U.M., and Robertson, J.D., 1995, Micro-PIXE analysis of framboidal pyrite and associated maceral yypes in oil-shale: Fuel, v. 74, p. 530–535, doi: 10.1016/0016-2361(95)98355-I.

Grimes, S.T., Brock, F., Rickard, D., Davies, K.L., Edwards, D., Briggs, D.E.G., and Parkes, R.J., 2001, Understanding fossilization: Experimental pyritization of plants: Geology, v. 29, p. 123–126, doi: 10.1130/0091-7613(2001)0292.0.CO;2.

Griscom, S.B., Fisher, N.S., Aller, R.C., and Lee, B.G., 2002, Effects of gut chemistry in marine bivalves on the assimilation of metals from ingested sediment particles: Journal of Marine Research, v. 60, p. 101–120, doi: 10.1357/002224002762341267.

Griscom, S.B., Fisher, N.S., and Luoma, S.N., 2000, Geochemical influences on assimilation of sediment-bound metals in clams and mussels: Environmental Science & Technology, v. 34, p. 91–99, doi: 10.1021/ES981309+.

Gronvold, F., and Westrum, E.F., 1976, Heat capacities of iron disulfides thermodynamics of marcasite from 5 to 700 K, pyrite from 300 to 780 K, and the transformation of marcasite to pyrite: Journal of Chemical Thermodynamics, v. 8, p. 1039–1048.

Guevremont, J., Bebié, J., Elsetinow, A.R., Strongin, D.R., and Schoonen, M.A.A., 1998a, Reactivity of the (100) plane of pyrite in oxidizing gaseous and aqueous environments: effects of surface imperfections: Environmental Science & Technology, v. 32, p. 3743–3748, doi: 10.1021/ES980298H.

Guevremont, J., Strongin, D.R., and Schoonen, M.A.A., 1997, Effects of surface imperfections on the binding of CH_3OH and H_2O on FeS_2(100): using adsorbed Xe as a probe of mineral structure: Surface Science, v. 391, p. 109–124, doi: 10.1016/S0039-6028(97)00461-5.

Guevremont, J.M., Elsetinow, A.R., Strongin, D.R., Bebié, J., and Schoonen, M.A.A., 1998b, Structure sensitivity of pyrite oxidation: Comparison of the (100) and (111) planes: American Mineralogist, v. 83, p. 1353–1356.

Guevremont, J.M., Strongin, D.R., and Schoonen, M.A.A., 1998c, Photoemission of adsorbed xenon, X-ray photoelectron spectroscopy, and temperature-programmed desorption studies of H_2O on FeS_2(100): Langmuir, v. 14, p. 1361–1366, doi: 10.1021/LA970722C.

Guevremont, J.M., Strongin, D.R., and Schoonen, M.A.A., 1998d, Thermal chemistry of H_2S and H_2O on the striated (100) plane of pyrite: Unique reactivity of defect sites: American Mineralogist, v. 83, p. 1246–1255.

Hamilton-Taylor, J., Smith, E.J., Davison, W., and Zhang, H., 1999, A novel DGT-sediment trap device for the in situ measurement of element remobilization from settling particles in water columns and its application to trace metal release from Mn and Fe oxides: Limnology and Oceanography, v. 44, p. 1772–1781.

Harmandas, N.G., Fernandez, E.N., and Koutsoukos, P.G., 1998, Crystal growth of pyrite in aqueous solutions. Inhibition by organophosphorus compounds: Langmuir, v. 14, p. 1250–1255, doi: 10.1021/LA970354C.

Harrison, P.M., and Arosio, P., 1996, The ferritins: Molecular properties, iron storage function and cellular regulation: Biochimica et Biophysica Acta (BBA)—Bioenergetics, v. 1275, p. 161–203, doi: 10.1016/0005-2728(96)00022-9.

Heinen, W., and Lauwers, A.-M., 1997, The iron-sulfur world and the origins of life: Abiotic thiol synthesis from metallic iron, H_2S and CO_2; a comparison of the thiol generating FeS/HCl $(H_2S)/CO_2$ system and its $Fe^0/H_2S/CO_2$ counterpart: Proceedings of the Koninglijke Nederlandse Akademie van Wetenschappen, v. 100, p. 11–25.

Heinen, W., and Lauwers, A.-M., 1996, Organic sulfur compounds resulting from the interaction of iron sulfide, hydrogen sulfide and carbon dioxide in an anaerobic environment: Origins of Life and Evolution of the Biosphere, v. 26, p. 131–150.

Holland, H.D., 2002, Volcanic gases, black smokers, and the Great Oxidation Event: Geochimica et Cosmochimica Acta, v. 66, p. 3811–3826, doi: 10.1016/S0016-7037(02)00950-X.

Huerta-Diaz, M.A., and Morse, J.W., 1990, A quantitative method for determination of trace metal concentration in sedimentary pyrite: Marine Chemistry, v. 29, p. 119–144.

Huerta-Diaz, M.A., and Morse, J.W., 1992, Pyritization of trace metals in anoxic sediments: Geochimica et Cosmochimica Acta, v. 56, p. 2681–2702.

Huerta-Diaz, M.A., Carignan, R., and Tessier, A., 1993, Measurement of trace metals associated with acid volatile sulfides and pyrite in organic freshwater sediments: Environmental Science & Technology, v. 27, p. 2367–2372.

Hurtgen, M.T., Lyons, T.W., Ingall, E.D., and Cruse, A.M., 1999, Anomalous enrichments of iron monosulfides in euxinic marine sediments and the role of H_2S in iron sulfide transformations: examples from Effingham Inlet, Orca Basin, and the Black Sea: American Journal of Science, v. 299, p. 100–101.

Jiang, W.T., Horng, C.S., Roberts, A.P., and Peacor, D.R., 2001, Contradictory magnetic polarities in sediments and variable timing of neoformation of authigenic greigite: Earth and Planetary Science Letters, v. 193, p. 1–12, doi: 10.1016/S0012-821X(01)00497-6.

Jørgenson, B.B., 1982, Mineralization of organic-matter in the sea bed—The role of sulfate reduction: Nature, v. 296, p. 643–645.

Kaag, N.H.B.M., Foekema, E.M., and Scholten, M.C.T., 1998, Ecotoxicity of contaminated sediments, a matter of bioavailability: Water Science and Technology, v. 37, p. 225–231, doi: 10.1016/S0273-1223(98)00202-9.

Kimblin, R.T., and Johnson, A.C., 1992, Recent localised sulphate reduction and pyrite formation in a fissured Chalk aquifer: Chemical Geology, v. 100, p. 119–127, doi: 10.1016/0009-2541(92)90105-E.

Kolker, A., and Finkelman, R.B., 1998, Potentially hazardous elements in coal: Modes of occurrence and summary of concentration data for coal components: Coal Preparation, v. 19, p. 133–157.

Kolker, A., Haack, S.K., Cannon, W.F., Westjohn, D.B., Kim, M.-J., Nriagu, J., and Woodruff, L.G., 2003, Arsenic in southeastern Michigan, *in* Welch, A.H., and Stollenwerk, K.G., eds., Arsenic in ground water: Geochemistry and occurrence: Norwell, Massachusetts, Kluwer Academic, p. 281–294.

Kolker, A., Huggins, F.E., Palmer, C.A., Shah, N., Crowley, S.S., Huffman, G.P., and Finkelman, R.B., 2000, Mode of occurrence of arsenic in four U.S. coals: Fuel Processing Technology, v. 63, p. 167–178, doi: 10.1016/S0378-3820(99)00095-8.

Kolker, A., and Koeppen, R.B., 1998, Arsenic variation in coal pyrite, *in* Proceedings of the Fifteenth Annual International Pittsburgh Coal Conference: Pittsburgh, Pennsylvania, International Pittsburgh Coal Conference, CD-ROM.

Kolker, A., Mroczkowski, S.J., Palmer, C.A., Dennen, K.O., Finkelman, R.B., and Bullock, J.H., 2002, Toxic substances from coal combustion—A comprehensive assessment, Phase II: Element modes of occurrence for the Ohio 5/6/7, Wyodak, and North Dakota coal samples: U.S. Geological Survey Open File Report 02–224, 79 p.

Kolker, A., and Nordstrom, D.K., 2001, Occurrence and micro-distribution of arsenic in pyrite, *in* Proceedings of the USGS Arsenic Workshop: Denver, U.S. Geological Survey, p. 3.

Konhauser, K.O., 1998, Diversity of bacterial iron mineralization: Earth-Science Reviews, v. 43, p. 91–121, doi: 10.1016/S0012-8252(97)00036-6.

Kriegmann-King, M.R., and Reinhard, M., 1994, Transformation of carbon tetrachloride by pyrite in aqueous solution: Environmental Science & Technology, v. 28, p. 692–700.

Large, D.J., Sawlowicz, Z., and Spratt, J., 1999, A cobaltite-framboidal pyrite association from the Kupferschiefer: Possible implications for trace element behaviour during the earliest stages of diagenesis: Mineralogical Magazine, v. 63, p. 353–361, doi: 10.1180/002646199548574.

Lee, B.G., Griscom, S.B., Lee, J.S., Choi, H.J., Koh, C.H., Luoma, S.N., and Fisher, N.S., 2000, Influences of dietary uptake and reactive sulfides on metal bioavailability from aquatic sediments: Science, v. 287, p. 282–284, doi: 10.1126/SCIENCE.287.5451.282.

Lennie, A.R., Redfern, S.A.T., Champness, P.E., Stoddart, C.P., Schofield, P.F., and Vaughan, D.J., 1997, Transformation of mackinawite to greigite: An in situ X-ray powder diffraction and transmission electron microscope study: American Mineralogist, v. 82, p. 302–309.

Lennie, A.R., Redfern, S.A.T., Schofield, P.F., and Vaughan, D.J., 1995, Synthesis and Rietveld crystal structure refinement of mackinawite, tetragonal FeS: Mineralogical Magazine, v. 59 p. 677–684.

Lin, S., Huang, K.-M., and Chen, S.-K., 2000, Organic carbon deposition and its control on iron sulfide formation of the southern East China Sea continental shelf sediments: Continental Shelf Research, v. 20, p. 619–635, doi: 10.1016/S0278-4343(99)00088-6.

Liu, C.-Y., and Bard, A.J., 1989, Irradiation-induced adsorption edge shifts in colloidal particles of FeS_2 (pyrite): Journal of Physical Chemistry, v. 93, p. 7047–7049.

Luepke, J.J., and Lyons, T.W., 2001, Pre-Rodinian (Mesoproterozoic) supercontinental rifting along the western margin of Laurentia: Geochemical evidence from the Belt-Purcell Supergroup: Precambrian Research, v. 111, p. 79–90, doi: 10.1016/S0301-9268(01)00157-7.

Luther, G.W., 1991, Pyrite synthesis via polysulfide compounds: Geochimica et Cosmochimica Acta, v. 55, p. 2839–2850, doi: 10.1016/0016-7037(91)90449-F.

Lyons, T.W., 1997, Sulfur isotopic trends and pathways of iron sulfide formation in upper Holocene sediments of the anoxic Black Sea: Geochimica et Cosmochimica Acta, v. 61, p. 3367–3382, doi: 10.1016/S0016-7037(97)00174-9.

Lyons, T.W., Luepke, J.J., Schreiber, M.E., and Zieg, G.A., 2000, Sulfur geochemical constraints on Mesoproterozoic restricted marine deposition: Lower Belt Supergroup, northwestern United States: Geochimica et Cosmochimica Acta, v. 64, p. 427–437, doi: 10.1016/S0016-7037(99)00323-3.

MacCrehan, W., and Shea, D., 1995, Temporal relationship of thiols to inorganic sulfur compounds in anoxic Chesapeake Bay sediment porewater, *in* Vairavamurthy, M.A., and Schoonen, M.A.A., eds., Geochemical transformations of sedimentary sulfur: Washington, D.C., American Chemical Society Symposium Series 612, p. 294–310.

Marnette, E.C.L., Van Breemen, N., Hordijk, K.A., and Cappenberg, T.E., 1993, Pyrite formation in two freshwater systems in the Netherlands: Geochimica et Cosmochimica Acta, v. 57, p. 4165–4177, doi: 10.1016/0016-7037(93)90313-L.

Mayer, L.M., Chen, Z., Findlay, R.H., Fang, J., Sampson, S., Self, R.F.L., Jumars, P.A., Quetel, C., and Donard, O.F.X., 1996, Bioavailability of sediment contaminants subject to deposit-feeders digestion: Environmental Science & Technology, v. 30, p. 2641–2645, doi: 10.1021/ES960110Z.

Mercone, D., Thomson, J., Croudace, I.W., and Troelstra, S.R., 1999, A coupled natural immobilisation mechanism for mercury and selenium in deep-sea sediments: Geochimica et Cosmochimica Acta, v. 63, p. 1481–1488, doi: 10.1016/S0016-7037(99)00063-0.

Meysman, F.J.R., Middelburg, J.J., Herman, P.M.J., and Heip, C.H.R., 2003, Reactive transport in surface sediments. II. Media: an object-oriented problem-solving environment for early diagenesis: Computers & Geosciences, v. 29, p. 301–318, doi: 10.1016/S0098-3004(03)00007-4.

Mikac, N., Niessen, S., Ouddane, B., and Fischer, J.C., 2000, Effects of acid volatile sulfides on the use of hydrochloric acid for determining solid-phase associations of mercury in sediments: Environmental Science & Technology, v. 34, p. 1871–1876, doi: 10.1021/ES990477E.

Morse, J.W., 1999, Sulfides in sandy sediments; new insights on the reactions responsible for sedimentary pyrite formation: Aquatic Geochemistry, v. 5, p. 75–85, doi: 10.1023/A:1009620021442.

Morse, J.W., and Cornwell, J.C., 1987, Analysis and distribution of iron sulfide minerals in recent anoxic marine sediments: Marine Chemistry, v. 22, p. 55–69, doi: 10.1016/0304-4203(87)90048-X.

Morse, J.W., and Luther, G.W., 1999, Chemical influences on trace metal-sulfide interactions in anoxic sediments: Geochimica et Cosmochimica Acta, v. 63, p. 3373–3378, doi: 10.1016/S0016-7037(99)00258-6.

Morse, J.W., Millero, F.J., Cornwell, J.C., and Rickard, D., 1987, The chemistry of the hydrogen sulfide and iron sulfide systems in natural waters: Earth-Science Reviews, v. 24, p. 1–42, doi: 10.1016/0012-8252(87)90046-8.

Morse, J.W., and Wang, Q., 1997, Pyrite formation under conditions approximating those in anoxic sediments: II. Influence of precursor iron minerals and organic matter: Marine Chemistry, v. 57, p. 187–193, doi: 10.1016/S0304-4203(97)00050-9.

Motelica-Heino, M., Naylor, C., Zhang, N., and Davison, W., 2003, Simultaneous release of metals and sulfides in lacustrine sediments: Environmental Science & Technology, v. 37, p. 4374–4381, doi: 10.1021/ES030035+.

Murowchick, J.B., 1992, Marcasite inversion and the petrographic determination of pyrite ancestry: Economic Geology and the Bulletin of the Society of Economic Geologists, v. 87 p. 1141–1152.

Murowchick, J.B., and Barnes, H.L., 1986, Marcasite precipitation from hydrothermal solutions: Geochimica et Cosmochimica Acta, v. 50, p. 2615–2629, doi: 10.1016/0016-7037(86)90214-0.

Neal, A.L., Techkarnjanaruk, S., Dohnalkova, A., McCready, D., Peyton, B.M., and Geesey, G.G., 2001, Iron sulfides and sulfur species produced at hematite surfaces in the presence of sulfate-reducing bacteria: Geochimica et Cosmochimica Acta, v. 65, p. 223–235, doi: 10.1016/S0016-7037(00)00537-8.

Odzak, N., Kistler, D., Xue, H., and Sigg, L., 2002, In situ trace metal speciation in a eutrophic lake using the technique of diffusion gradients in thin films (DGT): Aquatic Sciences, v. 64, p. 292–300.

Paktunc, A.D., and Dave, N.K., 2002, Formation of secondary pyrite and carbonate minerals in the Lower Williams Lake tailings basin, Elliot Lake, Ontario, Canada: American Mineralogist, v. 87, p. 593–602.

Parkhurst, D.L., and Appelo, C.A.J., 1999, User's guide to PHREEC (Version 2)—A computer program for speciation, batch-reaction, one-dimensional transport, and inverse geochemical calculations: Denver, U.S. Geological Survey—Water Resources Division, 310 p.

Passier, H.F., Dekkers, M.J., and de Lange, G.J., 1998, Sediment chemistry and magnetic properties in an anomalously reducing core from the eastern Mediterranean Sea: Chemical Geology, v. 152, p. 287–306, doi: 10.1016/S0009-2541(98)00121-1.

Passier, H.F., Middelburg, J.J., de Lange, G.J., and Boettcher, M.E., 1997, Pyrite contents, microtextures, and sulfur isotopes in relation to formation of the youngest eastern Mediterranean sapropel: Geology, v. 25, p. 519–522, doi: 10.1130/0091-7613(1997)0252.3.CO;2.

Raiswell, R., and Canfield, D.E., 1996, Rates of reactions between silicate iron and dissolved sulfide in Peru Margin sediments: Geochimica et Cosmochimica Acta, v. 60, p. 2777–2787.

Raiswell, R., 1998, Sources of iron for pyrite formation in marine sediments: American Journal of Science, v. 298, p. 219–245.

Raiswell, R., Canfield, D.E., and Berner, R.A., 1994, A comparison of iron extraction methods for determination of degree of pyritization and recognition of iron-limited pyrite formation: Chemical Geology, v. 111, p. 101–110.

Rakovan, J., Schoonen, M.A.A., Reeder, R.J., Tyrna, P., and Nelson, D.O., 1995, Epitaxial overgrowths of marcasite on pyrite from the tunnel and reservoir project, Chicago, Illinois, USA: Implications for marcasite growth: Geochimica et Cosmochimica Acta, v. 59, p. 343–346, doi: 10.1016/0016-7037(94)00320-L.

Reeder, R.J., and Rakovan, J., 1998, Surface structural controls on trace element incorporation during crystal growth, *in* Jamtveit, B., and Meakin, P., eds., Growth, dissolution, and pattern formation in geosystems: Dordrecht, Kluwer Academic, p. 143–162.

Reynolds, R.L., Tuttle, M.L., Rice, C.A., Fishman, N.S., Karrachewski, J.A., and Sherman, D.M., 1994, Magnetization and geochemistry of greigite-bearing Cretaceous strata, North Slope, Alaska: American Journal of Science, v. 294, p. 485–528.

Rickard, D., 1997, Kinetics of pyrite formation by the H2S oxidation of iron (II) monosulfide in aqueous solutions between 25 and 125°C: The rate equation: Geochimica et Cosmochimica Acta, v. 61, p. 115–134, doi: 10.1016/S0016-7037(96)00321-3.

Rickard, D., and Luther, G.W., 1997, Kinetics of pyrite formation by the H2S oxidation of iron (II) monosulfide in aqueous solutions between 25 and 125°C: The mechanism: Geochimica et Cosmochimica Acta, v. 61, p. 135–147, doi: 10.1016/S0016-7037(96)00322-5.

Rickard, D., Oldroyd, A., and Cramp, A., 1999, Voltammetric evidence for soluble FeS complexes in anoxic estuarine muds: Estuaries, v. 22, p. 693–701.

Rickard, D., Schoonen, M.A.A., and Luther, G.W., 1995, Chemistry of iron sulfides in sedimentary environments, *in* Vairavamurthy, M.A., and Schoonen, M.A.A., eds., Geochemical Transformations of Sedimentary Sulfur: Washington, D.C., American Chemical Society Symposium Series 612, p. 168–193.

Rickard, D.T., 1975, Kinetics and mechanism of pyrite formation at low temperatures: American Journal of Science, v. 275, p. 636–652.

Roberts, A.P., and Turner, G.M., 1993, Diagenetic formation of ferrimagnetic iron sulphide minerals in rapidly deposited marine sediments, South Island, New Zealand: Earth and Planetary Science Letters, v. 115, p. 257–273, doi: 10.1016/0012-821X(93)90226-Y.

Robie, R.A., Hemingway, B.S., and Fisher, J.R., 1978, Thermodynamic properties of minerals and related substances at 298 K and 1 Bar (10^5 Pascals) pressure and at higher temperatures: U.S. Geological Survey Bulletin 1452, 456 p.

Roychoudhury, A.N., Kostka, J.E., and Van Cappellen, P., 2003, Pyritization: A paleoenvironmental and redox proxy reevaluated: Estuarine, Coastal and Shelf Science, v. 57, p. 1183–1193, doi: 10.1016/S0272-7714(03)00058-1.

Rozan, T.F., Theberge, S.M., and Luther, G., 2000, Quantifying elemental sulfur (S^0), bisulfide (HS^-) and polysulfides (Sx^{2-}) using a voltammetric method: Analytica Chimica Acta, v. 415, p. 175–185.

Schaufuss, A.G., Nesbitt, H.W., Kartio, I., Kartio, I., Laajalehto, K., Bancroft, G.M., and Szargan, R., 1998, Incipient oxidation of fractured pyrite surfaces in air: Journal of Electron Spectroscopy and Related Phenomena, v. 96, p. 69–82, doi: 10.1016/S0368-2048(98)00237-0.

Schoonen, M.A.A., 1989, Mechanisms of pyrite and marcasite formation from solution between 25 and 300C [Ph.D. thesis]: University Park, Pennsylvania State University, 215 p.

Schoonen, M.A.A., and Barnes, H.L., 1991a, Reactions forming pyrite and marcasite from solution: I. Nucleation of FeS_2 below 100°C: Geochimica et Cosmochimica Acta, v. 55, p. 1495–1504, doi: 10.1016/0016-7037(91)90122-L.

Schoonen, M.A.A., and Barnes, H.L., 1991b, Reactions forming pyrite and marcasite from solution: II. Via FeS precursors below 100°C: Geochimica et Cosmochimica Acta, v. 55, p. 1505–1514, doi: 10.1016/0016-7037(91)90123-M.

Schoonen, M.A.A., and Barnes, H.L., 1991c, Mechanisms of pyrite and marcasite formation from solution: III. Hydrothermal processes: Geochimica et Cosmochimica Acta, v. 55, p. 3491–3504, doi: 10.1016/0016-7037(91)90050-F.

Schoonen, M.A.A., and Xu, Y., 2001, Nitrogen reduction under hydrothermal vent conditions: Implications for the prebiotic synthesis of C-H-O-N compounds: Astrobiology, v. 1, p. 133–140, doi: 10.1089/153110701753198909.

Schoonen, M.A.A., Xu, Y., and Strongin, D.R., 1998, An introduction to geocatalysis: Journal of Geochemical Exploration, v. 62, p. 201–215, doi: 10.1016/S0375-6742(97)00069-1.

Senior, C.L., Bool, L.E., III, Morency, G.P., Huggins, F., Huffman, G.P., Shah, N., Wendt, J.O.L., Shadman, F., Peterson, T., Seames, W., Wu, B., Sarofim, A.F., Olmez, I., Zeng, T., Crowley, S.S., Kolker, A., Palmer, C.A., Finkelman, R.B., Helble, J.J., and Wornat, M.J., 1997, Toxic substances from coal combustion—a comprehensive assessment: Final Report, DOE Contract DE-AC22–95PC95101: Pittsburgh, U.S. Department of Energy, Pittsburgh Energy Technology Center, 12 p.

Soetaert, K., Herman, P.M.J., and Middelburg, J.J., 1996, A model of early diagenetic processes from the shelf to abyssal depths: Geochimica et Cosmochimica Acta, v. 60, p. 1019–1040, doi: 10.1016/0016-7037(96)00013-0.

Sohnel, O., 1982, Electrolyte crystal-aqueous solution interfacial tension from crystallization data: Journal of Crystal Growth, v. 57, p. 101–108, doi: 10.1016/0022-0248(82)90254-8.

Stirling, A., Bernasconi, M., and Parrinello, M., 2003, Ab initio simulation of H_2S adsorption on the (100) surface of pyrite: The Journal of Chemical Physics, v. 119, p. 4934–4939, doi: 10.1063/1.1595632.

Strechie, C., Andre, F., Jelinowska, A., Tucholka, P., Guichard, F., Lericolais, G., and Panin, N., 2002, Magnetic minerals as indicators of major environmental change in Holocene black sea sediments: preliminary results: Physics and Chemistry of the Earth, Parts A/B/C, v. 27, p. 1363–1370, doi: 10.1016/S1474-7065(02)00119-5.

Suits, N.S., and Arthur, M.A., 2000, Sulfur diagenesis and partitioning in Holocene Peru shelf and upper slope sediments: Chemical Geology, v. 163, p. 219–234.

Suits, N.S., and Wilkin, R.T., 1998, Pyrite formation in the water column and sediments of a meromictic lake: Geology, v. 26, p. 1099–1102, doi: 10.1130/0091-7613(1998)0262.3.CO;2.

Sweeney, R.E., and Kaplan, I.R., 1973, Pyrite framboid formation: Laboratory synthesis and marine sediments: Economic Geology and the Bulletin of the Society of Economic Geologists, v. 68, p. 618–634.

Theberge, S.M., and Luther, G.W., 1997, Determination of the electrochemical properties of a soluble aqueous FeS species present in sulfidic solutions: Aquatic Geochemistry, v. 3, p. 191–211, doi: 10.1023/A:1009648026806.

Torre, M.C.A.D., Beaulieu, P. Y., and Tessier, A., 2000, In situ measurement of trace metals in lakewater using the dialysis and DGT techniques: Analytica Chimica Acta, v. 418, p. 53–69.

Trapp, H., and Meisler, H., 1992, The regional aquifer system underlying the Northern Atlantic Coastal Plain, in arts of North Carolina, Virginia, Maryland, Delaware, New Jersey, and New York: Summary: Denver, United States Geological Survey, USGS-WRI Investigations Report 92–4217, 59 p.

van den Berg, G.A., Buykx, S.E.J., van den Hoop, M., and van der Heijdt, L.M., 2001, Vertical profiles of trace metals and acid-volatile sulphide in a dynamic sedimentary environment: Lake Ketel, The Netherlands: Applied Geochemistry, v. 16, p. 781–791, doi: 10.1016/S0883-2927(00)00076-7.

van den Berg, G.A., Loch, J.P.G., van der Heijdt, L.M., and Zwolsman, J.J.G., 1998, Vertical distribution of acid-volatile sulfide and simultaneously extracted metals in a recent sedimentation area of the River Meuse in The Netherlands: Environmental Toxicology and Chemistry, v. 17, p. 758–763.

van den Berg, G.A., Lock, J.P.G., van Der Heijdt, L.M., and Zwolsman, J.J.G., 1999, Mobilisation of heavy metals in contaminated sediments in the river Meuse, The Netherlands: Water, Air, and Soil Pollution, v. 116, p. 567–586, doi: 10.1023/A:1005146927718.

van Griethuysen, C., Meijboom, E.W., and Koelmans, A.A., 2003, Spatial variation of metals and acid volatile sulfide in floodplain lake sediment: Environmental Toxicology and Chemistry, v. 22, p. 457–465.

Wang, F.Y., and Chapman, P.M., 1999, Biological implications of sulfide in sediment—A review focusing on sediment toxicity: Environmental Toxicology and Chemistry, v. 18, p. 2526–2532.

Wang, Q., and Morse, J.W., 1995, Laboratory simulation of pyrite formation in anoxic sediments, *in* Vairavamurthy, M.A., and Schoonen, M.A.A., eds., Geochemical Transformations of Sedimentary Sulfur: Washington, D.C., American Chemical Society Symposium Series 612, p. 206–223.

Wang, Q., and Morse, J.W., 1996, Pyrite formation under conditions approximating those in anoxic sediments: I. Pathway and morphology: Marine Chemistry, v. 52, p. 99–121, doi: 10.1016/0304-4203(95)00082-8.

Wang, W.X., Stupakoff, I., and Fisher, N.S., 1999, Bioavailability of dissolved and sediment-bound metals to a marine deposit-feeding polychaete: Marine Ecology Progress Series, v. 178, p. 281–293.

Weerasooriya, R., and Dharmasena, B., 2001, Pyrite-assisted degradation of trichloroethene (TCE): Chemosphere, v. 42, p. 389–396, doi: 10.1016/S0045-6535(00)00160-0.

Werne, J.P., Sageman, B.B., Lyons, T.W., and Hollander, D.J., 2002, An integrated assessment of a "type euxinic" deposit: Evidence for multiple controls on black shale deposition in the middle Devonian Oatka Creek formation: American Journal of Science, v. 302, p. 110–143.

Wharton, M.J., Atkins, B., Charnock, J.M., Livens, F.R., Pattrick, R.A.D., and Collison, D., 2000, An X-ray absorption study of the coprecipitation of Tc and Re with mackinawite (FeS): Applied Geochemistry, v. 15, p. 347–354, doi: 10.1016/S0883-2927(99)00045-1.

Wijsman, J.W.M., Middelburg, J.J., Herman, P.M.J., Böttcher, M.E., and Heip, C.H.R., 2001, Sulfur and iron speciation in surface sediments along the northwestern margin of the Black Sea: Marine Chemistry, v. 74, p. 261–278, doi: 10.1016/S0304-4203(01)00019-6.

Wilcoxon, J.P., Newcomer, P.P., and Samara, G.A., 1996, Strong quantum confinement effects in semiconductors: FeS2 nanoclusters: Solid State Communications, v. 98, p. 581–585, doi: 10.1016/0038-1098(95)00822-5.

Wilkin, R.T., and Arthur, M.A., 2001, Variations in pyrite texture, sulfur isotope composition, and iron systematics in the Black Sea: Evidence for Late Pleistocene to Holocene excursions of the O_2-H_2S redox transition: Geochimica et Cosmochimica Acta, v. 65, p. 1399–1416, doi: 10.1016/S0016-7037(01)00552-X.

Wilkin, R.T., Arthur, M.A., and Dean, W.E., 1997, History of water-column anoxia in the Black Sea indicated by pyrite framboid size distributions: Earth and Planetary Science Letters, v. 148, p. 517–525, doi: 10.1016/S0012-821X(97)00053-8.

Wilkin, R.T., and Barnes, H.L., 1996, Pyrite formation by reactions of iron monosulfides with dissolved inorganic and organic sulfur species: Geochimica et Cosmochimica Acta, v. 60, p. 4167–4179, doi: 10.1016/S0016-7037(97)81466-4.

Wilkin, R.T., and Barnes, H.L., 1997, Formation processes of framboidal pyrite: Geochimica et Cosmochimica Acta, v. 61, p. 323–339, doi: 10.1016/S0016-7037(96)00320-1.

Wilkin, R.T., Barnes, H.L., and Brantley, S.L., 1996, The size distribution of framboidal pyrite in modern sediments: An indicator of redox conditions: Geochimica et Cosmochimica Acta, v. 60, p. 3897–3912, doi: 10.1016/0016-7037(96)00209-8.

Wilkin, R.T., and Ford, R.G., 2002, Use of hydrochloric acid for determining solid-phase arsenic partitioning in sulfidic sediments: Environmental Science & Technology, v. 36, p. 4921–4927, doi: 10.1021/ES025862+.

Wolthers, M., 2003, Geochemistry and environmental mineralogy of the iron-sulphur-arsenic system [Ph.D. thesis]: Utrecht, Utrecht University, 185 p.

Xu, Y., and Schoonen, M.A.A., 1995, The stability of thiosulfate in the presence of pyrite in low-temperature aqueous solutions: Geochimica et Cosmochimica Acta, v. 59, p. 4605–4622, doi: 10.1016/0016-7037(95)00331-2.

Xu, Y., and Schoonen, M.A.A., 2000, The absolute energy position of conduction and valence bands of selected semiconducting minerals: American Mineralogist, v. 85, p. 543–556.

Xu, Y., Schoonen, M.A.A., and Strongin, D.R., 1996, Thiosulfate oxidation: Catalysis of synthetic sphalerite doped with transition metals: Geochimica et Cosmochimica Acta, v. 60, p. 4701–4710, doi: 10.1016/S0016-7037(96)00279-7.

Yu, K.C., Tsai, L.J., Chen, S.H., and Ho, S.T., 2001, Chemical binding of heavy metals in anoxic river sediments: Water Research, v. 35, p. 4086–4094, doi: 10.1016/S0043-1354(01)00126-9.

Zeng, T., Sarofim, A.F., and Senior, C.L., 2001, Vaporization of arsenic, selenium and antimony during coal combustion: Combustion and Flame, v. 126, p. 1714–1724, doi: 10.1016/S0010-2180(01)00285-1.

Zhang, H., Davison, W., Knight, B., and McGrath, S., 1998, In situ measurements of solution concentrations and fluxes of trace metals in soils using DGT: Environmental Science & Technology, v. 32, p. 704–711.

Zhang, H., Davison, W., Miller, S., and Tych, W., 1995, In situ high resolution measurements of fluxes of Ni, Cu, Fe, and Mn and concentrations of Zn and Cd in porewaters by DGT: Geochimica et Cosmochimica Acta, v. 59, p. 4181–4192.

Zhang, J., Ren, D., Zheng, C., Zeng, R., Chou, C.-L., and Liu, J., 2002, Trace element abundances in major minerals of Late Permian coals from southwestern Guizhou province, China: International Journal of Coal Geology, v. 53, p. 55–64, doi: 10.1016/S0166-5162(02)00164-7.

Zhang, S., Wang, S., and Shan, Q., 2001, Effect of sample pretreatment upon the metal speciation in sediments by sequential extraction procedure: Chemical Speciation and Bioavailability, v. 13, p. 69–74.

Manuscript Accepted by the Society March 28, 2004

Geological Society of America
Special Paper 379
2004

Organic sulfur biogeochemistry: Recent advances and future research directions

Josef P. Werne*
Large Lakes Observatory and Department of Chemistry, University of Minnesota Duluth, 10 University Drive, Duluth, Minnesota 55812, USA

David J. Hollander*
College of Marine Sciences, University of South Florida, St. Petersburg, Florida 33701, USA

Timothy W. Lyons*
Department of Geological Sciences, University of Missouri, Columbia, Missouri 65211, USA

Jaap S. Sinninghe Damsté*
Department of Marine Biogeochemistry and Toxicology, Royal Netherlands Institute for Sea Research (NIOZ), P.O. Box 59, 1790 AB Den Burg, Texel, The Netherlands

ABSTRACT

The biogeochemistry of sulfur is of widespread interest in the earth science community due to its impact on many different biogeochemical processes. Organic sulfur is of particular interest due to its impact on petroleum formation and refining and its relationship to microbial sedimentary processes, organic carbon accumulation, and the overall integrity of paleoenvironmental proxy records. This paper reviews many of the advances in organic sulfur biogeochemical research spanning the past ~15 years. These advances include (1) an improved mechanistic understanding of why sulfur-rich organic deposits form petroleum products earlier during diagenesis than sulfur-poor deposits, (2) constraints on the timing and pathways of organic sulfur formation as well as the forms of organic sulfur present in the environment, and (3) recognition of the impacts of organic matter sulfurization on organic carbon preservation at bulk and molecular scales and the implications of this enhanced preservation for paleoenvironmental studies.

Keywords: Organic sulfur, sedimentary sulfur, sulfur isotopes, macromolecular sulfur, global sulfur cycle.

INTRODUCTION

The sulfurization of organic matter is a globally significant biogeochemical process that has long been a topic of investigation. Among the reasons for this longstanding interest are the relationships between organic sulfur and (1) petroleum formation and quality, (2) the coupled global biogeochemical cycles of carbon, sulfur, and oxygen, (3) sedimentary microbial activity, and (4) organic matter preservation and molecularly based paleoenvironmental reconstructions. Despite the importance of

*E-mails: jwerne@d.umn.edu, davidh@seas.marine.usf.edu, lyonst@missouri.edu, damsté@nioz.nl.

Werne, J.P., Hollander, D.J., Lyons, T.W., and Sinninghe Damsté, J.S., 2004, Organic sulfur biogeochemistry: Recent advances and future research directions, *in* Amend, J.P., Edwards, K.J., and Lyons, T.W., eds., Sulfur biogeochemistry—Past and present: Boulder, Colorado, Geological Society of America Special Paper 379, p. 135–150. For permission to copy, contact editing@geosociety.org.

organic sulfur, our understanding of the processes surrounding its formation remains incomplete. One reason that the sulfurization of organic matter is poorly constrained is the wide variety of organic sulfur compounds found in nature, which appear to form via a range of possible mechanisms. For example, sulfur can be incorporated intramolecularly into organic compounds, forming a cyclo-sulfur group such as a thiophene or thiane, or it can be incorporated intermolecularly, resulting in organic compounds linked via C-S_x-C bonds into a macromolecular matrix. Table 1 illustrates some of the varieties of organic sulfur compounds identified in natural systems.

A second reason for gaps in our knowledge of organic sulfur formation is the complexity of biological and abiological sedimentary sulfur cycling (Fig. 1). Pore-water sulfide, which can be incorporated into organic matter, can react with iron to form iron monosulfides and ultimately pyrite. Additional reactions include oxidation of sulfide to sulfate and partial oxidation of sulfide to "reactive intermediates" such as polysulfides, elemental sulfur, thiosulfate, polythionates, and sulfite (Table 2). Sulfide can be oxidized by many pathways involving, for example, oxygen diffusing down from the water column, nitrate, (iron) oxide or oxyhydroxide minerals, or microbes (e.g., *Beggiatoa* sp.). Reactive intermediates produced through partial oxidation can then undergo a number of reactions, including disproportionation to sulfate and sulfide, complete reduction, or complete oxidation. Finally, sulfide can also diffuse upward into the overlying water

TABLE 1. EXAMPLES OF TYPICALLY ENCOUNTERED ORGANIC SULFUR COMPOUNDS

OSC type	General structure	Example	Reference
Intramolecular S incorporation			
thiolane	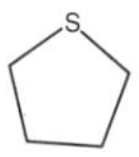	2-methyl-5-tridecylthiolane 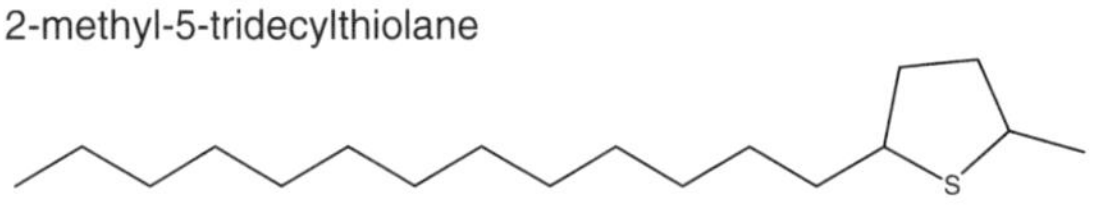	Kohnen et al., 1991b
thiane		malabarica-thiane 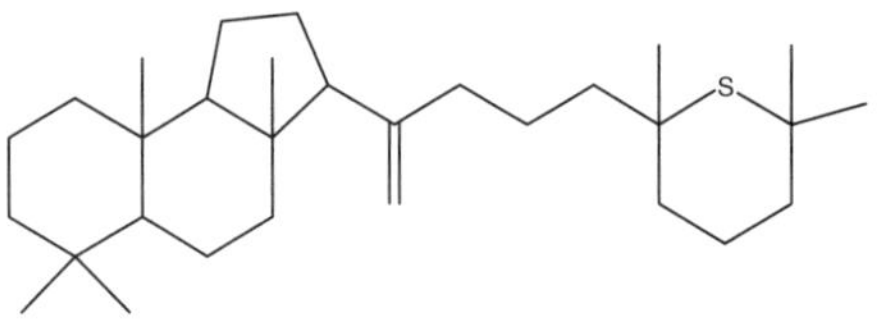	Werne et al., 2000
thiophene		highly-branched isoprenoid thiophene 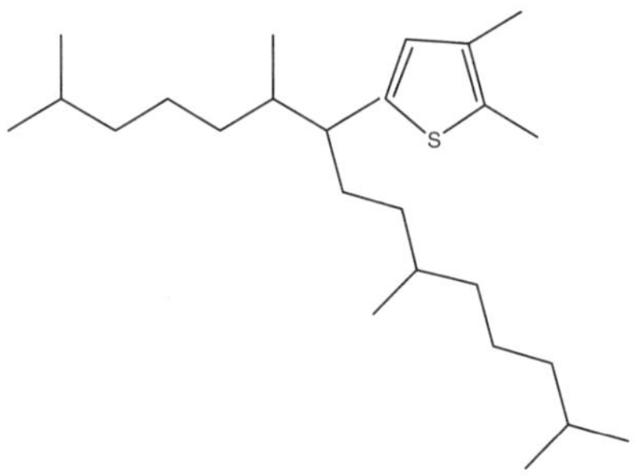	Respondek et al., 1997
Intermolecular S incorporation			
sulfide linked	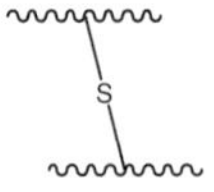	intermolecularly bound bacteriohopanoid 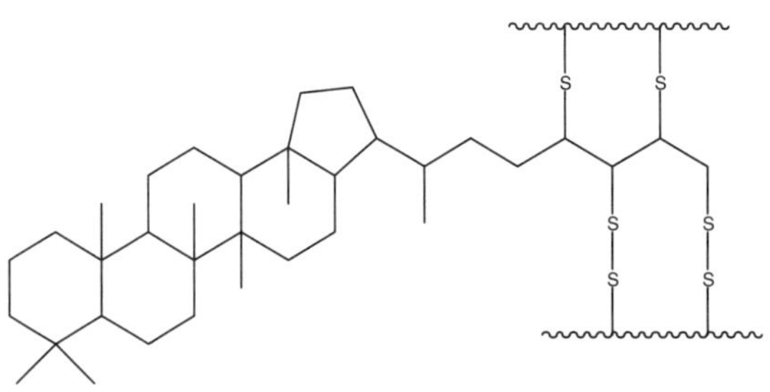	de Leeuw and Sinninghe Damsté, 1990
polysulfide linked	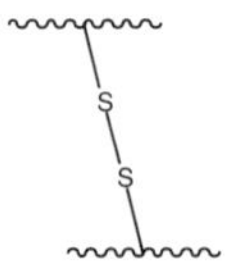		

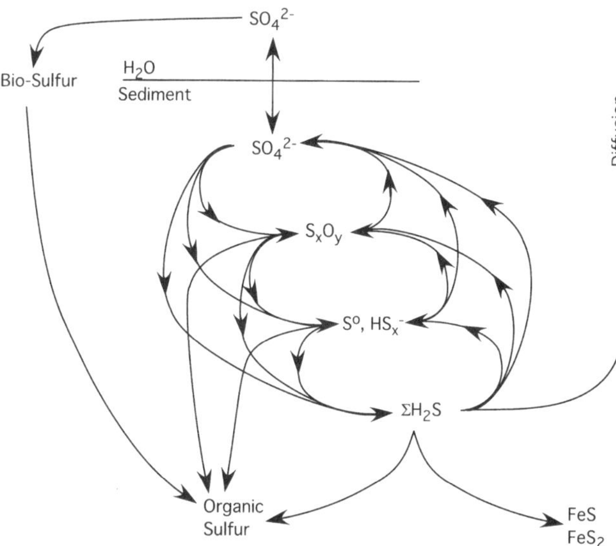

Figure 1. Schematic diagram illustrating the complex cycling of sulfur in sediments. The left side of the diagram shows reduction processes, including bacterial sulfate reduction and the reductive portion of disproportionation. The right side of the diagram shows oxidative processes, including microbial and/or abiotic (partial) sulfide oxidation as well as the oxidative portion of disproportionation. Arrows indicate directions of sulfur "flow" during sulfur cycling, whether chemical, as in the case of sulfate reduction or disproportionation reactions, or physical, as in diffusion of sulfide back into the overlying water column. Also included is the biological assimilation of sulfate into biomass and its deposition in sediments, and formation of iron sulfides. Possible immediate sources of organic sulfur include bio-sulfur, sulfides, and partially oxidized reactive intermediates such as elemental sulfur (S^0), polysulfides (HS_x^-), polythionates, thiosulfate, and sulfite (collectively represented as S_xO_y in the figure).

column. It is not known with certainty whether the sulfur that is incorporated into organic matter is pore-water sulfide, some other pool of reduced sulfur, such as polysulfides or elemental sulfur, or a mixture of different sources.

Dissimilatory bacterial reduction of dissolved sulfate is the primary means by which reduced sulfur (e.g., H_2S) is produced in the natural environment. There are many proposed chemical pathways to explain the incorporation of aqueous sulfides into organic matter, and in actuality, several are likely to be operating simultaneously. The dominant mechanisms would vary according to specific environmental conditions.

In order for organic sulfur formation to occur, certain depositional conditions must be met. First, there must be an adequate supply of reduced sulfur species, which implies the presence of anoxic conditions and sufficient sulfate reduction to produce the sulfide from which all other reduced sulfur species are derived. Second, there must be significant quantities of reactive organic matter present. This organic matter serves two purposes: it acts as a substrate supporting bacterial sulfate reduction, thus providing the reduced sulfur, and it reacts directly with the reactive reduced

TABLE 2. REACTIVE (R) AND NON-REACTIVE (NR) SULFUR SPECIES INVOLVED IN SEDIMENTARY CYCLE OF SULFUR AND THEIR OXIDATION STATES.

Compound	Formula	Oxidation state of sulfur	R/NR
sulfate	SO_4^{2-}	+6	R
sulfite	SO_3^{2-}	+4	R
tetrathionate	$S_4O_6^{2-}$	+2.5	R
thiosulfate	$S_2O_3^{2-}$	+2	R
elemental sulfur	S_8 (S^0)	0	R?
polysulfides	HS_x^-, S_x^{2-}	~–0.5	R
organic sulfur (disulfide)	R-S-S-R	–1	NR
pyrite	FeS_2	–1	NR
iron monosulfide	FeS	–2	R?
organic sulfur (thiol)	R-SH	–2	NR
bisulfide	HS^-	–2	R
hydrogen sulfide	H_2S	–2	R?

sulfur to form organic sulfur compounds. Organic matter content in sedimentary environments is typically enhanced by the presence of anoxic to euxinic conditions due to the exclusion of burrowing macrofauna. Finally, because pyrite formation is believed to be a kinetically favored process relative to organic matter sulfurization (Gransch and Posthuma, 1974), the environment must have a limited availability of reactive iron species (iron oxides and oxyhydroxides; Canfield, 1989; Canfield et al., 1992, 1996).

This final condition of the environment may not be absolutely required, as various studies have demonstrated the simultaneous formation of organic sulfur and iron sulfides (Brüchert and Pratt, 1996; Bates et al., 1995; Urban et al., 1999; Filley et al., 2002). Indeed, it may be that rapid input of iron oxides can actually promote both the sulfurization of organic matter and the formation of iron sulfides through the rapid production of polysulfides, as suggested by Filley et al. (2002). It has also been pointed out that frequent oscillations between oxidizing and reducing conditions may act as a catalyst for formation of both iron sulfides and organic sulfur through production of reactive intermediate sulfur species such as polysulfides, thiosulfate, and polythionates (Boulègue et al., 1982; Luther and Church, 1988; Ferdelman et al., 1991).

A full review of organic sulfur biogeochemistry is beyond the scope of this paper (for a thorough review of organic sulfur research up to 1990, see Sinninghe Damsté and de Leeuw [1990]). The focus of this paper is therefore on significant advances over the past 15 years.

ORGANIC SULFUR AND PETROLEUM FORMATION

All fossil fuels contain sulfur, ranging from trace amounts to more than 14% by weight (Orr and Sinninghe Damsté, 1990). In fact, the largest petroleum systems in the world are carbonate-evaporite sequences, which are typically high in sulfur (Vairavamurthy et al., 1995). Few crude oils with more than 4% sulfur are produced industrially, however, because organic sulfur compounds react during the refining process to produce substances such as sulfuric acid, which is corrosive and poisons catalysts, and sulfur dioxide, which contributes to numerous environmental problems such as acid rain (Orr, 1978; Tissot and Welte, 1984). In fact, the importance of organic sulfur in petroleum products has been recognized by a number of special sessions sponsored by the American Chemical Society Geochemistry Division, resulting in published proceedings volumes devoted to understanding the geochemistry of sulfur in fossil fuels (Orr and White, 1990) and the sedimentary environment (Vairavamurthy and Schoonen, 1995). Numerous research papers and reviews have been written about the geochemistry of sulfur in fossil fuels, including oils (Gransch and Posthuma, 1974; Orr and Sinninghe Damsté, 1990; Sinninghe Damsté et al., 1994), oil shales (Sinninghe Damsté et al., 1993; Barakat and Rüllkötter, 1995, 1999; de las Heras et al., 1997), and coals (Sinninghe Damsté and de Leeuw 1992; Sinninghe Damsté et al., 1999a; Sandison et al., 2002). Because of the large number of reviews on organic sulfur in petroleum systems in recent years, the focus here will be on a few studies that highlight the impact of organic sulfur on petroleum formation.

The quantity and forms of sulfur in petroleum systems are directly related to the properties of the source rocks from which their petroleum products were generated and therefore to the environment of deposition (Gransch and Posthuma, 1974). As mentioned previously, the organic sulfur content of a sedimentary deposit is controlled to a large extent by the availability of reduced sulfur species and by the availability of reactive iron (Canfield, 1989; Canfield et al., 1992, 1996). The formation of iron sulfides (e.g., pyrite, FeS_2) in sedimentary systems is generally believed to be kinetically favored relative to the formation of organic sulfur in the presence of readily available reactive iron species (Gransch and Posthuma, 1974; Hartgers et al., 1997). Certain iron minerals, specifically the iron oxides and oxyhydroxides, are known to be highly reactive with respect to hydrogen sulfide (Canfield et al., 1992; Larson and Postma, 2001), and reactivities of different iron minerals have been shown to vary as a function of particle size distributions, among other factors (Larson and Postma, 2001). Thus, delivery of reactive iron minerals, which varies with climate and source area, could impact the availability of reactive iron in natural systems (G. Mora and L. Hinnov, 2003, personal commun.).

Sedimentary organic sulfur is often present as polysulfide linkages between compounds in the macromolecular matrix of organic-rich deposits (Aizenshtat et al., 1983; Kohnen et al., 1991a), though a number of different types of organic sulfur compounds have recently been identified in kerogens (see below). Sulfur-sulfur bonds are cleaved more easily than carbon-sulfur or carbon-carbon bonds (Aizenshtat et al., 1995). A number of recent studies have utilized artificial maturation of natural organic sulfur-rich kerogens and shales via hydrous pyrolysis (Lewan, 1993) to quantify the effect of these S-S bonds on petroleum formation (Krein and Aizenshtat, 1995; Nelson et al., 1995; Tomic et al., 1995; Koopmans et al., 1996; Putschew et al., 1998; Sinninghe Damsté et al., 1998a). All of these studies clearly demonstrated that sulfur-rich kerogens produced petroleum under appreciably lower temperatures than their low sulfur counterparts, thus highlighting the impact of sulfur content on the kinetics of petroleum formation.

Of particular interest among the studies of hydrous pyrolysis is the investigation by Putschew et al. (1998), in which the remnant kerogens and pyrolysis products were also closely examined by molecular methods (stepwise selective chemical degradation) after artificial maturation. Although petroleum formation begins earlier due to the presence of organic sulfur, this study demonstrated that not all of the organic sulfur compounds were released. Indeed, much of the organic sulfur remained in the form of nonextractable, macromolecular organic matter (Putschew et al., 1998). Because this sulfur is no longer present in the kerogen, or in the petroleum products, it would be easily missed without molecular-level investigations.

These and many other studies implicate the relative weakness of S-S and C-S bonds compared to C-C bonds in petroleum formation. Alternatively, a recent study by Lewan (1998) suggested that it is not actually the relative weakness of S-S and C-S bonds that contributes to the early formation of high sulfur petroleum products. Instead, the presence of sulfur radicals may control petroleum formation rates. More specifically, the concentration of sulfur radicals generated during the initial stages of thermal maturation may be the critical factor, with greater concentration of sulfur radicals leading to more rapid petroleum generation (Lewan 1998). A second implication of the sulfur radical model is that once the petroleum products are generated and migrate away from the initial site of production, they will no longer be in contact with the sulfur radicals, which would slow the production of natural gas formed by continued cracking of C-C bonds. Lewan's (1998) study therefore provides an organic sulfur-dependent explanation for observed variation in petroleum formation rates (cf. Nelson et al., 1995; Tomic et al., 1995; Koopmans et al., 1996; Putschew et al., 1998; Sinninghe Damsté et al., 1998a) as well as the overall composition of the resulting petroleum.

ORGANIC SULFUR AND GLOBAL BIOGEOCHEMICAL CYCLES

The global biogeochemical cycles of carbon, sulfur, and oxygen are linked through a complex system of oxidation-reduction reactions that occur at the surface of the Earth. These reactions moderate the balance between reduced and oxidized forms of C and S on time scales of millions of years. Modeling studies have demonstrated that the sequestration of reduced sulfur in sediments affects the biogeochemical cycles of sulfur, carbon, and oxygen and therefore the evolution of atmospheric CO_2 and O_2 concentrations over geologic time (Garrels and Lerman, 1981, 1984; Kump and Garrels, 1986; Berner, 1987; Petsch and Berner, 1998; Canfield et al., 2000). Pyrite is clearly the most quantitatively significant sink for reduced sulfur in the sedimentary environment (Garrels and Lerman, 1984; Berner and Raiswell, 1983), and global biogeochemical models have typically approximated the total reduced sulfur pool as entirely pyrite sulfur. Recent studies, however, have shown that organic sulfur is also a significant pool of reduced sulfur (Anderson and Pratt, 1995; Vairavamurthy et al., 1995). Indeed, as much as 80% of the total reduced sulfur in some environments is present as organic sulfur (e.g., the Miocene Monterey Formation; see Fig. 2; Zaback and Pratt, 1992; Anderson and Pratt, 1995). A second potential source for error in model-based reconstructions of ancient atmospheric compositions is the fact that most models are dependent on assumptions made about the sulfur isotope offset between oxidized and reduced forms of sulfur (Garrels and Lerman, 1984; Kump and Garrels, 1986; Petsch and Berner, 1998). In general, a constant offset is assumed between the sulfur isotope composition of the total oxidized sulfur pool (estimated as sulfate in evaporite deposits) and the total reduced sulfur pool (estimated as pyrite). In a series of papers, however, Canfield and Teske (1996), Canfield (1998), Canfield and Raiswell (1999), Canfield et al. (2000), and Shen et al. (2001) argued that the offset between the sulfur isotope composition of sulfate and sulfides has changed significantly over geologic time. In addition, organic sulfur is generally enriched in ^{34}S relative to pyrite (Fig. 2; Anderson and Pratt, 1995; Werne et al., 2003). Thus, because the percent of total reduced sulfur that is organic can vary with time, the actual isotopic offset between oxidized and reduced forms of sulfur in the natural environment is not constant. These factors suggest that organic sulfur formation and burial may have affected the evolution of the atmosphere over geologic time scales, and our ability to model such changes is dependent on a better understanding of organic sulfur formation processes (Werne, 2000).

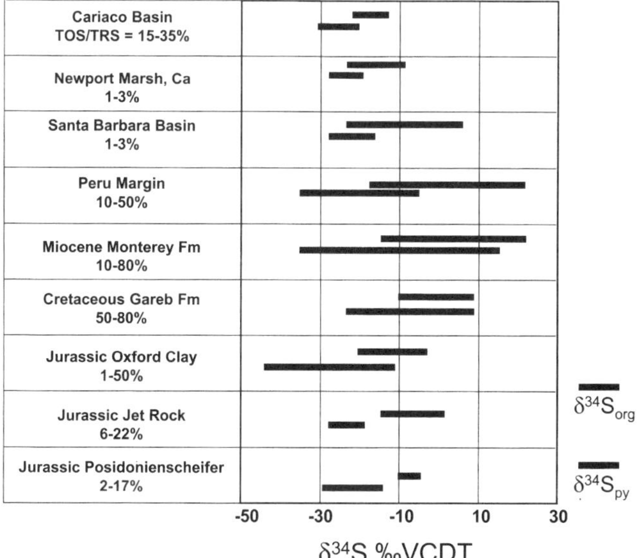

Figure 2. Summary diagram showing ranges of sulfur isotope compositions for pyrite sulfur ($\delta^{34}S_{py}$, black bars) and organic sulfur ($\delta^{34}S_{org}$, gray bars) in several environments. Beneath the name of each locality is the percent of the total reduced sulfur present as organic sulfur. Data are from Werne et al. (2003) and Anderson and Pratt (1995) and references therein. VCDT—Vienna Canyon Diablo Troilite.

While the full suite of reactions involved in the coupled C, S, and O biogeochemical cycles is extremely complex, the reactions directly related to reduced sulfur burial and its relationship to atmospheric O_2 and CO_2 can be simplified. The first critical reaction is the production of organic matter via photosynthesis:

$$CO_2 + H_2O \Leftrightarrow CH_2O + O_2 \quad (1)$$

This reaction in reverse (noted by the bi-directional arrow) is aerobic respiration (organic oxidation). The second critical reaction is the formation and burial of reduced sulfur as pyrite.

$$4Fe(OH)_3 + 8SO_4^{2-} + 15CH_2O \Leftrightarrow 4FeS_2 + 15HCO_3^- + 11H_2O + (5OH^-) \quad (2)$$

Lumped into this overall reaction is both the production of hydrogen sulfide via bacterial sulfate reduction and concomitant organic matter oxidation and the reaction of the resulting H_2S with iron oxides and oxyhydroxides to form pyrite. By adding Equations (1) and (2), we see that the coupling of organic matter production with pyrite burial represents a net sink of CO_2 and a net source of O_2 to the atmosphere.

$$4Fe(OH)_3 + 15CO_2 + 4H_2O + 8SO_4^{2-} \Leftrightarrow 4FeS_2 + 15HCO_3^- + (5OH^-) + 15O_2 \quad (3)$$

Alternatively, we can consider the formation and burial of organic sulfur. The relevant simplified equation for this process is a combination of production of hydrogen sulfide via bacterial sulfate reduction and organic matter oxidation and the reaction of sulfide with organic matter to produce organic sulfur.

$$SO_4^{2-} + 3CH_2O \Leftrightarrow CH_2S + 2HCO_3^- + H_2O \quad (4)$$

Coupling this equation with organic matter production yields a net equation for organic matter production and organic sulfur burial.

$$SO_4^{2-} + 3CO_2 + 2H_2O \Leftrightarrow CH_2S + 2HCO_3^- + 3O_2 \quad (5)$$

This equation demonstrates that, as was the case for pyrite burial, organic sulfur burial is a net sink of CO_2 and source for O_2.

There are two critical observations related to organic sulfur formation and burial that affect model-based reconstructions of atmospheric O_2 and CO_2. The first is the change in molar stoichiometry from 15/8 (O_2/pyrite-S, Equation 3) to 15/5 (O_2/Org-S, Equation 5). This change in the molar ratio between O_2 and S in itself implies a greater flux of O_2 to the atmosphere per mole of sulfur buried. The second critical observation is related to the isotopic composition of reduced sulfur species. Models predicting atmospheric composition based on biogeochemical cycling of C and S typically assume an offset between oxidized sulfur (gypsum) and reduced sulfur (pyrite) of 35‰, based on the assumption that all reduced sulfur is present as pyrite. Organic sulfur is typically enriched in ^{34}S by ~10‰ relative to pyrite sulfur in coeval sediments (Anderson and Pratt, 1995), which would give an isotopic offset of 25‰ if all reduced sulfur were present as organic sulfur. Thus, by simple isotopic mass balance calculation it is clear that if all reduced sulfur is buried as organic sulfur, nearly 40% more sulfur must be buried in sediments in order to generate the sulfur isotope composition of seawater sulfate expressed in evaporite deposits over time (cf. Garrels and Lerman, 1981, 1984; Kump and Garrels, 1986; Berner, 1987; Petsch and Berner, 1998). This increased sulfur burial would have increased the flux of CO_2 and O_2 to the atmosphere during periods of organic sulfur burial.

While organic sulfur burial may be volumetrically significant in some environments such as the Miocene Monterey Formation, where up to 80% of the total reduced S is organic-bound (Fig. 2; Anderson and Pratt, 1995), the paucity of data available for organic sulfur abundance and $\delta^{34}S$ through the Phanerozoic makes its impact difficult to assess. Because burial of organic sulfur is largely controlled by the relative availability of reduced inorganic sulfides and reactive iron species, it is likely that its influence would be most pronounced during times of widespread oceanic anoxia or euxinia, when sulfide is abundant and reactive Fe supplies are often limited. For example, mid-Cretaceous sediments from the proto-North Atlantic have very high organic sulfur contents of up to 80% of the total reduced sulfur (M.M. Kuypers, 2002, personal commun.).

ORGANIC SULFUR AND MICROBIAL ACTIVITY

The sulfur isotope composition of organic sulfur is dependent on (1) the sulfur isotope composition of the source sulfur, specifically reactive inorganic sulfur species, and (2) any isotopic fractionations associated with organic matter sulfurization. Because isotope fractionations associated with incorporation of inorganic sulfur species into organic matter are generally believed to be small (Price and Shieh, 1979; Fry et al., 1984, 1986, 1988), we will focus our discussion on the factors controlling the sulfur isotope composition of the sulfur source (though this belief is being challenged; Amrani and Aizenshtat, 2003). All of the possible inorganic sulfur sources, including sulfate (SO_4^{2-}), sulfide (ΣH_2S), and reactive intermediates such as elemental sulfur (S^0), polysulfides (HS_x^-), and thiosulfate ($S_2O_3^{2-}$) can be produced in the natural environment by microbial processes. Many can also be produced by abiotic processes, such as the formation of polysulfides via reactions between iron oxides and bisulfide or between dissolved sulfide and elemental sulfur (Pyzik and Sommer, 1981). Thus, the fundamental control on the biogeochemistry of sulfur isotopes, and specifically on the sulfur isotope composition of organic matter, is the microbial oxidative and reductive cycling of different forms of sulfur. Microbial fractionations of sulfur have recently been reviewed by Canfield (2001a) and are also discussed in several papers in this volume. We will therefore present only a brief summary here as required for our discussion of organic sulfur isotopes below and refer the interested reader to these other papers for a more detailed treatment of the subject.

The distribution of sulfur isotopes in the natural environment is controlled primarily by the fractionation imparted by dissimilatory bacterial sulfate reduction, which results in sulfide that is depleted in ^{34}S relative to the source sulfate (Chambers and Trudinger, 1979; Canfield, 2001a, 2001b, and references therein). This fractionation is highly variable but generally lies between 19‰ and 46‰, with values observed as low as 2‰ (Habicht and Canfield, 1997, 2001; Detmers et al., 2001; Brüchert et al., 2001). A second major control on the distribution of sulfur isotopes in sediments, which has only recently been identified, is the sedimentary cycle of microbial sulfide oxidation and subsequent disproportionation of intermediate phases of sulfur (e.g., elemental sulfur, thiosulfate) to sulfide and sulfate (Jørgensen, 1990; Canfield and Thamdrup, 1994; Canfield et al., 1998a; Habicht et al., 1998; Böttcher and Thamdrup, 2001; Böttcher et al., 2001). While the fractionations associated with sulfide oxidation are generally small (Fry et al., 1986), those associated with microbial disproportionation can be quite large. For example, the fractionations during elemental sulfur disproportionation have been shown to produce sulfide that is 6‰ depleted and sulfate that is 18‰ enriched in ^{34}S relative to the precursor elemental sulfur (Canfield et al., 1998a). In addition, disproportionation of sulfite will produce sulfide that is up to 37‰ depleted and sulfate that is up to 12‰ enriched in ^{34}S relative to the precursor sulfite (Habicht et al., 1998). Finally, disproportionation of thiosulfate in cultures has produced sulfide that is ^{34}S depleted relative to the sulfane sulfur in the precursor thiosulfate, though by extremely variable amounts ranging from ~2‰ to 20‰ (Habicht et al., 1998; Cypionka et al., 1998).

Discovery of the disporportionation pathway for sulfur cycling (Bak and Cypionka, 1987; Bak and Pfennig, 1987) and the substantial sulfur isotope fractionations associated with it (Canfield et al., 1998a; Habicht et al., 1998; Cypionka et al., 1998) ranks among the most significant discoveries in sulfur biogeochemistry in recent years. Among other implications, the disproportionation pathway provides an explanation for the discrepancy between fractionations observed in nature and those occurring experimentally during bacterial sulfate reduction. It has been proposed that the sedimentary cycling of sulfur, comprised of repeating cycles of sulfate reduction, sulfide oxidation, and disproportionation of elemental sulfur and other intermediates, is responsible for offsets between the sulfur isotope composition of sulfate and sulfide of up to and exceeding 65‰ (Canfield and Thamdrup, 1994), such as those observed in the modern Cariaco Basin (Werne et al., 2003). When compared to fractionations during one step sulfate reduction, which don't exceed ~46‰, it is clear that such a mechanism is required to produce the large values observed in nature. Furthermore, this refined understanding of microbial fractionations has spawned a proposal that the global ocean was insufficiently oxidizing to support disproportionation reactions until the Neoproterozoic (Canfield and Teske, 1996; Canfield, 1998).

There are two basic pathways by which organic sulfur is formed. The first is assimilatory sulfate reduction, which is the active uptake of sulfate into the cell, followed by its reduction to produce amino acids and other sulfur-requiring cellular components. There is generally little sulfur isotope fractionation associated with assimilatory sulfate reduction (Kaplan and Rittenberg, 1964; Trust and Fry, 1992), so this primary biogenic sulfur typically has an isotope composition similar to the ambient dissolved sulfate, which is ~+21‰ for modern seawater (Rees et al., 1978; Böttcher et al., 2000). The second and more significant pathway is the incorporation of reduced sulfur into organic matter during diagenesis. The mechanisms of diagenetic sulfur incorporation into organic matter are still debated (see discussion below addressing pathways of organic matter sulfurization) but fundamentally require that the sulfur is derived from pore-water sulfide, either directly or via reactive intermediates. Thus, the sulfur isotope composition of diagenetic organic sulfur is typically ^{34}S depleted relative to primary biogenic sulfur by 20‰–60‰. Mass balance modeling suggests that biogenic sulfur typically accounts for ~20%–25% of the total sedimentary organic sulfur in most marine settings (Anderson and Pratt, 1995; Werne et al., 2003).

TIMING OF ORGANIC MATTER SULFURIZATION

One of the most perplexing issues plaguing organic sulfur research is the timing of diagenetic S incorporation by organic compounds. It has been generally accepted that inorganic sulfides are incorporated into organic matter during early diagenesis, based on the identification of organic sulfur compounds in apolar fractions of organic extracts from near-surface sediments (Brassell et al., 1986; Kohnen et al., 1990, 1991b). Timing estimates range from several thousand years to only a few decades following initial deposition of the organic matter (Wakeham et al., 1995). Formation of macromolecular organic sulfur has also been identified in near-surface sediments (Francois, 1987; Eglinton et al., 1994). The organic sulfur compounds identified were typically algal lipids such as highly branched isoprenoids (Kohnen et al., 1990; Wakeham et al., 1995; Hartgers et al., 1997), other isoprenoids such as phytol derivatives (Peakman et al., 1989; Kenig and Huc, 1990), and steroids (Sinninghe Damsté et al., 1999b; Schouten et al., 1995a; Adam et al., 2000; Kok et al., 2000a), as well as sulfurized bacterial hopanoids (de las Heras et al., 1997) and fatty acids (Russell et al., 2000). Many lipids, such as phytane, are released from the nonextractable macromolecular organic matter (kerogen) from near-surface sediments upon treatment with desulfurizing agents like Raney Nickel or nickel boride (Schouten et al., 1993a). These compounds have typically been interpreted to represent the incorporation of sulfur into organic matter during early diagenesis, primarily because the sulfur compounds have not been identified in extant organisms. Unfortunately, due to a lack of clearly identifiable precursor-product relationships, constraining the timing of organic matter sulfurization any more precisely than "early diagenesis" was impossible in these studies.

A number of recent results, however, have more precisely bracketed the timing of early diagenetic sulfurization of organic matter. The first-ever precursor-product relationship for a diagenetic organic matter sulfurization reaction was identified in the sediments of the Cariaco Basin (Werne et al., 2000a). In this study, the near complete conversion of a tricyclic triterpenoid, (17E)-13β(H)-malabarica-14(27),17,21-triene, to a triterpenoid thiane was observed to span the upper ~3 m of sediment (Werne et al., 2000a). This depth interval represents ~10,000 yr, thus the timing of organic matter sulfurization, at least for this reaction, is constrained to be occurring over a 10 k.y. period. Other organic sulfur compounds, such as highly branched isoprenoid thiophenes and thiolanes, were identified in shallower sediments, suggesting that sulfurization of these compounds occurs more quickly than the 10 k.y. time period identified for the malabaricatriene to triterpenoid thiane conversion (Werne et al., 2000a). Similar results were obtained in a study of sediments from Ace Lake, Antarctica (Kok et al., 2000a). In their study, Kok et al. (2000a) found that steroids in Ace Lake sediments were sulfurized on a time scale of 1–3 k.y. Although they did not unambiguously identify a precursor-product relationship, they presented a convincing argument that their sulfurized steroids were formed through sulfurization of steroidal ketones deriving from biohydrogenation of Δ^5 sterols (sterols with a double bond at the 5 carbon position) by anaerobic bacteria (Kok et al., 2000a).

These two studies are consistent in that it appears that the order of magnitude of early diagenetic sulfurization of organic matter is 10^3 yr, but the specific rates of sulfurization of individual compounds may vary substantially. Studies in other settings with less well constrained timing have suggested a similar timeframe for sulfurization (e.g., Kohnen et al., 1990; Wakeham et al. 1995). Additional studies have attempted to investigate the timing and precursor-product relationships by using stable carbon isotope techniques. By comparing the carbon isotope composition of the organic sulfur compounds with suspected precursor compounds, these studies were able to reduce the number of potential precursors, which was adequate to support inferences about the timing of sulfurization (Filley et al., 1996, 2002). These studies also suggested a sulfurization time scale of 3–5 k.y. (Filley et al., 2002).

Much shorter time scales have also been suggested for the sulfurization. One study of lacustrine organic matter identified significant sulfurization in sediments that were all younger than ~60 yr (Urban et al., 1999), indicating that time scales of 10^3 yr are not always required for organic matter sulfurization (unless this material was reworked). A study by Adam et al. (1998) suggests that organic matter could become sulfurized through photochemically induced reactions in the water column. These findings are based on a laboratory study in which many different organic compounds were shown to be readily sulfurized under high light conditions following the addition of elemental sulfur, producing compounds very similar to those observed in nature. Unfortunately, while photochemical sulfurization appears to be likely in nature, particularly given the high concentrations of organic sulfur compounds present in some surface sediments (Wakeham et al., 1995), this rapid process has not yet been identified in a field study.

Support for a rapid sulfurization pathway can also be found in the results of Poinsot et al. (1998), who found sulfurization of polprenoids occurring in <1000 yr. Studies in the sediments of Lake Cadagno, Switzerland, indicate significant sulfurization of phytol derivatives (phytane) in sediments (Putschew et al., 1996). All of the sediments analyzed by Putschew et al. (1995, 1996) were younger than 100 yr. The very high concentrations of sulfur-bound phytane in the upper 2 cm are particularly intriguing given the carbon isotope data from the phytane and phytol compounds in these sediments. Putschew et al. (1995, 1996) showed that the sulfur-bound phytane has a carbon isotope composition that most closely matches that of the phytane in the microbes living in the chemocline portion of the water column. In contrast, the phytol and free phytane in the sediments are an isotopic match with the algae and higher plants in and around the lake (Putschew et al., 1995, 1996). While Putschew et al. (1996) attributed the differences in the depth profiles of concentration and carbon isotope composition of these phytyl compounds to differences in the stabilities of microbial versus algal organic matter, it is also possible that they result from a temporal change in the relative inputs of microbial versus algal phytyl compounds. An increase in the inputs of microbial phytanes in recent years could be a consequence of intensification of chemocline activity coupled with photochemical sulfurization of microbial organic matter in the water column.

The different time scales discussed above are not necessarily mutually exclusive. It is quite possible that both photochemical and sedimentary diagenetic sulfurization reactions are occurring in some environments, making the total time scale of sulfurization span from days to 10^3 yr. Furthermore, it is also likely that sulfurization could be continuing after the currently identified upper limit of 10^3 yr, particularly given the high concentrations of organic sulfur identified in ancient sedimentary environments worldwide (Kenig et al., 1995; Schaeffer et al., 1995; Schouten et al., 2001.

ORGANIC SULFUR AND ORGANIC MATTER PRESERVATION

In recent years, two new models have been proposed to explain enhanced preservation of organic matter in the sedimentary environment. The first implicates the selective preservation of biomacromolecules such as algaenans (de Leeuw and Largeau, 1993; Gelin et al., 1996a, 1996b, 1997, 1999; Blokker et al., 1998, 2000). The second model calls for the sulfurization of natural organic matter as a preservation mechanism (Valisolalao et al., 1984; Brassell et al., 1986; Sinninghe Damsté and de Leeuw, 1990). Reactions between sedimentary organic matter and reduced inorganic sulfur lead to a combination of low molecular weight organic sulfur compounds (through intramolecular sulfur incorporation) and high molecular weight macromolecules (through intermolecular sulfur incorporation). The nature of the macromolecular organic sulfur compounds formed, for example, through sulfide cross-linking (Schouten et al., 1995b), prevents or at least significantly hinders microbial attack and degradation by binding individual functionalized compounds into a macromolecular matrix, thereby preserving the organic compounds on geologic time scales.

Biomarker Sequestration

One of the most informative pathways of paleoenvironmental investigation is through the use of molecular biomarkers, which are organic compounds in the geologic records that can be linked unambiguously to a source organism and its associated ecological context (Eglinton and Hamilton, 1967). Furthermore, we can use changes in the relative distribution of different biomarkers to identify past variability in biological communities (Werne et al., 2000b).

Organic matter sulfurization has been shown to preserve the carbon skeletons of functionalized organic compounds in sediments (Brassell et al., 1986; Sinninghe Damsté and de Leeuw, 1990). The implications of this observation are twofold. First, organic matter sulfurization has the potential to preserve biomarkers that would otherwise be lost to degradation. For example, Sinninghe Damsté et al. (1990) used thiophenic biomarkers for paleoenvironmental assessment of the Jurf ed Darawish oil shale in Jordan. These compounds would likely not have been present without sulfurization. Other studies have investigated the compounds released from macromolecular organic matter via desulfurization, rather than investigating "free" organic sulfur compounds as in the study by Sinninghe Damsté et al. (1990). For example, Grice et al. (1998) found biomarkers for freshwater algae in the macromolecular organic matter of an ancient hypersaline euxinic ecosystem, which were released only upon cleavage of the sulfur bonds.

The second implication of biomarker preservation through sulfurization is that paleoenvironmental reconstructions can be significantly biased if selective sulfurization is not considered (Kohnen et al., 1991c, 1992). Specific compounds are not necessarily sulfurized to completion in the natural environment. Thus, because intramolecularly incorporated sulfur leads to organic sulfur compounds with a mass spectrum that is different from the original biomarker compound (cf. Werne et al., 2000a), the sulfurized portion of a particular biomarker compound could easily be overlooked in a molecularly based paleoenvironmental study. Furthermore, the biomarker compounds that are intermolecularly bound into the macromolecular matrix via sulfurization are generally not released by traditional organic extraction techniques, and many common organic geochemical techniques, such as pyrolysis, are destructive to the original (precursor) compound (Sinninghe Damsté and de Leeuw, 1990). These precursor compounds would therefore not be analyzed in a molecular paleoenvironmental study that did not consider the possibility of biomarkers becoming sulfur bound and requiring release from the kerogen or macromolecular bitumen.

Carbohydrate Preservation

One of the most interesting hypotheses to result from organic sulfur geochemical studies is that sulfurization of carbohydrates may substantially enhance organic carbon preservation in the geological environment. Carbohydrates are known to be labile compounds with high potential for rapid loss in sedimentary and aquatic environments. In a recent series of papers, however, Sinninghe Damsté and colleagues demonstrated that reaction with sulfur could sequester carbohydrates in sediments. Van Kaam-Peters et al. (1998) showed that intervals of high organic carbon content in the Jurassic Kimmeridge Clay Formation were characterized by high sulfur content and ^{13}C-enriched carbon isotope values. Because carbohydrates are known to be enriched in ^{13}C relative to total cell material (e.g., van Dongen et al., 2002), Van Kaam-Peters et al. (1998) hypothesized that these relationships reflected preservation of carbohydrate carbon through sulfurization during early diagenesis. Further support for this hypothesis was provided by experiments in which glucose and algae were sulfurized in the laboratory, resulting in sulfur-rich macromolecular organic matter (Sinninghe Damsté et al., 1998b; Kok et al., 2000b; van Dongen et al., 2003a). Upon pyrolysis, this macromolecular organic matter was found to have a molecular distribution very similar to that found in the Kimmeridge Clay (Sinninghe Damsté et al., 1998b; Kok et al., 2000b; van Dongen et al., 2003a).

The hypothesis that sulfurization leads to enhanced carbohydrate preservation was confirmed by detailed studies of the macromolecular organic matter in kerogen pyrolysates and through comparison with macromolecular organic matter produced from laboratory sulfurization of carbohydrates (van Dongen et al., 2003b; van Dongen 2003). These findings are critical because it was previously thought that carbohydrates are degraded much more readily than other classes of organic matter in the natural environment, and their preservation in sulfur-rich environments could potentially affect interpretations based on bulk organic carbon concentrations and isotope compositions alone (Sinninghe Damsté et al., 1998b).

Macromolecular Organic Sulfur

The combination of intermolecular and intramolecular sulfurization of organic matter can lead to a complex set of organic sulfur compounds in macromolecular organic matter. Detailed understanding of this pool of organic sulfur is still lacking, but significant steps have been made in recent years. The classical method of investigating macromolecular organic matter is through pyrolysis gas chromatography–mass spectrometry (pyGC-MS; Sinninghe Damsté and de Leeuw, 1990). This method involves the flash combustion of kerogens, followed by gas chromatographic separation and mass spectrometric identification of the pyrolysis products. pyGC-MS is useful for identifying the carbon skeletons bound to macromolecular organic matter by sulfur linkages (Eglinton et al., 1992; Sinninghe Damsté et al., 1998a; Luckge et al., 2002). For example, studies using pyGC-MS have identified molecular fossils of the alga *Gloeocapsomorpha prisca* in Ordovician kerogens (Douglas et al., 1991). However, because the technique is destructive, it does not yield information about the chemical form (e.g., polysulfide, sulfonate, etc.) of the organic sulfur itself. Researchers have therefore turned to other methods in addition to pyGC-MS to study the composition of macromolecular organic sulfur.

One method that has proven to be useful is the stepwise selective chemical degradation of the kerogen, which involves treating the kerogen with a sequence of different reagents and analyzing the released compounds. Early studies emphasized "total desulfurization" techniques, such as Raney-Nickel or nickel-boride (Perakis, 1986; Sinninghe Damsté et al., 1988; Schouten et al., 1993a). Studies of this type have confirmed that sulfurization can enhance the preservation of both macromolecular organic matter and specific biomarkers in sediments, and many types of organic matter vulnerable to sulfurization have been identified (Grice et al., 1998; Hefter et al., 1995; Putschew et al., 1996; Russell et al., 2000). Unfortunately, the sulfur is lost with these methods, eliminating our ability to identify the forms of sulfur present.

Other chemical degradation techniques have proven more fruitful (e.g., Schaeffer-Reiss et al., 1998). Specifically, cleavage of only S-S bonds using methyl lithium/methyl iodide (Kohnen et al., 1991a, 1993), superheated methyl iodide (Schouten et al., 1993b), and $LiAlH_4$ (Adam et al., 1991, 1992, 1993; Schaeffer et al., 1995) proved the presence of polysulfide (or disulfide) linkages in organic sulfur-rich macromolecular organic matter. These findings support the polysulfide pathway of organic matter sulfurization (cf. Aizenshtat et al., 1983; LaLonde et al., 1987).

There are also several nondestructive spectroscopic methods for investigating the forms of sulfur present in macromolecular organic matter. One of the most informative in recent years has been X-ray absorption near-edge structure spectroscopy (XANES), which was first applied in studies of petroleum products (Spiro et al., 1984; George and Gorbaty, 1989; Huffman et al., 1991, 1995; Waldo et al., 1991). Due to its sensitivity to the electronic structure, oxidation state, and geometry of neighboring atoms, this spectroscopic method has the capability of providing specific information about the functional groups of sulfur present (Vairavamurthy et al., 1994; 1997). One very intriguing result from the XANES studies is that 20%–40% of the total organic sulfur is actually present as sulfonates, an oxidized form of organic sulfur that was previously not known to exist in sediments (Vairavamurthy et al., 1994). Other sulfur-bearing compound types identified by XANES include reduced forms such as thiols, thiophenes, and disulfides and polysulfides, moderate oxidation state forms such as sulfones, and oxidized forms such as organic sulfates and sulfonates (Vairavamurthy et al., 1995, 1997). Other recent studies using XANES to investigate sulfur in ancient systems found that reduced forms such as thiophenes dominate (Sarret et al., 1999,

2002). It remains unclear what factors may be controlling the relative distribution of different forms of oxidized and reduced sulfur in modern and ancient organic-rich sediments.

PATHWAYS OF ORGANIC MATTER SULFURIZATION: CONSTRAINTS FROM LABORATORY SIMULATIONS AND SULFUR ISOTOPES

Laboratory Sulfurization Experiments

Despite intensive study, the geochemical pathways of organic matter sulfurization remain unclear. This uncertainty is compounded by the likely existence of multiple pathways in natural systems, which vary as a function of the specific conditions present. Most researchers currently favor the idea that the preferred pathway for the diagenetic formation of organic sulfur is the reaction of organic matter with inorganic polysulfides (Aizenshtat et al., 1983), but other mechanisms such as reaction directly with H_2S, elemental sulfur, or other intermediates such as thiosulfate cannot be eliminated. It is also not clear whether the sulfur reacts preferentially at sites of unsaturation (Sinninghe Damsté et al., 1989) or with functional groups such as ketones (Schneckenburger et al., 1998). Among the evidence in support of the polysulfide pathway, in addition to the presence of polysulfide linkages in sulfur-rich macromolecular organic matter (Kohnen et al., 1991a; Adam et al., 1993; Schaeffer et al., 1995), are the many laboratory experiments in which organic matter has been artificially sulfurized. All these studies were carried out under mild laboratory conditions (e.g., 50 °C) intended to be as similar to natural conditions as possible while still increasing the rate of sulfurization enough to enable laboratory study. These studies resulted in the sulfurization of phytol (de Graaf et al., 1992; Rowland et al., 1993), ketones and aldehydes (Schouten et al., 1994a, 1994b), olefins (de Graaf et al., 1995), algae (Gelin et al., 1998), and carbohydrates (Kok et al., 2000b; van Dongen et al., 2003a), yielding compounds similar to those identified in natural sedimentary systems.

Additional support for the polysulfide pathway of organic sulfur formation was provided in a study by Vairavamurthy et al. (1992) in which natural sediment samples were reacted with acrylic acid in a slurry to investigate the distribution and reactivity of polysulfides in sediments. This study determined that polysulfides, in addition to being dissolved in the aqueous phase, can be present in the solid fraction. As solid particles, polysulfides were bound to sediment grains and organic matter, which helped to bind the organic matter to the sediment particles (Vairavamurthy et al., 1992). This process may play an important role in the preservation of organic matter in sediments (Vairavamurthy et al., 1992).

Sulfur Speciation and Isotopic Studies

One of the most promising areas of organic sulfur research is detailed sulfur speciation coupled with sulfur isotope measurements. The theory underlying these studies is that environments conducive to the formation of organic sulfur typically also favor the formation of inorganic sulfides such as pyrite (Mossmann et al., 1991; Brüchert and Pratt, 1996). Furthermore, because of the wide range of sulfur isotope fractionations observed in nature (Canfield, 2001; Bottrell and Raiswell, 2000), isotope measurements can be used in combination with concentration data to trace pathways of sulfur cycling in sedimentary environments (Mossmann et al., 1991; Zaback and Pratt, 1992; Anderson and Pratt 1995; Henneke et al., 1997; Passier et al., 1997; Bates et al., 1995, 1998; Canfield et al., 1998b; Werne et al., 2003). Figure 3 shows depth trends of the $\delta^{34}S$ values of the major reduced sulfur species (pore-water sulfide, pyrite sulfur, and organic sulfur) in the Cariaco Basin (data from Werne et al., 2003) and the Peru Margin (data from Mossmann et al., 1991). Environments in which the supply of sulfate is restricted in deeper sediments lead ultimately to down core enrichment in ^{34}S in all the sulfur species.

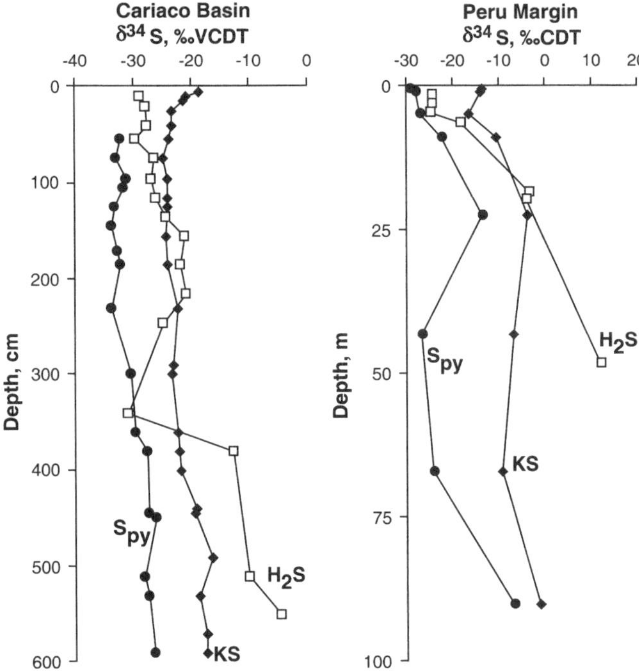

Figure 3. Depth trends of the sulfur isotope composition of the major reduced sulfur species in the Cariaco Basin and the Peru Margin (data from Werne et al. [2003] and Mossmann et al. [1991], respectively). Note the consistency among the trends, which are suggestive of restricted environments in which the consumption of the pore-water sulfur species, primarily sulfate via sulfate reduction to form sulfide, is faster than the supply through diffusion. VCDT—Vienna Canyon Diablo Troilite.

Speciation studies have shown clearly that organic sulfur represents a significant portion of the total reduced sulfur in the sediments of many different environments. For example, organic sulfur is up to 50% of the total sulfur in a Delaware salt marsh (Ferdelman et al., 1991), 50% in St. Andrew Bay, Florida (Brüchert and Pratt, 1996), 17%–43% in the hypersaline Tyro and Bannock Basins in the Mediterranean (Henneke et al., 1997), 85%–90% in Mangrove Lake, Bermuda (Canfield et al., 1998b), 50%–75% in the Everglades (Bates et al., 1998), 40% in the Peru Margin (Mossmann et al., 1991; Suits and Arthur, 2000), and up to 30% in the Cariaco Basin, Venezuela (Werne et al., 2003). Anderson and Pratt (1995) summarized results from a number of other environments in which organic sulfur is found to be as much as 80% of the total reduced sulfur.

Brüchert (1998) investigated organic sulfur bound in humic and fulvic acids and found that a portion of the fulvic acid sulfur in the upper centimeters of the sediments of St. Andrew Bay, Florida, was evidently recycled back to pore-water sulfur species. An additional portion was transferred into humic acids and protokerogens (the macromolecular organic matter discussed above). Furthermore, because of the characteristic isotopic signature of assimilated primary biogenic sulfur compared to organic sulfur incorporated during diagenesis, a number of studies have been able to demonstrate that this biogenic sulfur makes up 20%–25% of the sedimentary organic sulfur in many environments (Anderson and Pratt, 1995; Brüchert and Pratt, 1996; Werne et al., 2003). The remaining organic sulfur appears to be derived from pore-water sulfide or associated reactive intermediate, such as polysulfides (Werne et al., 2003; Brüchert, 1998; Canfield et al., 1998b).

Some of the strongest recent support for the polysulfide pathway of organic sulfur formation in fact comes from sulfur isotope analyses. In a recent study, Werne et al. (2003) measured the sulfur isotope composition of bulk organic matter. Based on the identification of a sulfurization reaction in Cariaco Basin sediments (Werne et al., 2000a), they then used mass balance considerations to model the isotope composition of sulfur incorporated diagenetically into a specific organic compound (Werne et al., 2003). Assuming that sulfur was incorporated into organic matter directly from pore-water sulfide with no isotopic fractionation, and using the depth profile for the isotopic composition of pore-water sulfide, they estimated values for their model organic sulfur compound that were very similar to those measured for bulk organic sulfur. This result suggested that their assumptions were correct and gave support to the direct reaction of organic matter with H_2S (Werne et al., 2003). Due to recent technical advances, however, it is now possible to measure the sulfur isotope composition of specific organic compounds. In a follow-up study, Werne et al. (2001) measured the sulfur isotope composition of the actual compound they had modeled and found that its sulfur isotope composition was in fact 10‰ heavier (^{34}S enriched) relative to model values and bulk organic sulfur.

These results indicate one or both of two possibilities: either the assumptions about the lack of sulfur isotope fractionation accompanying organic matter sulfurization were incorrect, or the sulfur is not derived directly from H_2S but rather from some other inorganic sulfur species that is ^{34}S enriched relative to H_2S, such as polysulfides. Studies of isotope fractionation associated with incorporation of S into organic matter are sparse and contradictory. Brüchert and Pratt (1996) proposed a kinetic fractionation (based on carbon isotope work of Schouten et al., 1995c) that would result in organic sulfur that was ^{34}S depleted relative to sulfide or polysulfides. More recently, Amrani and Aizenshtat (2003) hypothesized that organic matter sulfurization is an equilibrium process, with equilibrium S isotope fractionation leading to polysulfides that are ^{34}S enriched relative to sulfide. This second hypothesis fits well with the observations that elemental sulfur is enriched relative to solid phase sulfides (e.g., FeS_2) in sediments (Anderson and Pratt, 1995; Werne et al., 2003). Furthermore, ^{35}S radiolabel experiments have shown that there is rapid isotopic mixing between the various inorganic sulfur species (e.g., H_2S, S^0, S_x^{2-}, FeS; Fossing et al., 1992). Fractionations associated with microbial sulfide oxidation are generally small (Kaplan and Rittenberg, 1964; Fry et al., 1984, 1986, 1988); however, those associated with disproportionation can be large, with the reduced sulfur resulting from disproportionation being much more depleted than the oxidized sulfur (Canfield and Thamdrup, 1994; Canfield et al., 1998a; Habicht et al., 1998; Böttcher and Thamdrup, 2001; Böttcher et al., 2001). Thus, successive cycles of partial oxidation and disproportionation *could* lead to intermediate forms of sulfur that are enriched in ^{34}S relative to aqueous sulfide, though this has never been unequivocally observed.

Based on the above S isotope observations, elemental sulfur and polysulfides are the most likely candidates for direct incorporation in organic matter that could produce a ^{34}S enriched organic sulfur product. Both polysulfides and elemental sulfur are strong nucleophiles, with longer chain length polysulfides being stronger nucleophiles than the shorter chains (LaLonde et al., 1987). Because the pH values of most marine sedimentary systems are slightly basic, they favor the incorporation of polysulfides (via nucleophilic additions) over incorporation of elemental sulfur (Giggenbach, 1972; LaLonde et al., 1987; Aizenshtat et al., 1995). Thus, Werne et al. (2001) proposed that their isotopic data are best explained by the incorporation of inorganic sulfur into organic matter as polysulfides.

SUMMARY AND FUTURE RESEARCH DIRECTIONS

Our goal has been to review the most significant advances in organic sulfur biogeochemistry over the last decade. The highlights include:

1. the possibility that sulfur radical formation may play a central role in early petroleum formation from organic-rich sediments;

2. the potential of organic sulfur burial on the coupled global biogeochemical cycles of C, S, and O resulting in an underestimate in the reduced sulfur burial flux and thus on the fluxes of CO_2 and O_2 to the atmosphere on geological time scales;

3. the constraints placed on the timing of early diagenetic sulfurization of organic matter (e.g., 10s to 1000s of years);

4. the ability for sulfurization to enhance the preservation of organic matter, both specific biomarkers as well as classes of labile organic matter such as carbohydrates;

5. the identification of many different forms of organic sulfur in macromolecular organic matter, including partially oxidized forms such as sulfonates; and

6. the constraints placed on organic sulfur formation by studies combining sulfur speciation and parallel measurements of organic sulfur isotope compositions, particularly the recent advance of being able to measure the $\delta^{34}S$ of organic S at the level of individual organic compounds.

Most important, these results advance our comparatively weak understanding of the overall roles played by organic sulfur in the global biogeochemical cycles for C and S.

To continue moving forward, we need to identify the many forms of organic sulfur present in sediments, both specific organic sulfur compounds and macromolecular organic sulfur. There are many methods by which such goals can be achieved. Integrated approaches—e.g., a combination of selective chemical degradations and mass spectrometric analyses with nondestructive spectroscopic methods such as XANES—are likely to yield the best results. It is also clear that we have not quantified the extent to which organic matter preservation can be enhanced via sulfurization, nor do we have a good understanding of the kinetics of OM sulfurization.

Recent advances in the field of sulfur isotope geochemistry are increasing the sensitivity of sulfur isotope measurements and thus resolution of analyses now performed at the level of individual organic compounds (Fry et al., 2002; Studley et al., 2002). Eventually, such advances will make compound-specific sulfur isotope analyses commonplace.

ACKNOWLEDGMENTS

We would like to thank J. Amend and K. Edwards for inviting us to contribute to this volume. We also thank M. Kuypers for unpublished data and T. Filley and an anonymous reviewer for critical comments on an earlier version of this manuscript. Partial financial support for this project was provided by National Science Foundation award EAR-9875961 (TWL).

REFERENCES CITED

Adam, P., Schmid, J.C., Albrecht, P., and Connan, J., 1991, 2α and 3β steroid thiols from reductive cleavage of macromolecular petroleum fraction: Tetrahedron Letters, v. 32, p. 2955–2958, doi: 10.1016/0040-4039(91)80661-O.

Adam, P., Mycke, B., Schmid, J.C., Connan, J., and Albrecht, P., 1992, Steroid moieties attached to macromolecular petroleum fraction via di- or polysulfide bridges: Energy & Fuels, v. 6, p. 553–559.

Adam, P., Schmid, J.C., Mycke, B., Strazielle, C., Connan, J., Huc, A., Riva, A., and Albrecht, P., 1993, Structural investigation of nonpolar sulfur cross-linked macromolecules in petroleum: Geochimica et Cosmochimica Acta, v. 57, p. 3395–3419, doi: 10.1016/0016-7037(93)90547-A.

Adam, P., Phillippe, E., and Albrecht, P., 1998, Photochemical sulfurization of sedimentary organic matter: A widespread process occurring at early diagenesis in natural environments?: Geochimica et Cosmochimica Acta, v. 62, p. 265–271, doi: 10.1016/S0016-7037(97)00332-3.

Adam, P., Schneckenburger, P., Schaeffer, P., and Albrecht, P., 2000, Clues to early diagenetic sulfurization processes from mild chemical cleavage of labile sulfur-rich geomacromolecules: Geochimica et Cosmochimica Acta, v. 64, p. 3485–3503.

Aizenshtat, Z., Stoler, A., Cohen, Y., and Nielsen, H., 1983, The geochemical sulphur enrichment of recent organic matter by polysulfides in the Solar-Lake, *in* Bjorøy, M., et al., eds., Advances in Organic Geochemistry 1981: New York, John Wiley & Sons Limited, p. 279–288.

Aizenshtat, Z., Krein, E., Vairavamurthy, M., and Goldstein, T., 1995, Role of sulfur in the transformations of sedimentary organic matter: A mechanistic overview, *in* Vairavamurthy, M.A., and Schoonen, M.A.A., eds., Geochemical transformations of sedimentary sulfur: Washington, D.C., American Chemical Society Symposium Series 612, p. 378–396.

Amrani, A., and Aizenshtat, Z., 2003, Mechanisms of sulfur introduction chemically controlled: $\delta^{34}S$ imprint: Proceedings, 21st IMOG meeting, Krakow, Poland, p. 39–40.

Anderson, T.F., and Pratt, L.M., 1995, Isotope evidence for the origin of organic sulfur and elemental sulfur in marine sediments, *in* Vairavamurthy, M.A., and Schoonen, M.A.A., eds., Geochemical transformations of sedimentary sulfur: Washington, D.C., American Chemical Society Symposium Series 612, p. 378–396.

Bak, F., and Cypionka, H., 1987, A novel type of energy metabolism involving fermentation of inorganic sulphur compounds: Nature, v. 326, p. 891–892, doi: 10.1038/326891A0.

Bak, F., and Pfennig, N., 1987, Chemolithotrophic growth of *Desulfovibrio sulfodismutans* sp. nov. by disproportionation of inorganic sulfur compounds: Archives of Microbiology, v. 147, p. 184–189.

Barakat, A.O., and Rüllkötter, J., 1995, The distribution of free organic sulfur compounds in sediments from the Nördlinger Ries, Southern Germany, *in* Vairavamurthy, M.A., and Schoonen, M.A.A., eds., Geochemical transformations of sedimentary sulfur: Washington, D.C., American Chemical Society Symposium Series 612, p. 311–331.

Barakat, A.O., and Rüllkötter, J., 1999, Origin of organic sulfur compounds in sediments from the Nördlinger Ries (Southern Germany): Journal of Petroleum Science and Engineering, v. 22, p. 103–119, doi: 10.1016/S0920-4105(98)00060-6.

Bates, A., Spiker, E., Hatcher, P., Stout, S., and Weintraub, V., 1995, Sulfur geochemistry of organic-rich sediments from Mud Lake, Florida, USA: Chemical Geology, v. 121, p. 245–262, doi: 10.1016/0009-2541(94)00122-O.

Bates, A., Spiker, E., and Holmes, C., 1998, Speciation and isotopic composition of sedimentary sulfur in the Everglades, Florida, USA: Chemical Geology, v. 146, p. 155–170, doi: 10.1016/S0009-2541(98)00008-4.

Berner, R.A., 1987, Models for carbon and sulfur cycles and atmospheric oxygen: application to Paleozoic geologic history: American Journal of Science, v. 287, p. 177–196.

Berner, R.A., and Raiswell, R., 1983, Burial of organic carbon and pyrite sulfur in sediments over Phanerozoic time: a new theory: Geochimica et Cosmochimica Acta, v. 47, p. 855–862, doi: 10.1016/0016-7037(83)90151-5.

Blokker, P., Schouten, S., van den Ende, H., de Leeuw, J.W., Hatcher, P.G., and Sinninghe Damsté, J.S., 1998, Chemical structure of algaenans from the fresh water algae Tetraedron minimum, Scenedesmus communis and Pediastrum boryanum: Organic Geochemistry, v. 29, p. 1453–1468, doi: 10.1016/S0146-6380(98)00111-9.

Blokker, P., Schouten, S., de Leeuw, J.W., Sinninghe Damsté, J.S., and van den Ende, H., 2000, A comparative study of fossil and extant algaenans using ruthenium tetroxide degradation: Geochimica et Cosmochimica Acta, v. 64, p. 2055–2065, doi: 10.1016/S0016-7037(00)00367-7.

Böttcher, M.E., and Thamdrup, B., 2001, Anaerobic sulfide oxidation and stable isotope fractionation associated with bacterial sulfur disproportionation in the presence of MnO_2: Geochimica et Cosmochimica Acta, v. 65, p. 1573–1581, doi: 10.1016/S0016-7037(00)00622-0.

Böttcher, M.E., Thamdrup, B., and Vennemann, T.W., 2001, Oxygen and sulfur isotope fractionation during anaerobic bacterial disproportionation of elemental sulfur: Geochimica et Cosmochimica Acta, v. 65, no. 10, p. 1601–1609, doi: 10.1016/S0016-7037(00)00628-1.

Böttcher, M.E., Schale, H., Schnetger, B., Wallmann, K., and Brumsack, H.-J., 2000, Stable sulfur isotopes indicate net sulfate reduction in near-surface sediments of the deep Arabian Sea: Deep-Sea Research II, v. 47, p. 2769–2783.

Bottrell, S.H., and Raiswell, R., 2000, Sulphur isotopes and microbial sulphur cycling, *in* Riding, R.E., and Awramik, S.M., eds., Microbial sediments: Berlin, Springer-Verlag, p. 96–104.

Boulègue, J., Lord, C.J., III, and Church, T.M., 1982, Sulfur speciation and associated trace metals (Fe, Cu) in the pore waters of Great Marsh, Delaware: Geochimica et Cosmochimica Acta, v. 46, p. 453–464, doi: 10.1016/0016-7037(82)90236-8.

Brassell, S.C., Lewis, C.A., de Leeuw, J.W., de Lange, F., and Sinninghe Damsté, J.S., 1986, Isoprenoid thiophenes: Novel diagenetic products in sediments?: Nature, v. 320, p. 160–162.

Brüchert, V., 1998, Early diagenesis of sulfur in estuarine sediments: The role of sedimentary humic and fulvic acids: Geochimica et Cosmochimica Acta, v. 62, p. 1567–1586, doi: 10.1016/S0016-7037(98)00089-1.

Brüchert, V., and Pratt, L.M., 1996, Contemporaneous early diagenetic formation of organic and inorganic sulfur in estuarine sediments from St. Andrew Bay, Florida, USA: Geochimica et Cosmochimica Acta, v. 60, no. 13, p. 2325–2332, doi: 10.1016/0016-7037(96)00087-7.

Brüchert, V., Knoblauch, C., and Jørgensen, B.B., 2001, Controls on stable sulfur isotope fractionation during bacterial sulfate reduction in Arctic sediments: Geochimica et Cosmochimica Acta, v. 65, p. 763–776, doi: 10.1016/S0016-7037(00)00557-3.

Canfield, D.E., 1989, Reactive iron in marine sediments: Geochimica et Cosmochimica Acta, v. 53, p. 619–632, doi: 10.1016/0016-7037(89)90005-7.

Canfield, D.E., 1998, A new model for Proterozoic ocean chemistry: Nature, v. 396, p. 450–453, doi: 10.1038/24839.

Canfield, D.E., 2001a, Biogeochemistry of sulfur isotopes, *in* Valley, J.W., and Cole, D.R., eds., Stable isotope geochemistry: Washington, D.C., The Mineralogical Society of America Reviews in Mineralogy and Geochemistry 43, p. 607–636.

Canfield, D.E., 2001b, Isotope fractionation by natural populations of sulfate-reducing bacteria: Geochimica et Cosmochimica Acta, v. 65, no. 7, p. 1117–1124, doi: 10.1016/S0016-7037(00)00584-6.

Canfield, D.E., and Raiswell, R., 1999, The evolution of the sulfur cycle: American Journal of Science, v. 299, p. 697–723.

Canfield, D.E., and Teske, A., 1996, Late Proterozoic rise in atmospheric oxygen concentration inferred from phylogenetic and sulphur-isotope studies: Nature, v. 382, p. 127–132, doi: 10.1038/382127A0.

Canfield, D.E., and Thamdrup, B., 1994, The production of ^{34}S-depleted sulfide during bacterial disproportionation of elemental sulfur: Science, v. 266, p. 1973–1975.

Canfield, D.E., Raiswell, R., and Bottrell, S., 1992, The reactivity of sedimentary iron minerals toward sulfide: American Journal of Science, v. 292, p. 818–834.

Canfield, D.E., Lyons, T.W., and Raiswell, R., 1996, A model for iron deposition to euxinic Black Sea sediments: American Journal of Science, v. 296, p. 818–834.

Canfield, D.E., Thamdrup, B., and Fleischer, S., 1998a, Isotope fractionation and sulfur metabolism by pure and enrichment cultures of elemental sulfur-disproportionating bacteria: Limnology and Oceanography, v. 43, p. 253–264.

Canfield, D.E., Boudreau, B.P., Mucci, A., and Gundersen, J.K., 1998b, The early diagenetic formation of organic sulfur in the sediments of Mangrove Lake, Bermuda: Geochimica et Cosmochimica Acta, v. 62, p. 767–781, doi: 10.1016/S0016-7037(98)00032-5.

Canfield, D.E., Habicht, K.S., and Thamdrup, B., 2000, The Archean sulfur cycle and the early history of atmospheric oxygen: Science, v. 288, p. 658–661.

Chambers, L.A., and Trudinger, P.A., 1979, Microbiological fractionation of stable sulfur isotopes: a review and critique: Geomicrobiology Journal, v. 1, p. 249–293.

Cypionka, H., Smock, A., and Böttcher, M., 1998, A combined pathway of sulfur compound disproportionation in *Desulfovivrio desulfuricans*: FEMS Microbiology Letters, v. 166, p. 181–186, doi: 10.1016/S0378-1097(98)00330-9.

de Graaf, W., Sinninghe Damsté, J., and de Leeuw, J., 1992, Laboratory simulation of natural sulphurization: I. Formation of monomeric and oligomeric isoprenoid polysulphides by low-temperature reactions of inorganic polysulphides with phytol and phytadienes: Geochimica et Cosmochimica Acta, v. 56, p. 4321–4328, doi: 10.1016/0016-7037(92)90275-N.

de Graaf, W., Sinninghe Damsté, J., and de Leeuw, J., 1995, Low-temperature addition of hydrogen polysulfides to olefins: Formation of 2.2′-dialkyl polysulfides from alk-1-enes and cyclic (poly)sulfides and polymeric organic sulfur compounds from α,ω-dienes: Journal of the Chemical Society, Perkin Transactions 1: Organic and Bio-Organic Chemistry, v. I, p. 634–640.

de las Heras, F.X.C., Grimalt, J.O., Lopez, J.F., Albaiges, J., Sinninghe Damsté, J.S., Schouten, S., and de Leeuw, J.W., 1997, Free and sulphurized hopanoids and highly branched isoprenoids in immature lacustrine oil shales: Organic Geochemistry, v. 27, p. 41–63.

de Leeuw, J.W., and Largeau, C., 1993, A review of macromolecular organic compounds that comprise living organisms and their role in kerogen, coal and petroleum formation, *in* Engel, M.H., and Macko, S.A., eds., Organic geochemistry, principles and applications—Topics in Geobiology: New York, Plenum, v. 11, p. 23–72.

de Leeuw, J.W., and Sinninghe Damsté, J.S., 1990, Organic sulphur compounds and other biomarkers as indicators of Palaeosalinity: A critical evaluation, *in* Orr, W.L., and White, C.M., Geochemistry of sulfur in fossil fuels: Washington, D.C., American Chemical Society Symposium Series 249, p. 417–443.

Detmers, J., Brüchert, V., Habicht, K.S., and Kuever, J., 2001, Diversity of sulfur isotope fractionations by sulfate-reducing prokaryotes: Applied and Environmental Microbiology, v. 67, p. 888–894, doi: 10.1128/AEM.67.2.888-894.2001.

Douglas, A., Sinninghe Damsté, J., Fowler, M., Eglinton, T., and de Leeuw, J., 1991, Unique distributions of hydrocarbons and sulphur compounds released by flash pyrolysis from the fossilized alga *Gloecapsomorpha prisca*, a major constituent in one of four Ordovician kerogens: Geochimica et Cosmochimica Acta, v. 55, p. 275–291, doi: 10.1016/0016-7037(91)90417-4.

Eglinton, G., and Hamilton, R.J., 1967, Leaf epicuticular waxes: Science, v. 156, p. 1322–1335.

Eglinton, T.I., Irvine, J.E., Vairavamurthy, A., Zhou, W., and Manowitz, B., 1994, Formation and diagenesis of macromolecular organic sulfur in Peru margin sediments, *in* Telnaes, N., van Graas, G., and Øygard, K., eds., Advances in Organic Geochemistry 1993: Organic Geochemistry, v. 22, p. 781–799.

Eglinton, T.I., Sinninghe Damsté, J.S., Pool, W., de Leeuw, J.W., Eijkel, G., and Boon, J.J., 1992, Organic sulphur in macromolecular sedimentary organic matter. II. Analysis of distributions of sulphur-containing pyrolysis products using multivariate techniques: Geochimica et Cosmochimica Acta, v. 56, p. 1545–1560, doi: 10.1016/0016-7037(92)90224-7.

Ferdelman, T.G., Church, T.M., and Luther, G.W., III, 1991, Sulfur enrichment of humic substances in a Delaware salt marsh sediment core: Geochimica et Cosmochimica Acta, v. 55, p. 979–988, doi: 10.1016/0016-7037(91)90156-Y.

Filley, T., Freeman, K., Wilkin, R., and Hatcher, P., 2002, Biogeochemical controls on reaction of sedimentary organic matter and aqueous sulfides in Holocene sediments of Mud Lake, Florida: Geochimica et Cosmochimica Acta, v. 66, p. 937–954, doi: 10.1016/S0016-7037(01)00829-8.

Filley, T.R., Freeman, K.H., and Hatcher, P.G., 1996, Carbon isotope relationships between sulfide-bound steroids and proposed functionalized lipid precursors in sediments from the Santa Barbara Basin, California: Organic Geochemistry, v. 25, p. 367–377, doi: 10.1016/S0146-6380(96)00142-8.

Fossing, H., Thode-Andersen, S., and Jørgensen, B.B., 1992, Sulfur isotope exchange between ^{35}S-labeled inorganic sulfur compounds in anoxic marine sediments: Marine Chemistry, v. 38, p. 117–132, doi: 10.1016/0304-4203(92)90071-H.

Francois, R., 1987, A study of sulphur enrichment in the humic fraction of marine sediments during early diagenesis: Geochimica et Cosmochimica Acta, v. 51, p. 17–27, doi: 10.1016/0016-7037(87)90003-2.

Fry, B., Gest, H., and Hayes, J.M., 1984, Isotope effects associated with the anaerobic oxidation of sulfide by the purple photosynthetic bacterium, *Chromatium vinosum*: FEMS Microbiology Letters, v. 22, p. 283–287, doi: 10.1016/0378-1097(84)90025-9.

Fry, B., Cox, J., Gest, H., and Hayes, J., 1986, Discrimination between ^{34}S and ^{32}S during bacterial metabolism of inorganic sulfur compounds: Journal of Bacteriology, v. 165, p. 328–330.

Fry, B., Gest, H., and Hayes, J.M., 1988, $^{34}S/^{32}S$ fractionation in sulfur cycles catalyzed by anaerobic bacteria: Applied and Environmental Microbiology, v. 54, no. 1, p. 250–256.

Fry, B., Silva, S.R., Kendall, C., and Anderson, R.K., 2002, Oxygen isotope corrections for online $\delta^{34}S$ analysis: Rapid Communications in Mass Spectrometry, v. 16, p. 854–858, doi: 10.1002/RCM.651.

Garrels, R.M., and Lerman, A., 1981, Phanerozoic cycles of sedimentary carbon and sulfur: Proceedings of the National Academy of Sciences of the United States of America, v. 78, no. 8, p. 4652–4656.

Garrels, R.M., and Lerman, A., 1984, Coupling of the sedimentary sulfur and carbon cycles—An improved model: American Journal of Science, v. 284, p. 989–1007.

Gelin, F., Boogers, I., Noordeloos, A.A.M., Sinninghe Damsté, J.S., Hatcher, P.G., and de Leeuw, J.W., 1996a, Novel, resistant microalgal polyethers: an important sink of organic carbon in the marine environment?: Geochimica et Cosmochimica Acta, v. 60, p. 1275–1280, doi: 10.1016/0016-7037(96)00038-5.

Gelin, F., Sinninghe Damsté, J.S., Harrison, W.N., Reiss, C., Maxwell, J.R., and de Leeuw, J.W., 1996b, Variations in origin and composition of kerogen constituents as revealed by analytical pyrolysis of immature kerogens before and after desulphurization: Organic Geochemistry, v. 24, p. 705–714, doi: 10.1016/0146-6380(96)00061-7.

Gelin, F., Boogers, I., Noordeloos, A.A.M., Sinninghe Damsté, J.S., Riegman, R., and de Leeuw, J.W., 1997, Resistant biomacromolecules in five marine microalgae of the classes Eustigmatophyceae and Chlorophyceae: Geochemical implications: Organic Geochemistry, v. 26, p. 659–675, doi: 10.1016/S0146-6380(97)00035-1.

Gelin, F., Kok, M., de Leeuw, J., and Sinninghe Damsté, J., 1998, Laboratory sulfurisation of the marine microalga *Nannochloropsis salina*: Organic Geochemistry, v. 29, p. 1837–1848, doi: 10.1016/S0146-6380(98)00171-5.

Gelin, F., Volkman, J.K., Largeau, C., Derenne, S., Sinninghe Damsté, J.S., and de Leeuw, J.W., 1999, Distribution of aliphatic, nonhydrolyzable biopolymers in marine microalgae: Organic Geochemistry, v. 30, p. 147–159, doi: 10.1016/S0146-6380(98)00206-X.

George, G.N., and Gobaty, M.L., 1989, Sulfur K-edge x-ray absorption spectroscopy of petroleum asphaltenes and model compounds: Journal of the American Chemical Society, v. 111, p. 3182–3186.

Giggenbach, W., 1972, Optical spectra and equilibrium distribution of polysulfide ions in aqueous solution at 20 °C: Inorganic Chemistry, v. 11, p. 1201–1207.

Gransch, J.A., and Posthuma, J., 1974, On the origin of sulphur in crudes, *in* Tissot, B., and Bienner, F., Advances in Organic Geochemistry 1973: Paris, Editions Technip, p. 727–739.

Grice, K., Schouten, S., Nissenbaum, A., Charrach, J., and Sinninghe Damsté, J., 1998, A remarkable paradox: Sulfurised freshwater algal (*Botryococcus braunii*) lipids in an ancient hypersaline euxinic ecosystem: Organic Geochemistry, v. 28, p. 195–216, doi: 10.1016/S0146-6380(97)00127-7.

Habicht, K.S., and Canfield, D.E., 1997, Sulfur isotope fractionation during bacterial sulfate reduction in organic-rich sediments: Geochimica et Cosmochimica Acta, v. 61, p. 5351–5361.

Habicht, K.S., and Canfield, D.E., 2001, Isotope fractionation by sulfate-reducing natural populations and the isotopic composition of sulfide in marine sediments: Geology, v. 29, p. 555–558, doi: 10.1130/0091-7613(2001)0292.0.CO;2.

Habicht, K.S., Canfield, D.E., and Rethmeier, J., 1998, Sulfur isotope fractionation during bacterial reduction and disproportionation of thiosulfate and sulfite: Geochimica et Cosmochimica Acta, v. 62, p. 2585–2595, doi: 10.1016/S0016-7037(98)00167-7.

Hartgers, W.A., Lopez, J.F., Sinninghe Damsté, J.S., Reiss, C., Maxwell, J.R., and Grimalt, J.O., 1997, Sulfur-binding in recent environments: II. Speciation of sulfur andiron and implications for the occurrence of organo-sulfur compounds: Geochimica et Cosmochimica Acta, v. 61, p. 4769–4788, doi: 10.1016/S0016-7037(97)00279-2.

Hefter, J., Hauke, V., Richnow, H.H., and Michaelis, W., 1995, Alkanoic subunits in sulfur-rich geomacromolecules, *in* Vairavamurthy, M.A., and Schoonen, M.A.A., eds., Geochemical transformations of sedimentary sulfur: Washington, D.C., American Chemical Society Symposium Series 612, p. 91–109.

Henneke, E., Luther, G., de Lange, G., and Hoefs, J., 1997, Sulphur speciation in anoxic hypersaline sediments from the eastern Mediterranean Sea: Geochimica et Cosmochimica Acta, v. 61, p. 307–321, doi: 10.1016/S0016-7037(96)00355-9.

Huffman, G.P., Mitra, S., Huggins, F.E., Shah, N.S., Vaidya, N., and Lu, F., 1991, Quantitative analysis of all major forms of sulfur in coal by x-ray absorption fine structure spectroscopy: Energy & Fuels, v. 5, p. 574–581.

Huffman, G.P., Shah, N., Huggins, F.E., Stock, L.M., Chatterjee, K., Kilbane, J.J., Chou, M., and Buchanan, D.H., 1995, Sulfur speciation of desulfurized coals by XANES spectroscopy: Fuel, v. 74, p. 549–555, doi: 10.1016/0016-2361(95)98358-L.

Jørgensen, B.B., 1990, A thiosulfate shunt in the sulfur cycle of marine sediments: Science, v. 249, p. 152–154.

Kaplan, I., and Rittenberg, S., 1964, Microbiological fractionation of sulphur isotopes: Journal of General Microbiology, v. 34, p. 195–212.

Kenig, F., and Huc, A.Y., 1990, Incorporation of sulfur into recent organic matter in a carbonate environment (Abu Dhabi, United Arab Emirates), *in* Orr, W.L., and White, C.M., Geochemistry of sulfur in fossil fuels: Washington, D.C., American Chemical Society Symposium Series 249, p. 170–185.

Kenig, F., Sinninghe Damsté, J.S., Frewin, N.L., Hayes, J.M., and de Leeuw, J.W., 1995, Molecular indicators for paleoenvironmental change in a Messinian evaporitic sequence (Vena del Gesso, Italy). II. High-resolution variations in abundances in $\delta^{13}C$ contents of free and sulphur-bound carbon skeletons in a single marl bed: Organic Geochemistry, v. 23, p. 485–526, doi: 10.1016/0146-6380(95)00049-K.

Kohnen, M.E.L., Sinninghe Damsté, J.S., Kock-van Dalen, A.C., ten Haven, H.L., Rüllkötter, J., and de Leeuw, J.W., 1990, Origin and diagenetic transformations of C_{25} and C_{30} highly branched isoprenoid sulphur compounds: Further evidence for the formation of organically bound sulphur during early diagenesis: Geochimica et Cosmochimica Acta, v. 54, p. 3053–3063, doi: 10.1016/0016-7037(90)90121-Z.

Kohnen, M.E.L., Sinninghe Damsté, J.S., Kock-van Dalen, A.C., and de Leeuw, J.W., 1991a, Di- or polysulphide-bound biomarkers in sulphur-rich geomacromolecules as revealed by selective chemolysis: Geochimica et Cosmochimica Acta, v. 55, p. 1375–1394, doi: 10.1016/0016-7037(91)90315-V.

Kohnen, M.E.L., Sinninghe Damsté, J.S., ten Haven, H.L., Kock-van Dalen, A.C., Schouten, S., and de Leeuw, J.W., 1991b, Identification and geochemical significance of cyclic di- and trisulphides with linear and acyclic isoprenoid skeletons in immature sediments: Geochimica et Cosmochimica Acta, v. 55, p. 3685–3695, doi: 10.1016/0016-7037(91)90067-F.

Kohnen, M.E.L., Sinninghe Damsté, J.S., and de Leeuw, J.W., 1991c, Biases from natural sulphurization in palaeoenvironmental reconstruction based on hydrocarbon biomarker distributions: Nature, v. 349, p. 775–778, doi: 10.1038/349775A0.

Kohnen, M., Schouten, S., Sinninghe Damsté, J., de Leeuw, J., Merrit, D., and Hayes, J., 1992, The combined application of organic sulphur and isotope geochemistry to assess multiple sources of palaeobiochemicals with identical carbon skeletons: Organic Geochemistry, v. 19, p. 403–419, doi: 10.1016/0146-6380(92)90008-L.

Kohnen, M.E.L., Sinninghe Damsté, J.S., Baas, M., Kock-van Dalen, A.C., and de Leeuw, J.W., 1993, Sulphur-bound steroid and phytane carbon skeletons in geomacromolecules: Implications for the mechanism of incorporation of sulphur into organic matter: Geochimica et Cosmochimica Acta, v. 57, p. 2515–2528, doi: 10.1016/0016-7037(93)90414-R.

Kok, M.D., Rijpstra, W.I.C., Robertson, L., Volkman, J., and Sinninghe Damsté, J.S., 2000a, Early steroid sulfurisation in surface sediments of a permanently stratified lake (Ace Lake, Antarctica): Geochimica et Cosmochimica Acta, v. 64, p. 1425–1436, doi: 10.1016/S0016-7037(99)00430-5.

Kok, M.D., Schouten, S., and Sinninghe Damsté, J.S., 2000b, Formation of insoluble, nonhydrolyzable, sulfur-rich macromolecules via incorporation of inorganic sulfur species into algal carbohydrates: Geochimica et Cosmochimica Acta, v. 64, p. 2689–2699, doi: 10.1016/S0016-7037(00)00382-3.

Koopmans, M.P., de Leeuw, J.W., Lewan, M.D., and Sinninghe Damsté, J.S., 1996, Impact of dia- and catagenesis on sulphur and oxygen sequestration of biomarkers as revealed by artificial maturation of an immature sedimentary rock: Organic Geochemistry, v. 25, p. 391–426, doi: 10.1016/S0146-6380(96)00144-1.

Krein, E.B., and Aizenshtat, Z., 1995, Proposed thermal pathways for sulfur transformations in organic macromolecules: laboratory simulation experiments, *in* Vairavamurthy, M.A., and Schoonen, M.A.A., eds., Geochemical transformations of sedimentary sulfur: Washington, D.C., American Chemical Society Symposium Series 612, p. 110–137.

Kump, L.R., and Garrels, R.M., 1986, Modeling atmospheric O_2 in the global sedimentary redox cycle: American Journal of Science, v. 286, p. 337–360.

LaLonde, R., Ferrara, L., and Hayes, M., 1987, Low-temperature, polysulfide reactions of conjugated ene carbonyls: A reaction model for the geologic origin of S-heterocycles: Organic Geochemistry, v. 11, p. 563–571, doi: 10.1016/0146-6380(87)90010-6.

Larsen, O., and Potsma, D., 2001, Kinetics of reductive bulk dissolution of lepidocrocite, ferrihydrite, and goethite: Geochimica et Cosmochimica Acta, v. 65, p. 1367–1379.

Lewan, M., 1993, Laboratory simulation of petroleum formation: Hydrous Pyrolysis, *in* Engel M.H., and Macko, S.A., eds., Organic geochemistry, principles and applications—Topics in geobiology: New York, Plenum, v. 11, p. 419–444.

Lewan, M.D., 1998, Sulphur-radical control on petroleum formation rates: Nature, v. 391, p. 164–166, doi: 10.1038/34391.

Luckge, A., Horsfield, B., Littke, R., and Scheeder, G., 2002, Organic matter preservation and sulfur uptake in sediments from the continental margin off Pakistan: Organic Geochemistry, v. 33, p. 477–488, doi: 10.1016/S0146-6380(01)00171-1.

Luther, George W. III, Church, Thomas M., 1988, Seasonal cycling of sulfur and iron in pore-waters of a Delaware salt marsh: Marine Chemistry, v. 23, p. 295–309.

Mossmann, J.R., Aplin, A.C., Curtis, C.D., and Coleman, M.L., 1991, Geochemistry of inorganic and organic sulphur in organic-rich sediments from the Peru Margin: Geochimica et Cosmochimica Acta, v. 55, p. 3581–3595, doi: 10.1016/0016-7037(91)90057-C.

Nelson, B.C., Eglinton, T.I., Seewald, J.S., Vairavamurthy, M.A., and Miknis, F.P., (1995) Transformations in organic sulfur speciation during maturation of Monterey Shale: constraints from laboratory experiments, *in* Vairavamurthy, M.A., and Schoonen, M.A.A., eds., Geochemical transformations of sedimentary sulfur: Washington, D.C., American Chemical Society Symposium Series 612, p. 138–166.

Orr, W., 1978, Sulphur in heavy oils, oils sands and oil shales, *in* Strausz, O., and Lown, E., eds., Oil sand and oil shale chemistry: Berlin, Verlag Chemie Int., p. 223–243.

Orr, W., and White, C., editors, 1990, Geochemistry of sulfur in fossil fuels: Washington, D.C., American Chemical Society Symposium Series 429, 708 p.

Orr, W.L., and Sinninghe Damsté, J.S., 1990, Geochemistry of sulfur in petroleum systems, *in* Orr, W.L., and C.M. White, eds., Geochemistry of sulfur in fossil fuels: Washington, D.C., American Chemical Society Symposium Series 429, p. 2–29.

Passier, H.F., Luther, G.W., III, and de Lange, G.J., 1997, Early diagenesis and sulphur speciation in sediments of the Oman Margin, northwestern Arabian Sea: Deep-Sea Research II, v. 44, p. 1361–1380, doi: 10.1016/S0967-0645(97)00014-3.

Peakman, T.M., Sinninghe Damsté, J.S., and de Leeuw, J.W., 1989, The identification and geochemical significance of a second series of alkylthiophene comprising a linearly extended phytane skeleton in sediments and oils: Geochimica et Cosmochimica Acta, v. 53, p. 3317–3322, doi: 10.1016/0016-7037(89)90111-7.

Perakis, N., 1986, Séparation et détection sélective de composés soufrés dans les fractions Lourdes de pétroles: Géochimie des benzo[b]thiophénes [Ph.D. Dissertation]: University of Strasbourg.

Petsch, S.T., and Berner, R.A., 1998, Coupling the geochemical cycles of C, P, Fe, and S: The effects on atmospheric O^2 and the isotopic records of carbon and sulfur: American Journal of Science, v. 298, p. 246–262.

Poinsot, J.P., Schneckenburger, P., Adam, P., Schaeffer, P., Trendel, J.M., Riva, A., and Albrecht, P., 1998, Novel polycyclic sulfides derived from regular polyprenoids in sediments: Characterization, distribution and geochemical significance: Geochimica et Cosmochimica Acta, v. 62, p. 805–814, doi: 10.1016/S0016-7037(98)00022-2.

Price, F.T., and Shieh, Y.N., 1979, Fractionation of sulfur isotopes during laboratory synthesis of pyrite at low temperatures: Chemical Geology, v. 27, p. 245–253, doi: 10.1016/0009-2541(79)90042-1.

Putschew, A., Scholz-Böttcher, B.M., and Rullkötter, J., 1995, Organic geochemistry of sulphur-rich surface sediments of meromictic Lake Cadagno, Swiss Alps, *in* Vairavamurthy, M.A., and Schoonen, M.A.A., eds., Geochemical transformations of sedimentary sulfur: Washington, D.C., American Chemical Society Symposium Series 612, p. 59–79.

Putschew, A., Scholz-Böttcher, B.M., and Rullkötter, J., 1996, Early diagenesis of organic matter and related sulphur incorporation on surface sediments of meromictic Lake Cadagno in the Swiss Alps: Organic Geochemistry, v. 25, p. 379–390, doi: 10.1016/S0146-6380(96)00143-X.

Putschew, A., Schaeffer-Reiss, C., Schaeffer, P., Koopmans, M.P., de Leeuw, J.W., Lewan, M.D., Sinninghe Damsté, J.S., and Maxwell, J.R., 1998, Release of sulfur- and oxygen-bound components from a sulfur-rich kerogen during simulated maturation by hydrous pyrolysis: Organic Geochemistry, v. 29, p. 1875–1890., doi: 10.1016/S0146-6380(98)00191-0

Pyzik, A.J., and Sommer, S.E., 1981, Sedimentary iron monosulfides: Kinetics and mechanism of formation: Geochimica et Cosmochimica Acta, v. 45, p. 687–698, doi: 10.1016/0016-7037(81)90042-9.

Rees, C.E., Jenkins, W.J., and Monster, J., 1978, The sulfur isotopic composition of ocean water sulfate: Geochimica et Cosmochimica Acta, v. 42, p. 377–381, doi: 10.1016/0016-7037(78)90268-5.

Rospondek, M.J., Köster, J., and Sinninghe Damsté, J.S., 1997, Novel C_{26} highly branched isoprenoid thiophenes and alkane from the Menilite Formation, Outer Carpathians, SE Poland: Organic Geochemistry, v. 26, p. 295–304, doi: 10.1016/S0146-6380(97)00021-1.

Rowland, S., Rockey, C., al-Lihaibi, S.S., and Wolff, G.A., 1993, Incorporation of sulphur into phytol derivatives during simulated early diagenesis: Organic Geochemistry, v. 20, p. 1–5, doi: 10.1016/0146-6380(93)90075-M.

Russell, M., Hartgers, W., and Grimalt, J., 2000, Identification and geochemical significance of sulphurized fatty acids in sedimentary organic matter from the Lorca Basin, SE Spain: Geochimica et Cosmochimica Acta, v. 64, p. 3711–3723, doi: 10.1016/S0016-7037(00)00470-1.

Sandison, C.M., Alexander, R., Kagi, R.I., and Boreham, C.J., 2002, Sulfurisation of lipids in a marine-influenced lignite: Organic Geochemistry, v. 33, p. 1053–1077, doi: 10.1016/S0146-6380(02)00083-9.

Sarret, G., Connan, J., Kasrai, M., Bancroft, G.M., Charrié-Duhaut, A., Lemoine, S., Adam, P., Albrecht, P., and Eybert-Bérard, L., 1999, Chemical forms of sulfur in geological and archeological asphaltenes from Middle East, France, and Spain determined by sulfur K- and L-edge X-ray absorption near-edge structure spectroscopy: Geochimica et Cosmochimica Acta, v. 63, p. 3767–3779, doi: 10.1016/S0016-7037(99)00205-7.

Sarret, G., Mongenot, T., Connan, J., Derenne, S., Kasrai, M., Bancroft, G., and Largeau, C., 2002, Sulfur speciation in kerogens of the Orbagnoux deposit (Uper Kimmeridgian, Jura) by XANES spectroscopy and pyrolysis: Organic Geochemistry, v. 33, p. 877–895, doi: 10.1016/S0146-6380(02)00066-9.

Schaeffer, P., Reiss, C., and Albrecht, P., 1995, Geochemical study of macromolecular organic matter from sulfur-rich sediments of evaporitic origin (Messinian of Sicily) by chemical degradations: Organic Geochemistry, v. 23, p. 567–581, doi: 10.1016/0146-6380(95)00045-G.

Schaeffer-Reiss, C., Schaeffer, P., Putschew, A., and Maxwell, J.R., 1998, Stepwise chemical degradation of immature S-rich kerogens from Vena del Gesso (Italy): Organic Geochemistry, v. 29, p. 1857–1873, doi: 10.1016/S0146-6380(98)00175-2.

Schneckenburger, P., Adam, P., and Albrecht, P., 1998, Thioketones as key intermediates in the reduction of ketones to thiols by HS^- in natural environments: Tetrahedron Letters, v. 39, p. 447–450, doi: 10.1016/S0040-4039(97)10572-X.

Schouten, S., Pavlovic, D., Sinninghe Damsté, J., and de Leeuw, J., 1993a, Nickel boride: An improved desulphurizing agent for sulphur-rich geomacromolecules in polar and asphaltene fractions: Organic Geochemistry, v. 20, p. 901–909, doi: 10.1016/0146-6380(93)90101-G.

Schouten, S., Pavlovic, D., Sinninghe Damsté, J.S., and de Leeuw, J.W., 1993b, Selective cleavage of acyclic sulphide moieties of sulphur-rich geomacromolecules by superheated methyl iodide: Organic Geochemistry, v. 20, p. 911–916, doi: 10.1016/0146-6380(93)90102-H.

Schouten, S., de Graaf, W., Sinninghe Damsté, J.S., van Driel, G.B., and de Leeuw, J.W., 1994a, Laboratory simulation of natural sulphurization. II. Reaction of multifunctionalized lipids with inorganic polysulphides at low temperatures, *in* Telnæs, N., van Graas, G., and Øygard, K., Advances in organic geochemistry 1993: Organic Geochemistry, v. 22, p. 825–834, doi: 10.1016/0146-6380(94)90142-2.

Schouten, S., van Driel, G.B., Sinninghe Damsté, J.S., and de Leeuw, J.W., 1994b, Natural sulphurization of ketones and aldehydes: A key reaction in the formation of organic sulphur compounds: Geochimica et Cosmochimica Acta, v. 58, p. 5111–5116.

Schouten, S., Sinninghe Damsté, J., Baas, M., Kock-van Dalen, A., Kohnen, M., and de Leeuw, J., 1995a, Quantitative assessment of mono- and polysulphide-linked carbon skeletons in sulphur-rich macromolecular aggregates present in bitumens and oils: Organic Geochemistry, v. 23, p. 765–775, doi: 10.1016/0146-6380(95)00055-J.

Schouten, S., Eglinton, T.I., and Sinninghe Damsté, J.S., 1995b, Influence of sulphur cross-linking on the molecular-size distribution of sulphur-rich macromolecules in bitumen, *in* Vairavamurthy, M.A., and Schoonen, M.A.A., eds., Geochemical transformations of sedimentary sulfur: Washington, D.C., American Chemical Society Symposium Series 612, p. 80–92.

Schouten, S., Sinninghe Damsté, J.S., Kohnen, M.E.L., and de Leeuw, J.W., 1995c, The effect of hydrosulphurization on stable carbon isotopic compositions of free and sulphur-bound lipids: Geochimica et Cosmochimica Acta, v. 59, p. 1605–1609, doi: 10.1016/0016-7037(95)00066-9.

Schouten, S., de Loureiro, M.R.B., Sinninghe Damsté, J.S., and de Leeuw, J.W., 2001, Molecular biogeochemistry of Monterey sediments (Naples Beach, USA). I. distributions of hydrocarbons and organic sulphur compounds, *in* Isaacs, C.M., and Rüllkötter, J., eds., The Monterey Formation: From rocks to molecule: Columbia University Press, p. 150–174.

Shen, Y., Buick, R., and Canfield, D.E., 2001, Isotopic evidence for microbial sulphate reduction in the early Archean era: Nature, v. 410, p. 77–81, doi: 10.1038/35065071.

Sinninghe Damsté, J.S., and de Leeuw, J.W., 1990, Analysis, structure and geochemical significance of organically-bound sulphur in the geosphere: State of the art and future research, *in* Durand B., and Behar, F., Advances in organic geochemistry 1989: Organic Geochemistry, v. 16, p. 1077–1101, doi: 10.1016/0146-6380(90)90145-P.

Sinninghe Damsté, J.S., and de Leeuw, J.W., 1992, Organically bound sulphur in coal: A molecular approach: Fuel Processing Technology, v. 30, p. 109–178, doi: 10.1016/0378-3820(92)90045-R.

Sinninghe Damsté, J.S., Rijpstra, W.I.C., de Leeuw, J.W., and Schenck, P.A., 1988, Origin of organic sulphur compounds and sulphur-containing high molecular weight substances in sediments and immature crude oils, *in* Mattavelli, L., and Novelli, L., Advances in organic geochemistry 1987: Organic Geochemistry, v. 13, p. 593–606, doi: 10.1016/0146-6380(88)90079-4.

Sinninghe Damsté, J.S., Rijpstra, W.I.C., Kock-van Dalen, A.C., de Leeuw, J.W., and Schenck, P.A., 1989, Quenching of labile functionalised lipids by inorganic sulphur species: Evidence for the formation of sedimentary organic sulphur compounds at the early stages of diagenesis: Geochimica et Cosmochimica Acta, v. 53, p. 1343–1355, doi: 10.1016/0016-7037(89)90067-7.

Sinninghe Damsté, J.S., Kohnen, M.E.L., and de Leeuw, J.W., 1990, Thiophenic biomarkers for palaeoenvironmental assessment and molecular stratigraphy: Nature, v. 345, p. 609–611, doi: 10.1038/345609A0.

Sinninghe Damsté, J.S., de las Heras, X.F.C., van Bergen, P.F., and de Leeuw, J.W., 1993, Characterization of Tertiary Catalan lacustrine oil shales: Discovery of extremely organic sulphur-rich Type I kerogens: Geochimica et Cosmochimica Acta, v. 57, p. 389–415.

Sinninghe Damsté, J.S., Rijpstra, W.I.C., de Leeuw, J.W., and Lijmbach, G.W.M., 1994, Molecular characterization of organically-bound sulfur in crude oils: A feasibility study for the application of Raney Ni desulfurization as a new method to characterize crude oils: Journal of High Resolution Chromatography, v. 17, p. 489–500.

Sinninghe Damsté, J.S., Kohnen, M.E.L., and Horsfield, B., 1998a, Origin of low-molecular-weight alkylthiophenes in pyrolysates of sulphur-rich kerogens as revealed by micro-scale sealed vessel pyrolysis: Organic Geochemistry, v. 29, p. 1891–1903, doi: 10.1016/S0146-6380(98)00166-1.

Sinninghe Damsté, J., Kok, M., Köster, J., and Schouten, S., 1998b, Sulfurized carbohydrates: an important sedimentary sink for organic carbon?: Earth and Planetary Science Letters, v. 164, p. 7–13, doi: 10.1016/S0012-821X(98)00234-9.

Sinninghe Damsté, J.S., White, C.M., Green, J.B., and de Leeuw, J.W., 1999a, Organosulfur compounds in sulfur-rich Rasa Coal: Energy & Fuels, v. 13, p. 728–738, doi: 10.1021/EF980236C.

Sinninghe Damsté, J.S., Schouten, S., de Leeuw, J.W., van Duin, A.C.T., and Geenevasen, J.A.J., 1999b, Identification of novel sulfur-containing steroids in sediments and petroleum: Probably incorporation of sulfur into D5,7-sterols during early diagenesis: Geochimica et Cosmochimica Acta, v. 63, p. 31–38, doi: 10.1016/S0016-7037(98)00295-6.

Spiro, C.L., Wong, J., Lytle, F.W., Greegor, R.B., Maylotte, D.H., and Lamson, S.H., 1984, X-ray absorption spectroscopic investigation of sulfur sites in coals: organic sulfur identification: Science, v. 226, p. 48–50.

Studley, S.A., Ripley, E.M., Elswick, E.R., Dorais, M.J., Fong, J., Finkelstein, D., and Pratt, L.M., 2002, Analysis of sulfides in whole rock matrices by elemental analyzer–continuous flow isotope ratio mass spectrometry: Chemical Geology, v. 192, p. 141–148, doi: 10.1016/S0009-2541(02)00162-6.

Suits, N., and Arthur, M., 2000, Sulfur diagenesis and partitioning in Holocene Peru shelf and upper slope sediments: Chemical Geology, v. 163, p. 219–234, doi: 10.1016/S0009-2541(99)00114-X.

Tissot, B., and Welte, D., 1984, Petroleum formation and occurrence, 2nd edition: Heidelberg, Springer, 699 p.

Tomic, J., Behar, F., Vandenbrouke, M., and Tang, Y., 1995, Artificial maturation of Monterey kerogen (Type II-S) in a closed system and comparison with Type II kerogen: implications on the fate of sulfur: Organic Geochemistry, v. 23, p. 647–660, doi: 10.1016/0146-6380(95)00043-E.

Trust, B.A., and Fry, B., 1992, Sulphur isotopes in plants: a review: Plant, Cell & Environment, v. 15, p. 1105–1110.

Urban, N., Ernst, K., and Bernasconi, S., 1999, Addition of sulfur to organic matter during early diagenesis of lake sediments: Geochimica et Cosmochimica Acta, v. 63, p. 837–853, doi: 10.1016/S0016-7037(98)00306-8.

Vairavamurthy, M., and Schoonen, M., editors, 1995, Geochemical transformations of sedimentary sulfur: Washington, D.C., American Chemical Society Symposium Series 612, 467 p.

Vairavamurthy, A., Mopper, K., and Taylor, B., 1992, Occurrence of particle-bound polysulfides and significance of their reaction with organic matters in marine sediments: Geophysical Research Letters, v. 19, p. 2043–2046.

Vairavamurthy, A., Zhou, W., Eglinton, T., and Manowitz, B., 1994, Sulfonates: A novel class of organic sulfur compounds in marine sediments: Geochimica et Cosmochimica Acta, v. 58, p. 4681–4687, doi: 10.1016/0016-7037(94)90200-3.

Vairavamurthy, M., Orr, W., and Manowitz, B., 1995, Geochemical transformations of sedimentary sulfur: An introduction, *in* Vairavamurthy, M.A., and Schoonen, M.A.A., eds., Geochemical transformations of sedimentary sulfur: Washington, D.C., American Chemical Society Symposium Series 612, p. 1–15.

Vairavamurthy, M., Maletic, D., Want, S., Manowitz, B., Eglinton, T., and Lyons, T., 1997, Characterization of sulfur-containing functional groups in sedimentary humic substances by X-ray absorption near-edge structure spectroscopy: Energy & Fuels, v. 11, p. 546–553, doi: 10.1021/EF960212A.

Valisolalao, J., Perakis, N., Chappe, B., and Albrecht, P., 1984, A novel sulfur containing C_{35} hopanoid in sediments: Tetrahedron Letters, v. 25, p. 1183–1186, doi: 10.1016/S0040-4039(01)91555-2.

van Dongen, B.E., 2003, Natural sulfurization of carbohydrates in marine sediments: consequences for the chemical and carbon isotopic composition of sedimentary organic matter [Ph.D. thesis]: University of Utrecht, 149 p.

van Dongen, S., Schouten, S., and Sinninghe Damsté, J., 2002, Carbon isotope variability in monosaccharides and lipids of aquatic algae and terrestrial plants: Marine Ecology Progress Series, v. 232, p. 83–92.

van Dongen, B.E., Schouten, S., Baas, M., Geenevasen, J.A.J., and Sinninghe Damsté, J.S., 2003a, An experimental study of the low-temperature sulfurization of carbohydrates: Organic Geochemistry, v. 34, p. 1129–1144, doi: 10.1016/S0146-6380(03)00060-3.

van Dongen, B.E., Schouten, S., and Sinninghe Damsté, J.S., 2003b, Sulfurization of carbohydrates results in n S-rich, unresolved complex mixture in kerogen pyrolysates: Energy & Fuels, v. 17, p. 1109–1118, doi: 10.1021/EF0202283.

Van Kaam-Peters, H.M.E., Schouten, S., Köster, J., and Sinninghe Damsté, J.S., 1998, Controls on the molecular and carbon isotopic composition of organic matter deposited in a Kimmeridgian euxinic shelf sea: Evidence for preservation of carbohydrates through sulfurisation: Geochimica et Cosmochimica Acta, v. 62, p. 3259–3283, doi: 10.1016/S0016-7037(98)00231-2.

Wakeham, S.G., Sinninghe Damsté, J.S., Kohnen, M.E.L., and de Leeuw, J.W., 1995, Organic sulfur compounds formed during early diagenesis in Black Sea sediments: Geochimica et Cosmochimica Acta, v. 59, no. 3, p. 521–533, doi: 10.1016/0016-7037(94)00361-O.

Waldo, G.S., Carlson, M.K., Moldowan, J.M., Peters, K.E., and Penner-Hahn, J.E., 1991, Sulfur speciation in heavy petroleums: Information from x-ray absorption near edge structure: Geochimica et Cosmochimica Acta, v. 55, p. 801–814, doi: 10.1016/0016-7037(91)90343-4.

Werne, J., 2000, A geochemical evaluation of depositional controls and paleoenvironmental reconstructions in organic rich sedimentary deposits: Evidence from the modern Cariaco Basin, Venezuela, and application to the Devonian Appalachian Basin [Ph.D. dissertation]: Northwestern University, 310 p.

Werne, J.P., Hollander, D.J., Behrens, A., Schaeffer, P., Albrecht, P., and Sinninghe Damsté, J.S.S., 2000a, Timing of early diagenetic sulfurization of organic matter: A precursor-product relationship in Holocene sediments of the anoxic Cariaco Basin, Venezuela: Geochimica et Cosmochimica Acta, v. 64, p. 1741–1751.

Werne, J.P., Hollander, D.J., Lyons, T.W., and Peterson, L.C., 2000b, Climate-induced variations in productivity and planktonic ecosystem structure from the Younger Dryas to Holocene in the Cariaco Basin, Venezuela: Paleoceanography, v. 15, no. 1, p. 19–29.

Werne, J.P., Hollander, D.J., Lyons, T.W., and Sinninghe Damsté, J.S., 2001, Compound-specific sulfur isotope constraints on the pathway(s) of diagenetic sulfurization of organic matter, Geological Society of America Abstracts with Programs, v. 33, no. 6, p. A-94.

Werne, J.P., Lyons, T.W., Hollander, D.J., Formolo, M., and Sinninghe Damsté, J.S., 2003, Reduced sulfur in euxinic sediments of the Cariaco Basin: Sulfur isotope constraints on organic sulfur formation: Chemical Geology, v. 195, p. 159–179, doi: 10.1016/S0009-2541(02)00393-5.

Zaback, D.A., and Pratt, L.M., 1992, Isotope composition and speciation of sulfur in the Miocene Monterey Formation: Reevaluation of sulfur reactions during early diagenesis in marine environments: Geochimica et Cosmochimica Acta, v. 56, p. 763–774, doi: 10.1016/0016-7037(92)90096-2.

Manuscript Accepted by the Society March 28, 2004

Geological Society of America
Special Paper 379
2004

Using sulfur isotopes to elucidate the origin of barite associated with high organic matter accumulation events in marine sediments

A. Paytan
Department of Geological and Environmental Science, Stanford University, Stanford, California 94305-2115, USA

F. Martinez-Ruiz
Instituto Andaluz de Ciencias de la Tierra (CSIC-UGR), Campus Fuentenueva, 18002 Granada, Spain

M. Eagle
A. Ivy
S.D. Wankel
Department of Geological and Environmental Science, Stanford University, Stanford, California 94305-2115, USA

ABSTRACT

Events of widespread deposition of organic-carbon–rich marine sediments, identified as ocean anoxic events, occurred in the middle of the Cretaceous. Similar deposits termed *sapropels* occurred during the Pliocene and Pleistocene in the Mediterranean Basin. High biological productivity and/or anoxia have been invoked as possible causes for these widespread high organic carbon deposition events. We use the S isotopic composition of barite associated with these events to confirm that high barite accumulation rates are a result of elevated marine biological productivity and not a diagenetic artifact. The accumulation and good preservation of biogenic barite, which dissolves when pore-water sulfate concentrations are low, in association with high organic matter and authigenic pyrite, indicates that the rate of bacterial sulfate reduction was low enough for downward diffusion of seawater sulfate to replenish the pore water and prevent depletion of sulfate. The organic C to S burial ratio in samples with high barite accumulation is typically high (>5 wt ratio), supporting burial in high-productivity open-ocean regions, where pyrite formation is restricted.

Keywords: ocean anoxic events, sapropels, sulfur isotopes, marine productivity, barite.

INTRODUCTION

Time intervals in the geological record during which ocean conditions promoted accumulation of organic rich sediments, such as black shales, have usually been referred to as ocean anoxic events (OAE) (Arthur et al., 1990). The enhanced organic matter burial rates could have led to a significant drop in atmospheric CO_2 concentrations (Freeman and Hayes, 1992; Arthur et al., 1988; Kuypers et al., 1999), thereby providing negative feedback to "greenhouse" climates. Widespread occurrences of such black shale deposits are found in the mid-Cretaceous. Although restricted to the Mediterranean, and spanning shorter time intervals, Mediterranean sapropels—defined as discrete layers >1 cm thick and containing >2% total organic carbon (Kidd et

*apaytan@pangea.stanford.edu

Paytan, A., Martinez-Ruiz, F., Eagle, M., Ivy, A., and Wankel, S.D., 2004, Using sulfur isotopes to elucidate the origin of barite associated with high organic matter accumulation events in marine sediments *in* Amend, J.P., Edwards, K.J., and Lyons, T.W., eds., Sulfur biogeochemistry—Past and present: Boulder, Colorado, Geological Society of America Special Paper 379, p. 151–160. For permission to copy, contact editing@geosociety.org.

al., 1978)—have been considered as possible younger analogues of black shales (Calvert, 1983; Nijenhuis et al., 1999). The increased burial rates of organic matter during both Cretaceous OAE and Mediterranean sapropel deposition periods have usually been attributed to two different mechanisms (Arthur et al., 1990; Calvert and Pedersen, 1993): (1) decreased organic matter mineralization promoted by decreased oxygen content in seawater as a result of ocean stagnation (Stanley, 1978; Bralower and Thierstein, 1987; Sarmiento et al., 1988; Nolet and Corliss, 1990; Aksu et al., 1995; Barron et al., 1995; Erbacher et al., 2001), and/or (2) increased export production and rapid supply of organic matter to the sediment, which overwhelmed mineralization (Schlanger and Jenkyns, 1976; Calvert, 1983; Weissert et al., 1985; Weissert, 1989; Calvert and Pedersen, 1992). These two models imply different causes for the high organic matter accumulation and ocean anoxia. In the "ocean stagnation" model, external physical processes (temperature, evaporation, runoff, continent configuration) caused intense vertical gradients of temperature and salinity, which resulted in stable stratification, reduced ventilation of deep water, and finally, oceanic anoxia. In contrast, the "high productivity" model invokes biogeochemical processes internal to the ocean; extensive use of oxygen for (partial) organic matter mineralization results in lower oceanic dissolved oxygen content and even anoxia. In the latter model, changes in the carbon cycle caused the lower oxygen levels and are not merely a result of anoxic conditions induced by reduced deepwater circulation. The causes for increased productivity, however, are ultimately driven by nutrient availability in the euphotic zone. These nutrients are supplied via intensified upwelling or continental runoff, which, in turn, are controlled by tectonic and climatic changes such as uplift, precipitation, and wind stress (Parrish, and Curtis, 1982; Rossignol-Strick, 1985; Weissert, 1989; Schmidt and Mysak, 1996). Changes in hydrothermal activity (Larson and Erba, 1999) and water column redox state (VanCappellen and Ingall, 1996; Slomp et al., 2002; Filippelli et al., 2003) may also have contributed to nutrient availability.

High organic carbon accumulation in marine sediments is considered an indicator of increased biological production (Berger et al., 1988, 1989); however, a number of parameters, including seawater oxygen content, influence organic carbon preservation in marine sediments (Berger et al., 1988, 1989; Anderson et al., 2001; Sarmiento et al., 1988; Pedersen and Calvert, 1990; Canfield, 1994). At very low dissolved oxygen concentrations, organic carbon preservation increases (Sarmiento et al., 1988; Canfield, 1994); accordingly, the high organic carbon accumulation in marine sediments during mid-Cretaceous OAE and Mediterranean sapropel deposition cannot be interpreted unequivocally as an indication of increased export production or of low oxygen conditions induced by water stagnation. Here we provide evidence from S isotope analyses of barite, which is associated with these events (i.e., separated from the black shale or sapropel layers), that support increased export production as the dominant cause of organic matter accumulation in the open ocean during these time intervals.

Marine barite forms in microenvironments in association with decaying organic matter (Bishop, 1988; Dehairs et al., 1980; Ganeshram et al., 2003); therefore, barite fluxes measured in sediment traps and in suspended particles correlate well with biological productivity in the overlying water column, and barite accumulation rates (BaAR) in marine sediments are related to export production (Dehairs et al., 1980; Bishop, 1988; Dymond et al., 1992; Paytan et al., 1996; Eagle et al., 2003). Barite is relatively resistant to diagenetic alteration after burial in sediments where pore waters are sulfate rich and has been used to reconstruct paleoproductivity at various oceanic locations and time intervals (Schmitz, 1987; Gingele and Dahmke, 1994; Rutsch et al., 1995; Paytan et al., 1996; Dean et al., 1997; Nürnberg et al., 1997; Bonn and Gingele, 1998; Schroeder et al., 1997; Thompson and Schmitz, 1997; Murray et al., 2000; Bains et al., 2000; Klump et al., 2000). In particular, high excess Ba content (total barium minus the fraction associated with terrigenous material) has long been recognized in many eastern Mediterranean sapropel layers and is considered the most reliable proxy for the identification of sapropel layers and for determining the original sapropel thickness (e.g., Van Santvoort et al., 1996; Nijenhuis et al., 1998; Wehausen and Brumsack, 1998; Thomson et al., 1999; Martinez-Ruiz et al., 2000; Weldeab et al., 2003). Similarly, high barite accumulation rates have recently been reported for mid-Cretaceous OAE in several deep-sea cores (Paytan, 2002). These high excess Ba or barite accumulations that coincide with the organic carbon–rich layers have been interpreted as evidence for high oceanic productivity during these events (based on the relation between excess Ba, barite, and productivity in the present day ocean). This interpretation assumes that the excess Ba (which is presumed to be associated with barite) and/or the barite observed in these sediments originated in the water column and precipitated in association with decaying organic matter and therefore could be directly related to export production.

In sulfate-reducing sediments, where pore waters have low sulfate concentrations, however, barite is remobilized and is not preserved (Dean, and Schreiber, 1977; Torres et al., 1996a, 1996b; McManus et al., 1998, Bréhéret and Brumsack, 2000; Schenau et al., 2001). In such sediments, the remobilized Ba may diffuse within the sediment column, and upon contact with sulfate-containing pore waters, barite may precipitate diagenetically as "barite fronts" within sediments (Bolze et al., 1973; Dean and Schreiber, 1977; Brumsack and Gieskes, 1983; Cecile et al., 1983; Brumsack, 1986; Pruysers et al., 1991; van Os et al., 1991; von Breymann et al., 1992; Falkner et al., 1992; Torres et al., 1996b; Bréhéret and Brumsack, 2000) (see also the schematic diagram in Fig. 1). As demonstrated in Figure 1, when sulfate is depleted in pore waters as a result of sulfate reduction, pore waters become significantly undersaturated with respect to barite, and barite is remobilized, releasing Ba to pore waters. This process also enriches the residual pore-water sulfate in ^{34}S. As pore-water Ba diffuses upward in the sediments and encounters solutions with sulfate, barite will precipitate diagenetically within the sediment in distinct fronts. This barite will be enriched

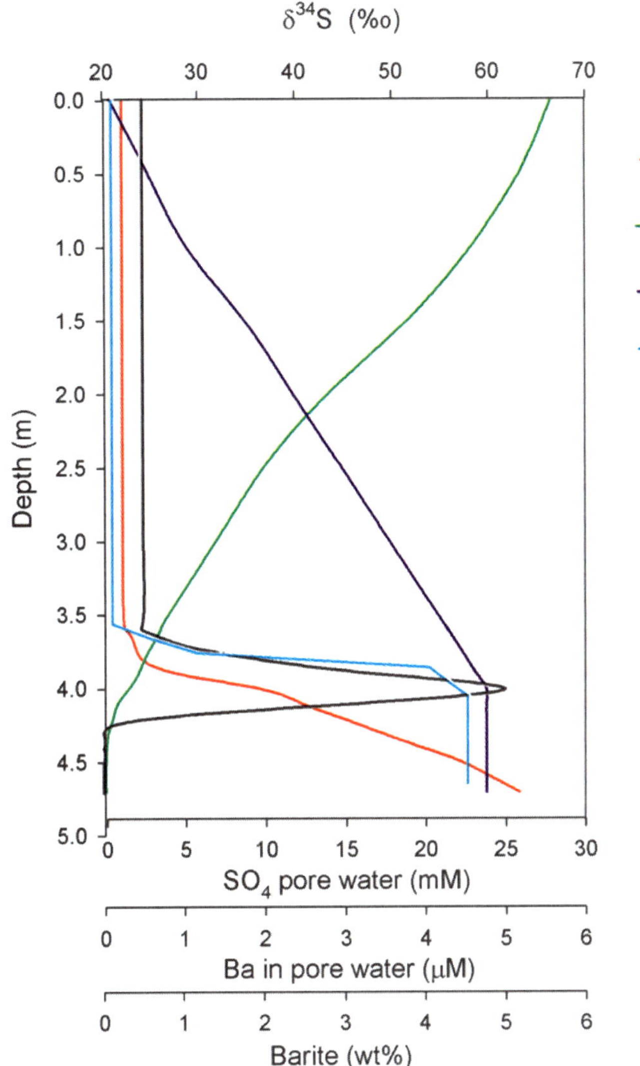

Figure 1. Schematic diagram of concentration-depth profiles for pore-water sulfate, sulfide, dissolved Ba, the isotopic composition of pore-water sulfate, and barite concentrations in the solid phase. Modified from Brumsack (1986) and Torres et al. (1996b).

in ^{34}S as well. Accordingly, the interpretation that increased barite accumulation suggests high biological productivity is contingent upon proof that indeed the barite in these organic rich layers has formed in the water column and is not a postdepositional diagenetic artifact. We use the S isotopic composition of barite separated from OAE and sapropel deposits to distinguish between the two potential barite sources. If the barite associated with the high organic matter accumulation event has precipitated in the water column and has been preserved in the sediments since the time of deposition, the S isotope ratio recorded in this barite should reflect contemporaneous seawater sulfate S isotope ratios (Paytan et al., 1998; Paytan et al., 2002). If, on the other hand, the barite has precipitated diagenetically within the sediment due to extensive sulfate reduction and Ba remobilization, the S isotopic composition of this diagenetic barite is expected to be significantly heavier than that of contemporaneous seawater sulfate, due to the preferential removal of light S in the process of sulfate reduction, which increases the isotopic composition of the residual sulfate in pore waters (Torres et al., 1996b; Aquilina et al., 1997; Naehr et al., 2000; Paytan et al., 2002)

METHODS

Barite was separated from the sediment using a sequential leaching procedure that includes reaction with 6N hydrochloric, sodium hypochlorite, hydroxylamine, and an HF-HNO_3 mixture (Table 1; Collier and Edmond, 1984; Paytan et al., 1998; Eagle et al., 2003). During barite extraction, each of the leaching steps targets a major (operationally defined) sedimentary fraction, leaving a final residue composed of barite and a few other refractory minerals. The extraction yield was determined to be better than 90% (Eagle et al., 2003). This method was modified from the procedure described in Eagle et al. (2003) by using 6N HCl in a glove bag under a N_2 atmosphere to remove the carbonate and sulfides and prevent oxidation of S and potential reprecipitation of barite during preparation. The insoluble residue of the sequential leaching procedure is examined under a scanning electron microscope (SEM) to determine barite content. Rutile and anatase (TiO_2) were the most common minerals other than barite in the residue. Only samples with no S bearing phases other than barite present in the residue were analyzed.

We used samples from several Deep Sea Drilling Project (DSDP) and Ocean Drilling Program (ODP) cores. The Cretaceous deep oceanic sediments spanning OAE time intervals are from ODP Sites 305, 417, 418, 551, 766, and 1049. Barite accumulation rates for these cores have been published previously (Paytan et al., 2002) and indicate increased accumulation rates in all OAE present in these cores. Barite was also extracted from sapropel samples of Pliocene to Pleistocene age recovered from eastern Mediterranean cores (ODP Sites 964, 966 and 967). Ba enrichments associated with sapropel layers in some of these cores have been previously reported (Thomson et al., 1995, 1999; Nijenhuis et al., 1998; Wehausen and Brumsack, 1998; Warning and Brumsack, 2000; Calvert and Fontugne, 2001; Weldeab et al., 2003), and it has also been shown and generally accepted that the Ba excess in sapropel layers is derived from marine barite (Thomson et al., 1995; Martinez-Ruiz et al., 2000).

Sulfur isotope analyses were done by continuous-flow mass spectrometry using a Carlo Erba NA 1500 elemental analyzer connected to a Micromass Isoprime mass spectrometer. Samples of 4–8 mg were introduced in tin boats with ~5 mg vanadium pentoxide mixed with each sample. A commercial tank of SO_2 was used as a reference gas for $\delta^{34}S$ measurements, and results are reported relative to the Canyon Diablo Troilite standard, with a standard deviation (2σ) of ±0.3‰.

TABLE 1. BARITE SEPARATION SEQUENTIAL LEACHING

1. Weigh ~10 g dry sediment.
2. Remove carbonates and sulfides with 6 N hydrochloric acid under a N_2 atmosphere (room temperature, ~12 h).
3. Wash three times with DI water (repeat after each step).
4. Remove organic matter in 5% sodium hypochlorite (50 °C, ~12 h).
5. Remove Fe-Mn oxyhydroxides with 0.2 N hydroxylamine in 25% acetic acid (by volume) (80 °C, ~12 h).
6. Digest in 1:2 40% hydrofluoric acid:1 N nitric acid (room temperature, ~12 h).
7. Digest in 1:1 40% hydrofluoric acid:1 N nitric acid (room temperature, ~12 h).
8. Digest in 2:1 40% hydrofluoric acid:1 N nitric acid (room temperature, ~12 h).
9. Rinse residue in saturated $AlCl_3$ in 0.1 N HNO_3 to remove fluorides (90 °C, 1 h).
10. Ash sample at 700 °C for 2 h.
11. Weigh residue and check purity with SEM or XRD.

Note: After Paytan et al. (1998) and Eagle et al. (2003). DI—distilled water; SEM—scanning electron microscope; XRD—X-ray diffraction.

RESULTS AND DISCUSSION

In all the cores used, the time interval identified as the time frame of sapropel or OAE deposition is characterized by higher barite or excess Ba deposition compared to sections below or above (Wehausen and Brumsack, 1998; Thomson et al., 1995, 1999; Martinez-Ruiz et al., 2000; Warning and Brumsack, 2000; Calvert and Fontugne, 2001; Paytan, 2002). This enrichment, at least in sapropels, has been recognized for quite some time (Calvert, 1983; Calvert and Pedersen, 1992; Thomson et al., 1995, 1999; Van Santvoort et al., 1996; Weldeab et al., 2003) and has been interpreted as indicating high biological productivity. However, postdepositional barite precipitation has not been ruled out (Dean, and Schreiber, 1977; Brumsack, 1986; Bréhéret and Brumsack, 2000), although based on the peak shape of the Ba enrichment, it has been suggested that in most cases, at least for the sapropels, these are not diagenetic features (Weldeab et al., 2003). In some sapropel layers, however, the Ba enrichment has been interpreted as a post burial redistribution of Ba feature (van Os et al., 1991, 1994; Weldeab et al., 2003). The S isotope signature of the barite could serve as an independent and conclusive indicator for the origin of the barite associated with the sapropel and black shale deposits studied here (in addition to the model based peak shape characteristics).

The S isotopic composition of barite separated from sapropel and black shale sections is presented in Table 2. As can be seen, barite samples analyzed here record the S isotopic composition of contemporaneous seawater sulfate (e.g., ~21–22‰ for the Pliocene and Pleistocene and ~16–19‰, depending on the exact age, for the mid Cretaceous [Fig. 2; Paytan et al., 1998; Nielsen 1978; Claypool et al., 1980]). It should be emphasized that although the S isotopic compositions of sapropel and OAE barite samples are compared with a seawater isotope curve derived from barite, the seawater S isotope curve was constructed from multiple cores from a wide range of sites, is consistent with evaporite based records (Claypool, et al., 1980), and has been shown to record contemporaneous seawater sulfate S isotopic composition (Paytan et al., 2004).

It is expected that barite of diagenetic origin that has precipitated from pore fluids that have encountered some degree of sulfate loss due to bacterial sulfate reduction will not record the open seawater isotopic composition. Sulfate reduction leads to enrichment of the heavy S isotope (^{34}S) in the residual sulfate of pore fluids (Harrison and Thode, 1958; Hartmann and Nielsen, 1969; Jørgensen, 1979; Habicht and Canfield, 1997). Barite precipitation within the sedimentary column occurs when Ba-rich fluids (from barite dissolution by the sulfate-reduction process) migrate by diffusion or advection toward sections in the sediment where sulfate is available (Dean and Schreiber, 1977; Brumsack, 1986; Torres et al., 1996a, 1996b; Bréhéret and Brumsack, 2000). The S isotopic composition of such fluids is significantly enriched in ^{34}Scompared to seawater (see Fig. 1). Indeed, at one Cretaceous

TABLE 2. SULFUR ISOTOPES IN BARITE AND C/S WEIGHT RATIOS IN BULK SEDIMENT OF SAPROPEL AND OCEAN ANOXIC EVENT DEPOSITS

Core					Depth below surface (m)	Longitude	Latitude	Water depth (m)	Age	$\delta^{34}S$ (‰)	C/S wt. ratio
Leg	Site	Core	Section	Interval							
Mediterranean Sapropels											
160	964A	9H	3	42–44	76.72	36°15′N	17°44′E	3657	Pliocene	21.12	11.2
160	964B	4H	3	31–37	30.43	36°15′N	17°45′E	3658	mid. Pleistocene	21.55	na
160	964B	4H	3	117–122	31.28	36°15′N	17°45′E	3658	mid. Pleistocene	21.32	na
160	964B	6H	2	90–98	48.53	36°15′N	17°45′E	3658	early Pleistocene	20.12	7.5
160	964C	9H	3	86–92	75.46	36°15′N	17°45′E	3660	Pliocene	21.46	na
160	967B	6H	7	14–20	52.44	34°04′N	32°43′E	2555	Pliocene	20.97	na
160	967C	8H	6	48–50	74.48	34°04′N	32°43′E	2552	late Pliocene	21.03	10.6
160	969A	4H	2	34–37	28.54	33°50′N	24°53′E	2200	early Pleistocene	20.22	6.5
160	969A	4H	2	41–48	28.61	33°50′N	24°53′E	2200	early Pleistocene	19.92	na
160	969E	1H	3	80–82	3.80	33°50′N	24°52′E	2201	late Pleistocene	19.82	12
TTR	GL94	–	–	25–30	0.265	36°04′N	21°58′E	2687	Holocene	21.32	4
Cretaceous ocean anoxic events											
32	305	42	1	108–112	383.58	32°00′N	157°51′E	2921	OAE 2	18.90	23.4
32	305	59	1	88–91	551.38	32°00′N	157°51′E	2921	OAE 1b	16.81	na
51	417D	21	2	110–113	337.10	25°06′N	68°02′W	5482	OAE 1a	16.40	12.7
51	418B	30	2	112–115	275.00	25°02′N	68°03′W	5514	OAE 1b	15.35	na
80	551	5	1	145–148	134.91	48°54′N	13°30′W	3887	OAE 2	19.10	31.1
80	551	5	2	91–94	133.95	48°54′N	13°30′W	3887	OAE 2	18.84	21.9
80	551	5	CC	1–4	135.21	48°54′N	13°30′W	3887	OAE 2	19.02	4.2
123	766A	14	CC	7–10	130.9	19°55′S	110°27′E	3997	OAE 2	19.04	10.5
123	766A	16	4	55–59	147.95	19°55′S	110°27′E	3997	OAE 1d	16.32	na
123	766A	21	1	2226	191.22	19°55′S	110°27′E	3997	OAE 1b	16.14	na
123	766A	27	1	40–44	249.30	19°55′S	110°27′E	3997	OAE 1a	17.23	8.5
170	1049C	12	1	42–44	139.72	30°08′N	76°06′W	2670	OAE 1b	16.33	na
170	1049C	12	2	127–129	142.07	30°08′N	76°06′W	2670	OAE 1b	16.29	5.1
170	1049C	12	3	100–102	143.30	30°08′N	76°06′W	2670	OAE 1b	16.26	na

Note: C/S—organic carbon to sulfur ratio; OAE—ocean anoxic event.

site investigated (DSDP Site 765) barite with S isotopic composition significantly higher than expected for the appropriate age was found; this sample (765C-30-CC at 635.3 m depth, age ca. 108 Ma) was not high in organic matter and is not from an OAE interval. It is possible that similar diagenetic barite is present in association with some sapropel and black shale sections. Although such barite deposits cannot be used to infer past levels of productivity, it must be kept in mind that the ultimate source of the Ba in these deposits is marine barite (Torres et al., 1996a).

In addition to the S isotope signature, diagenetic barite crystals are typically large (20–700 μm), flat, tabular-shaped crystals and tend to appear as barite beds in the sedimentary column (Torres et al., 1996a, 1996b; Paytan et al., 2002). Figure 3A and B are scanning electron micrographs of the barite crystals separated from Cretaceous black shales and from sapropel deposits, respectively, and Figure 2C is barite from Site 765C-30-CC (diagenetic). The barite crystals observed in the OAE and sapropel sections are all elliptical crystals or aggregates, ranging in size from 0.5 to 5 μm, and differ from the diagenetic barite shown in Figure 3C. These crystals are similar in shape and size to barite observed in the water column and extracted from sediment trap samples (Dehairs et al., 1980; Bishop, 1988; Paytan et al., 2002) and support a water column origin. Similar barite morphology has been observed in all of the core samples presented here. The identification of barite that forms in the water column in association with organic matter export in these sections reaffirms the "increased productivity" model as the main cause for the high organic matter accumulation in these time intervals at the studied sites.

The occurrence and preservation of marine barite in these sediments suggest that the water column and pore waters were not significantly depleted of sulfate (e.g., sulfate reduction rates were not high relative to sulfate diffusion into the sediment from the overlying seawater). Solubility calculations (Rushdi et al., 2000) indicate that for typical pore water in the open ocean with Ba concentrations of ~200 nM (Paytan and Kastner, 1996) at 2 °C and 3500 m depth, when sulfate concentrations are lower than 22 mM, undersaturation conditions exist, and barite will dissolve. On the other hand, when Ba-rich pore fluids with concentrations of over 2000 nM, as seen in some sediments where barite is remobilized (Brumsack, 1986; Torres et al., 1996a), diffuse and reach pore waters with sulfate concentrations as low as 3 mM, super saturation is achieved and barite may precipitate (Rushdi et al., 2000). Such low sulfate concentrations at the site of diagenetic barite precipitation (the "barite front") are likely to have high $\delta^{34}S$ values (Brumsack, 1986) (see Fig. 1). The abundance of marine and not diagenetic barite in association with the

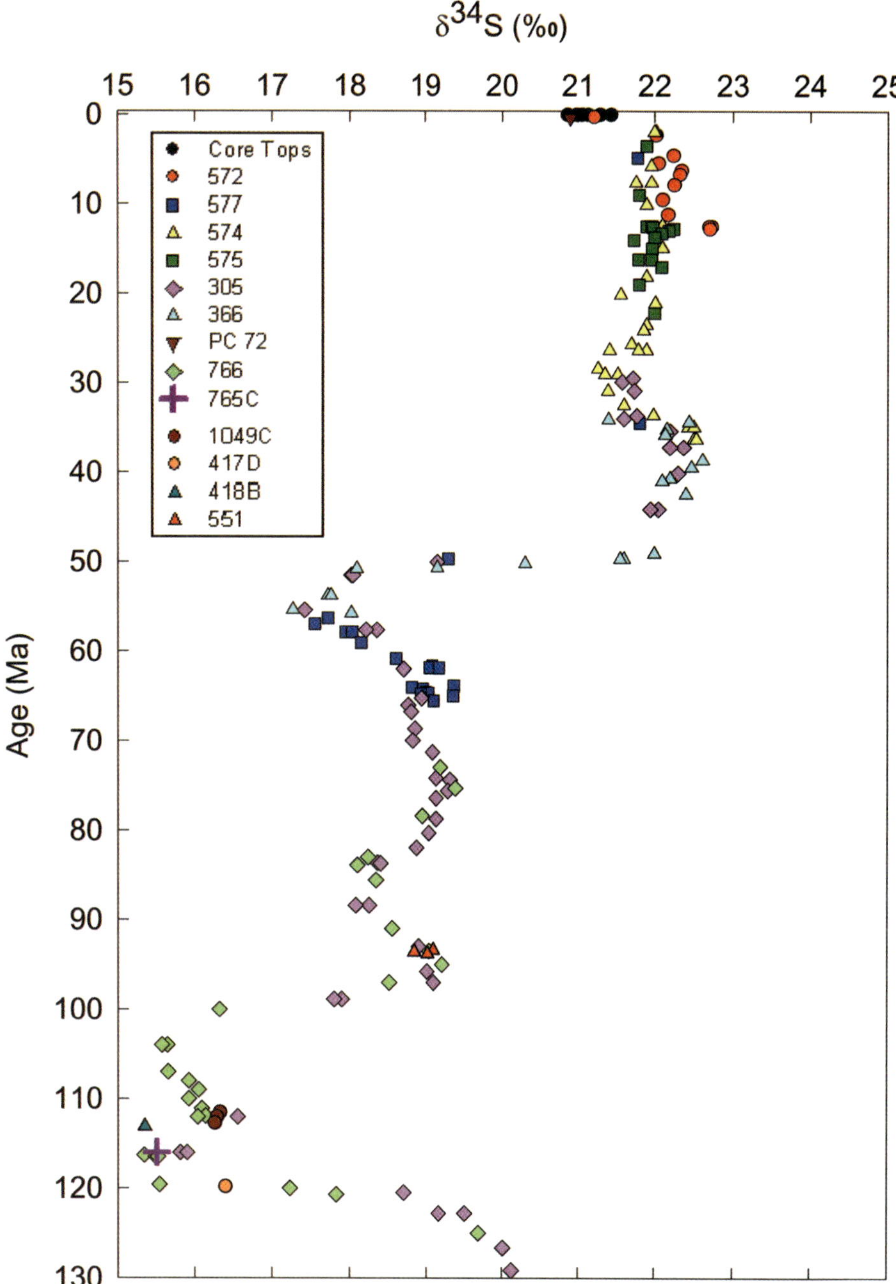

Figure 2. Seawater sulfate S isotope curve for the past 130 million years. From Paytan et al. (1998, 2004)

sapropel and OAE deposits investigated here suggests that sufficient sulfate was present in the pore fluids of these sediments. These results are in agreement with Passier et al. (1996, 1999), who used different lines of evidence (pyrite distribution and morphology, $\delta^{34}S$ of pyrite) to show that sapropel sulfate reduction rates were not very high and took place in an open system, where sulfate supply via diffusion or advection was large relative to sulfate reduction rates, implying that the pore water was not significantly depleted of sulfate during and after pyrite deposition.

It is evident, however, that at least some sulfate reduction has taken place in some of these sediments, as indicated by the presence of pyrite (e.g., Passier et al., 1999, and references therein). Even sediments lacking high pyrite concentrations most likely have experienced sulfate reduction, but pyrite precipitation may have been suppressed due to iron limitation, in particular in the carbonate rich sediments of sites 305 and 766 (Canfield, 1989, 1994; Canfield et al., 1992, 1996). Sedimentary settings conducive to sulfate reduction coupled with barite preservation

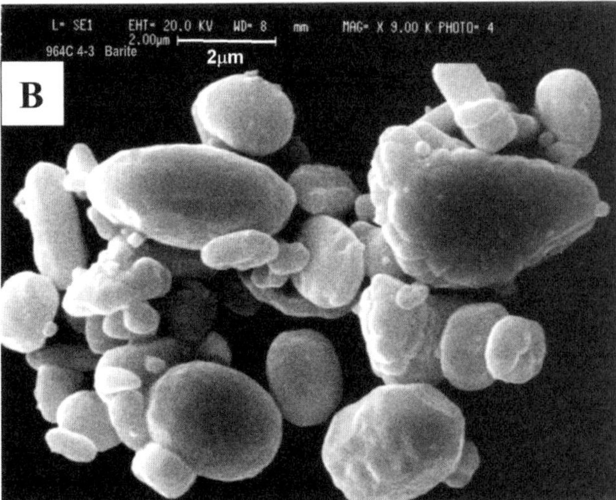

Figure 3. Scanning electron microscope micrographs. A. Primary, unaltered barite microcrystals from Leg 80 site 551 section CC, 0–7 cm. B. Primary, unaltered barite microcrystals from Leg 160 site 964B section 4–3 31–37 cm. C. Diagenetically formed barite crystals from section 765C-30-CC at 635.3 m, 108 Ma.

are those where high rates of downward diffusion of sulfate from deep water into pore water prevents barite dissolution, while high organic content consumes other oxidants, promoting sulfate reduction. Such sedimentary conditions, where high accumulation and preservation of both barite and pyrite are found, are not prevalent in the present day ocean. At present in settings where pyrite is abundant (e.g., continental margins, anoxic and suboxic basins, etc.), marine barite microcrystals are not preserved beyond the upper few centimeters of the sediments (Paytan, personal observation). This is due to the dependence of pyrite formation on high organic matter burial, which, at present, is more widespread in continental margins, and to the low barite preservation at these settings (McManus et al., 1998). The co-occurrence of these two phases simultaneously indicates that continental margins (at least as represented today) are not good analogues for the deposition environment of sapropels and black shales investigated here and implies that a significant fraction of organic carbon burial during these events occurred in open ocean settings.

As an additional test to determine the oceanic/sedimentary setting where high organic matter, barite, and pyrite burial will persist (as during sapropels and OAE), we have determined the organic C to total S ratio in some of our samples (Table 2). In the modern ocean, carbon and sulfur burial rates are coupled through burial of organic C and pyrite in marine environments (Holland, 1973; Berner, 1987; Kump and Garrels, 1986; Canfield et al., 2000). Pyrite forms in sediments by the reduction of seawater sulfate at the expense of sedimentary organic carbon, and this is a strictly anaerobic process. Sedimentary sulfide formation is more abundant in shelf, deltaic, estuarine, and hemipelagic muds than in the deep ocean (Berner, 1982). Berner (1982) noted that sediments accumulating in shelf and deltaic environments tend to have a remarkably constant organic C to pyrite ratios (Corg/Spy) (~7.5 molar ratio; 2.8 wt ratio). Analyses of shales indicate that the ratio of 2.8 was maintained throughout the Phanerozoic (Raiswell and Berner, 1986). The relationship between organic carbon and sulfide burial can change when the locus of carbon burial shifts away from normal shelf-deltaic environments (Berner and Raiswell, 1983). Several environments inhibit the burial of pyrite (Berner, 1984; Lyons and Berner, 1992; Calvert et al., 1996; Canfield et al., 1996; Wilkin et al., 1997; Raiswell and Canfield, 1998). Among these are high-productivity open-ocean regions and shallow water calcareous sediments, where pyrite formation may be limited by the availability of dissolved iron, and terrestrial environments (soils, swamps, coal basins), where sulfate is in limited supply. In contrast, pyrite burial rates are high in euxinic environments, where sulfides may form in the water column (Raiswell and Berner, 1985). The Corg/Spy burial, therefore, could be used to establish the burial conditions (e.g., normal marine, euxinic, and freshwater burial environments). As seen in Table 2, the organic C to total S ratio varies significantly among our core samples but is always higher than a 2.8 weight ratio (this is despite the use of total, not pyrite, S). The high Corg/S burial ratio in most of our samples suggests a net shift of organic carbon sedimentation to settings where the rates of bacterial sulfate reduction and pyrite burial were low enough for downward diffusion of sulfate to prevent significant pore-water sulfate depletion. This could be accomplished in highly productive pelagic settings where the overall sedimentation rates are relatively low (compared to continental margin settings), while organic C and barite formation and burial are high. Similar C/S ratios have previously been recorded in some organic rich sapropels (Passier and de Lange, 1998) and have been interpreted as reflecting the limitation of pyrite formation by the availability of reactive iron oxides.

Results obtained here, therefore, suggest that the barite associated with sapropel and Cretaceous OAE deposits (at least at the sites investigated here) did not formed diagenetically within sulfate-poor sediments but rather is of seawater origin. These results confirm the marine origin of the barite deposits and thus the association of sapropel and OAE with periods of high biological productivity. The above conclusion that high organic carbon deposition events resulted from increased productivity and were not necessarily associated with changes in oceanic circulation (e.g., stagnation) has important implications to the understanding and modeling of the global sulfur, carbon, and oxygen cycles. In particular, during the mid-Cretaceous when such events were relatively frequent, widespread, and persisted for a long time (in the order of a million years), changes in organic matter burial may have exerted feedbacks in the global carbon cycle (e.g., lower atmospheric CO_2 and reduced greenhouse effect). Increased C burial without ocean stagnation has two implications: (1) the supply of nutrients to the surface ocean through upwelling is maintained, supporting high productivity, and (2) re-supply of deep water oxygen at a fast enough rate that it would be available for effective oxidation of sulfides. This would result in an overall relatively lower burial of pyrite and other reduced S minerals and thus may have acted as negative feedback in the coupled C-S-O cycle (e.g., maintaining a general negative correlation between the burial of reduced C and S acts to sustain atmospheric oxygen at a relatively constant level). Indeed, the mid-Cretaceous $\delta^{34}S$ values are considerably lower than Cenozoic values (~16‰ and 22‰, respectively; Fig. 2) and a decrease in $\delta^{34}S$ in the last million years of Earth's history is also observed (Paytan et al., 1998), suggesting lower pyrite burial.

ACKNOWLEDGMENTS

This work was supported by the National Science Foundation grant OCE 0095754 to AP. We thank the ODP core repositories for the sediment samples.

REFERENCES CITED

Aksu, A.E., Yasar, D., and Mudie, P.J., 1995, Paleoclimatic and paleoceanographic conditions leading to development of sapropel layer S1 in the Aegean Sea: Palaeogeography, Palaeoclimatology, Palaeoecology, v. 116, p. 71–101, doi: 10.1016/0031-0182(94)00092-M.

Anderson, L.D., Delaney, M.L., and Faul, K.L., 2001, Carbon to phosphorus ratios in sediments: Implications for nutrient cycling: Global Biogeochemical Cycles, v. 15, p. 65–79, doi: 10.1029/2000GB001270.

Aquilina, L., Dia, A.N., Boulegue, J., Bourgois, J., and Fouiliac, A.M., 1997, Massive barite deposits in the convergent margin off Peru: Implications for fluid circulation within subduction zones: Geochimica et Cosmochimica Acta, v. 61, p. 1233–1245.

Arthur, M.A., Dean, W.E., and Pratt, L.M., 1988, Geochemical and climatic effects of increased marine organic-carbon burial at the Cenomanian Turonian boundary: Nature, v. 335, p. 714–717, doi: 10.1038/335714A0.

Arthur, M.A., Jenkyns, H.C., Brumsack, H., and Schlanger, S.O., 1990, Stratigraphy, geochemistry, and paleoceanography of organic-rich mid-Cretaceous sequences, *in* Ginsburg, R.N., and Beaudoin, B., eds., Cretaceous resources, events, and rhythms: Dordrecht, Kluwer Academic, NATO ASI Series C, 304, p. 75–119.

Bains, S., Norris, R.D., Corfield, R.M., and Faul, K.L., 2000, Termination of global warmth at the Palaeocene/Eocene boundary through productivity feedback: Nature, v. 407, p. 171–174, doi: 10.1038/35025035.

Barron, E.J., Fawcett, P.J., Peterson, W.H., Pollard, D., and Thompson, S.L., 1995, A simulation of Mid-Cetaceous climate: Paleocenography, v. 10, p. 953–962, doi: 10.1029/95PA01624.

Berger, W.H., Fischer, K., Lai, C., and Wu, G., 1988, Ocean carbon flux: Global maps of primary production and export production, *in* Agegian, C.R., ed., Biogeochemical cycling and fluxes between the deep euphotic zone and other oceanic realms: NOAA National Undersea Research Program, Research Report 88–1, p. 131–176.

Berger, W.H., Smetacek, V.S., and Wefer, G., 1989, Productivity of the ocean present and past: New York, John Wiley & Sons, 471 p.

Berner, R.A., 1982, Burial of organic carbon and pyrite sulfur in the modern ocean: Its geochemical and environmental significance: American Journal of Science, v. 282, p. 451–473.

Berner, R.A., 1984, Sedimentary pyrite formation: an update: Geochimica et Cosmochimica Acta, v. 48, p. 605–615, doi: 10.1016/0016-7037(84)90089-9.

Berner, R.A., 1987, Models for carbon and sulfur cycles and atmospheric oxygen: Application to Paleozoic geologic history: American Journal of Science, v. 287, p. 177–196.

Berner, R.A., and Raiswell, R., 1983, Burial of organic carbon and pyrite sulfur in sediments over Phanerozoic time: A new theory: Geochimica et Cosmochimica Acta, v. 47, p. 855–862, doi: 10.1016/0016-7037(83)90151-5.

Bishop, J.K.B., 1988, The barite-opal-organic carbon association in oceanic particulate matter: Nature, v. 332, p. 341–343, doi: 10.1038/332341A0.

Bolze, C.E., Malone, P.G., and Smith, M.J., 1973, Microbial mobilization of barite: Chemical Geology, v. 13, p. 341–343.

Bonn, W.J., and Gingele, F.X., 1998, Palaeoproductivity at the Antarctic continental margin: Opal and barium records for the last 400 ka: Palaeogeography, Palaeoclimatology, Palaeoecology, v. 139, p. 195–211, doi: 10.1016/S0031-0182(97)00144-2.

Bralower, T.J., and Thierstein, R.H., 1987, Organic carbon and metal accumulation rates in Holocene and Mid-Cretaceous sediments: Paleoceanographic significance, *in* Brooks, J., and Fleet, A.J., eds., Marine petroleum source rocks: London, Geological Society Special Publication 26, p. 345–369.

Bréhéret, J.-G., and Brumsack H.-J. 2000, Barite concretions as evidences of pauses in sedimentation in the Marnes Bleues Formation of the Vocontian Basin (SE France): Sedimentary Geology, v. 130, p. 205–228, doi: 10.1016/S0037-0738(99)00112-8.

Brumsack, H., and Gieskes, J.M., 1983, Interstitial water trace metal chemistry of laminated sediments from the Gulf of California, Mexico: Marine Chemistry, v. 14, p. 89–106, doi: 10.1016/0304-4203(83)90072-5.

Brumsack, H.-J., 1986, The inorganic geochemistry of Cretaceous black shales (DSDP Leg 41) in comparison to modern upwelling sediments from the Gulf of California, *in* Shackelton, N.J., Summerhayes, C.P., eds., North Atlantic paleoceanography: London, Geological Society Special Publication 21, p. 447–462.

Calvert, S.E., 1983, Geochemistry of Pleistocene sapropels and associated sediments from the Eastern Mediterranean: Oceanologica Acta, v. 6, p. 255–267.

Calvert, S.E., and Fontugne, M.R., 2001, On the late Pleistocene-Holocene sapropel record of climatic and oceanographic variability in the eastern Mediterranean: Paleoceanography, v. 16, no. 1, p. 78–94, doi: 10.1029/1999PA000488.

Calvert, S.E., and Pedersen, T.F., 1992, Organic carbon accumulation and preservation in marine sediments: How important is anoxia? *in* Whelan, J.K., Farrington, J.W., eds., Productivity, accumulation and preservation of organic matter in recent and ancient sediments: New York, Columbia University Press, p. 231–263.

Calvert, S.E., and Pedersen, T.F., 1993, Geochemistry of Recent oxic and anoxic marine sediments: Implications for the geological record: Marine Geology, v. 113, p. 67–88, doi: 10.1016/0025-3227(93)90150-T.

Calvert, S.E., Thode, H.G., Yeung, D., and Karlin, R.E., 1996, A stable isotope study of pyrite formation in the Late Pleistocene and Holocene sediments of the Black Sea: Geochimica et Cosmochimica Acta, v. 60, p. 1261–1270, doi: 10.1016/0016-7037(96)00020-8.

Canfield, D.E., 1989, Reactive iron in marine sediments: Geochimica et Cosmochimica Acta, v. 53, p. 619–632, doi: 10.1016/0016-7037(89)90005-7.

Canfield, D.E., 1994, Factors influencing organic carbon preservation in marine sediments: Chemical Geology, v. 114, p. 315–329, doi: 10.1016/0009-2541(94)90061-2.

Canfield, D.E., Raiswell, R., and Bottrell, S., 1992, The reactivity of sedimentary iron minerals toward sulfide: American Journal of Science, v. 292, p. 818–834.

Canfield, D.E., Lyons, T.W., and Raiswell, R., 1996, A model for iron deposition to euxinic Black Sea sediments: American Journal of Science, v. 292, p. 818–834.

Canfield, D.E., Habicht, K.S., and Thamdrup, B., 2000, The Archean sulfur cycle and the early history of atmospheric oxygen: Science, v. 288, p. 658–661, doi: 10.1126/SCIENCE.288.5466.658.

Cecile, M.P., Shakur, M.A., and Krouse, H.R., 1983, The isotopic composition of western Canadian barites and the possible derivation of oceanic sulfate $\delta^{34}S$ and $\delta^{18}O$ age curves: Canadian Journal of Earth Sciences, v. 20, p. 1528–1535.

Claypool, G.E., Holser, W.T., Kaplan, I.R., Sakai, H., and Zak, I., 1980, The age curve of sulfur and oxygen isotopes in marine sulfate and their mutual interpretation: Chemical Geology, v. 28, p. 199–260, doi: 10.1016/0009-2541(80)90047-9.

Collier, R., and Edmond, J.M., 1984, The trace elements geochemistry of marine biogenic particulate matter: Progress in Oceanography, v. 13, p. 113–199, doi: 10.1016/0079-6611(84)90008-9.

Dean, W.E., and Schreiber, B.C., 1977, Authigenic barite, *in* Lancelot, Y., Seibold, E., et al., eds., Proceedings of the Deep Sea Drilling Project, Initial reports: Washington (U.S. Government Printing Office), v. 41, p. 915–931.

Dean, W.E., Gardner, J.V., and Piper, D.A., 1997, Inorganic geochemical indicators of glacial-interglacial changes in productivity and anoxia on the California continental margin: Geochimica et Cosmochimica Acta, v. 61, p. 4507–4518, doi: 10.1016/S0016-7037(97)00237-8.

Dehairs, F., Chesselet, R., and Jedwab, J., 1980, Discrete suspended particles of barite and the barium cycle in the open ocean: Earth and Planetary Science Letters, v. 49, p. 528–550, doi: 10.1016/0012-821X(80)90094-1.

Dymond, J., Suess, E., and Lyle, M., 1992, Barium in deep sea sediments: A geochemical proxy for paleoproductivity: Paleoceanography, v. 7, p. 163–181.

Eagle, M., Paytan, A., Arrigo, K.R., van Dijken, G., and Murray, R.W., 2003, A comparison between excess barium and barite as indicators of carbon export: Paleoceanography, v. 18, p. 21.1–21.13, doi: 10.1029/2002PA000793.

Erbacher, R.D., Huber, B.T., Norris, R.D., and Markey, M., 2001, Increased thermohaline stratification as a possible cause for a Cretaceous anoxic event: Nature, v. 409, p. 325–327, doi: 10.1038/35053041.

Filippelli, G.M., Sierro, F.J., Flores, J.A., Vazquez, A., Utrilla, R., Perez-Folgado, M., and Latimer, J.C., 2003, A sediment-nutrient-oxygen feedback responsible for productivity variations in Late Miocene sapropel sequences of the western Mediterranean: Palaeogeography, Palaeoclimatology, Palaeoecology, v. 190, p. 335–348, doi: 10.1016/S0031-0182(02)00613-2.

Freeman, K.H., and Hayes, J.M., 1992, Fractionation of carbon isotopes by phyto-plankton and estimates of ancient CO_2 levels: Global Biogeochemical Cycles, v. 6, p. 185–198.

Ganeshram, R.S., Francois, R., Commeau, J., and Brown-Leger, S.L., 2003, An experimental investigation of barite formation in seawater: Geochimica et Cosmochimica Acta, v. 67, p. 2599–2605, doi: 10.1016/S0016-7037(03)00164-9.

Gingele, F., and Dahmke, A., 1994, Discrete barite particles and barium as a tracer of paleoproductivity in south Atlantic sediments: Paleoceanography, v. 9, p. 151–168, doi: 10.1029/93PA02559.

Habicht, K.S., and Canfield, D.E., 1997, Sulfur isotope fractionation during bacterial sulfate reduction in organic rich sediments: Geochimica et Cosmochimica Acta, v. 61, p. 5351–5361, doi: 10.1016/S0016-7037(97)00311-6.

Harrison, A.G., and Thode, H.G., 1958, Mechanism of bacterial reduction of sulfate from isotope fractionation studies: Faraday Society Transactions, v. 54, p. 84–92.

Hartmann, M., and Nielsen, H., 1969, $\delta^{34}S$-Werte in rezenten Meeressedimenten und ihre Deutung am Beispiel einiger Sedimentprofile aus der westlichen Ostsee: Geologische Rundschau, v. 58, p. 621–655.

Holland, H.D., 1973, Systematics of the isotope composition of sulfur in the oceans during the Phanerozoic and its implications for atmospheric oxygen: Geochimica et Cosmochimica Acta, v. 37, p. 2605–2616, doi: 10.1016/0016-7037(73)90268-8.

Jørgensen, B.B., 1979, A theoretical model of the stable sulfur isotope distribution in marine sediments: Geochimica et Cosmochimica Acta, v. 43, p. 363–374, doi: 10.1016/0016-7037(79)90201-1.

Kidd, R.B., Cita, M.B., and Ryan, W.B.F., 1978, Stratigraphy of eastern Mediterranean sapropel sequences recovered during DSDP Leg 142 and their paleoenvironmental significance: Initial Reports of the Deep Sea Drilling Project, v. 42, no. 1, p. 421–443.

Klump, J., Hebbeln, D., and Wefer, G., 2000, The impact of sediment provenance on barium-based productivity estimates: Marine Geology, v. 169, p. 259–271, doi: 10.1016/S0025-3227(00)00092-X.

Kump, L.R., and Garrels, R.M., 1986, Modeling atmospheric O_2 in the global sedimentary redox cycle: American Journal of Science, v. 286, p. 337–360.

Kuypers, M.M.M., Pancost, R.D., and Sinninghe Damsté, J.S., 1999, A large and abrupt fall in atmospheric CO_2 concentration during Cretaceous times: Nature, v. 399, p. 342–345, doi: 10.1038/20659.

Larson, R.L., and Erba, E., 1999, Onset of the mid-Cretaceous greenhouse in the Barremian-Aptian: Igneous events and the biological, sedimentary, and geological responses: Paleoceanography, v. 14, p. 663–678, doi: 10.1029/1999PA900040.

Lyons, T.W., and Berner, R.A., 1992, Carbon-sulfur-iron systematics of uppermost deep-water sediments of the Black Sea: Chemical Geology, v. 99, p. 1–27, doi: 10.1016/0009-2541(92)90028-4.

Martinez-Ruiz, F., Kastner, M., Paytan, A., Ortega-Huertas, M., and Bernasconi, S.M., 2000, Geochemical evidence for enhanced productivity during S1 sapropel deposition in the eastern Mediterranean: Paleoceanography, v. 15, p. 200–209, doi: 10.1029/1999PA000419.

McManus, J., Berelson, W.M., Klinkhammer, G.P., Johnson, K.S., Coale, K.H., Anderson, R.F., Kumar, N., Burdige, D.J., Hammond, D.E., Brumsack, H.J., McCorkle, D.C., and Rushdi, A., 1998, Geochemistry of barium in marine sediments: Implications for its use as a paleoproxy: Geochimica et Cosmochimica Acta, v. 62, p. 3453–3473, doi: 10.1016/S0016-7037(98)00248-8.

Murray, R.W., Knowlton, C., Leinen, M., Mix, A.C., and Polsky, C.H., 2000, Export production and carbonate dissolution in the central equatorial Pacific Ocean over the past 1 Ma: Paleoceanography, v. 15, p. 570–592, doi: 10.1029/1999PA000457.

Naehr, T.H., Stakes, D.S., and Moore, W.S., 2000, Mass wasting, ephemeral fluid flow, and barite deposition on the California continental margin: Geology, v. 28, p. 315–318.

Nielsen, H., 1978, Sulfur isotopes in nature, *in* Wedepohl, K.H., ed., Handbook of geochemistry, Part 16B: Heidelberg, Springer-Verlag, p. 1–40.

Nijenhuis, I.A., Brumsack, H.J., and De Lange., G.J., 1998, The trace element budget of the eastern Mediterranean during Pliocene sapropel formation, *in* Robertson, et al., eds., Proceedings ODP, Leg 160, Scientific Results, Mediterranean I: College Station, Texas, Ocean Drilling Program, p. 199–206.

Nijenhuis, I.A., Bosch, H.J., Sinnighe Damsté, J.S., Brumsack, H.-J., and de Lange, G.J., 1999, Organic matter and trace element rich sapropels and black shales: a geochemical comparison: Earth and Planetary Science Letters, v. 169, p. 277–290.

Nolet, G.J., and Corliss, B.H., 1990, Benthic foraminiferal evidence for reduced deep-water circulation during sapropel deposition in the eastern Mediterranean: Marine Geology, v. 94, p. 109–130, doi: 10.1016/0025-3227(90)90106-T.

Nürnberg, C.C., Bohrmann, G., and Schlüter, M., 1997, Barium accumulation in the Atlantic sector of the Southern Ocean: Results from 190,000-year record: Paleoceanography, v. 12, p. 594–603, doi: 10.1029/97PA01130.

Parrish, J.T., and Curtis, R.L., 1982, Atmospheric circulation, upwelling, and organic-rich rocks in the Mesozoic and Cenozoic eras: Palaeogeography, Palaeoclimatology, Palaeoecology, v. 40, p. 31–66, doi: 10.1016/0031-0182(82)90085-2.

Passier, H.F., and de Lange, G.J., 1998, Sedimentary sulfur and iron chemistry in relation to the formation of Eastern Mediterranean sapropels, *in* Robertson, A.H.F., et al., eds., Proceedings ODP, Leg 160, Scientific Results, Mediterranean I: College Station, Texas, Ocean Drilling Program, p. 249–259.

Passier, H.F., Middelburg, J.J., van Os, B.J.H., and de Lange, G.J., 1996, Diagenetic pyritisation under eastern Mediterranean sapropels caused by downward sulfide diffusion: Geochimica et Cosmochimica Acta, v. 60, p. 751–763, doi: 10.1016/0016-7037(95)00419-X.

Passier, H.F., Middelburg, J.J., de Lange, G.J., and Böttcher, M.E., 1999, Modes of sapropel formation in the eastern Mediterranean: Some constraints based on pyrite properties: Marine Geology, v. 153, p. 199–219, doi: 10.1016/S0025-3227(98)00081-4.

Paytan, A., 2002, Tales of black shales., *in* White, K., and Urquhart, E., eds., ODP Greatest Hits, Volume 2, available from the World Wide Web http://www.joiscience.org/greatesthits2/pdfs/Paytanshales.pdf.

Paytan, A., and Kastner, M., 1996, Benthic Ba fluxes in the central Equatorial Pacific, implications for the oceanic Ba cycle: Earth and Planetary Science Letters, v. 142, p. 439–450, doi: 10.1016/0012-821X(96)00120-3.

Paytan, A., Kastner, M., and Chavez, F.P., 1996, Glacial to interglacial fluctuations in productivity in the Equatorial Pacific as indicated by marine barite: Science, v. 274, p. 1355–1357, doi: 10.1126/SCIENCE.274.5291.1355.

Paytan, A., Kastner, M., Campbell, D., and Thiemens, M.H., 1998, Sulfur isotope composition of Cenozoic seawater sulfate: Science, v. 282, p. 1459–1462, doi: 10.1126/SCIENCE.282.5393.1459.

Paytan, A., Mearon, S.A., Cobb, K., and Kastner, M., 2002, Origin of marine barite deposits: Sr and S isotope characterization: Geology, v. 30, p. 747–750, doi: 10.1130/0091-7613(2002)0302.0.CO;2.

Paytan, A., Kastner, M., Campbell, D., and Thiemens, M.H., 2004, Seawater sulfur isotope fluctuations in the Cretaceous: Science (in press).

Pedersen, T.F., and Calvert, S.E., 1990, Anoxia vs. productivity: What controls the formation of organic-carbon-rich sediments and sedimentary rocks?: American Association of Petroleum Geologists Bulletin, v. 74, p. 454–466.

Pruysers P.A., de Lange, G.J., and Middelburg, J.J., 1991, Geochemistry of eastern Mediterranean sediments: Primary sediment composition and diagenetic alterations: Marine Geology, v. 100, p. 137–154.

Raiswell, R., and Berner, R.A., 1985, Pyrite formation in euxinic and semi-euxinic sediments: American Journal of Science, v. 285, p. 710–724.

Raiswell, R., and Berner, R.A., 1986, Pyrite and organic matter in Phanerozoic normal marine shale: Geochimica et Cosmochimica Acta, v. 50, p. 1967–1976, doi: 10.1016/0016-7037(86)90252-8.

Raiswell, R., and Canfield, D.E., 1998, Sources of iron for pyrite formation in marine sediments: American Journal of Science, v. 298, p. 219–245.

Rossignol-Strick, M., 1985, Mediterranean Quaternary sapropels: An immediate response of the African monsoon to variation of insolation: Palaeogeography, Palaeoclimatology, Palaeoecology, v. 49, p. 237–265, doi: 10.1016/0031-0182(85)90056-2.

Rushdi, A., McManus, J., and Collier, R., 2000, Marine barite and celestite saturation in seawater: Marine Chemistry, v. 69, p. 19–31, doi: 10.1016/S0304-4203(99)00089-4.

Rutsch, H.J., Mangini, A., Bonai, G., Dittrich-Hannen, B., Kubik, P.W., Suter, M., and Segl, M., 1995, ^{10}Be and Ba concentrations in West African sediments trace productivity in the past: Earth and Planetary Science Letters, v. 133, p. 129–143, doi: 10.1016/0012-821X(95)00069-O.

Sarmiento, J.L., Herbert, T.D., and Toggweiler, J.R., 1988, Causes of anoxia in the world ocean: Global Biogeochemical Cycles, v. 2, p. 115–128.

Schenau, S.J., Prins, M.A., De Lange, G.J., and Monnin, C., 2001, Barium accumulation in the Arabian Sea: Controls on barite preservation in marine sediments: Geochimica et Cosmochimica Acta, v. 65, p. 1545–1556, doi: 10.1016/S0016-7037(01)00547-6.

Schlanger, S.O., and Jenkyns, H.C., 1976, Cretaceous anoxic events: Cause and consequences: Geologie en Mijnbouw, v. 55, p. 179–184.

Schmidt, G.A., and Mysak, L.A., 1996, Can increased pole-ward oceanic heat flux explain the warm Cretaceous climate?: Paleoceanography, v. 11, p. 579–593, doi: 10.1029/96PA01851.

Schmitz, B., 1987, Barium, equatorial high productivity, and the wandering of the Indian continent: Paleoceanography, v. 2, p. 63–77.

Schroeder, J.O., Murray, R.W., Leinen, M., Pflaum, R.C., and Janecek, T.R., 1997, Barium in equatorial Pacific carbonate sediment: Terrigenous, oxide, and biogenic associations: Paleoceanography, v. 12, p. 125–146, doi: 10.1029/96PA02736.

Slomp, C.P., Thomson, J., and de Lange, G.J., 2002, Enhanced regeneration of phosphorus during formation of the most recent eastern Mediterranean sapropel (S1): Geochimica et Cosmochimica Acta, v. 66, p. 1171–1184, doi: 10.1016/S0016-7037(01)00848-1.

Stanley, D.J., 1978, Ionian Sea sapropel distribution and late Quaternary palaeoceanography in the eastern Mediterranean: Nature, v. 274, p. 149–151.

Thompson, E.I., and Schmitz, B., 1997, Barium and late Paleocene δ^{13}C maximum: Evidence of increased marine surface productivity: Paleoceanography, v. 12, p. 239–254, doi: 10.1029/96PA03331.

Thomson, J., Higgs, N.C., Wilson, T.R.S., Croudace, I.W., De Lange, G.J., and Van Santvoort, P.J.M., 1995, Redistribution and geochemical behaviour of redox-sensitive elements around S1, the most recent eastern Mediterranean sapropel: Geochimica et Cosmochimica Acta, v. 59, p. 3487–3501, doi: 10.1016/0016-7037(95)00232-O.

Thomson, J., Mercone, D., De Lange, G.J., and Van Santvoort, P.J.M., 1999, Review of recent advances in the interpretation of eastern Mediterranean sapropel S1 from geochemical evidence: Marine Geology, v. 153, p. 77–89, doi: 10.1016/S0025-3227(98)00089-9.

Torres, M.E., Bohrmann, G., and Suess, E., 1996a, Authigenic barites and fluxes of barium associated with fluid seeps in the Peru subduction zone: Earth and Planetary Science Letters, v. 144, p. 469–481, doi: 10.1016/S0012-821X(96)00163-X.

Torres, M.E., Brumsack, H.J., Bohrmann, G., and Emeis, K.C., 1996b, Barite fronts in continental sediments: A new look at barium remobilization in the zone of sulfate reduction and formation of heavy barites in authigenic fronts: Chemical Geology, v. 127, p. 125–139, doi: 10.1016/0009-2541(95)00090-9.

Van Cappellen, P., and Ingall, E.D., 1996, Redox stabilization of atmosphere and oceans by phosphorus-limited marine productivity: Science, v. 271, p. 493–496.

van Os, B.J.H., Middelburg, J.J., and deLange, G.J., 1991, Possible diagenetic mobilization of barium in sapropelic sediments from the eastern Mediterranean: Marine Geology, v. 100, p. 125–136, doi: 10.1016/0025-3227(91)90229-W.

van Os, B.J.H., Lourens, L.J., Hilgen, F.J., de Lange, G.J., and Beauford, L., 1994, The formation of Pliocene sapropels and carbonate cycles in the Mediterranean: diagenesis, dilution, and productivity: Paleoceanography, v. 9, p. 601–617, doi: 10.1029/94PA00597.

Van Santvoort, P.J.M., De Lange, G.J., Thomson, J., Cussen, H., Wilson, T.R.S., Krom, M.D., and Strohle, K., 1996, Active post-depositional oxidation of the most recent sapropel (S1) in sediments of the eastern Mediterranean Sea: Geochimica et Cosmochimica Acta, v. 60, p. 4007–4024, doi: 10.1016/S0016-7037(96)00253-0.

von Breymann, M.T., Brumsack, H.-J., and Emeis, K.C., 1992, Depositional and diagenetic behavior of barium in the Japan Sea, *in* Tamaki K., et al., eds., Proceedings of the Ocean Drilling Program Scientific Results, Legs 127 and 128: College Station, Texas, Ocean Drilling Program, v 127/128, Pt. 1, p. 651–665.

Warning, B., and Brumsack, H.J., 2000, Trace metal signatures of eastern Mediterranean sapropels: Palaeogeography, Palaeoclimatology, Palaeoecology, v. 158, p. 293–309, doi: 10.1016/S0031-0182(00)00055-9.

Wehausen, R., and Brumsack, H.J., 1998, The formation of Pliocene Mediterranean sapropels: constraints from high-resolution major and minor elements studies, *in* Robertson, A.A.F., et al., eds., Proceedings ODP, Leg 160, Scientific Results, Mediterranean I: College Station, Texas, Ocean Drilling Program, p. 207–218.

Weissert, H., 1989, C-isotope stratigraphy, a monitor of paleoenvironmental change: A case study from the early Cretaceous: Surveys in Geophysics, v. 10, p. 1–61.

Weissert, H., McKenzie, J.A., and Channell, J.E.T., 1985, Natural variations in the carbon cycle during the Early Cretaceous, *in* Sundquist, E. T. and Broecker, W. S., eds., The Carbon Cycle and Atmospheric CO_2: Natural Variations Archean to Present: Washington D.C., American Geophysical Union Geophysical Monograph 32, p. 531–545.

Weldeab, S., Emeis, K.C., Hemleben, C., Schmiedl, G., and Schulz, H., 2003, Spatial productivity variations during sapropel S5 and S6 in the Mediterranean Sea: Evidence from Ba contents: Palaeogeography, Palaeoclimatology, Palaeoecology, v. 191, p. 169–190, doi: 10.1016/S0031-0182(02)00711-3.

Wilkin, R.T., Atrur, M.A., and Dean, W.E., 1997, History of water-column anoxia in the Black Sea indicated by pyrite framboid size distributions: Earth and Planetary Science Letters, v. 148, p. 517–525, doi: 10.1016/S0012-821X(97)00053-8.

Manuscript Accepted by the Society March 28, 2004

Printed in the USA

Geological Society of America
Special Paper 379
2004

Sites of anomalous organic remineralization in the carbonate sediments of South Florida, USA: The sulfur cycle and carbonate-associated sulfate

Timothy W. Lyons
Department of Geological Sciences, University of Missouri, Columbia, Missouri 65211, USA

Lynn M. Walter
Department of Geological Sciences, University of Michigan, Ann Arbor, Michigan 48109, USA

Anne M. Gellatly
Department of Geological Sciences, University of Missouri, Columbia, Missouri 65211, USA

Anna M. Martini
Department of Geology, Amherst College, Amherst, Massachusetts 01002, USA

Ruth E. Blake
Department of Geology and Geophysics, Yale University, New Haven, Connecticut 06511, USA

ABSTRACT

The modern shallow-platform, calcium-carbonate–dominated sediments of the Florida Keys (Florida Bay and Atlantic reef tract) are diverse in their biological, sedimentological, and geochemical properties. Sites of intense bioturbation and thick seagrass cover are pervasive within Florida Bay and are often characterized by appreciable early diagenetic aragonite dissolution. Additional, less common sites show atypical diagenetic profiles that suggest strong reworking and/or very rapid deposition of the upper sediment layer extending to a depth of at least 20 cm. Diagenesis in these seagrass-free areas is dominated by rapid burial of labile organic matter that would otherwise be degraded aerobically under conditions of slower burial. Correspondingly, these oozy, water-rich sediments display anomalously high rates of microbial decomposition as recorded in ^{35}S-sulfate reduction rates and patterns of sulfate depletion, high dissolved sulfide concentrations in excess of several millimolar (mM), and elevated alkalinities. Unlike many sites in Florida Bay where solute concentrations suggest volumetrically significant net dissolution of metastable carbonate phases, dramatic increases in carbonate alkalinity from organic matter oxidation during bacterial sulfate reduction support net precipitation of $CaCO_3$ in the highly reactive surface layer. This early carbonate mineralization is indicated by measured depletions in Ca approaching 4 mM relative to overlying seawater. Geochemical signatures of sediment reworking or rapid sedimentation are corroborated by porosity

Lyons, T.W., Walter, L.M., Gellatly, A.M., Martini, A.M., and Blake, R.E., 2004, Sites of anomalous organic remineralization in the carbonate sediments of South Florida, USA: The sulfur cycle and carbonate-associated sulfate, *in* Amend, J.P., Edwards, K.J., and Lyons, T.W., eds., Sulfur Biogeochemistry—Past and Present: Geological Society of America Special Paper 379, p. 161–176, For permission to copy, contact editing@geosociety.org.

trends; visual evaluations, including X-radiography; and an interval of essentially constant ^{210}Pb activity.

Rapid burial within the reactive layer gives rise to restricted-system diagenetic behavior that is recorded in the sulfur isotope compositions of dissolved sulfate and sulfide. Nevertheless, despite strong ^{34}S enrichments in the pore-water sulfate and clear evidence for diagenetic calcium carbonate precipitation, carbonate-associated sulfate (CAS) trapped within the muds (at concentrations from ~2400 to 4200 ppm) preserves the original $^{34}S/^{32}S$ ratio of the overlying seawater. Such preservation of the $\delta^{34}S$ of seawater sulfate in bulk lime mud samples—even in the presence of appreciable diagenetic overprinting—confirms the broad utility of the CAS approach in reconstructing ancient ocean chemistry.

Keywords: carbonate sediments, geochemistry, diagenesis, carbonate-associated sulfate, sulfur isotopes, seawater proxy.

INTRODUCTION

Unlike their shallow siliciclastic equivalents, carbonate platform settings receive only minimal fluxes of detritally delivered iron and, correspondingly, show only low levels of iron-sulfide formation. As a result of limited Fe availability (typical reactive Fe contents are on the order of hundreds of ppm or less), the biogeochemical patterns and pathways for sulfur are far less studied than those in Fe-rich siliciclastic settings. In the absence of Fe, millimolar levels of hydrogen sulfide build up in the interstitial waters, and sulfur reaction pathways are dominated by both biotic and abiotic oxidation and by uptake within sedimentary organic matter (Werne et al., this volume). Furthermore, because of the extensive biological production of calcium carbonate in the shallow waters of the open ocean and dissolution at greater depths (Archer et al., 1989), platform carbonate sediments have been comparatively neglected in models for the global carbon cycle, despite evidence for dissolution even at these shallow depths. For example, Walter and Burton (1990) and Ku et al. (1999) argued for early dissolution of roughly half of the gross annual carbonate production in the shallow platform sediments of South Florida. Consequently, present-day carbonate recycling within shelf pore fluid is not insignificant given that platform carbonates comprise ~5% of the oceanic carbonate inventory (Milliman, 1974). Shallow recycling, however, would have been even more significant in the geologic past when the locus of carbonate deposition, driven largely by biological production, had not yet shifted from shallow to deep settings and when vast, shallow, carbonate-depositing seaways covered large portions of the continents.

Walter and Burton (1990) observed concentrations of dissolved calcium in pore waters from Florida Bay in excess of the overlying seawater and attributed at least some of this to model-predicted aragonite undersaturation during progressive, closed-system sulfate reduction in Fe-poor sediments. More specifically, conditions of aragonite undersaturation were predicted (Ben-Yaakov, 1973; Gardner, 1973; Canfield and Raiswell, 1991; Boudreau and Canfield, 1993) and observed (Walter and Burton, 1990) at low degrees of sulfate reduction (<~3 mM; i.e., up to 10% reduction of original seawater sulfate). During the initial stages of bacterial sulfate reduction under Fe-limited conditions, the absence of volumetrically significant Fe-sulfide precipitation favors the accumulation of H_2S, a weak acid, in the pore waters. Elevated levels of dissolved sulfide and concomitant production of carbonic acid as a consequence of organic matter oxidation during sulfate reduction lead to lower pH values and, correspondingly, aragonite undersaturation. However, as sulfate reduction progresses to levels in excess of 3 mM, increasing alkalinity ultimately promotes aragonite supersaturation (see Walter and Burton [1990] and Canfield and Raiswell [1991] for details).

While sulfate reduction may play a critical, direct role in the development of carbonate undersaturation, studies over the past ten years have emphasized that it alone cannot provide a sufficient source of acid to account for the observed levels of carbonate dissolution (Walter et al., 1993). For example, data from South Florida fail to show the positive trend between degree of sulfate reduction and amount of excess Ca (i.e., Ca contributions via dissolution) expected at low degrees of sulfate reduction (<~3 mM). Further challenge was provided by closed-system anoxic incubations in which increased concentrations of dissolved Ca and thus dissolution were not observed with progressive sulfate reduction at low levels of total sulfate consumption (Walter et al., 1993).

Using detailed inventories of dissolved CO_2, Walter et al. (1993) delineated a suite of sediments characterized by excess levels of metabolic dissolved carbon (CO_2 derived directly from the oxidation of organic matter). More specifically, they observed ΣCO_2 levels in excess of the concentrations predicted from the observed degree of sulfate depletion and the corresponding 2-to-1 reaction stoichiometry of moles of organic carbon oxidized to moles of sulfate reduced. The most significant elevation of dissolved Ca was observed at sites with excess metabolic CO_2. A detailed solute mass balance showed that levels of dissolved CO_2 track the full extent of sulfate reduction, while apparent degrees of sulfate reduction indicated by observed levels of interstitial sulfate have been heavily overprinted by the reoxidation of dissolved sulfide back to sulfate. Walter et al. (1993) could not dismiss aerobic respiration as a significant contributor to the observed excess metabolic CO_2, but by invoking volumetrically important sulfide oxidation, they were able to reconcile the rapid rates of sulfate reduction measured in Florida Bay with the low "apparent" degrees of sulfate reduction. Oxidation of sulfide by O_2 represents a potentially important source of protons and, as such, a critical component in carbonate dissolution.

Based on the above model, Walter et al. (1993) defined an intensely burrowed subtidal end member within Florida Bay characterized by appreciable H_2S production and high levels of sulfide oxidation and carbonate dissolution (Fig. 1, Type 1 sediment). Rapid oxygen transport is required in this model to explain sulfate concentrations far greater than those predicted from the rates of sulfate reduction as recorded in the concentrations of metabolic CO_2 and as measured via incubation experiments. Furthermore, in order for the system to sustain high levels of oxidation of sulfide by O_2, while at the same time maintaining pore-water concentrations that deviate significantly from those of the overlying water for Ca, CO_2, etc., Walter et al. (1993) argued for a "rapid and selective mass transport of oxygen." This mechanism is unlike simple mixing or mass transport during advection, diffusion, or bioirrigation (Berner, 1980; Boudreau, 1997; Aller, 2001). Walter et al. (1993) further suggested enhanced non-stoichiometric O_2 transport in the pore waters as mediated by sulfide oxidizing bacteria lining the oxic-anoxic interfaces defined by the abundant burrow walls.

Most recently, Ku et al. (1999) investigated these highly bioturbated, pervasively grass-covered muds in Florida Bay. As outlined above, mass balance considerations, such as sediments with excess metabolic CO_2 relative to observed sulfate depletion, delineated sulfide oxidation as a source of appreciable carbonate dissolution. Beyond the concentration relationships, Ku et al. used sulfur and oxygen isotopes to further constrain the system. Among the principal findings, the $\delta^{18}O$ mass balance for sulfate verifies the importance of sulfide oxidation and shows that O_2 dissolved in seawater and O_2 released by seagrass roots are essential oxidants. The latter—that is, photosynthetically derived O_2 released directly into the H_2S-containing sediments from the extensive rhizome networks of the seagrass *Thalassia testudinum*—is a key mechanism of enhanced and selective transport of O_2 into the subsurface.

A second end member in the South Florida system dominates much of the reef tract—specifically the sandy muds and muddy sands on the shelf seaward of the Keys. Here, the environments lack the thick seagrass cover of the bay, and the organic inputs are sufficiently low (compared to the bay) that pore waters are largely unevolved chemically relative to overlying seawater (Walter et al., 1993; Ku et al., 1999). These unreactive sediments show low H_2S concentrations and little evidence for calcium carbonate dissolution and precipitation (Fig. 1, Type 2 sediment).

The third end member, and the focus of the present study, represents sites of net calcium carbonate precipitation. Sediments encountered at a representative seagrass-free area, such as those commonly observed at Bob Allen Keys Bank, are characterized by extreme rates of microbial decomposition and, correspondingly, a Ca trend that records net carbonate precipitation as a product of the high level of alkalinity production (Fig. 1, Type 3 sediment). In these settings, either rapid burial or sediment reworking favor the transport of reactive organic phases into the subsurface where they support the high rates of anaerobic respiration—most notably, sulfate reduction. Despite evidence for appreciable diagenetic carbonate precipitation at these sites, data for carbonate-associated sulfate (CAS), an emerging proxy for the $\delta^{34}S$ of seawater sulfate and a central theme of this ongoing study, preserve the S isotope composition of the overlying seawater. Our data also speak to the origins of at least a subset of the seagrass-free areas that abound in the region and the possible anthropogenic causes. Many barren areas in Florida Bay have been linked to an epidemic of widespread grass mortality driven by human activities.

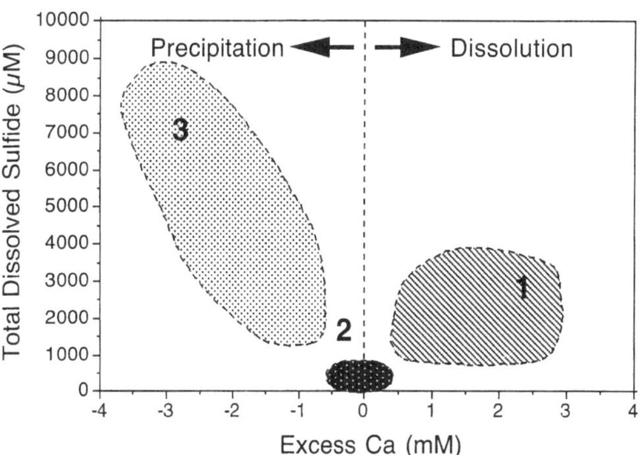

Figure 1. Schematic representation of the dissolved sulfide versus excess Ca relationships for three fundamental sediment types in South Florida (see text and Walter et al. [1993] for details). Concentrations are approximated and show large intra- and intersite variability. While variation among the three geochemical facies reflects basic differences in the depositional regimes and the corresponding sediment properties, data at a given site can vary substantially as a function of the sampling method (box core versus push core versus peeper).

BACKGROUND

Carbonate Sediments of South Florida

Metastable polymorphs of calcium carbonate—aragonite and high-Mg calcite—dominate the Holocene shallow-water sediments of the Florida Keys (Florida Bay and the adjacent Atlantic reef tract) (Walter and Burton, 1990; Hover et al., 2001). Low-Mg calcite occurs in minor amounts. Many past studies have addressed the depositional conditions, interstitial chemistries, and solid-phase/pore-water interactions recorded in the classic shallow-platform muddy sands to sandy lime muds of South Florida and the Bahamas (Rosenfeld, 1979; Bosence, 1989; Swart et al., 1989; Wanless and Tagett, 1989; Burns and Swart, 1992; Randazzo and Jones, 1997), including a few that have argued for volumetrically significant dissolution of the metastable phases (Morse et al., 1985; Walter and Burton, 1990; Rude and Aller, 1991; Walter et al., 1993; Ku et al., 1999).

Sources of the calcareous mud that dominates Florida Bay and that is also found in great abundance in the sandier sediment seaward of the Keys have been debated with renewed intensity over the past 15 years. Historically, the prevailing paradigm called for benthic calcareous green algae, such as *Halimeda* sp. and *Penicillus* sp., as the primary producers of the aragonite needle muds (Stockman et al., 1967; Neumann and Land, 1975). In more recent years, however, the relative roles of biotic and abiotic pathways of $CaCO_3$ precipitation have been revisited, particularly as related to the whiting phenomena commonly observed in Florida Bay and the Bahamas and, more generally, as controlling the sources of the lime mud that dominates these platform settings (Shinn et al., 1989; Robbins and Blackwelder, 1992; Boss and Neumann, 1993; Robbins et al., 1997; Broecker et al., 2000, 2001; Morse et al., 2003). While the debate persists, some workers are reviving an earlier idea and recasting the suspended sediment of whitings as more strongly a product of spontaneous precipitation than resuspension of algal mud (compare, for example, Robbins et al., 1997, and Broecker et al., 2000).

Bacterial Sulfate Reduction—Sulfur Isotope Effects

The kinetic isotope effect associated with dissimilatory bacterial sulfate reduction (BSR) results in hydrogen sulfide that is depleted in ^{34}S relative to the $^{34}S/^{32}S$ ratios of residual, coexisting sulfate (Goldhaber and Kaplan, 1974). Dissimilatory sulfate reduction under pure-culture laboratory conditions can produce sulfide depleted in ^{34}S by roughly 2‰–46‰ relative to the parent sulfate (Chambers et al., 1975; Canfield, 2001; Detmers et al., 2001). Although this range is generally accepted, controls on the magnitude of fractionation are less well known. For example, the relationship during BSR between rates of sulfate reduction and the isotopic offset between parent sulfate and product HS^- ($\Delta^{34}S$) is complex and not fully understood (compare Kaplan and Rittenberg, 1964; Canfield, 2001; Detmers et al., 2001; Habicht and Canfield, 2001). Nevertheless, it seems that isotope fractionations during BSR are unaffected by sulfate concentration at levels >1–2 mM (Canfield, 2001) and perhaps as low as to 200 μM (Habicht et al., 2002).

In light of the significantly smaller isotope effects attributable to BSR under pure-culture conditions, and assuming that the experiments mimic nature (compare Habicht and Canfield, 2001, and Wortmann et al., 2001), recent studies have addressed the large fractionations of up to and exceeding 60‰ observed in the modern Black Sea (Lyons, 1997) and throughout the Phanerozoic. One model invokes bacterial disproportionation of elemental sulfur and other S intermediates as a means of exacerbating the ^{34}S depletions observed in HS^- and in pyrite formed through reaction of the dissolved sulfide with Fe (Canfield and Thamdrup, 1994; Habicht and Canfield, 2001).

Ultimately, net isotopic fractionations reflect both the collective magnitude of bacterial effects and the properties of the sulfate reservoir (Zaback et al., 1993). Even in the presence of large fractionations during BSR and coupled disproportionation, high $\delta^{34}S_{sulfide}$ and $\delta^{34}S_{sulfate}$ values occur in pore-water systems with restricted renewal of sulfate relative to the rate of bacterial consumption (e.g., under conditions of rapid sediment accumulation or in systems with high rates of BSR driven by anaerobic oxidation of methane; Aharon and Fu, 2003; Jørgensen et al., 2004). Conversely, low $\delta^{34}S$ values typically represent marine systems where sulfate availability does not limit BSR. As a result of these multiple controlling factors, bacterial sulfide can display a broad range of $\delta^{34}S$ values that are often very low (^{34}S-depleted) relative to coeval sulfate. These broad ranges and common ^{34}S depletions are the oft-cited fingerprints of BSR.

Carbonate-Associated Sulfate

Although the presence of CAS has long been known, the pioneering study of Burdett et al. (1989) was the first to comprehensively demonstrate that this sulfate (as recorded in modern and fossil micro- and macro-skeletal grains) can be isotopically similar to modern seawater and ancient, coeval evaporite deposits. Sulfate in modern and ancient sedimentary carbonate minerals typically occurs in concentrations ranging from a few hundred ppm to extremes of 10^4 ppm (Staudt and Schoonen, 1995). Recent studies of Precambrian limestones and dolostones show CAS concentrations typically ranging from 10^0 to 10^2 ppm, which are consistent with a hypothesized low-sulfate Proterozoic ocean (Kah et al., 2001; Hurtgen et al., 2002; Pavlov et al., 2003; Shen et al., 2003; Lyons et al., 2004).

The position of the sulfate within the minerals is imprecisely known, with models proposing structural substitution within the carbonate-ion lattice position (the structurally substituted sulfate, or SSS, of some authors), sulfate-containing fluid inclusions, trace sulfate mineral-phase inclusions, or incorporation into organic matrices (see Burdett et al., 1989; Staudt and Schoonen, 1995; Strauss, 1999; and references therein for a detailed historical account). A variety of spectroscopic techniques, as reviewed in Staudt and Schoonen (1995), confirm structural substitution within the carbonate lattice as the essential CAS reservoir. To date, skeletal carbonate has been the primary focus for studies of modern and ancient settings (Burdett et al., 1989; Kampschulte and Strauss, 1998; Kampschulte et al., 2001). The isotopic integrity of bulk lime mud samples—the emphasis of the present study—is less well known, although bulk micrite and dolomicrite samples are already proving useful in the interpretation of ancient sequences (Kampschulte et al., 2001; Hurtgen et al., 2002; Kampschulte and Strauss, 2004; Lyons et al., 2004). Sulfate partitioning into specific carbonate mineralogies (i.e., distribution coefficients) are poorly known. However, I. Gavrieli (2001, personal commun.) and others (e.g., Staudt and Schoonen, 1995) have shown very high concentrations of CAS in aragonite relative to calcite. Preliminary results from modern settings suggest that isotopic fractionations during CAS incorporation into $CaCO_3$ are minor to negligible (this study; Burdett et al., 1989), which is corroborated by comparisons between ancient $\delta^{34}S_{CAS}$ data and $\delta^{34}S$ results from coeval gypsum/anhydrite deposits (Burdett et al., 1989; Strauss, 1999; Kampschulte and Strauss, 2004; Lyons et al., 2004).

SAMPLES AND METHODS

Location, Sampling Protocol, and Sample Descriptions

This paper emphasizes a broad range of complementary geochemical data from a single, representative location—a grass-free, highly reactive site in the Bob Allen Keys area of Florida Bay. Here, diagenetic carbonate (re)precipitation is occurring today in the presence of a highly evolved pore-water sulfate reservoir. In recent years, barren (grass-free) areas in Florida Bay have come under close scrutiny as large patches of seagrass mortality have been observed and attributed to human perturbations of the broader ecosystem. This is not, however, meant to be a comprehensive study of this phenomenon or a survey of geochemistry throughout the Bay. Instead, we emphasize the details of a particular and somewhat anomalous facies characterized by extreme rates of microbial activity with important implications for pathways of early carbonate precipitation, seagrass ecology, the integrity of ancient CAS records, and patterns of organic remineralization in shallow carbonate-platform settings.

The shallow water depths of our study site (generally ≤~1 m), and throughout much of Florida Bay, generally allowed us to collect cores by wading short distances from the boat. Samples were collected at the Bob Allen Keys site using three methods: (1) plexiglass box cores pressed into the sediments by the operator (~25 × 35 cm), (2) hand-inserted butyrate push cores (~7.5 cm diameter), and (3) peepers—which are in situ pore-water sampling devices whereby vertical chemical profiles of interstitial species are generated through collection from small sample reservoirs (2.5 mL) following diffusional equilibration across a permeable filtration membrane. Because solid-phase constituents are central to this report, peeper data are discussed only briefly. The push core from the Bob Allen Keys site "compacted" by almost 40% during collection. This extreme length reduction may be an artifact of the unique sediment properties at this site, as described below. Immediately following collection, the box cores and push cores were processed in nitrogen-filled glovebags. The cores were sectioned into 2 cm intervals for the box cores and thicker intervals for the push cores (Tables 1 and 2) and sealed in centrifuge tubes. The pore-water fractions were then isolated by centrifugation, filtered (0.45 µm) and subdivided under an N_2 atmosphere for a variety

TABLE 1. BOX CORE DATA FROM F-6

Depth (cm)	Porosity (%)	xs ^{210}Pb (dpm/g)	TOC (wt%)	SRR nmoles/cc/d	mM SO_4/M Cl	ΣH_2S (µM)	xs Ca (mM)	Alk (meq/l)	δ^{34}S-ΣH_2S (‰)	δ^{34}S-SO_4 (‰)
0–2	88.2	2.73	3.43	643.3	47.7	1787	0.24	7.47	–13.5	+24.0
2–4	85.8	2.07	3.73	621.2	42.2	3106	–0.34	12.76	–9.3	+27.7
4–6	85.3	2.54	3.47		32.3	6505	–1.22	21.47	–7.5	+36.0
6–8	84.4		3.94	317.1	26.7	5687	–1.28	23.03		
8–10	83.5	2.41	3.76		25.2	6287	–2.43	25.60	–6.3	+46.6
10–12	82.4		3.42		25.6	6382	–2.95	25.24		
12–14	80.7	2.63	3.15	194.5	25.6	6410	–2.82	23.13	–8.8	+49.6
16–18	73.3		2.83		29.1	6235	–2.15	22.03	–12.9	+46.9
18–20	73.5		2.58	104.3	31.4	5124	–2.28	19.74		
20–22	73.4	1.97	2.64		32.0	6008	–2.30	19.20	–15.9	+44.2

TOC—total organic carbon. SRR—Sulfate reduction rate.

TABLE 2. PUSH CORE DATA FROM F-6

Depth (cm)	Porosity (%)	xs ^{210}Pb (dpm/g)	TOC (wt%)	SRR nmoles/cc/d	mM SO_4/M Cl	ΣH_2S (µM)	xs Ca (mM)	Alk (meq/l)	δ^{34}S-ΣH_2S (‰)	δ^{34}S-SO_4 (‰)
0–6			3.76		23.5	5897	–3.29	20.72	+8.5	+32.5
6–12			3.92		3.1			38.67	+15.0	+31.5
12–18			3.34		9.7	8099	–3.61	39.10		
18–24			2.47		29.1		–3.23	22.52		
24–30		0.66	2.58		37.3	4472	–1.81	18.22	–13.7	+39.2
30–36		0.33	2.43		40.1		–1.04	16.82		
36–42		0.32	2.29		41.5	3375	–1.88	11.41		
42–54		0.22	2.73		43.4		–0.74	12.87	–26.4	+32.3
54–66			2.32		45.0	2203	–0.88	9.98	–30.4	+29.9
66–78			2.07		47.2		–0.21	8.74	–32.1	+27.5
78–86			1.78		48.3	983	–0.12	6.49		

TOC—total organic carbon. SRR—Sulfate reduction rate.

of field and later laboratory analyses (as outlined under "Analytical Procedures"). Plexiglass-enclosed sediment slabs (~4 cm thick, 28 cm long) were collected for field-based X-radiography, photography, and visual description. Sediment samples for later analysis were immediately frozen.

The site referred to here as F-6 is the focus of this paper. The F-6 study area, with a water depth of <1 m, is located near the Bob Allen Keys (Fig. 2) and is characterized by numerous irregularly shaped seagrass-free patches ranging from a few to several meters in their longest dimensions. All the samples were collected from a small area within one of these patches. (Rude and Aller [1991] described dimensions of 10–50 m for analogous features in the same general area of Bob Allen Keys Bank.) These areas, which are completely devoid of seagrass, are bordered by muds with thick grass cover, with sharp boundaries between the two facies. The grass-free areas are characterized by anomalously oozy textures and medium gray colors below a few millimeters of oxidized sediment at the sediment-water interface. Our recent work has shown analogous features to also occur within the Captain Key region east-southeast of Bob Allen Keys. Despite some smearing along the edges during collection, sediments collected at the F-6 site suggested a sharp downcore color change from the medium gray surface interval to lighter tan-gray mud at ~18 cm below the sediment-water interface. This interface is also visible in the X-radiograph, where it is expressed as a sharp but irregular transition from "bright" (relatively X-ray transparent) mud rich in what appear to be cm-scale fragments of seagrass to the underlying denser, more X-ray absorbent mud with less abundant grass (Fig. 3). The comparatively lower X-ray absorption in the upper 18 cm is consistent with the soupy, highly porous fabric that characterizes the upper layer.

Rude and Aller (1991) focused on similar grass-free muds in the Bob Allen Keys area. They described these features as "blowout areas" but stated explicitly that no mode of formation was

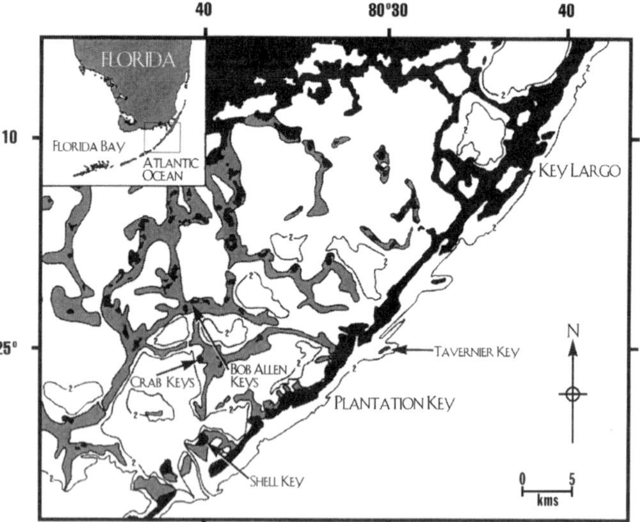

Figure 2. Map showing general locations of Bob Allen Keys (F-6) and Crab Key (F-4) sites (after Rude and Aller, 1991).

implied by their use of that term. Models for the origins of these grass-free patches are discussed below in the context of our work and the results of previous studies.

Sediments at the F-4 site, discussed only briefly here for comparison, were collected at Crab Key mudbank (Fig. 2) and are more typical of the seagrass-covered sediments that dominate much of Florida Bay. Here, beneath the moderate grass cover, the diverse floral and faunal assemblage, and the surface-most (several cm) winnowed grainy layer, the uniform sulfidic muds show a carbonate chemistry very different than that observed at the macrobiologically lifeless F-6 site. More specifically, F-4 is an ideal example of the sites in Florida Bay dominated by $CaCO_3$ dissolution. Net carbonate dissolution typifies much of the grassy, burrowed mud of Florida Bay (Fig. 1; Ku et al., 1999) but stands in strong contrast to the pronounced net carbonate precipitation recorded at F-6.

In addition to the Florida muds emphasized in this study, a small suite of identifiable skeletal fragments and mixed-faunal skeletal sands was collected adjacent to Heron Island, southern Great Barrier Reef, Australia. The results from these carbonate grains are included here for comparison and to specifically demonstrate the robustness of the CAS method beyond the bulk lime mud samples from South Florida. There is further significance because the genus of calcareous algae analyzed from Australia (*Halimeda*) is an important source of carbonate mud throughout the world.

Analytical Procedures

Pore waters (2.75 mL per interval) were fixed with 1 mL of a 3% Zn acetate solution for later analysis of total dissolved sulfide (ΣH_2S) using the methylene blue technique of Cline (1969) with a 3 μM detection limit. An additional aliquot was fixed with a few grains of granular Zn acetate for later analysis of dissolved sulfate. Anion ratios (SO_4^{2-}/Cl^-) were determined by ion chromatography with a precision of ±0.5% (2σ) using IAPSO as a check standard. As a further check, chlorinity was measured by potentiometric titration with a precision of ±0.2% (2σ) (results not shown). Dissolved Ca concentrations and total alkalinity were also determined by potentiometric titration with precisions of ±0.25% and ±0.5%, respectively. Methane concentrations, which will be discussed only briefly here, were measured by gas chromatographic analysis of headspace gas. Additional analytical details for the pore-water concentration analyses are available in Walter and Burton (1990) and Ku et al. (1999).

Dissolved sulfate was also quantitatively precipitated as $BaSO_4$ by the addition of $BaCl_2$. The $BaSO_4$ was combusted offline in the presence of Cu_2O and SiO_2 to SO_2 for sulfur isotope analysis. Pore-water HS^- (ΣH_2S) was initially fixed for isotope analysis by addition of excess Cd acetate (solid) to varying volumes of pore water in the N_2-filled glovebag. The CdS samples were later converted to Ag_2S as follows. The CdS was separated from seawater through multiple rinses with de-ionized H_2O;

each rinse was followed by centrifugation and decantation. Following the rinses, a 3% silver nitrate and 10% NH_4OH solution (~25 mL) was added to each centrifuge tube, leading to spontaneous conversion to Ag_2S at room temperature. The precipitate was filtered, dried, and later combusted with cuprous oxide to SO_2. Sulfur isotope analyses were performed using a VG 903 mass spectrometer with a precision of ±0.2‰ (2σ). All S isotope data are reported relative to Canyon Diablo Troilite (CDT) using the standard delta notation.

Samples for solid-phase analysis were either dried at low temperature (~40 °C) or freeze dried and then gently disaggregated by mortar and pestle. Porosities were calculated from the weight loss during drying and assuming a dry bulk sediment density of ~2.8 g/cm^3. The dried samples that were weighed for porosity determinations and used for the full range of solid-phase analyses were not salt corrected. Total carbon and inorganic ($CaCO_3$) carbon were determined by coulometric titration of the CO_2 liberated during combustion and HCl digestion, respectively. Organic carbon was calculated by difference. Analyses of $CaCO_3$ standards were generally accurate within 1%–2%. Activities of unsupported (excess) ^{210}Pb were quantified via gamma-ray detection using dried sediments in sealed containers and correcting for the activity supported by the decay of ^{226}Ra in the sediment (Moore and O'Neill, 1991). Solid-phase Fe was extracted using the buffered (pH 4.8) dithionite procedure outlined by Canfield (1989) and Raiswell et al. (1994). Sediment samples (0.2–0.3 g) were shaken at room temperature for ~2 h in 50 mL of the buffered (0.35 M acetic acid and 0.2 M sodium citrate) sodium dithionite (50 g/L) solution.

Rates of sulfate reduction were determined by $^{35}SO_4^{2-}$-labeled sediment incubations (Jørgensen, 1978; Fossing, 1995; Hurtgen et al., 1999). The rates were calculated from the $H_2{}^{35}S$/$^{35}SO_4^{2-}$ activity ratios measured by liquid scintillation counting using the known incubation time (~11 h) and an assumed fractionation factor (α) of 1.06 (Jørgensen, 1978). The incubations were terminated by freezing the sediments, and the radiolabeled sulfate and hydrogen sulfide were separated by a 6N HCl/15% $SnCl_2$ extraction under N_2 (Chanton and Martens, 1985); the liberated H_2S was trapped in a 2% Zn acetate and 6% NH_4OH solution. Because of pronounced Fe deficiencies within the carbonate sediments, we assumed that the ^{35}S-labeled sulfide produced by bacterial sulfate reduction during the incubation was preserved in the sediment as ΣH_2S and was thus quantitatively liberated during the HCl extraction. Triplicate rate measurements for a given interval were quite variable but often agreed within 10%–20%.

Skeletal grains (mixed sands and individual fragments) from Australia and bulk mud samples from South Florida were analyzed for their CAS concentrations and sulfur isotope compositions. Dried, powdered samples (4–20 g; specifically 10–20 g for

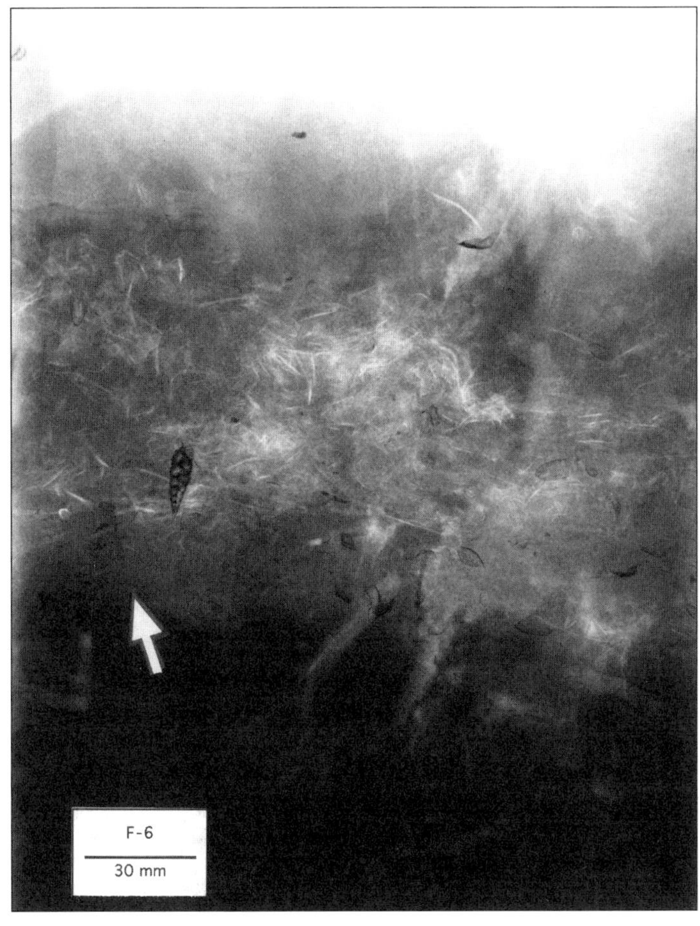

Figure 3. X-radiograph of core from Bob Allen Keys (F-6) site. The arrow delineates the interface ~18 cm below the sediment-water interface corresponding with the transition in porosity (Fig. 5) and a broad range of geochemical properties (Fig. 6). The darker character of the lower portion of the X-radiograph (below the 18 cm interface) reflects the higher density of the less-soupy mud. Note the abundance of grass fragments in the upper portion of the core and the relatively sharp interface at the base of the reactive layer.

the Florida mud samples) were initially rinsed in de-ionized H_2O for at least one to two days and sonicated to remove soluble salts. The samples were then soaked in acetone for ~12 h to remove any elemental sulfur that was originally in the sediment or that formed during sample oxidation. Following a 1-day water rinse, the samples were treated with 5.25% bleach solution (NaOCl) to remove labile organic phases and carefully rinsed again for 1 day with de-ionized H_2O. The sonication, acetone treatment, and repeated water rinses are beyond what is necessary with ancient carbonate samples. Our concern, however, was that removal of seawater salts precipitated during the drying of unrinsed samples would require additional care. Samples were then dissolved slowly at room temperature in 4N HCl until reaction stopped and filtered to remove the insoluble residues. The solution was brought to a volume of 1 L and filtered to remove the insoluble residue. After a 15 mL aliquot was removed for concentration measurements, the trace sulfate was precipitated as barium sulfate ($BaSO_4$) through addition of 125 mL of saturated barium chloride ($BaCl_2$) solution (~250 g/L). The barium sulfate was allowed to precipitate at room temperature over a period of ~3 days and then filtered.

Barium sulfate precipitates were homogenized, combined with an excess of V_2O_5, and analyzed for sulfur isotope compositions at Indiana University—Bloomington using a Finnigan MAT 252 gas source mass spectrometer fitted with an elemental analyzer (EA) for online sample combustion. This protocol permitted high $\delta^{34}S$ precision for very small samples (<0.5 mg of pure $BaSO_4$). All sulfur isotope compositions for CAS are expressed in standard delta notation as per mil (‰) deviations from Vienna Canyon Diablo Troilite (V-CDT), with an analytical error of 0.2‰. (The difference in $\delta^{34}S$ values expressed relative to CDT versus V-CDT is small [<1‰; Ding et al., 2001].)

Concentrations of carbonate-associated sulfate were measured at the University of Missouri using a Perkin Elmer inductively coupled plasma–optical emission spectrometer (ICP-OES) fitted with a micro-concentric nebulizer and calibrated through a series of standard and repetitive sample runs. Precision was better than 10%. Certified reference materials were used for the standard calibration, with the addition of HCl and dissolved Ca to simulate sample matrix conditions. The detection limit for dissolved sulfate was calculated at 0.7 ppm solution concentration (as sulfate). Reliability of the ICP-OES approach was further tested by measuring sulfate concentrations gravimetrically (as $BaSO_4$). The strong agreement between the two methods even at concentrations <10^3 ppm (CAS in solid) validates the use of ICP-OES (Fig. 4).

Because of historical differences, both seasonal and longer-term, in the extent of evaporation and freshwater input to the bay, downcore variations in reactive dissolved species—such as calcium—may be less related to pathways of calcium carbonate dissolution and precipitation than to dilution and evaporative enrichment. To minimize the risk of spurious interpretations, the concentrations of potentially reactive species are often normalized to the corresponding concentration of a dissolved species, such as

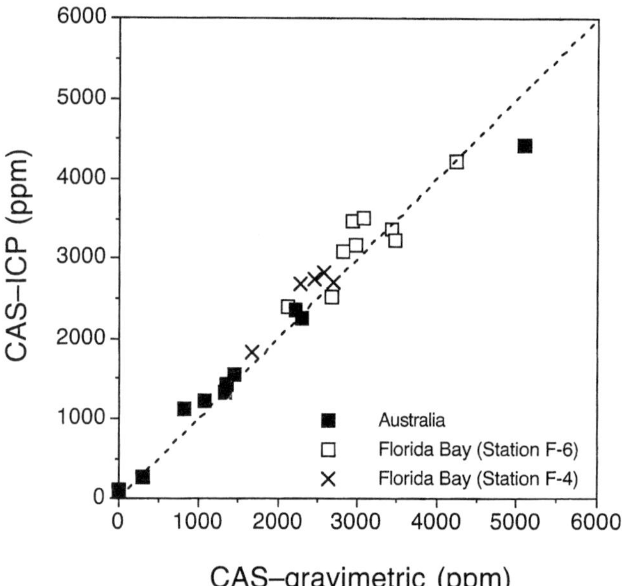

Figure 4. Comparison of replicate carbonate-associated sulfate (CAS) concentration data as determined by ICP-OES (inductively coupled plasma–optical emission spectrometer) and gravimetry. Samples from Australia and Florida Bay are shown (Tables 3 and 4).

chloride, assumed to behave conservatively over the conditions of interest. This way, the ratio of the calcium concentration, for example, to the chloride content remains constant, despite historical trends in evaporation and freshwater dilution, unless reaction is occurring. An approach of even greater value is to calculate the "excess" concentration of the reactive species using Cl^- as a conservative tracer. The measured pore-water concentration is compared to a chloride-corrected initial concentration:

$$\text{Excess } [\,]_{pw} = [\,]_{pw} - ([\,]_{ow} \times [Cl^-]_{pw}/[Cl^-]_{ow}),$$

where the subscripts *pw* and *ow* refer to pore water and present overlying water, respectively. This way, a negative or positive Ca excess unambiguously records dissolved Ca uptake or release during calcium carbonate precipitation or dissolution, respectively (Fig. 1).

RESULTS

The complete results are provided in Tables 1–6 and are summarized graphically in Figures 4 through 6. The transition at 18 cm—visible as a color change in the cores and as juxtaposed contrasting sediment fabrics in the X-radiograph—is also recorded in the porosity relationships (Fig. 5). The oozy, gray muds of the upper 18 cm are characterized by anomalously high porosities relative to the F-4 core and the lighter tan-gray muds below 18 cm at Site F-6. The porosity within the upper 18 cm shows a systematic downcore decrease, followed by an abrupt drop to the uniform values observed below the transition.

TABLE 3. CARBONATE-ASSOCIATED SULFATE CONCENTRATION DATA FOR SKELETAL CARBONATES*, HERON ISLAND, GREAT BARRIER REEF, AUSTRALIA

Sample type	ppm
Halimeda discoidea	1322 (1332)
Skeletal sand #1	1550 (1446)
Halimeda sp.	1437 (1356)
Halimeda opuntia	1116 (813)
Skeletal sand #2	2355 (2232)
Skeletal sand #3	2268 (2309)
Halimeda macroloba	1221 (1080)
Acripora sp.	4426 (5086)
Clam	283 (305)
Gastropod	118 (12†)

*Determined by inductively coupled plasma–optical emission spectrometry (ICP-OES) and gravimetry (in parentheses).
†This gravimetric analysis was compromised by the small sample size (4 g). ICP-OES analysis yielded greater sensitivity, particularly for small samples.

TABLE 5. CARBONATE-ASSOCIATED SULFATE $\delta^{34}S$ DATA FOR SKELETAL CARBONATES, HERON ISLAND, GREAT BARRIER REEF, AUSTRALIA

Sample type	$\delta^{34}S$ (‰, V-CDT)
Halimeda discoidea	+19.7
Skeletal sand #1	+21.4
Halimeda sp.	+21.6
Halimeda opuntia	+20.1
Skeletal sand #2	+20.8
Skeletal sand #3	+21.3
Halimeda macroloba	+19.6
Halimeda macroloba (duplicate analysis)	+19.8
Acripora sp.	+21.5
Clam	+19.4
Gastropod	+20.4
mean	+20.5 ± 0.8‰

Note: Corresponding concentration data are provided in Table 3. V-CDT—Vienna Canyon Diablo Troilite.

TABLE 4. DOWNCORE CARBONATE-ASSOCIATED SULFATE CONCENTRATION DATA* FOR ARAGONITIC MUDS OF FLORIDA BAY AT SITES OF DIAGENETIC $CaCO_3$ DISSOLUTION (F-4) AND PRECIPITATION (F-6)

F-4 (cm)	ppm	F-6 (cm)	ppm
0–2	1835 (1676)	0–2	3380 (3435)
2–4	2689 (2288)	2–4	4235 (4245)
6–8	2820 (2578)	4–6	– –
12–14	2700 (2700)	8–10	3512 (3078)
18–20	2753 (2453)	12–14	2529 (2674)
		20–22	3176 (2981)
		24–30	3092 (2824)
		30–36	3239 (3476)
		36–42	3470 (2945)
		42–54	2397 (2119)

*By inductively coupled plasma–optical emission spectrometry and gravimetry (in parentheses).

TABLE 6. DOWNCORE CARBONATE-ASSOCIATED SULFATE $\delta^{34}S$ DATA FOR ARAGONITIC MUDS OF FLORIDA BAY AT SITES OF DIAGENETIC $CaCO_3$ DISSOLUTION (F-4) AND PRECIPITATION (F-6)

F-4 (cm)	$\delta^{34}S$ (‰)	F-6 (cm)	$\delta^{34}S$ (‰)
0–2	+19.8	0–2	+19.5
2–4	+20.0	2–4	+19.9
6–8	+19.9	4–6	+18.9
12–14	+19.4	8–10	+19.5
18–20	+19.5	12–14	+19.8
		20–22	+19.6
		24–30	+19.6
		30–36	+19.7
		36–42	+19.5
		42–54	+18.6
mean	+19.7 ± 0.3‰	mean	+19.5 ± 0.4‰

In concert with the contrasting physical characteristics, the sediments at F-6 from above and below the ~18 cm transition show striking geochemical contrasts (Fig. 6). Sediments within the upper 18 cm, labeled tentatively as "reworked layer," display C_{org} concentrations approaching 4 wt%, which drop off to values generally between 2 wt% and 3 wt% below the transition. Activities of unsupported ^{210}Pb are essentially uniform within the upper 18 cm and decrease exponentially below. Rates of sulfate reduction in excess of 600 nmole cm^{-3} d^{-1} are highest at the surface and decrease down core over the upper 18 cm. These surface-most rates are roughly equivalent to the value for the 0–2 cm interval at F-4 (results not shown). Compared to F-6, however, the rates drop off more precipitously at F-4, and surface values at these two sites are roughly a factor of three greater than those observed at other sites in the region (sediment types 1 and 2 in Fig. 1). As predicted from the general persistence of higher rates at F-6, the alkalinity values for the box core are roughly a factor of three to five times higher than those of F-4, and $\delta^{34}S$ values for sulfate exceed those of F-4 by as much as ~30‰ over the same interval. It is important to remember, however, that the grassy F-4

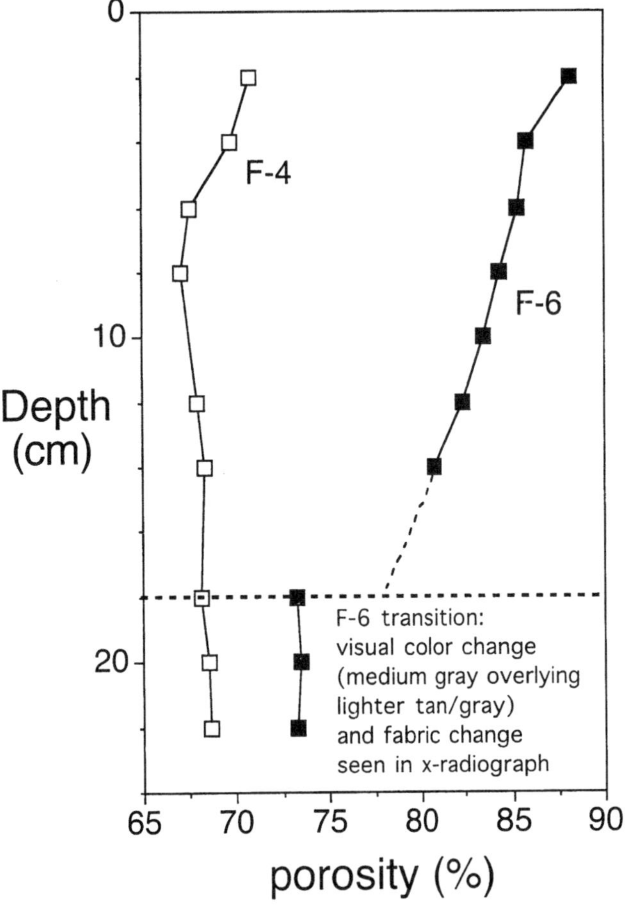

Figure 5. Downcore porosity trends at the Florida Bay F-4 and F-6 sites. The ~18 cm transition between the more-reactive and less-reactive muds at F-6 is indicated.

site is by no means unreactive and that pore-water relationships are likely complicated by sulfide oxidation and burrow-related irrigation.

Ratios of dissolved sulfate-to-chloride (mM/M) at F-6 decrease below the surface to a minimum of 3.1 in the 6–12 cm interval of the push core and then increase sharply toward the interface at 18 cm. Below 18 cm the ratio increases gradually and approaches the overlying seawater value of ~52 by the base of the push core. The box core SO_4^{2-}/Cl^- ratios (Table 1) show the same general trend as the push core data displayed in Table 2 and Figure 6, but the minimum measured ratio is only 25.2 in the box core. Given the comparatively high extents of sulfate depletion in the surface interval at F-6 compared to F-4 and to other sites throughout the region, F-6 sediments are generally the most rich in methane, although headspace concentrations are still only in the single to low double-digit μM range. Sulfate concentrations over the upper 18 cm in the F-6 box core are roughly a factor of two lower than those over the same interval in the box core from F-4. This difference manifests even more strongly in the F-6 push core and in intersite comparisons of peeper data.

Alkalinity data for the push core shown in Figure 6 increase dramatically within the upper layer and then decrease gradually to roughly the seawater value by the base of the core. The box core alkalinity data show the same general trend for the upper interval, but the values are substantially lower in the box core. The excess Ca data for the push core (Fig. 6) show strongly negative values in the upper layer, which increase gradually below 18 cm to essentially zero by the base of the push core. The higher-resolution Ca data of the box core (Table 1) are higher (less negative) than those from the gray layer in the push core but still decrease in the upper 12 cm from a maximum value of +0.24 in the 0–2 cm interval to a minimum of –2.95 (10–12 cm).

Concentrations of ΣH_2S are provided in Tables 1 and 2 for the box core and push core collected at F-6. The concentrations are high (in excess of 8 mM) relative to a maximum value of ~2 mM measured in a peeper from F-4 (and compared to values of << 1 mM in the F-4 box core) and generally increase down core within the upper intervals of the gray layer and decrease abruptly below 18 cm. Like many of the other data from F-6, the sulfur isotope results for ΣH_2S and dissolved SO_4^{2-} show unusual, nonsteady-state characteristics. Sulfate $\delta^{34}S$ values for the box core increase abruptly to almost +50‰ by 12–14 cm and then decrease slightly to +44.2‰ at 20–24 cm. The +24.0‰ measured for the 0–2 cm interval is close to the seawater value of +20–21‰ (Rees et al., 1978). The push core SO_4^{2-} data increase to a maximum of ~+39‰ at 24–30 cm and then decrease down core. The push core $\delta^{34}S$ for ΣH_2S increases to +15.0‰ by 6–12 cm and then decreases to –32.1‰ in the deepest measured interval at 66–78 cm. The high-resolution box core data are significantly more depleted in ^{34}S compared to the push core results seen in Figure 6 and show a subtle downcore increase followed by decrease over the top 18 cm (Table 1).

Data for CAS concentrations and $\delta^{34}S$ for the skeletal grains collected from the Great Barrier Reef are presented in Tables 3 and 5. The concentrations, with two notably low exceptions, are in excess of 1000 ppm. The S isotope data, with a mean of +20.5 ± 0.8‰ (±1σ) cluster very close to the $\delta^{34}S$ value of modern seawater sulfate. Concentrations of CAS (as determined by ICP-OES) in bulk Florida muds at the site of net dissolution (F-4) and net precipitation (F-6) show ranges of 1835–2820 ppm and 2397–4235 ppm, respectively (Table 4). The Florida muds also cluster near the $\delta^{34}S$ value of seawater sulfate, with mean values of +19.7 ± 0.3‰ and +19.5 ± 0.4‰ at sites F-4 and F-6, respectively (Table 6). The summary of CAS $\delta^{34}S$ values in Figure 6 for a combination of push core and box core samples highlights their downcore uniformity and the pronounced offset with respect to the dissolved sulfate from the same interval.

Concentrations of Na-dithionite extractable Fe average 671 ppm in the upper 18 cm at F-6 and 451 ppm below this interface, compared to a mean value of 263 ppm at F-4. Raiswell et al. (1994) confirmed that the Na dithionite extraction used here is specific to highly reactive oxide phases, with only minimal contributions from less reactive Fe-bearing silicates. These numbers are roughly 20%–50% of the total Fe concentrations reported by

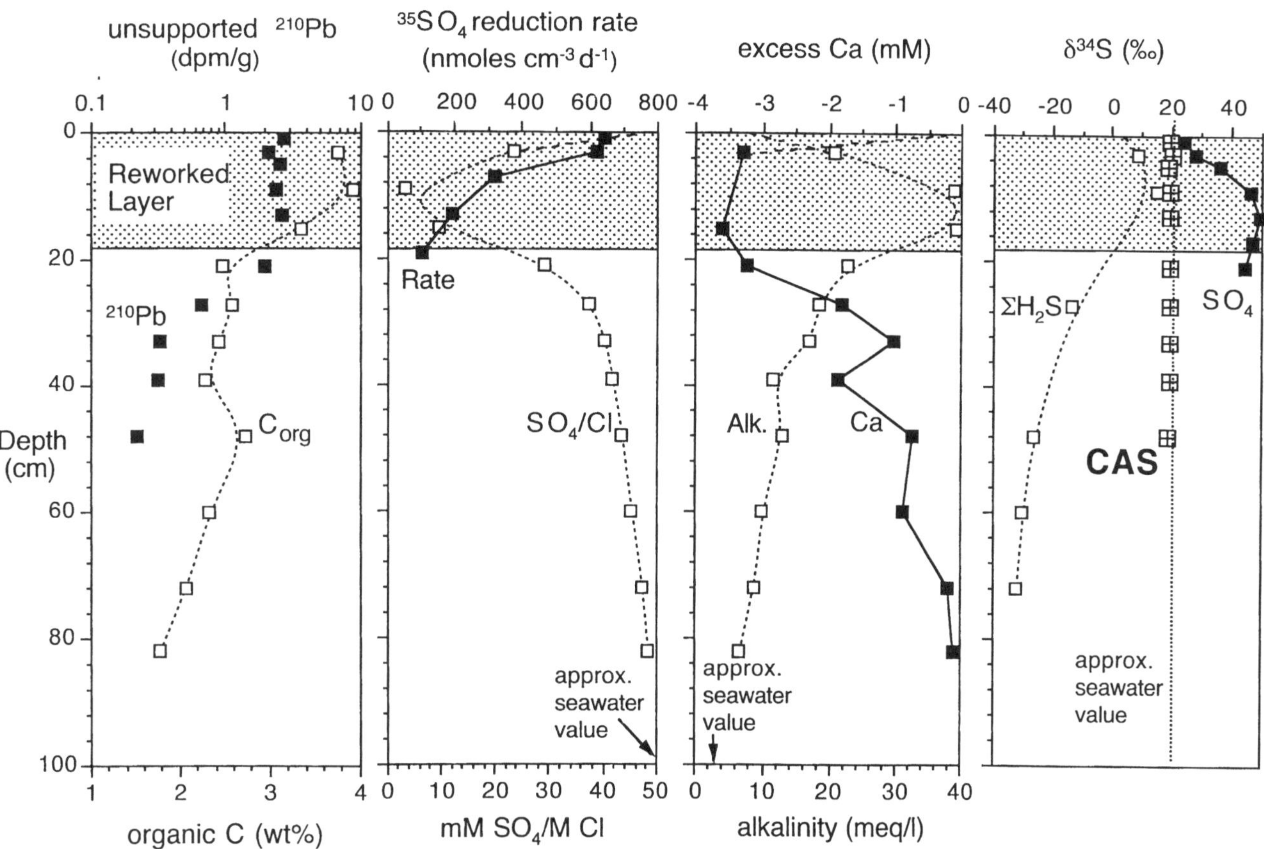

Figure 6. Downcore trends at the F-6 site for unsupported ^{210}Pb (mixed push-core and box-core data); organic C (C_{org}, push core); ^{35}S sulfate reduction rates (box core); sulfate-to-chloride ratios (push core); excess Ca (push core; negative and positive values record net $CaCO_3$ precipitation and dissolution, respectively); total alkalinity (push core); and the $\delta^{34}S$ values of total dissolved sulfide (ΣH_2S, push core), carbonate-associated sulfate (CAS; mixed push core and box core), and dissolved sulfate (box core). The CAS $\delta^{34}S$ data are also summarized in Table 6; the remainder of the data are provided in Tables 1 and 2. The interval labeled as the "reworked layer" corresponds to the highly reactive sediment (above ~18 cm) discussed in the text. While this layer may reflect physical (storm) reworking, rapid deposition may also be the cause of the anomalously high rates of microbial activity and the corresponding net carbonate precipitation. These two models are thoroughly discussed in the text.

Rude and Aller (1991) for Florida Bay (0.12%–0.14%) and are thus consistent with a typical ratio of highly reactive Fe-to-total Fe of ~0.3 reported by Raiswell and Canfield (1998) for average oxic to dysoxic marine sediments. It is worth noting that the highly sulfidic interval at F-6 does show enrichment in reactive Fe. The total Fe concentrations in the carbonate-dominated sediments of Florida Bay are, however, only 2%–3% of those in average crust (Taylor and McLennan, 1985). The Fe results are provided here to highlight the prevailing low Fe conditions in Florida Bay, which favor limited iron sulfide formation and the potential for high levels of dissolved sulfide in these sediments.

DISCUSSION

Among the most striking results of this study are the differences among the push core, box core, and peeper geochemical data sets. Peepers and box cores from a given location yield comparable values for chloride and total dissolved silica, suggesting that the peepers were deployed for sufficient duration to permit equilibration with the ambient pore fluids. Despite this equilibration, however, the same comparison indicated dramatic differences in the concentrations of other dissolved species, including total alkalinity, total dissolved sulfide, and excess calcium. While similar downcore trends are observed for the peeper and box core data, the box core alkalinities and dissolved sulfide concentrations are consistently lower than those of the peepers by as much as a factor of 2 to 3. These disparities highlight the potential for artifacts that are specific to the method of sampling, including pronounced CO_2 and H_2S degassing, concomitant calcium carbonate precipitation, and possible pore water mixing during the collection and processing of the box cores. We cannot, however, rule out the possibility of enhanced microbial activity linked in some way to the emplacement of the peepers.

The push core and box core samples also show dramatic differences, including the results from F-6 (Tables 1 and 2). Although the trends are generally very similar, the box cores appear decidedly more vulnerable to CO_2 and H_2S degassing, $CaCO_3$ precipitation, and sample mixing during collection and processing. This contrast is not surprising given the respective surface areas exposed in the glovebag during extrusion and the times required to process each, as well as the comparative degrees of disruption during processing and collection. Nevertheless, these inter- and intrasite sampling complications are expressed primarily in the absolute values of the data; the overall trends and the fundamental conclusions of this paper are sound regardless of the sampling method emphasized.

What is most clear in Figure 6 is the decidedly nonsteady-state nature of deposition and diagenesis at F-6 and the anomalously high rates of BSR and carbonate diagenesis in the upper ~18 cm. Alkalinity and sulfate concentration are inversely related at this site because of the bicarbonate produced during sulfate reduction and corresponding organic matter oxidation:

$$2CH_2O + SO_4^{2-} \rightarrow H_2S + 2HCO_3^-,$$

and the pronounced negative value for excess Ca reflects the bicarbonate-driven net precipitation of calcium carbonate:

$$Ca^{2+} + 2HCO_3^- = CO_2 + H_2O + CaCO_3.$$

Beneath the reactive layer, alkalinity, sulfate, and calcium all return gradually to seawater values, supporting the notion of atypically high rates of bacterial sulfate reduction in the surface interval and a fundamental shift in the depositional and diagenetic regimes. These high rates are further indicated by the ^{35}S results, C_{org} enrichments of roughly a factor of two relative to the deeper sediments, and the high ΣH_2S concentrations (Tables 1 and 2). Although there is still significant ^{34}S depletion in the dissolved sulfide of the reactive layer compared to sulfate at the same depth, both data sets show the telltale ^{34}S enrichments that typify high rates of sulfate consumption during BSR that outpaces limited diffusional sulfate renewal. Below the reactive layer, sulfide shows the strong ^{34}S depletions that are more typical of low rates of consumption relative to renewal.

The most reasonable explanation for the geochemical relationships at the seagrass-free F-6 site at Bob Allen Keys (Fig. 6) requires either extremely high rates of sediment accumulation or rapid physical sediment mixing. These rapid processes, expressed over the upper ~18 cm in uniform ^{210}Pb profile, drive high rates of BSR and ultimately net $CaCO_3$ precipitation. By inference, our interpretations for F-6 can be extended to the many other grass-free patches in the vicinity of Bob Allen Keys, and while recognizing that not all grass-free areas have the same characteristics and origins, our conclusions may have relevance throughout the Bay. It is clear the high concentrations of sulfide preclude the recolonization by seagrass and other benthic flora and fauna. What is less clear, however, is what initiated the loss of grass cover, the rapid burial, and the high rates of bacterial activities.

Rude and Aller (1991, p. 2492) reported on similar areas lacking plant cover at Bob Allen Keys Bank, which "commonly occur within the otherwise continuous *Thalassia testudinum* beds covering mudbanks in the Bay." They referred to these as "blowout areas," but emphasized that the storm-related reworking and/or erosion traditionally implied by this term were not necessarily the mode of origin. Nevertheless, uniform ^{210}Pb activity in the upper ~20 cm suggested rapid accumulation or biological or physical mixing on time scales of 10 yr or less. Below 20 cm, their profile for excess ^{210}Pb is consistent with decay under slower rates of sedimentation (maximum 1 cm/yr) or an absence of pronounced mixing. Using a mass balance model incorporating carbonate mineral reactions, fluorapatite formation, and pore-water profiles for F, Sr, and Ca, Rude and Aller argued for aragonite and high-Mg calcite dissolution and low-Mg calcite precipitation, with net loss of $CaCO_3$ through dissolution. These results, when compared to our strongly negative Ca excess indicating net carbonate precipitation, highlight the possibility of variance among the grass-free muds of Bob Allen Keys. As an additional example of this variability, Robbins et al. (2000) measured ^{210}Pb at a barren site at Bob Allen Keys and failed to observe the essentially constant activity described here and by Rude and Aller (1991). Most importantly, we stress that these sites of extreme bacterial activity and potentially abundant diagenetic $CaCO_3$ precipitation are anomalous compared to the grassy expanses that dominated much of the Bay where net dissolution often prevails (Fig. 1; Walter and Burton, 1990; Walter et al., 1993; Ku et al., 1999).

We observed sharp boundaries between reactive barren sites at Bob Allen and adjacent seagrass-covered areas. Past workers in the Bob Allen Keys area have suggested burial of former seagrass beds by fine-grained sediment (Orem et al., 1999). However, there were no hints of topographic controls that might have localized very rapid sediment accumulation in the barren areas. Given these observations, the uniform excess ^{210}Pb profile, and the sharp interface observed at the base of the soupy, reactive layer, it is reasonable to imagine that the seagrass-free sites represent blowouts in the true sense—that is, rapidly reworked sediment associated with storm processes. By these processes, reactive organic phases that would normally be lost through aerobic degradation at the sediment-water interface under normal, steady-state depositional conditions were rapidly mixed into the subsurface (e.g., the abundant grass fragments in Fig. 3), where they supported anomalously high rates of BSR. The positive relationship between rapid burial and high rates of sulfate reduction has long been known (Toth and Lerman, 1977; Berner, 1980); such rapid burial can occur either through mixing or high sediment inputs.

Nevertheless, there are two arguments against the reworking-mixing model that instead favor very rapid accumulation of sediment newly transported to the barren sites of accumulation. First, the porosity profile of Figure 5 shows a downcore decrease within the reactive layer, suggesting progressive accumulation rather than instantaneous emplacement or mixing. (Note the porosity contrast across the interface.) Second, the excess ^{210}Pb profile in Figure 6 is not offset across the interface at 18 cm. A discontinuity would be expected with the reworking associated with storm processes. Also, the profile could show uniform values as a result of very rapid sedimentation over a short time interval rather than mixing.

We are left with seemingly contradictory observations. It is not clear why rapid sedimentation would have been highly localized, yet in situ reworking is also not straightforwardly indicated by the collective data. What is clear, however, is that physical processes are involved in the production of at least some of the sites of anomalous microbial activity within Florida Bay and that a chemical feedback—i.e., high H_2S resulting from degradation of rapidly buried organic phases—perpetuates the absence of seagrass (see also Carlson et al., 1994). The absence of the baffling and binding effects of grass cover could lead to further burial of labile organic material by rapid mixing or through scour followed by rapid sediment replacement.

Seagrass die-off is epidemic in parts of Florida Bay, and proposed causative factors (as reviewed in Carlson et al. [1994]) include salinity and nutrient perturbations, pathogens, and toxic compounds—all of which could have anthropogenic ties. Although the critical cause-and-effect relationships remain uncertain, it is unlikely that the highly reactive character of the F-6 muds and equivalent sites at Bob Allen Keys and elsewhere in the Bay initiated with seagrass die-off. Our ongoing work is further exploring this and related sites, including the character of seagrass-covered areas immediately adjacent to the barren patches.

While the F-6 site and the associated quantity and rate of $CaCO_3$ precipitation might be anomalous in the overall scheme of modern carbonate mud accumulation, and details remain uncertain regarding the origin of this highly reactive site, it is an ideal natural laboratory for testing the newly emerging CAS approach to ancient seawater chemistry. At F-6, appreciable net $CaCO_3$ precipitation is occurring in pore waters that show pronounced ^{34}S enrichments within the dissolved sulfate reservoir. Such diagenetic effects could shift the bulk CAS reservoir away from the $\delta^{34}S$ of the overlying seawater. Our work in Australia (Table 5) shows that a variety of skeletal grains, including calcareous green algae, initially record the sulfur isotope composition of modern seawater sulfate. By inference, we can assume the muds in Florida Bay begin with a primary seawater signal, but despite the unequivocal evidence for net $CaCO_3$ precipitation and perhaps additional cycling by dissolution and reprecipitation within a highly evolved pore-water environment, the muds at the F-6 site retain essentially the $\delta^{34}S$ of the primary sediment and seawater. In anticipation of additional data from our ongoing study of these sites, we tentatively suggest that the primary value is preserved as a result of a favorable mass balance. Specifically, the diagenetic carbonate would be volumetrically insufficient to perturb the $\delta^{34}S$ of the bulk lime mud away from the primary value.

Rude and Aller (1991) suggested that early carbonate-mineral reactivity might be more extreme than what the excess calcium data suggest. Excess Ca tracks only the extent of net $CaCO_3$ precipitation or dissolution. As a result, the dissolution-reprecipitation intrinsic to carbonate mineral recrystallization and mineral transformation (e.g., aragonite to low-Mg calcite stabilization) should also record the ambient pore-water $\delta^{34}S$. Hover et al. (2001) also argued for appreciable alteration of muddy sediment in Florida Bay based on fine-scale changes in the crystal morphologies of biogenic aragonite and high-Mg calcite. Although appreciable mineralogical change did not accompany the dissolution and reprecipitation that resulted in the recrystallization documented by Hover et al., shifts in trace and minor element compositions are a distinct possibility. From the perspective of CAS, however, the data of Rude and Aller and Hover et al. must exaggerate the extent of carbonate mineral transformation and/or recrystallization at F-6, or the primary inputs of mud must swamp the diagenetic overprint. We also need to explore the possibility that carbonate precipitation during diagenesis may sequester less sulfate than the original aragonite mud. Such differences could reflect varying $CaCO_3$ mineralogy, crystal size, or some unknown control linked specifically to the environmental conditions of carbonate precipitation. Despite remaining uncertainties, these results confirm that CAS can be a viable proxy for seawater sulfate even at sites of appreciable microbial activity. The implications of this important result are discussed in detail below.

IMPLICATIONS FOR THE CARBONATE-ASSOCIATED SULFATE PALEOCEANOGRAPHIC METHOD

The CAS method has already emerged as a tool of proven paleoceanographic value, but its full potential is not yet known. To date, a number of studies have documented the close match between $\delta^{34}S$ values for CAS, primarily from skeletal grains, and coeval evaporite deposits (Burdett et al., 1989; Strauss, 1999; Kampschulte et al., 2001; Kampschulte and Strauss, 2004; Lyons et al., 2004). Most recently, CAS results from whole-rock dolomicrites of Mesoproterozoic age, which match interbedded gypsum deposits by ~1‰–3‰ or better and thus approximate ~1.2 Ga seawater (Kah et al., 2001; Lyons et al., 2004), hint of the promise for bulk analyses even in samples of Precambrian age (see also Hurtgen et al., 2002; Lyons et al., 2004). What is less known, however, is the full integrity of the method under a broad range of depositional and diagenetic conditions. To date, studies incorporating bulk samples of ancient fine-grained carbonate have stressed C_{org}-deficient samples out of concern for diagenetic resetting or overprinting within subsurface environments characterized by intense BSR.

Despite the extreme rates of BSR at the F-6 site of the present study, the highly evolved $\delta^{34}S$ of the pore-water sulfate reservoir, and evidence for appreciable net subsurface $CaCO_3$ precipitation, the CAS in the bulk muds remains uniformly consistent with the value for seawater sulfate. We note that this is not an artifact of insufficient rinsing of seawater salts precipitated during sample drying. Recall that the $\delta^{34}S$ values of pore-water sulfate in these samples are significantly enriched in ^{34}S relative to the overlying seawater. It is these ^{34}S-enriched values that would be observed if secondary salts were insufficiently rinsed. Instead, the $\delta^{34}S$ of CAS bears the signature of the primary carbonate grains and the overlying seawater. These primary inputs likely swamp any secondary signal from $CaCO_3$ precipitation.

Mineralogical transformations from aragonite to calcite in the presence of a modified sulfate reservoir have the potential to shift the bulk mud CAS away from the seawater $\delta^{34}S$ value. If Rude and Aller (1991) are correct in their estimates of early carbonate reactivity at Bob Allen Keys and their results apply to our site, such carbonate mineral transformations do not appear to have a large effect on the $\delta^{34}S$ of CAS. We are in the process of addressing the effects of mineral stabilization through deeper coring in Florida Bay.

The effects of carbonate diagenesis on CAS concentrations are also a concern. I. Gavrieli (2001, personal commun.) and others (e.g., Staudt and Schoonen, 1995) have shown very high concentrations of CAS in aragonite relative to calcite. While the $\delta^{34}S$ of CAS should be buffered to primary values during dissolution-reprecipitation in the presence of sulfate deficient meteoric fluids or evolved seawater, dramatic concentration decreases are possible during transformations from aragonite to calcite. Despite the possibility of CAS loss during diagenesis, at least some Precambrian carbonates show the predicted relationships between CAS concentration and independent local facies constraints on sulfate levels in the early seawater (L. Kah, 2003, personal commun.). Furthermore, Hurtgen et al. (2002) and Pavlov et al. (2003) have used low CAS concentrations in Proterozoic carbonates as an indication of the sulfate-deficient global ocean that likely existed at this time (Canfield, 1998; Shen et al., 2003; Lyons et al., 2004).

These mineralogical considerations have particular relevance in light of models for calcite versus aragonite seas over geologic time (Sandberg, 1983; Wilkinson and Given, 1986) and our ability to use both CAS concentrations and isotopic relationships to reconstruct ancient environments. What appears certain, however, is that CAS in bulk micrite and dolomicrite samples has strong potential to preserve the primary $\delta^{34}S$ of seawater—even at sites of active early carbonate precipitation driven by high bacterial activity.

ACKNOWLEDGMENTS

Carol Nabelek and Ted Huston provided invaluable analytical assistance. Jon Fong, Mike Formolo, and Steve Studley assisted with the CAS isotope analyses at Indiana University. Billy Moore provided the ^{210}Pb data, and Max Coleman generated the $\delta^{34}S$ results for dissolved sulfate and sulfide. Tracy Frank made the collection of samples in Australia possible. She and Linda Kah are also thanked for many valuable discussions. We are grateful to Rolf Arvidson and two anonymous reviewers for their insightful comments. This work was supported by National Science Foundation grants OCE-9102545 (LMW) and EAR-9725538 and EAR-0207565 (TWL).

REFERENCES CITED

Aharon, P., and Fu, B., 2003, Sulfur and oxygen isotopes of coeval sulfate-sulfide in pore fluids of cold seep sediments with sharp redox gradients: Chemical Geology, v. 195, p. 201–218, doi: 10.1016/S0009-2541(02)00395-9.

Aller, R.C., 2001, Transport and reactions in the bioirrigated zone, *in* Boudreau B.P., and Jørgensen, B.B., eds., The benthic boundary layer: Oxford University Press, p. 269–301.

Archer, D., Emerson, S., and Reimers, C., 1989, Dissolution of calcite in deep-sea sediments: pH and O_2 microelectrode results: Geochimica et Cosmochimica Acta, v. 53, p. 2831–2845, doi: 10.1016/0016-7037(89)90161-0.

Ben-Yaakov, S., 1973, pH buffering of pore water of recent anoxic marine sediments: Limnology and Oceanography, v. 18, p. 86–94.

Berner, R.A., 1980, Early diagenesis—A theoretical approach: Princeton, Princeton University Press, 241 p.

Bosence, D., 1989, Surface sublittoral sediments of Florida Bay: Bulletin of Marine Science, v. 44, p. 434–453.

Boss, S.K., and Neumann, A.C., 1993, Physical versus chemical processes of "whiting" formation in the Bahamas: Carbonates and Evaporites, v. 8, p. 135–148.

Boudreau, B.P., 1997, Diagenetic models and their implementation—Modelling transport and reactions in aquatic sediments: New York, Springer-Verlag, 414 p.

Boudreau, B.P., and Canfield, D.E., 1993, A comparison of closed- and open-system models for porewater pH and calcite dissolution: Geochimica et Cosmochimica Acta, v. 57, p. 317–334, doi: 10.1016/0016-7037(93)90434-X.

Broecker, W.S., Langdon, C., Takahashi, T., and Peng, T.-H., 2001, Factors controlling the rate of $CaCO_3$ precipitation on Great Bahama Bank: Global Biogeochemical Cycles, v. 15, p. 589–596, doi: 10.1029/2000GB001350.

Broecker, W.S., Sanyal, A., and Takahashi, T., 2000, The origin of Bahamian whitings revisited: Geophysical Research Letters, v. 27, p. 3759–3760, doi: 10.1029/2000GL011872.

Burdett, J.W., Arthur, M.A., and Richardson, M., 1989, A Neogene seawater sulfate isotope age curve from calcareous pelagic microfossils: Earth and Planetary Science Letters, v. 94, p. 189–198, doi: 10.1016/0012-821X(89)90138-6.

Burns, S.J., and Swart, P.K., 1992, Diagenetic processes in Holocene carbonate sediments: Florida Bay mudbanks and islands: Sedimentology, v. 39, p. 285–304.

Canfield, D.E., 1989, Reactive iron in marine sediments: Geochimica et Cosmochimica Acta, v. 53, p. 619–632, doi: 10.1016/0016-7037(89)90005-7.

Canfield, D.E., 1998, A new model for Proterozoic ocean chemistry: Nature, v. 396, p. 450–453, doi: 10.1038/24839.

Canfield, D.E., 2001, Isotope fractionation by natural populations of sulfate-reducing bacteria: Geochimica et Cosmochimica Acta, v. 65, p. 1117–1124, doi: 10.1016/S0016-7037(00)00584-6.

Canfield, D.E., and Raiswell, R., 1991, Carbonate precipitation and dissolution: Its relevance to fossil preservation, *in* Allison, P.A., and Briggs, D.E.G., eds., Taphonomy: Releasing the data locked in the fossil record: Plenum Press, New York, p. 337–387.

Canfield, D.E., and Thamdrup, B., 1994, The production of ^{34}S-depleted sulfide during bacterial disproportionation of elemental sulfur: Science, v. 266, p. 1973–1975.

Carlson, P.R., Jr., Yarbro, L.A., and Barber, T.R., 1994, Relationship of sediment sulfide to mortality of *Thalassia testudinum* in Florida Bay: Bulletin of Marine Research, v. 54, p. 733–746.

Chambers, L.A., Trudinger, P.A., Smith, J.W., and Burns, M.S., 1975, Fractionation of sulfur isotopes by continuous cultures of *Desulfovibrio desulfuricans*: Canadian Journal of Microbiology, v. 21, p. 1602–1607.

Chanton, J.P., and Martens, C.S., 1985, The effects of heat and stannous chloride addition on the active distillation of acid volatile sulfide from pyrite-rich marine sediment samples: Biogeochemistry, v. 1, p. 375–383.

Cline, J.D., 1969, Spectrophotometric determination of hydrogen sulfide in natural water: Limnology and Oceanography, v. 14, p. 454–458.

Detmers, J., Brüchert, V., Habicht, K.S., and Kuever, J., 2001, Diversity of sulfur isotope fractionations by sulfate-reducing prokaryotes: Applied and Environmental Microbiology v. 67, p. 888–894, doi: 10.1128/AEM.67.2.888-894.2001.

Ding, T., Valkiers, S., Kipphardt, H., De Bièvre, P., Taylor, P.D.P., Gonfiantini, R., and Krouse, R., 2001, Calibrated sulfur isotope abundance ratios of three IAEA sulfur isotope reference materials and V-CDT with a reassessment of the atomic weight of sulfur: Geochimica et Cosmochimica Acta, v. 65, p. 2433–2437, doi: 10.1016/S0016-7037(01)00611-1.

Fossing, H., 1995, ^{35}S-radiolabeling to probe biogeochemical cycling of sulfur, *in* Vairavamurthy, M.A., and Schoonen, M.A.A., eds., Geochemical transformations of sedimentary sulfur: American Chemical Society Symposium Series 612, p. 348–364.

Gardner, L.R., 1973, Chemical models for sulfate reduction in closed anaerobic marine environments: Geochimica et Cosmochimica Acta, v. 37, p. 53–68, doi: 10.1016/0016-7037(73)90243-3.

Goldhaber, M.B., and Kaplan, I.R., 1974, The sulfur cycle, *in* E.D. Goldberg, ed., The sea (v. 5): Wiley, New York, p. 569–655.

Habicht, K.S., and Canfield, D.E., 2001, Isotope fractionation by sulfate-reducing natural populations and the isotopic composition of sulfide in marine sediments: Geology, v. 29, p. 555–558, doi: 10.1130/0091-7613(2001)0292.0.CO;2.

Habicht, K.S., Gade, M., Thamdrup, B., Berg, P., and Canfield, D., 2002, Calibration of sulfate levels in the Archean Ocean: Science, v. 298, p. 2372–2374, doi: 10.1126/SCIENCE.1078265.

Hover, V.C., Walter, L.M., and Peacor, D.R., 2001, Early marine diagenesis of biogenic aragonite and Mg-calcite: New constraints from high-resolution STEM and AEM analyses of modern platform carbonates: Chemical Geology, v. 175, p. 221–248, doi: 10.1016/S0009-2541(00)00326-0.

Hurtgen, M.T., Arthur, M.A., Suits, N., and Kaufman, A.J., 2002, The sulfur isotopic composition of Neoproterozoic seawater sulfate: Implications for a snowball Earth?: Earth and Planetary Science Letters, v. 203, p. 413–429, doi: 10.1016/S0012-821X(02)00804-X.

Hurtgen, M.T., Lyons, T.W., Ingall, E.D., and Cruse, A.M., 1999, Anomalous enrichments of iron monosulfide in euxinic sediments and the role of H_2S in iron sulfide transformations: Examples from Effingham Inlet, the Orca Basin and the Black Sea: American Journal of Science, v. 299, p. 556–588.

Jørgensen, B.B., 1978, A comparison of methods for the quantification of bacterial sulfate reduction in coastal marine sediments, I, Measurement with radiotracer techniques: Geomicrobiology Journal, v. 1, p. 11–28.

Jørgensen, B.B., Böttcher, M.E., Lüschen, H., Neretin, L.N., and Volkov, I.I., 2004, Anaerobic methane oxidation and a deep H_2S sink generate isotopically heavy sulfides in Black Sea sediments: Geochimica et Cosmochimica Acta, v. 68, p. 2095–2118.

Kah, L.C., Lyons, T.W., and Chesley, J.T., 2001, Geochemistry of a 1.2 Ga carbonate-evaporite succession, northern Baffin and Bylot Islands: Implications for Mesoproterozoic marine evolution: Precambrian Research, v. 111, p. 203–234.

Kampschulte, A., and Strauss, H., 1998, The isotopic composition of trace sulphates in Paleozoic biogenic carbonates: Implications for coeval seawater and geochemical cycles: Mineralogical Magazine, v. 62A, p. 744–745.

Kampschulte, A., and Strauss, H., 2004, The sulfur isotopic evolution of Phanerozoic seawater based on the analysis of structurally substituted sulfate in carbonates: Chemical Geology, v. 204, p. 255–286.

Kampschulte, A., Bruckschen, P., and Strauss, H., 2001, The sulphur isotopic composition of trace sulphates in Carboniferous brachiopods: Implications for coeval seawater, correlation with other geochemical cycles and isotope stratigraphy: Chemical Geology, v. 175, p. 149–173, doi: 10.1016/S0009-2541(00)00367-3.

Kaplan, I.R., and Rittenberg, S.C., 1964, Microbiological fractionation of sulphur isotopes: Journal of General Microbiology, v. 34, p. 195–212.

Ku, T.C.W., Walter, L.M., Coleman, M.L., Blake, R.E., and Martini, A.M., 1999, Coupling between sulfur recycling and syndepositional carbonate dissolution: Evidence from oxygen and sulfur isotope composition of pore water sulfate, South Florida Platform, U.S.A: Geochimica et Cosmochimica Acta, v. 63, p. 2529–2546, doi: 10.1016/S0016-7037(99)00115-5.

Lyons, T.W., 1997, Sulfur isotopic trends and pathways of iron sulfide formation in upper Holocene sediments of the anoxic Black Sea: Geochimica et Cosmochimica Acta, v. 61, p. 3367–3382, doi: 10.1016/S0016-7037(97)00174-9.

Lyons, T.W., Kah, L.C., and Gellatly, A.M., 2004, The Precambrian sulphur isotope record of evolving atmospheric oxygen, *in* Eriksson, P.G., et al., eds., The Precambrian Earth: Tempos and Events: Developments in Precambrian Geology: Amsterdam, Elsevier, p. 421–439.

Milliman, J.D., 1974, Marine carbonates: New York, Springer-Verlag, 375 p.

Moore, W.S., and O'Neill, D.J., 1991, Radionuclide distributions in recent Black Sea sediments, *in* Izdar, E., and Murray, J.W., eds., Black Sea Oceanography: NATO ASI Series: Kluwer, Dordrecht, Series C, v. 351, p. 257–270.

Morse, J.W., Gledhill, D.K., and Millero, F.J., 2003, $CaCO_3$ precipitation kinetics in waters from the Great Bahama Bank: Implications for the relationship between Bank hydrochemistry and whitings: Geochimica et Cosmochimica Acta, v. 67, p. 2819–2826, doi: 10.1016/S0016-7037(03)00103-0.

Morse, J.W., Zullig, J.J., Bernstein, L.D., Millero, F.J., Milne, P., Mucci, A., and Choppin, G.R., 1985, Chemistry of calcium carbonate-rich shallow water sediments in the Bahamas: American Journal of Science, v. 285, p. 147–185.

Neumann, A.C., and Land, L.S., 1975, Lime mud deposition and calcareous algae in the Bight of Abaco, Bahamas: A budget: Journal of Sedimentary Petrology, v. 45, p. 763–786.

Orem, W.H., Holmes, C.W., Kendall, C., Lerch, H.E., Bates, A.L., Silva, S.R., Boylan, A., Corum, M., Marot, M., and Hedgman, C., 1999, Geochemistry of Florida Bay sediments: Nutrient history at five sites in eastern and central Florida Bay: Journal of Coastal Research, v. 15, p. 1055–1071.

Pavlov, A.A., Hurtgen, M.T., Kasting, J.F., and Arthur, M.A., 2003, Methane-rich Proterozoic atmosphere?: Geology, v. 31, p. 87–90, doi: 10.1130/0091-7613(2003)0312.0.CO;2.

Raiswell, R., and Canfield, D.E., 1998, Sources of iron for pyrite formation in marine sediments: American Journal of Science, v. 298, p. 219–245.

Raiswell, R., Canfield, D.E., and Berner, R.A., 1994, A comparison of iron extraction methods for the determination of degree of pyritisation and the recognition of iron-limited pyrite formation: Chemical Geology, v. 111, p. 101–110, doi: 10.1016/0009-2541(94)90084-1.

Randazzo, A.F., and Jones, D.S., 1997, The geology of Florida: Gainesville, University Press of Florida, 327 p.

Rees, C.E., Jenkins, W.J., and Monster, J., 1978, The sulphur isotopic composition of ocean water sulphate: Geochimica et Cosmochimica Acta, v. 42, p. 377–381.

Robbins, J.A., Holmes, C., Halley, R., Bothner, M., Shinn, E., Graney, J., Keeler, G., tenBrink, M., Orlandini, K.A., and Rudnick, D., 2000, Time-averaged fluxes of lead and fallout radionuclides to sediment in Florida Bay: Journal of Geophysical Research, v. 105, p. 28805–28821, doi: 10.1029/1999JC000271.

Robbins, L.L., and Blackwelder, P.L., 1992, Biochemical and ultrastructural evidence for the origin of whitings: A biologically induced calcium carbonate precipitation mechanism: Geology, v. 20, p. 464–468.

Robbins, L.L., Tao, Y., and Evan, C.A., 1997, Temporal and spatial distribution of whitings on Great Bahama Bank and a new lime mud budget: Geology, v. 25, p. 947–950, doi: 10.1130/0091-7613(1997)0252.3.CO;2.

Rosenfeld, J.K., 1979, Interstitial water and sediment chemistry of two cores from Florida Bay: Journal of Sedimentary Petrology, v. 49, p. 989–994.

Rude, P.D., and Aller, R.C., 1991, Fluorine mobility during early diagenesis of carbonate sediment: An indicator of mineral transformations: Geochimica et Cosmochimica Acta, v. 55, p. 2491–2509, doi: 10.1016/0016-7037(91)90368-F.

Sandberg, P.A., 1983, An oscillating trend in Phanerozoic non-skeletal carbonate mineralogy: Nature, v. 305, p. 19–22.

Shen, Y., Knoll, A.H., and Walter, M.R., 2003, Evidence for low sulphate and anoxia in a mid-Proterozoic marine basin: Nature, v. 423, p. 632–635, doi: 10.1038/NATURE01651.

Shinn, E.A., Steinen, R.P., Lidz, B.H., and Swart, P.K., 1989, Whitings, a sedimentologic dilemma: Journal of Sedimentary Petrology, v. 59, p. 147–161.

Staudt, W.J., and Schoonen, M.A.A., 1995, Sulfate incorporation into sedimentary carbonates, *in* Vairavamurthy, M.A., and Schoonen, M.A.A., eds., Geochemical transformations of sedimentary sulfur: American Chemical Society Symposium Series 612, p. 332–345.

Stockman, K.W., Ginsburg, R.N., and Shinn, E.A., 1967, The production of lime mud by algae in South Florida: Journal of Sedimentary Petrology, v. 37, p. 633–648.

Strauss, H., 1999, Geological evolution from isotope proxy signals—Sulfur: Chemical Geology, v. 161, p. 89–101, doi: 10.1016/S0009-2541(99)00082-0.

Swart, P.K., Berler, D., McNeill, D., Guzikowski, M., Harrison, S.A., and Dedick, E., 1989, Interstitial water geochemistry and carbonate diagenesis in the sub-surface of a Holocene mud island in Florida Bay: Bulletin of Marine Science, v. 44, p. 490–514.

Taylor, S.R., and McLennan, S.M., 1985, The continental crust: Its composition and evolution: Oxford, Blackwell, 312 p.

Toth, D.J., and Lerman, A., 1977, Organic matter reactivity and sedimentation rates in the ocean: American Journal of Science, v. 277, p. 265–285.

Walter, L.M., and Burton, E.A., 1990, Dissolution of recent platform carbonate sediments in marine pore fluids: American Journal of Science, v. 290, p. 601–643.

Walter, L.M., Bischof, S.A., Patterson, W.P., and Lyons, T.W., 1993, Dissolution and recrystallization in modern shelf carbonates: evidence from pore water and solid phase chemistry: Philosophical Transactions of the Royal Society of London, Series A: Mathematical and Physical Sciences, v. 344, p. 27–36.

Wanless, H.R., and Tagett, M.G., 1989, Origin, growth and evolution of carbonate mudbanks in Florida Bay: Bulletin of Marine Science, v. 44, p. 454–489.

Wilkinson, B.H., and Given, R.K., 1986, Secular variation in abiotic marine carbonate: Constraints on Phanerozoic atmospheric carbon dioxide contents and oceanic Mg/Ca ratios: Journal of Geology, v. 94, p. 321–334.

Wortmann, U.G., Bernasconi, S.M., and Böttcher, M.E., 2001, Hypersulfidic deep biosphere indicates extreme sulfur isotope fractionation during single-step microbial sulfate reduction: Geology, v. 29, p. 647–650, doi: 10.1130/0091-7613(2001)0292.0.CO;2.

Zaback, D.A., Pratt, L.M., and Hayes, J.M., 1993, Transport and reduction of sulfate and immobilization of sulfide in marine black shales: Geology, v. 21, p. 141–144, doi: 10.1130/0091-7613(1993)0212.3.CO;2.

Manuscript Accepted by the Society March 28, 2004

Printed in the USA

Geological Society of America
Special Paper 379
2004

The sulfur isotope composition of carbonate-associated sulfate in Mesoproterozoic to Neoproterozoic carbonates from Death Valley, California

Matthew T. Hurtgen*
Michael A. Arthur
Penn State Astrobiology Research Center and Department of Geosciences, Pennsylvania State University, University Park, Pennsylvania 16802, USA

Anthony R. Prave
School of Geography and Geosciences, University of St. Andrews, St. Andrews, Fife, KY16 9AL, Scotland

ABSTRACT

We have analyzed the concentration and sulfur isotope composition of trace sulfate in carbonate from three Proterozoic formations in Death Valley, California. Trace sulfate concentrations for the Crystal Spring Formation and Beck Spring Dolomite, which were deposited in the late Mesoproterozoic and mid-Neoproterozoic and are not associated with glacial sediments, range from 0 to 144 ppm with $\delta^{34}S_{sulfate}$ values spanning 11.0‰–27.4‰. Within these formations, stratigraphic shifts in $\delta^{34}S_{sulfate}$ of up to ~9‰ occur over <50 m. Trace sulfate concentrations for the Noonday Dolomite, which was deposited in the late Neoproterozoic and directly overlies glacial sediments associated with the "snowball Earth" events, range from 2 to 272 ppm with $\delta^{34}S_{sulfate}$ values varying between 15‰ and 35‰. The ~17‰ $\delta^{34}S_{sulfate}$ increase at the base of the Noonday Dolomite is similar in magnitude and rate to the >20‰ positive $\delta^{34}S$ shifts recorded in Neoproterozoic postglacial carbonates from Namibia. The results indicate that the sulfur cycle behaved differently in the late versus early Neoproterozoic as a possible consequence of severe late Neoproterozoic glacial events. Furthermore, based on $\delta^{34}S_{sulfate}$ patterns and carbonate-associated sulfate concentrations recorded in the Crystal Spring and Beck Spring formations, we speculate that late Mesoproterozoic to mid-Neoproterozoic oceanic sulfate concentrations were ~10% of modern values (e.g., ~3 mM).

Keywords: Neoproterozoic, sulfur isotopes, carbonate-associated sulfate, Death Valley, sulfate concentration, snowball Earth.

INTRODUCTION

The late Neoproterozoic sulfur isotope record, recorded in trace sulfate in carbonates, exhibits anomalously large positive shifts (>20‰) over short stratigraphic intervals (<10 m). These large variations have been linked to a Neoproterozoic ocean with low sulfate concentrations and, consequently, a sulfate reservoir more susceptible to change (Hurtgen et al., 2002). The primary source of seawater sulfate is riverine delivery resulting from the oxidative weathering of pyrite and dissolution of calcium sulfate

*Present address: Department of Earth and Planetary Sciences, Harvard University, Cambridge, Massachusetts 02138, USA, mhurtgen@fas.harvard.edu.

Hurtgen, M.T., Arthur, M.A., and Prave, A.R., 2004, The sulfur isotope composition of carbonate-associated sulfate in Mesoproterozoic to Neoproterozoic carbonates from Death Valley, California, *in* Amend, J.P., Edwards, K.J., and Lyons, T.W., eds., Sulfur Biogeochemistry—Past and Present: Geological Society of America Special Paper 379, p. 177–194. For permission to copy, contact editing@geosociety.org.

(evaporite) minerals. Therefore, low levels of oceanic sulfate in the Neoproterozoic have been attributed to reduced oxygen concentrations within Earth's atmosphere-ocean system. Several lines of sedimentological and geochemical evidence indicate that oxygen concentrations in the atmosphere-ocean system were decreased (relative to today) through much of the Neoproterozoic—a period of time preceding the evolution of metazoans (Canfield and Teske, 1996; Knoll and Canfield, 1998).

Sulfate concentrations are believed to have been low through much of the Proterozoic (Canfield and Teske, 1996; Canfield 1998). However, while sulfur isotope measurements of carbonate-associated sulfate (CAS) in Mesoproterozoic carbonates exhibit large sulfur isotope shifts (~16‰) over stratigraphic intervals of ~140–400 m (Kah et al., 2002), none are as large or occur as "rapidly" as those within late Neoproterozoic postglacial sediments. In this context, positive $\delta^{34}S$ deviations in trace sulfate associated with carbonates (Hurtgen et al., 2002) and pyrite (Ross et al., 1995; Gorjan et al., 2000) in sedimentary rocks that were deposited subsequent to late Neoproterozoic glaciations suggest that, in addition to low oceanic sulfate concentrations, sulfur cycling may have been strongly influenced by changes in ocean chemistry accompanying severe glaciations.

We have analyzed the sulfur isotope composition of CAS from three formations in Death Valley, California: the late Mesoproterozoic to Neoproterozoic Crystal Spring Formation and the Neoproterozoic Beck Spring Dolomite and Noonday Dolomite. At least one of these, the Crystal Spring Formation, predate the proposed late Neoproterozoic "snowball Earth" events, whereas the Noonday Dolomite directly overlies presumed Marinoan-equivalent glacial deposits. Our objectives were to: (1) extend a protocol that has been used to assess the integrity of $\delta^{13}C_{carbonate}$ in Proterozoic carbonates to address diagenetic concerns and the possibility for retention of primary $\delta^{34}S_{CAS}$ values; (2) examine $\delta^{34}S_{sulfate}$ in a postglacial carbonate (Noonday Dolomite) in order to ascertain the global nature of sulfur cycling in the aftermath of severe Neoproterozoic glaciations; (3) examine $\delta^{34}S_{sulfate}$ in carbonates not directly associated with Neoproterozoic glacial deposits in order to determine whether the large $\delta^{34}S_{sulfate}$ variations were unique to postglacial rocks or, alternatively, represented a style of change that was characteristic of the entire Era; and (4) further assess the utility of CAS concentrations as a proxy for Neoproterozoic sulfate concentrations.

METHODS

The method of extraction and isotopic analysis of sulfate in carbonates used in this study is modified slightly from that described by Burdett et al. (1989). In short, carbonate rock samples ranging in mass from 15 to 80 g were ground and soaked in a 5.25% sodium hypochlorite solution for ~24 h to remove any non-CAS, organic sulfur compounds and metastable sulfides. The sediment was rinsed with deionized water, dissolved in 3 *N* HCl, and the insoluble residues were removed using 0.5 µm filters. Approximately 25–30 mL of saturated $BaCl_2$ was added to the filtrate and brought to a near boil for more than 4 h. The solution was allowed to cool, and the precipitated barium sulfate was filtered through Whatman No. 42 ashless filter papers.

The preparation of sulfate minerals for isotopic analysis was modified from Ueda and Krouse (1986). Approximately 2–10 mg of $BaSO_4$ sample (equivalent to 9–43 µmoles of S) was mixed with equal amounts of V_2O_5 and combusted in an Elemental Analyzer at 1000 °C to determine sulfate concentrations and isolate SO_2 for isotopic analysis. Sulfate yields of 95% and higher were achieved for the barite standard (NBS 127) using this technique. The purified SO_2 was sealed in pyrex tubing and analyzed on a VG Prism Series II isotope ratio mass spectrometer. Sulfur isotope ratios are expressed as per mil (‰) deviations from the S isotope composition of Cañon Diablo Troilite (CDT) using the conventional delta ($\delta^{34}S$) notation. Sulfur isotope results were generally reproducible within ±0.2‰.

Concentrations of Ca, Mg, Mn, and Sr in carbonate were determined on an inductively coupled plasma (ICP) spectrophotometer in the Pennsylvania State Materials Characterization Laboratory. Approximately 0.25 mg of sample was weighed, reacted with 5% acetic acid, and diluted with deionized water. Insoluble residues were removed via filtration, dried, and weighed. The mass of insoluble residue was subtracted from the original mass to calculate trace- and minor-element concentrations relative to 100% carbonate and reported in ppm.

SAMPLES AND RESULTS

In the Death Valley region, carbonates were collected from the middle and upper portion of the Crystal Spring Formation (Alexander Hills and Saddle Peak Hills localities), the Beck Spring Dolomite (Alexander Hills), and the Noonday Dolomite (southern Nopah Range) (Figure 1). The Crystal Spring Formation is a mixed siliciclastic and carbonate unit that has been interpreted to represent fluvial and near-shore marine environments (Roberts, 1982). A diabase sill intrudes the middle Crystal Spring Formation but is truncated by a major unconformity at the base of the upper Crystal Spring Formation (Prave, 1994). The sills have yielded a U-Pb date of 1.08 Ga (Heaman and Grotzinger, 1992).

The Beck Spring Dolomite occurs stratigraphically above the Crystal Spring Formation and is comprised mainly of microbially laminated dolostone that is slumped locally and contains varying amounts of rip-up clasts and sedimentary breccias (Fig. 2). Age constraints are poor, but Dehler et al. (2001) suggested that the top of the Beck Spring Dolomite is ca. 742 Ma based on lithologic correlations to a radiometrically dated ash layer at the top of the Chuar Group in the Grand Canyon, Arizona (Karlstrom et al., 2000). The temporal relationship between the Beck Spring Dolomite and the Sturtian (ca. 750–720 Ma) glacial episode is contentious. The Beck Spring Dolomite has not been found to occur stratigraphically above any glacial deposits. However, Corsetti and Kaufman (2003) have suggested that geochemical and textural similarities of the lower Beck Spring beds (e.g., organic-rich microbial laminites and roll-up structures)

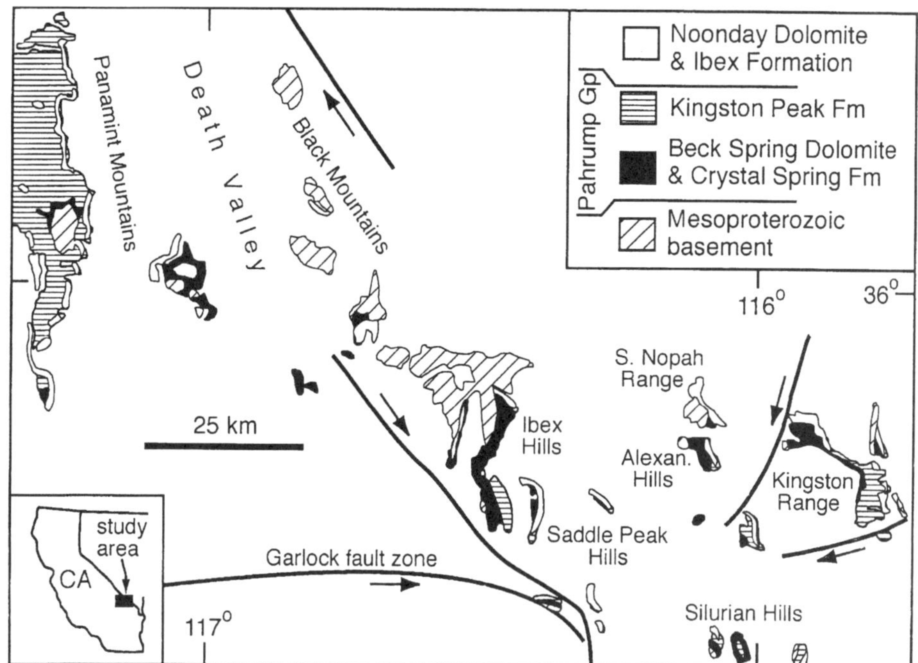

Figure 1. Pahrump Group outcrop belt, Death Valley region, California. Samples were collected from exposures in the Alexander Hills, Saddle Peak Hills, and southern Nopah Range.

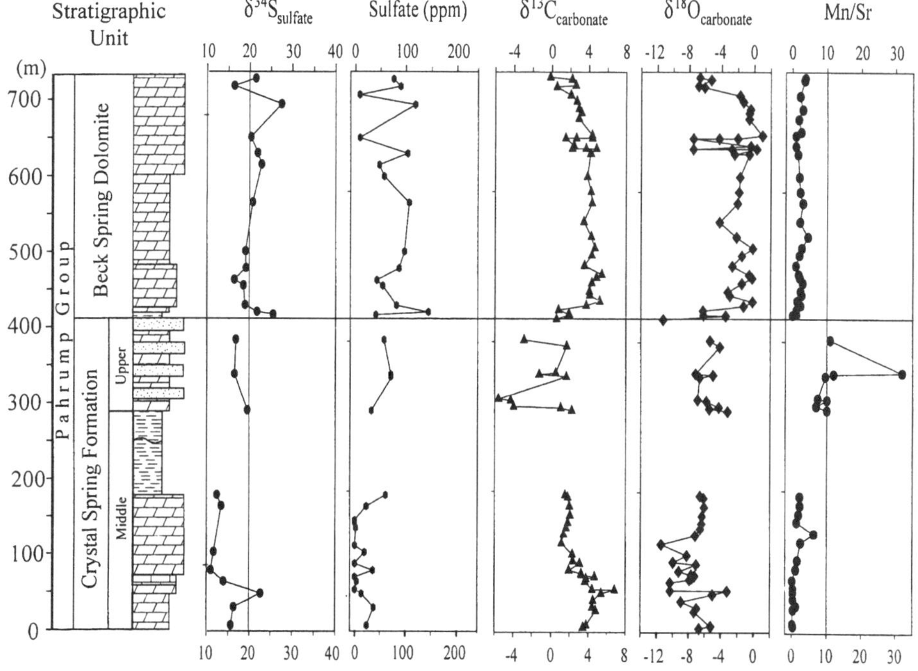

Figure 2. The sulfur isotope composition of carbonate-associated sulfate (CAS), CAS concentrations, $\delta^{13}C_{carbonate}$, $\delta^{18}O_{carbonate}$, and Mn/Sr plotted against late Mesoproterozoic to early Neoproterozoic carbonates collected from the middle and upper Crystal Spring Formation and Beck Spring Dolomite of Death Valley, California.

make them a possible equivalent to the Rasthof cap carbonate in Namibia—which overlies a Sturtian diamictite (Chuos Formation). By contrast, Prave (1999) suggested that the Kingston Peak Formation, which occurs stratigraphically above the Beck Spring Dolomite, contains both the Sturtian- and Marinoan- (ca. 600 Ma) equivalent glacial deposits and, therefore, the Beck Spring Dolomite predates the Sturtian glacial event. Regardless, if the top of the Beck Spring Dolomite is ca. 750 Ma, the middle Crystal Spring Formation and Beck Spring Dolomite together may encompass ~330 m.y. The contact between the upper Crystal Spring Formation and the overlying Beck Spring Dolomite has been identified as a disconformable sequence boundary and, while it is difficult to constrain, the upper Crystal Spring Formation is believed to be closer in age to the Beck Spring Dolomite than the middle Crystal Spring. At any rate, as Heaman and Grotzinger (1992) suggested, significant temporal gaps likely exist in this record.

The Noonday Dolomite rests depositionally on the upper glacial interval (Wildrose Diamictite) of the Kingston Peak Formation. The Noonday is a stratigraphically complex unit containing an intraformational unconformity, which separates a lower dolomicritic and microbial carbonate unit from an overlying mixed carbonate-siliciclastic package of rocks. The lower unit shares the lithologic (pinkish color, tubestone stromatolites, sheet-crack cements, and crystal fans) and C-isotope characteristics of Marinoan-equivalent cap carbonates on other continents (Prave, 1999; Kennedy et al., 1998; Corsetti and Kaufman, 2003). Corsetti and Kaufman (2004) have suggested that the intraformational unconformity that separates the lower from the upper Noonday is glacially related.

The sulfur isotope composition and concentrations of CAS within Crystal Spring and Beck Spring carbonates are presented in Figure 2 and Tables 1 and 2. Carbon and oxygen-isotope data and Mn/Sr ratios are also provided. Because of the length of time represented (ca. 330 Ma) and the existence of major unconformities within these two formations, it is important to consider the data in smaller subsets (i.e., middle Crystal Spring, upper Crystal Spring, and Beck Spring). $\delta^{34}S_{sulfate}$ values for the middle Crystal Spring vary between 11.0‰ and 22.7‰ (mean = 15‰), while CAS concentrations range from 0 to 61 ppm (mean = 24 ppm). $\delta^{34}S_{sulfate}$ varies between 16.4‰ and 19.5‰ (mean = 18‰) for the upper Crystal Spring, with CAS values between 32 and 71 ppm (mean = 58 ppm). $\delta^{34}S_{sulfate}$ values for the Beck Spring fall between 16.3‰ and 27.4‰ (mean = 21‰), and CAS concentrations vary between 8.6 and 144.3 ppm (mean = 73 ppm). Note that both average CAS concentrations and $\delta^{34}S_{sulfate}$ values increase from the middle Crystal Spring through the Beck Spring (see Table 3).

Three sulfur isotope excursions are expressed within the Crystal Spring and Beck Spring formations, two of which are not robust because they are represented by only single points (middle Crystal Spring and top of Beck Spring; Fig. 2). Nevertheless, taken at face value, the data indicate that a positive 6‰ shift, followed by a 9‰ decrease, occurs at the base of the middle Crystal Spring and is not accompanied by a systematic shift in $\delta^{13}C_{carbonate}$ or $\delta^{18}O_{carbonate}$. $\delta^{34}S_{sulfate}$ values are relatively stable in the upper Crystal Spring with values ranging from 16.4‰ to 19.5‰. Over this same interval, $\delta^{13}C$ and $\delta^{18}O$ values show significant scatter, with values ranging from –5.6‰ to 2.2‰ and –7.2‰ to –3.2‰, respectively; Mn/Sr values are elevated relative to other units (Fig. 2).

The second shift occurs above the unconformity that separates the upper Crystal Spring Formation from the Beck Spring Dolomite and is marked by an ~9‰ shift with values increasing from 16.7‰ to 25.5‰. At the base of the Beck Spring, $\delta^{34}S_{sulfate}$ values drop ~9‰ from 25.5‰ to 16.4‰ over 46 m, with the majority of that drop (~7‰) occurring over the initial 12 m. The scatter in $\delta^{13}C$ and $\delta^{18}O$ values in the upper Crystal Spring Formation makes it difficult to determine if isotopic values for these elements shifted across the unconformity; however, $\delta^{13}C$ and $\delta^{18}O$ values in the lower Beck Spring Dolomite increase systematically, coincident with a fall in $\delta^{34}S_{sulfate}$ from 25.5‰ to 16.4‰ (Fig. 2). Finally, a third $\delta^{34}S_{sulfate}$ excursion (~7‰)—again, defined by only a single data point—is expressed near the top of the Beck Spring Dolomite. $\delta^{13}C$ and $\delta^{18}O$ values decrease over this same interval.

Because the middle Crystal Spring is believed to be much older than the upper Crystal Spring and the Beck Spring (late Mesoproterozoic versus middle Neoproterozoic), we have delineated each by different symbols on the plots of Figure 3A–3F. For example, if all three rock units are considered together, a correlation between $\delta^{34}S_{sulfate}$ and $\delta^{18}O_{carbonate}$ (Fig. 3A) exists (r^2 = 0.6). However, if the individual units are considered separately, only a weak correlation exists between $\delta^{34}S_{sulfate}$ and $\delta^{18}O_{carbonate}$ for the middle Crystal Spring (r^2 = 0.33) and the Beck Spring (r^2 = 0.16). There is essentially no correlation between $\delta^{18}O_{carbonate}$ and CAS concentration, CAS and $\delta^{34}S_{sulfate}$, $\delta^{18}O_{carbonate}$ and Mn/Sr, $\delta^{34}S_{sulfate}$ and Mn/Sr, or CAS and Mn/Sr for middle Crystal Spring and Beck Spring carbonates (see Figure 3 caption for r^2 values). Too few points are available to test for correlations in the upper Crystal Spring; however, Mn/Sr are elevated (>10) compared to middle Crystal Spring and Beck Spring carbonates.

The sulfur isotope composition and concentrations of CAS within the Noonday Dolomite are shown in Figure 4 and Table 4. In the lower Noonday, $\delta^{34}S_{sulfate}$ values vary between 15.2‰ and 32.0‰ (mean = 25‰) as CAS concentrations range from 2 to 272 ppm (mean = 64 ppm)(Table 3). $\delta^{34}S_{sulfate}$ values are low initially at ~15‰ and rise sharply to ~32‰ within a few meters of the base of the unit and then decrease upsection to around 20‰. A second excursion with $\delta^{34}S_{sulfate}$ reaching 31‰ occurs further upsection near the top of the lower Noonday. $\delta^{34}S_{sulfate}$ values for the upper Noonday, represented by only three data points, are low initially at 23.3‰ and rise to 34.5‰ (mean = 30‰) upsection over ~50 m, while CAS concentrations range from 9 to 61 ppm (mean = 27 ppm) over the same interval (Table 3). $\delta^{34}S_{sulfate}$ excursions do not correspond to systematic variations in carbon and oxygen isotopes. Figure 5A illustrates that a weak correlation exists between $\delta^{34}S_{sulfate}$ and $\delta^{18}O_{carbonate}$ (r^2 = 0.11 for the lower

TABLE 1. GEOCHEMICAL CHARACTERISTICS OF THE CRYSTAL SPRING FORMATION

Sample	Formation	Depth (m)	$\delta^{34}S_{sulfate}$	Sulfate (ppm)	$\delta^{13}C_{carbonate}$	$\delta^{18}O_{carbonate}$	Ca (ppm)	Mg (ppm)	Mn (ppm)	Sr (ppm)
SP-CSU-10	upper Crystal Spring	602	16.74	57.62	–2.90	–5.37	209220	113475	780	71
SP-CSU-9	upper Crystal Spring	594			1.65	–4.17				
SP-CSU-8	upper Crystal Spring	558			0.46	–7.19	222816	124777	7130	223
SP-CSU-7	upper Crystal Spring	556	16.43	70.92	–1.26	–4.99	230812	125139	2002	167
SP-CSU-6	upper Crystal Spring	553		70.8	1.62	–6.67	214346	120968	4881	509
SP-CSU-5	upper Crystal Spring	523			–5.62	–6.91	172674	101906	7926	1076
SP-CSU-4A	upper Crystal Spring	521			–4.33	–5.80	197429	117080	3214	321
SP-CSU-3	upper Crystal Spring	513			–4.00	–4.28	155361	92951	1195	173
SP-CSU-2	upper Crystal Spring	511			1.01	–5.44				
SP-CSU-1	upper Crystal Spring	507.5	19.46	32.23	2.19	–3.21	209273	123418	1073	107
AH-CSM-47	middle Crystal Spring	395	12.39	61.18	1.51	–6.59				
AH-CSM-46	middle Crystal Spring	392			1.81	–6.13	281121	30865	434	192
AH-CSM-43	middle Crystal Spring	380	13.39	22.62	1.99	–6.07	311014	31438	415	180
AH-CSM-40	middle Crystal Spring	368			2.05	–6.34	346270	22665	435	227
AH-CSM-38A	middle Crystal Spring	361.5		0						
AH-CSM-37	middle Crystal Spring	358		0	1.82	–6.39	348869	25350	350	254
AH-CSM-35	middle Crystal Spring	351		2.02	1.64	–6.54				
AH-CSM-32	middle Crystal Spring	342			1.38	–7.19	295987	40732	1195	190
AH-CSM-28	middle Crystal Spring	330			1.14	–11.46	271709	19547	684	274
AH-CSM-27	middle Crystal Spring	327		0						
AH-CSM-24A	middle Crystal Spring	318	11.68	19.23						
AH-CSM-23	middle Crystal Spring	315			2.31	–8.24				
AH-CSM-20	middle Crystal Spring	306			2.36	–9.98	384372	30750	384	246
AH-CSM-19	middle Crystal Spring	303		0	3.09	–7.07				
AH-CSM-16	middle Crystal Spring	294	11.02	35.67	1.94	–9.25	407012	5945	351	320
AH-CSM-15	middle Crystal Spring	289.4			3.25	–7.68				
AH-CSM-14	middle Crystal Spring	288		0	3.31	–7.26				
AH-CSM-13	middle Crystal Spring	285.25		0	4.69	–7.58				
AH-CSM-12	middle Crystal Spring	283			3.76	–7.89				
AH-CSM-11	middle Crystal Spring	279.25	13.92	3.95	3.67	–10.34	414964	3064	89	919
AH-CSM-10	middle Crystal Spring	268.5		0	4.43	–10.33	371397	11364	291	998
AH-CSM-9	middle Crystal Spring	268			6.84	–3.24				
AH-CSM-8	middle Crystal Spring	263.25	22.66	13.43	5.39	–5.03	210840	118198	170	596
AH-CSM-6	middle Crystal Spring	254			4.53	–8.97	252027	32355	204	596
AH-CSM-4	middle Crystal Spring	245	16.28	37.25	4.44	–7.00	180727	111018	516	452
AH-CSM-3	middle Crystal Spring	240.45			4.81	–7.30	183292	113095	221	715
AH-CSM-2	middle Crystal Spring	221.25	15.63	23.58	3.79	–5.24	169884	103818	189	1416
AH-CSM-1	middle Crystal Spring	218.5			3.45	–6.68	209969	117255	382	1214

Noonday and 0.20 for the entire Noonday). There is no clear relationship between CAS concentration and $\delta^{18}O_{carbonate}$, $\delta^{34}S_{sulfate}$ and CAS concentration, Mn/Sr and $\delta^{18}O_{carbonate}$ or $\delta^{34}S_{sulfate}$ and CAS concentration (Fig. 5B–5F).

DISCUSSION

Diagenetic Considerations

The elemental and stable isotope compositions of carbonates may be affected by any or all of the following processes: early diagenesis, an assortment of fluid-rock interactions, dissolution of primary carbonate and reprecipitation of secondary carbonate, and metamorphism. Assessing the extent of geochemical alteration in carbonates provides critical constraints on the postdepositional history of sediments and their utility as primary recorders of ancient ocean chemistries. Some researchers have utilized petrographic and geochemical attributes to suggest that carbonates within the Beck Spring Dolomite have been overprinted by various postdepositional processes—including the dolomitization of primary calcite and aragonite in mixed marine-meteoric fluids—and therefore primary (e.g., marine) elemental and stable isotopic signals have been compromised (Zempolich et al. 1988; Kenny and Knauth, 2001).

In the present study, we extend a protocol that is used for assessing the integrity of $\delta^{13}C_{carbonate}$ to address diagenetic concerns and the integrity of primary $\delta^{34}S_{CAS}$ values. The oxygen-isotope composition of carbonate as recorded in ancient limestones and

TABLE 2. GEOCHEMICAL CHARACTERISTICS OF THE BECK SPRING DOLOMITE

Sample	Formation	Depth (m)	$\delta^{34}S_{sulfate}$	Sulfate (ppm)	$\delta^{13}C_{carbonate}$	$\delta^{18}O_{carbonate}$	Ca (ppm)	Mg (ppm)	Mn (ppm)	Sr (ppm)
AH-BS-55	Beck Spring	952			–0.18	–6.69	224984	88869	1237	337
AH-BS-54	Beck Spring	949	21.45	75.57	2.17	–5.26	170564	102339	478	136
AH-BS-53A	Beck Spring	941			2.58	–6.85				
AH-BS-52	Beck Spring	939	16.33	89.65	0.54	–6.08				
AH-BS-50	Beck Spring	928		8.62	2.06	–1.74	228851	128437	257	117
AH-BS-49	Beck Spring	920			2.69	–1.33	166489	102771	185	0.02
AH-BS-48	Beck Spring	915	27.44	118.12						
AH-BS-47	Beck Spring	910			2.91	–0.45	171400	107125	107	36
AH-BS-46A	Beck Spring	905			3.18	–0.58				
AH-BS-44	Beck Spring	897			2.91	–0.62	198883	116451	118	65
AH-BS-39	Beck Spring	875			4.33	0.99	204544	126445	93	37
AH-BS-38A	Beck Spring	871	20.29	9.07	4.36	–2.01	178630	119648	84	84
AH-BS-38B	Beck Spring	871			2.63	–4.27				
AH-BS-38C	Beck Spring	871			1.50	–7.48				
AH-BS-36	Beck Spring	861			2.37	–0.43				
AH-BS-35A	Beck Spring	857			2.20	–7.49	175993	110662	93	93
AH-BS-35B	Beck Spring	857			3.67	–2.74				
AH-BS-35C	Beck Spring	857			4.80	0.31				
AH-BS-33	Beck Spring	850	21.87	103.43	4.23	–2.40				
AH-BS-40	Beck Spring	850			4.15	–0.58	194986	119991	150	94
AH-BS-32A	Beck Spring	835	22.79	47.45						
AH-BS-31	Beck Spring	820		57.5	3.83	–1.73	174308	107778	106	53
AH-BS-30	Beck Spring	800			4.22	–1.84	223443	131070	112	50
AH-BS-29	Beck Spring	785	20.73	106.34	4.34	–2.02	179571	119328	139	46
AH-BS-28	Beck Spring	760			3.45	–4.22	200539	118212	232	106
AH-BS-27	Beck Spring	740			4.25	–2.15	122216	75294	98	22
AH-BS-26	Beck Spring	725			4.64	–0.16	205019	122107	121	45
AH-BS-25A	Beck Spring	720	18.97	97.46						
AH-BS-24	Beck Spring	715			4.30	–1.50	168965	104875	117	58
AH-BS-21	Beck Spring	701			3.53	–2.66	238779	138240	173	173
AH-BS-20A	Beck Spring	697	19.08	87						
AH-BS-18	Beck Spring	689			5.40	–0.59	169183	114991	119	66
AH-BS-17	Beck Spring	685			4.82	–0.25	195650	118492	179	83
AH-BS-16B	Beck Spring	681.5	16.36	43.1						
AH-BS-15	Beck Spring	678			4.32	–1.50	187572	116661	206	69
AH-BS-14A	Beck Spring	674	18.49	54.57						
AH-BS-12	Beck Spring	667			4.06	–3.20	183581	112476	155	65
AH-BS-11	Beck Spring	662			4.11	–2.96	214171	128230	218	82
AH-BS-9	Beck Spring	654			5.16	–0.20	184928	121160	112	80
AH-BS-7	Beck Spring	648	18.87	81.97	3.72	–1.29	165864	104756	153	65
AH-BS-5	Beck Spring	642			0.75	–6.26	212440	117033	89	76
AH-BS-4	Beck Spring	639	21.78	144.3						
AH-BS-3	Beck Spring	636			1.92	–6.19	194761	120255	78	65
AH-BS-2	Beck Spring	635.5	25.52	40.76						
AH-BS-1	Beck Spring	635			1.80	–3.47	6560	21470	119	895
SP-BS-0	Beck Spring	630			0.53	–11.27				

TABLE 3. AVERAGE $\delta^{34}S_{SULFATE}$ AND CAS VALUES

Formation/Unit	Average CAS (ppm)	CAS range (ppm)	Average $\delta^{34}S$ (‰)	$\delta^{34}S$ range (‰)
Upper Noonday	27	8.9–60.7	30	23.3–34.5
Lower Noonday	64	1.6–271.7	25	15.2–32.0
Beck Spring	73	8.6–144.3	21	16.3–27.4
Upper Crystal Spring	58	32–71	18	16.4–19.5
Middle Crystal Spring	24	2–61	15	11.0–22.7

Note: CAS—carbonate-associated sulfate.

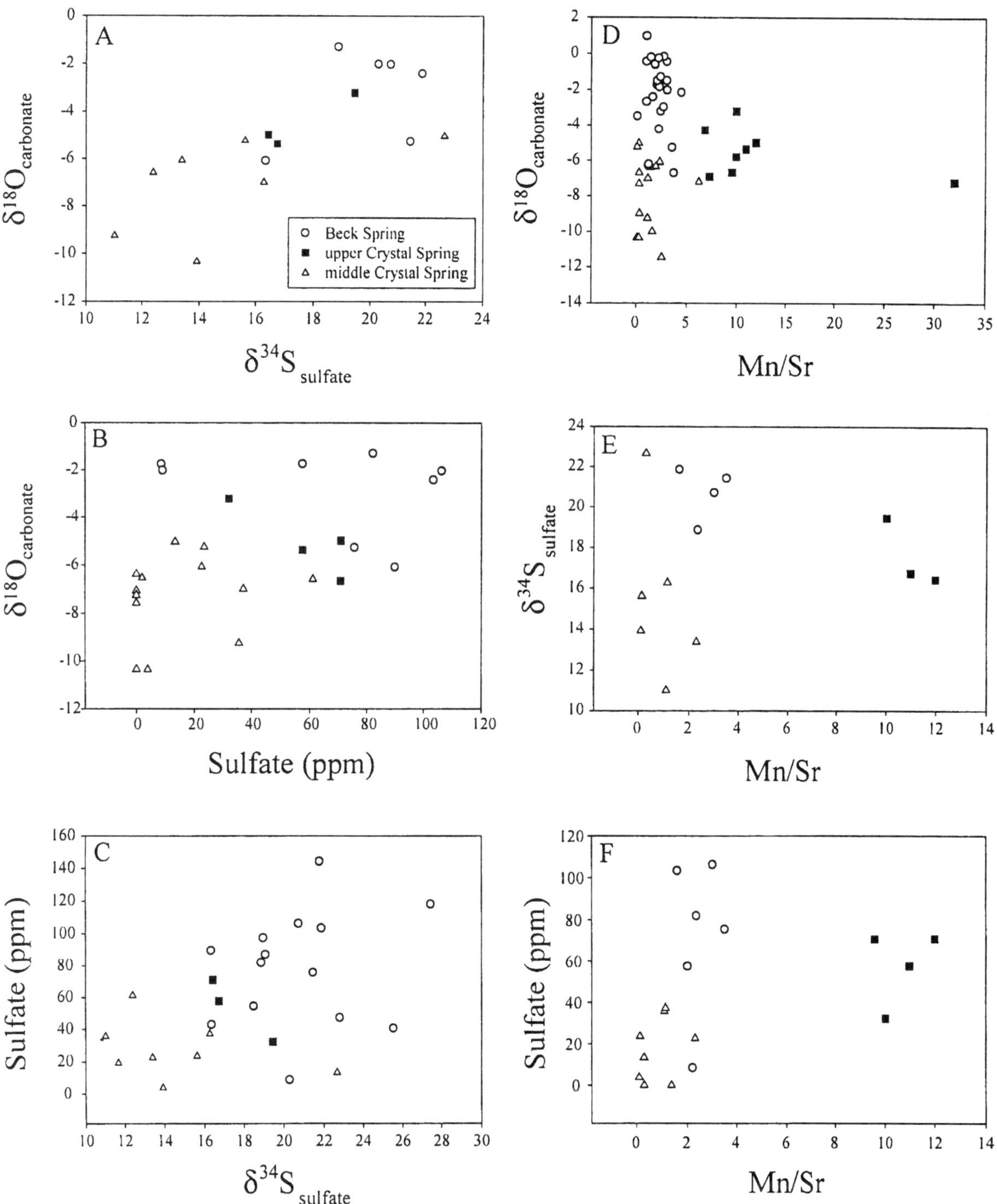

Figure 3. Crystal Spring Formation and Beck Spring Dolomite elemental and stable isotope data. (A) $\delta^{18}O_{carbonate}$ versus $\delta^{34}S_{sulfate}$. $R^2 = 0.33$ for middle Crystal Spring (MCS) and 0.16 for Beck Spring (BS). (B) $\delta^{18}O_{carbonate}$ versus carbonate-associated sulfate (CAS) concentration (ppm). $R^2 = 0.04$ for MCS and 0.09 for BS. (C) CAS concentration versus $\delta^{34}S_{sulfate}$. $R^2 = 0.12$ for MCS and 0.03 for BS. (D) $\delta^{18}O_{carbonate}$ versus Mn/Sr. $R^2 = 0.002$ for MCS, 0.20 for upper Crystal Spring (UCS) and 0.01 for BS. (E) $\delta^{34}S_{sulfate}$ versus Mn/Sr. $R^2 = 0.16$ for MCS. (F) CAS concentration versus Mn/Sr. $R^2 = 0.10$ for MCS, 0.11 for UCS and 0.02 for BS.

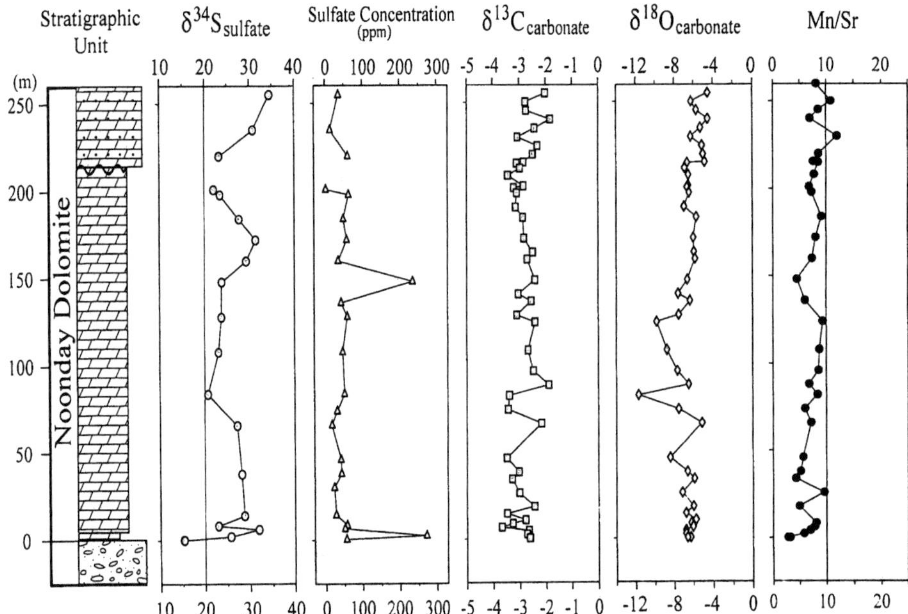

Figure 4. The sulfur isotope composition of carbonate-associated sulfate (CAS), CAS concentrations, $\delta^{13}C_{carbonate}$, $\delta^{18}O_{carbonate}$, and Mn/Sr plotted against Neoproterozoic carbonates collected from the Noonday Dolomite in Death Valley, California.

dolostones is a useful indicator of the extent of geochemical alteration. More specifically, ^{18}O-depleted signatures are often attributed to postdepositional alteration resulting from isotopic exchange with meteoric or hydrothermal fluids as a result of carbonate dissolution and reprecipitation. While ^{13}C-depleted signatures may also be found within meteoric or hydrothermal fluids, carbon concentrations within these fluids are typically low (relative to the carbonate host rock) and, therefore, $\delta^{13}C$ is often buffered to the rock values. Within this context, researchers have plotted $\delta^{18}O$ versus $\delta^{13}C$ from carbonate lithologies in order to evaluate the influence of post-depositional processes on carbon isotopes (Given and Lohmann, 1985; Kaufman et al., 1991; Kaufman and Knoll, 1995; Kennedy, 1996; Kah et al. 2001; Halverson et al., 2002). In instances where crossplots of $\delta^{18}O$ versus $\delta^{13}C$ for single formations show no clear relationship and, in particular, when $\delta^{13}C$ values show little or no variation as $\delta^{18}O$ varies significantly, then $\delta^{13}C$ values are buffered to the carbonate host rock values. However, positive correlations between $\delta^{18}O$ and $\delta^{13}C$ suggest that meteoric diagenesis may have influenced both $\delta^{18}O$ and $\delta^{13}C$.

A similar approach may be used to test the integrity of $\delta^{34}S_{CAS}$ values. Analogous to the carbon system, sulfate concentrations within meteoric or hydrothermal fluids may be low and, therefore, $\delta^{34}S_{CAS}$ may be buffered to the rock values even in the presence of reset $\delta^{18}O$. In extending this approach to evaluate $\delta^{34}S_{CAS}$, a correlation exists between $\delta^{34}S_{sulfate}$ and $\delta^{18}O_{carbonate}$ if the Crystal Spring and Beck Spring formations are considered together (Fig. 3A). However, this approach is not warranted given that $\delta^{18}O$ reconstructions through time show that progressively older samples are ^{18}O-depleted (Knauth and Epstein, 1976; Walker and Lohmann, 1989; Veizer et al., 1997; Frank and Lyons, 2000). Several explanations for this trend have been put forward, including temporal evolution of primary seawater, diagenetic overprinting, and ocean temperature. Additionally, the sulfur isotope composition of seawater sulfate is believed to generally increase through the Proterozoic and, in particular in the Neoproterozoic, as sulfate concentrations increased and bacterial sulfate reduction and the disproportionation of intermediate sulfur species became increasingly important processes as a result of the evolution of oxygen within the atmosphere-ocean system (Canfield and Teske, 1996).

Enough time separates the middle Crystal Spring from upper Crystal Spring and the Beck Spring that it is necessary to treat these units separately in order to avoid correlations between $\delta^{18}O$ and $\delta^{34}S$ that may have resulted from the temporal evolution of marine waters. It is worth noting that the oldest unit, the middle Crystal Spring, has the lowest $\delta^{18}O$ and $\delta^{34}S$ values. If these formations are considered separately, only a very weak to nonexistent correlation exists. As mentioned previously, the $\delta^{18}O$ of altered or replaced carbonate is decreased as a result of ^{18}O-depleted meteoric waters. If meteoric waters contained any sulfate, they should also have relatively depleted $\delta^{34}S_{sulfate}$ values (due to the oxidation of isotopically depleted sulfides), perhaps similar to $\delta^{34}S_{sulfate}$ of rivers (~6‰; Arthur, 2000). Therefore, the lack of any significant correlation between $\delta^{18}O$ and $\delta^{34}S$ of Crystal Spring and Beck Spring carbonates might suggest that primary $\delta^{34}S_{sulfate}$ values have not been altered as a result of carbonate dissolution and reprecipitation in the presence of fluids influenced by meteoric water. Additionally, correlations between $\delta^{18}O_{carbonate}$ and CAS concentration and CAS concentration versus $\delta^{34}S_{sulfate}$ are weak (Fig. 3B and 3C). Furthermore, samples were carefully screened based on carbonate fabric retention, and those samples showing neomorphic sparry textures were not analyzed.

TABLE 4. GEOCHEMICAL CHARACTERISTICS OF THE NOONDAY DOLOMITE.

Sample	Formation	Depth (m)	$\delta^{34}S_{sulfate}$	Sulfate (ppm)	$\delta^{13}C_{carbonate}$	$\delta^{18}O_{carbonate}$	Ca (ppm)	Mg (ppm)	Mn (ppm)	Sr (ppm)
SN-ND-76	Upper Noonday	2367			–3.34	–6.60	165618	101211	565	26
SN-ND-75	Upper Noonday	2362			–1.49	–4.84	203057	121323	447	38
SN-ND-74	Upper Noonday	2357		8.93	–1.90	–4.95	194797	118192	383	42
SN-ND-73	Upper Noonday	2352			–2.28	–5.38	183025	116470	480	59
SN-ND-72	Upper Noonday	2347	34.521	35.57	–2.01	–4.53				
SN-ND-71	Upper Noonday	2342			–2.76	–6.24	202859	121024	380	35
SN-ND-70	Upper Noonday	2337			–2.74	–5.68	193968	113918	349	41
SN-ND-69	Upper Noonday	2332			–1.81	–4.53	184034	114101	258	37
SN-ND-68	Upper Noonday	2327	30.885	13.54	–2.41	–5.26				
SN-ND-67	Upper Noonday	2322			–3.07	–6.28	210128	121443	384	32
SN-ND-66	Upper Noonday	2317			–2.30	–5.10				
SN-ND-65B	Upper Noonday	2312	23.3	60.74	–2.48	–5.00	170315	103942	326	38
SN-ND-64A	Upper Noonday	2307.5			–2.84	–4.84	168810	104974	213	28
SN-ND-63	Noonday cap carb.	2162.7			–3.11	–6.66	184726	113764	383	45
SN-ND-62	Noonday cap carb.	2159.7			–2.98	–6.92				
SN-ND-61	Noonday cap carb.	2155.7			–3.44	–6.53	209836	131573	442	57
SN-ND-60	Noonday cap carb.	2149.7			–2.84	–6.54				
SN-ND-59	Noonday cap carb.	2148.7	22.057	1.64	–3.22	–6.68	218698	125482	294	43
SN-ND-56	Noonday cap carb.	2145.7	23.521	62.87	–3.10	–6.45	182489	118143	306	42
SN-ND-54	Noonday cap carb.	2137.7			–3.15	–6.99				
SN-ND-52	Noonday cap carb.	2131.7	27.762	48.28	–2.86	–5.68	217636	124971	391	43
SN-ND-49	Noonday cap carb.	2119.7	31.456	57.48	–2.83	–6.01	169446	105614	650	81
SN-ND-47	Noonday cap carb.	2111.7			–2.49	–5.94				
SN-ND-46	Noonday cap carb.	2107.7	29.336	34.24	–2.70	–5.82	176675	111735	420	57
SN-ND-43	Noonday cap carb.	2095.7	23.872	234.86	–2.40	–6.66	170455	108262	415	92
SN-ND-41	Noonday cap carb.	2087.7			–3.05	–7.61				
SN-ND-40	Noonday cap carb.	2083.7		41.89	–2.55	–6.38	177478	109294	361	60
SN-ND-38	Noonday cap carb.	2075.7	23.781	58.84	–3.11	–7.55				
SN-ND-37	Noonday cap carb.	2071.7			–2.40	–9.84	186074	116674	1127	121
SN-ND-33	Noonday cap carb.	2055.7	23.086	45.74	–2.67	–8.77	219371	126305	643	74
SN-ND-31	Noonday cap carb.	2047.7		94.81						
SN-ND-30	Noonday cap carb.	2043.7			–2.47	–7.69	172981	111476	436	51
SN-ND-28	Noonday cap carb.	2035.7			–1.88	–6.52	176373	110099	802	118
SN-ND-27	Noonday cap carb.	2031.7	20.693	50.59						
SN-ND-26	Noonday cap carb.	2029.7			–3.40	–11.69	214409	120407	2112	253
SN-ND-24	Noonday cap carb.	2021.7		30.56	–3.45	–7.56	175579	109117	476	79
SN-ND-22	Noonday cap carb.	2013.7	27.257	16.67	–2.16	–5.14	186027	118592	337	47
SN-ND-17	Noonday cap carb.	1993.7		40.18	–3.50	–8.43	209018	124281	720	127
SN-ND-15	Noonday cap carb.	1985.7	28.241	41.52	–3.03	–6.66	180668	115509	385	74
SN-ND-14	Noonday cap carb.	1981.7			–3.30	–5.91	184889	112221	396	92
SN-ND-13	Noonday cap carb.	1977.7		22.44						
SN-ND-12	Noonday cap carb.	1973.7			–3.02	–7.20	209701	125821	544	57
SN-ND-10	Noonday cap carb.	1965.7			–2.44	–6.01	191995	114927	338	68
SN-ND-9A	Noonday cap carb.	1961.7	28.78	26.58	–3.50	–6.83				
SN-ND-9B	Noonday cap carb.	1961.7			–3.39	–13.19				
SN-ND-8A	Noonday cap carb.	1957.95			–2.78	–5.76				
SN-ND-8B	Noonday cap carb.	1957.95			–3.26	–8.92				
SN-ND-7	Noonday cap carb.	1955.95	22.919	56.61	–3.28	–6.26	208565	124012	338	42
SN-ND-6	Noonday cap carb.	1953.95	31.999	50.06	–3.70	–6.06	187893	118373	366	47
SN-ND-5	Noonday cap carb.	1951.95			–2.65	–6.83	169611	108363	440	63
SN-ND-4	Noonday cap carb.	1949.95	25.706	271.68	–2.73	–6.82	207642	123878	342	59
SN-ND-3A	Noonday cap carb.	1947.95	15.252	54.78	–2.59	–6.34				
SN-ND-2	Noonday cap carb.	1947.85			–2.61	–6.31	171247	109975	220	79
SN-ND-1	Noonday cap carb.	1947.75			–2.62	–6.67	175785	115315	225	70

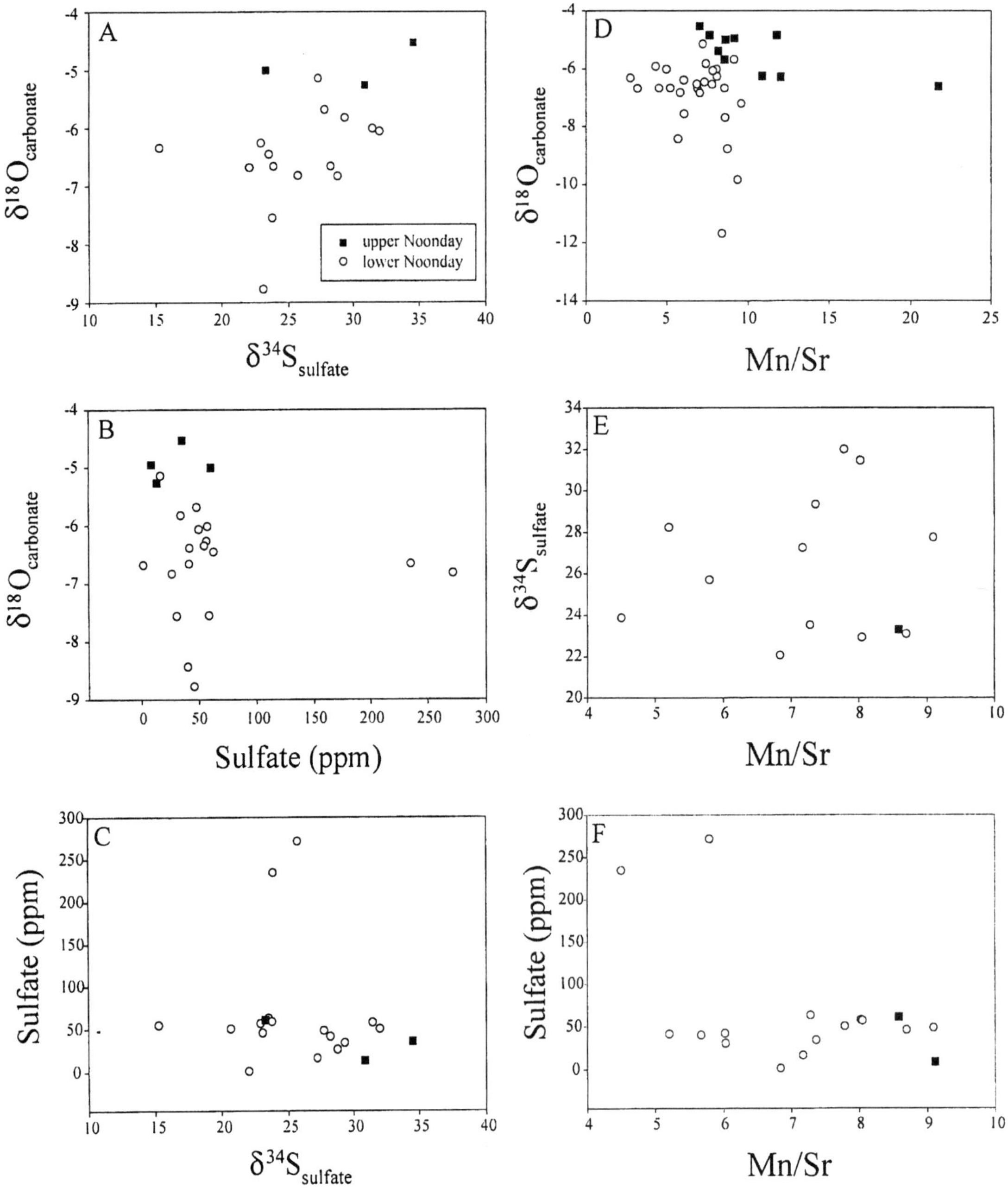

Figure 5. Elemental and stable isotopic data from the Noonday Dolomite. (A) $\delta^{18}O_{carbonate}$ versus $\delta^{34}S_{sulfate}$. $R^2 = 0.11$ for lower Noonday (LN). (B) $\delta^{18}O_{carbonate}$ versus carbonate-associated sulfate (CAS) (ppm). $R^2 = 0.002$ for LN and 0.076 for upper Noonday (UN). (C) CAS concentration versus $\delta^{34}S_{sulfate}$. $R^2 = 0.005$ for LN. (D) $\delta^{18}O_{carbonate}$ versus Mn/Sr. $R^2 = 0.09$ for LN. E) $\delta^{34}S_{sulfate}$ versus Mn/Sr. $R^2 = 0.02$ for LN. F) CAS concentration versus Mn/Sr. $R^2 = 0.02$ for LN.

Elemental concentrations of Mn and Sr have also been used by researchers to assess the effects of meteoric diagenesis and dolomitization in Neoproterozoic carbonates (Kaufman et al., 1991; Derry et al., 1992; Kaufman and Knoll, 1995). Strontium is typically lost in carbonates during meteoric diagenesis, while Mn is often enriched because oxic seawater contains relatively little Mn compared to freshwater (Brand and Veizer, 1981). Therefore, carbonates that contain high Mn to Sr ratios were likely affected by meteoric diagenesis. Neoproterozoic carbonates typically have low Sr concentrations and moderate to high Mn concentrations. Kaufman and Knoll (1995) suggested that Neoproterozoic limestones and dolostones with Mn/Sr (wt. ratio) <10 record near primary $\delta^{13}C$ values.

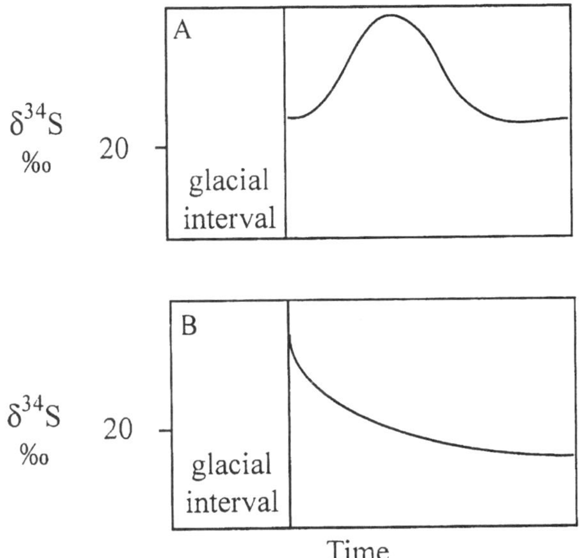

Figure 6. Schematic illustration of $\delta^{34}S_{sulfate}$ behavior (A) typically recorded in Neoproterozoic postglacial carbonates versus (B) not typically associated with postglacial sediments.

Mn/Sr values are well under 10 through the middle Crystal Spring and Beck Spring formations. The upper Crystal Spring Formation is an exception, with values ranging from 5 to 30 (Fig. 2). Figure 3D–3F illustrates that no correlation exists between Mn/Sr and $\delta^{18}O_{carbonate}$, or $\delta^{34}S_{sulfate}$ and CAS concentration for the middle Crystal Spring, upper Crystal Spring and Beck Spring units. Therefore, the elevated Mn/Sr values in the upper Crystal Spring Formation suggest that meteoric fluids may have influenced the elemental and isotopic composition of these carbonates. However, the sulfur isotope composition of the middle Crystal Spring and the Beck Spring, as indicated by relationships between elemental (Mn/Sr, CAS) and isotopic ($\delta^{34}S$, $\delta^{18}O$) compositions, may well reflect primary $\delta^{34}S_{sulfate}$ of Neoproterozoic seawater.

Elemental and stable isotopic compositions of the Noonday Dolomite suggest that significant meteoric diagenesis has not occurred in that formation. Figure 5A illustrates that a weak correlation exists between $\delta^{34}S_{sulfate}$ and $\delta^{18}O_{carbonate}$ ($r^2 = 0.11$ for the lower Noonday and 0.20 for the entire Noonday). There is no clear relationship between CAS concentration and $\delta^{18}O_{carbonate}$ or $\delta^{34}S_{sulfate}$ and CAS concentration (Fig. 5B–5C). Whereas Mn/Sr ratios are elevated relative to the middle Crystal Spring Formation and Beck Spring Dolomite, they are generally <10 and exhibit no correlation with $\delta^{18}O_{carbonate}$, $\delta^{34}S_{sulfate}$ or CAS concentration (Figs. 4 and 5D–5F).

Controls on the Sulfur Isotopic Composition of Seawater Sulfate

Sulfate is the second most abundant anion in the modern ocean, and the stable isotope composition of sulfate-sulfur reflects a balance between fractionations associated with the microbial processes that mediate the major inputs and outputs of seawater sulfate. The primary source of seawater sulfate is riverine delivery resulting from the oxidative weathering of pyrite and dissolution of calcium sulfate (evaporites) minerals. No isotopic fractionation is associated with the weathering of evaporites and only a negligible fractionation is associated with sulfide oxidation during weathering (Taylor et al., 1984). Sulfate is removed from the oceans primarily by calcium sulfate precipitation (evaporites and ridge-flank hydrothermal systems) and bacterial sulfate reduction (BSR) with associated Fe-sulfide burial. Sulfur removed during calcium sulfate precipitation in evaporite settings involves minimal isotopic fractionation (0–2.4‰; Ault and Kulp, 1959; Raab and Spiro, 1991). Additionally, it is believed that hydrothermal circulation of seawater sulfate at mid-ocean ridges (MOR) has a minimal effect on the sulfur isotope composition of seawater sulfate (Alt, 1995). As seawater is cycled through basalt at temperatures of 150–200 °C, sulfate is removed from solution through anhydrite precipitation (Bischoff and Dickson, 1975; Seyfried and Bischoff, 1979). Empirical evidence suggests that there is no isotopic fractionation associated with this process (Alt, 1995). This seawater-basalt interaction could result in the complete extraction of sulfate from the ocean inasmuch as the entire volume of seawater circulates through the MOR in 10^7 years, about the same length as the estimated residence time of sulfate in the ocean today (calculated in terms of riverine sulfate supply). Anhydrite that forms at such elevated temperatures likely dissolves when the upper oceanic crust experiences lower-temperature off-axis circulation, reintroducing sulfate with the same isotope composition as that which entered the oceanic crust (Alt, 1995). Interestingly, the overall effect of hydrothermal circulation of seawater sulfate through the MOR may be to buffer the ocean against large $\delta^{34}S_{sulfate}$ shifts.

In contrast to the minimal isotope fractionations associated with calcium sulfate precipitation and the overall effect of hydrothermal circulation of seawater sulfate through the MOR, a pronounced kinetic isotope effect occurs during dissimilatory BSR as obligate anaerobes preferentially dissimilate the lighter ^{32}S in the production of H_2S, which is then precipitated as sedimentary pyrite. Factors influencing the kinetic sulfur isotope effect during BSR have been studied both experimentally (Kaplan and Rittenberg, 1964; Goldhaber and Kaplan, 1974; Canfield, 2001; Detmers et al., 2001; Habicht et al. 2002) and in natural systems (Habicht and Canfield, 1997, 2001) for a phylogenetically diverse group of bacteria under a range of environmental conditions. These studies indicate that both pure bacterial cultures and natural populations of sulfate-reducing bacteria preferentially consume ^{32}S relative to ^{34}S leading to apparent fractionation of 2‰–46‰. Sulfate concentrations are an important consideration for sulfur isotope fractionations during BSR. For example, Habicht et al. (2002) determined that sulfur isotope fractionation during BSR decreased significantly (fractionation <6‰) when sulfate concentrations were <50–200 μM.

The maximum difference in isotopic composition between seawater sulfate and pyrite ($\Delta^{34}S_{sulfate\text{-}pyrite}$) appears to be 55‰ or greater for much of the Phanerozoic (Canfield and Teske, 1996). In order to attain $\Delta^{34}S_{sulfate\text{-}pyrite}$ values larger than 46‰, researchers have suggested an overprint resulting from a consortium of bacteria linked with the oxidative part of the sulfur cycle (Jørgensen 1990; Canfield and Thamdrup, 1994). Very depleted $\delta^{34}S_{sulfide}$ values are thought to reflect an initial fractionation by sulfate-reducing bacteria followed by an additional fractionation that results from the disproportionation of intermediate sulfur species such as S^0 and thiosulfate. The disproportionation of elemental S and thiosulfate produces sulfide depleted in ^{34}S relative to the initial reactant by 7‰–11‰ (Canfield and Thamdrup, 1994; Habicht et al., 1998). Through repeated cycles of sulfide oxidation to S^0 and thiosulfate and subsequent disproportionation, ΣH_2S becomes depleted in ^{34}S to a greater degree than that produced by initial BSR. Canfield and Teske (1996) suggested that the evolution of aerobic, non-photosynthetic sulfide-oxidizing bacteria occurred between 1.0 and 0.64 Ga (roughly coincident with the period of time represented by the Crystal Spring, Beck Spring and Noonday formations) as a result of increases in Earth's atmospheric oxygen content to ~10% present atmospheric levels (PAL). They argued that these bacteria were then responsible for the production of intermediate sulfur species, which thereby facilitated disproportionation reactions and larger $\Delta^{34}S$ values.

Sulfur Cycling in the Late Neoproterozoic (circa 750–590 Ma)

It should be emphasized that the hypotheses forwarded within this manuscript are speculative given the limited amount of data, and that more data will be necessary to differentiate between them. The sulfur cycle appears to have behaved differently in the late versus the early Neoproterozoic. The ~17‰ $\delta^{34}S_{sulfate}$ positive shift that occurs over ~5 m at the base of the Noonday Dolomite is similar in magnitude and stratigraphic thickness to the >20‰ positive $\delta^{34}S_{sulfate}$ variations from Neoproterozoic postglacial and nonglacial carbonates from Namibia (Hurtgen et al., 2002). By contrast, the early Neoproterozoic formations (Crystal Spring and Beck Spring; Fig. 2) display variations on the order of ~9‰ over tens of meters. Recall that two of the three excursions expressed within the Crystal Spring and Beck Spring formations are not robust because they are represented by single data points. Furthermore, recall that workers have suggested that the Beck Spring Dolomite may represent a Sturtian cap carbonate despite the absence of any underlying glacial deposits.

Nevertheless, one explanation has been offered by Hurtgen et al. (2002), who proposed that the >20‰ (and as high as 30‰) positive $\delta^{34}S_{sulfate}$ shifts in late Neoproterozoic, postglacial carbonates are consistent with details of the "snowball Earth" hypothesis (Kirschvink, 1992; Hoffman et al., 1998a) in that they appear to reflect nearly complete reduction of sulfate in an anoxic global ocean (Hurtgen et al., 2002). Kirschvink (1992) suggested that if the entire ocean were covered with ice, the hydrologic cycle would have essentially been disabled and oceanic anoxia would have developed, encouraging the build up of ferrous iron and the return of banded-iron formations (BIFs). The occurrence of BIFs with Neoproterozoic glacial deposits is evidence for widespread oceanic anoxia (Kirschvink, 1992; Hoffman et al., 1998a, Hoffman and Schrag, 2002). Sulfate removal would be favored if the oceans were covered with ice and the hydrologic cycle disabled, and the riverine delivery of relatively depleted $\delta^{34}S_{sulfate}$ from the continents via pyrite weathering and/or evaporite dissolution essentially ceased. Assuming that sufficient organic substrates existed to allow sulfate-reducing bacteria to continue to preferentially dissimilate ^{32}S and that virtually all of this sulfide was precipitated as pyrite (sufficient Fe^{2+}), the isolated oceanic reservoir of sulfate would become progressively enriched in ^{34}S (e.g., the ocean would have been a closed reservoir). Such a long period of isolation might facilitate a significant decrease in the mass of the oceanic sulfate reservoir. During deglaciation, deepwater overturn would supply ^{34}S-enriched sulfate to surface waters, albeit in low concentrations, and as a result, trace sulfate incorporated in postglacial carbonates would reflect the enriched $\delta^{34}S$ and low $[SO_4^{2-}]$.

The extent of $\delta^{34}S_{sulfate}$ enrichment would have depended on the duration of the snowball Earth event, initial sulfate concentrations and the availability of organic substrates to fuel bacterial sulfate reduction. In a simple numerical model, Hurtgen et al. (2002) produced a 20‰ $\delta^{34}S_{sulfate}$ excursion assuming that oceanic sulfate concentrations were half that of present values, the snowball event lasted 5 m.y., and BSR continued at preglacial levels (e.g., sufficient organic carbon availability) throughout the glaciation. It is very possible that BSR levels would have been reduced during a snowball Earth event due to a decreased availability of organic substrates. In order to produce a 20‰ $\delta^{34}S_{sulfate}$ shift under conditions of reduced BSR levels, the duration of the snowball event would have to increase and/or initial oceanic sulfate concentrations decreased (see discussion below).

While the snowball Earth hypothesis as envisioned by Kirschvink (1992) and Hoffman et al. (1998a) is consistent with this scenario, other explanations deserve consideration, especially given that positive $\delta^{34}S_{sulfate}$ excursions are known to occur in rocks that are not associated with glacial deposits, such as the upper Rasthof and Gruis Formations in Namibia (Hurtgen et al., 2002). Recent work has suggested that the hydrologic cycle may have been active at some level during the glacial episodes. For example, Hyde et al. (2000) and Crowley and Hyde (2001) used a coupled climate/ice-sheet model to simulate a snowball Earth event and suggested that ice-free zones may have been present in some equatorial regions, a finding echoed by Poulsen (2003). Additionally, Condon et al. (2002) and Leather et al. (2002) found evidence for episodicity in various Neoproterozoic glacial deposits and concluded that a dynamic glacial system—and therefore a hydrologic cycle—was active.

If a hydrologic cycle did exist as a result of an ice-free equatorial ocean, rivers would deliver some amount of relatively depleted $\delta^{34}S_{sulfate}$ from the continents. However, depending on the extent of riverine sulfate delivery—and more importantly the balance between riverine sulfate inputs and BSR outputs—the oceanic sulfate cycle probably was not operating under steady-state conditions. If BSR continued at near preglacial levels and riverine sulfate inputs were diminished during glacial episodes, oceanic sulfate concentrations would have decreased through time, and the residual sulfate reservoir would become progressively ^{34}S-enriched. Therefore, in terms of the sulfur cycle, the enriched $\delta^{34}S_{sulfate}$ values found in postglacial carbonates may also be consistent with severe Neoproterozoic glaciations in which equatorial oceans remained ice free. Thus, the main difference between a "hard" versus a "soft" glacial episode would be the amount of time required to attain the enriched $\delta^{34}S_{sulfate}$ signal (values > ~40‰). The farther out of balance riverine inputs and BSR outputs were—that is, outputs > inputs—the faster the oceanic sulfate pool would have become ^{34}S-enriched. Additionally, the lower preglacial sulfate concentrations, the more rapid $\delta^{34}S_{sulfate}$ values would rise.

Other possibilities exist. For example, the pattern of $\delta^{34}S$ in postglacial carbonates suggests that seawater sulfate may have been partially driven to more ^{34}S-enriched values as a result of intense BSR in the glacial aftermath (Gorjan et al., 2000; Hurtgen et al., 2002). In both the Noonday Dolomite (Fig. 4) and the Maieberg Formation (Namibia; Hurtgen et al., 2002), $\delta^{34}S_{sulfate}$ values begin relatively low and rise sharply upsection (i.e., Figure 6A versus 6B). However, both the Maieberg Formation and the Noonday dolomite (and the Marinoan events in general; Kennedy et al. 1998) contain low organic carbon concentrations, suggesting that enhanced postglacial BSR was not important during deposition of these carbonates. Alternatively, Varni et al. (2001) and Kaufman et al. (2002) suggest that CAS uptake during cap carbonate formation may have aided in driving the upsection increase in $\delta^{34}S_{sulfate}$ by further reducing the oceanic sulfate reservoir thereby making it more susceptible to closed-system Raleigh distillation effects.

Another possibility is that the entire oceanic sulfate reservoir was exhausted or nearly exhausted during the glacial event as a result of BSR. During deglaciation, strong weathering inputs and the possible oxidation of ΣH_2S within the ocean may have caused $\delta^{34}S_{sulfate}$ to vary significantly over short timescales depending on the regional importance of these processes (weathering inputs versus oxidation). The widespread occurrence of banded-iron formations (BIFs) for the Sturtian glacial event suggests that Fe availability exceeded sulfide supply—at least in areas of BIF formation—and that oceanic sulfate may have been nearly exhausted during that event. By contrast, BIF deposition is uncommon during the Marinoan glacial event (Kennedy et al., 1998), which could signal a greater availability of sulfide and therefore sulfate (Hurtgen et al., 2002).

It is clear that more detailed, high-resolution sampling for $\delta^{34}S_{sulfate}$, coupled with other sedimentological and geochemical characteristics on additional postglacial carbonates, is necessary to distinguish among hypotheses for $\delta^{34}S_{sulfate}$ excursions. Regardless, we suggest that Neoproterozoic glacial events, whether global or not, forced high amplitude sulfur isotope variations by substantially decreasing oceanic sulfate concentrations. The positive $\delta^{34}S_{sulfate}$ values reflect either the enriched ^{34}S ocean resulting from BSR during glaciation (as described above) or in the aftermath of glaciation as a result of BSR in an ocean with reduced sulfate concentrations *or* a combination of both.

Sulfur Cycling in the Early Neoproterozoic (circa 1000–750 Ma)

Patterns of $\delta^{34}S_{sulfate}$ evolution, as recorded in late Mesoproterozoic and early Neoproterozoic carbonate sequences (Crystal Spring and Beck Spring Formations), illustrate that maximum shifts of ~9‰ occur over stratigraphic distances of tens of meters (Fig. 2). Additionally, the late Mesoproterozoic $\delta^{34}S_{sulfate}$ trends for the middle Crystal Spring are consistent with, in both absolute value and magnitude of variation, $\delta^{34}S_{sulfate}$ trends for CAS in the ca. 1.2 Ga Apache Group, Arizona (Gellatly and Lyons, 2002). These excursions likely reflect changes in either the weathering/oxidation of sulfates and sulfides or the extent of BSR and subsequent pyrite burial (relative to calcium sulfate deposition) or both. However, as several researchers have suggested (Hurtgen et al., 2002; Lyons et al., 2004; Gellatly and Lyons, 2002), the high amplitude $\delta^{34}S_{sulfate}$ variations that occur over "rapid" timescales (tens of meters) may not be compatible with an ocean with sulfate concentrations comparable to modern values (28 mM and a residence time of 20 m.y.).

It is very difficult to define "rapid" timescales and assign durations to the $\delta^{34}S_{sulfate}$ excursions in the Crystal Spring and Beck Spring formations. Wilkinson et al. (1991) calculated long-term carbonate accumulation rates using meter- and epoch-scale Phanerozoic marine sequences and determined that average Phanerozoic accumulation rates ranged between 14 and 32 m/m.y. However, Neoproterozoic carbonate precipitation was largely abiotic (devoid of carbonate skeletal material) and Grotzinger and Kasting (1993) have suggested that Neoproterozoic seawater may have been highly oversaturated with respect to calcium carbonate. Therefore, it is possible that carbonate accumulations rates were in general higher in the Neoproterozoic relative to the Phanerozoic. Hoffman et al. (1998b) suggested that the average shallow-water carbonate accumulation rate for the Otavi Group (Neoproterozoic) in Namibia was ~52 m/m.y. Therefore, assuming that Crystal Spring and Beck Spring carbonates accumulated at roughly comparable rates, the ~9‰ shifts in $\delta^{34}S_{sulfate}$ that span stratigraphic distances of tens of meters, may have occurred in less than 1–2 m.y., or even more rapidly given that that these sediments were deposited at least intermittently during times of active tectonism and therefore sedimentation rates may have been higher (Prave, 1994).

How does this compare with $\delta^{34}S_{sulfate}$ variations in the Cenozoic (65–0 Ma) when sulfate concentrations ranged from ~18 to 28 mM (Lowenstein et al., 2003) and the residence time of oceanic sulfate was perhaps ~10–20 m.y. (Petsch and Berner, 1998)? Paytan et al. (1998) constructed a high-resolution (~1 m.y.) $\delta^{34}S_{sulfate}$ record for the Cenozoic using marine barite. The largest $\delta^{34}S_{sulfate}$ shift expressed during this time interval is only 5‰ over ~10 m.y. The larger sulfur isotope changes that occur over shorter timescales in the Mesoproterozoic and earlier Neoproterozoic (relative to the Cenozoic) are compatible with an ocean with significantly lower sulfate concentrations than today.

The Paleozoic and Mesozoic record of $\delta^{34}S_{sulfate}$, as well as oceanic sulfate concentrations and residence times, is more ambiguous. Holser (1977) recognized rapid (<5 m.y.) positive $\delta^{34}S_{sulfate}$ excursions (ranging from 5‰ to 15‰) in the late Neoproterozoic, late Devonian, and early Triassic and argued that it would be extremely difficult to modify the sulfur isotope composition of seawater sulfate by 15‰ in less than 5 m.y. assuming that the oceanic residence time of sulfate was 20 m.y. Holser (1977) suggested that anoxia in stratified basins allowed sulfide resulting from BSR to occur within an isolated deep-water mass causing $\delta^{34}S_{sulfate}$ to increase within those basins. Subsequent oceanic mixing forced a rapid increase in $\delta^{34}S_{sulfate}$ recorded in evaporites that formed from surface seawater. However, it is important to note that the isotope excursions identified by Holser (1977) were recorded in evaporitic gypsum deposits that occur sporadically in the geologic record, are difficult to date, and could represent local conditions under which global ocean signals may have been modified. Therefore, the actual timing, magnitude, and global extent of the sulfur isotope excursions need to be confirmed. A nice example of a high-resolution $\delta^{34}S_{sulfate}$ study for the Paleozoic is presented in Kampschulte et al. (2001). They found a 10‰ $\delta^{34}S_{sulfate}$ shift in Carboniferous sediments that occurred over ~20 m.y., a rate of change that is consistent with Cenozoic $\delta^{34}S_{sulfate}$ variation and with modern oceanic sulfate concentrations.

In any event, whereas the $\delta^{34}S_{sulfate}$ variations expressed in late Mesoproterozoic and early Neoproterozoic carbonates are not as large as those in the later part of the Neoproterozoic, the magnitude (~10‰) and best estimate of the rate-of-change (<2 m.y.) of the excursions appear to be consistent with an ocean with significantly lower sulfate concentrations than today's. However, an important question is whether $\delta^{34}S_{sulfate}$ changes in the late Mesoproterozoic–early Neoproterozoic reflect global shifts in the sulfur isotopic composition of seawater sulfate—as they appear to in the late Neoproterozoic (this study; Hurtgen et al. 2002)—or simply local perturbations to an ocean with low sulfate concentrations and therefore more susceptible to change. This can only be answered when more sections of equivalent age are assessed.

CAS Concentrations as a Proxy for Oceanic Sulfate Concentrations

Additional evidence for low sulfate concentrations in the late Mesoproterozoic to early Neoproterozoic may come from CAS concentrations. Busenberg and Plummer (1985) found that the amount of trace sulfate substituting for the carbonate ion is primarily dictated by the sulfate concentration in the original solution and the rate of crystal growth. Cenozoic carbonates commonly contain CAS concentrations on the order of 10^3 ppm (Burdett et al., 1989; Staudt et al., 1993) and the average concentration for modern carbonates is estimated to be between ~2400 and 4200 ppm (Lyons et al., this volume). By contrast, CAS concentrations for late Mesoproterozoic–early Neoproterozoic carbonates are an order of magnitude less, ranging from 0 to 144 ppm (Fig. 2). Therefore, assuming no diagenetic CAS loss, one might predict that late Mesoproterozoic–early Neoproterozoic oceanic sulfate concentrations were ~10% of modern values (e.g., ~3 mM). That being said, Neoproterozoic carbonate precipitation was largely abiotic (devoid of carbonate skeletal material) and Grotzinger and Kasting (1993) have suggested that Neoproterozoic seawater may have been highly oversaturated with respect to calcium carbonate. Thus, it is also possible that sulfate concentrations in Neoproterozoic carbonate were higher than predicted (based on CAS concentrations) because of rapid rates of precipitation; that is, unless diagenesis has altered the primary values.

It is certainly possible that a portion of the difference in CAS concentrations between modern versus Neoproterozoic carbonates may reflect diagenetic loss. For example, Staudt and Schoonen (1995) analyzed CAS concentrations in Mississippian dolomites (Burlington-Keokuk Formation) that experienced varying degrees of diagenesis. They found that early diagenetic dolomites (burial depths <500) contained CAS levels of 1000–3500 ppm. By contrast, dolomites of the same formation that experienced greater degrees of diagenesis (burial depths >500 m) generally contained <500 ppm CAS. Therefore, it is possible that the low CAS concentrations found in Death Valley carbonates, in part, reflect diagenetic loss of sulfate.

In nearly all cases, light-colored (organic-poor), fabric retentive, micritic limestones and dolostones were used for trace sulfate analyses. The fabric retentive micrite argues against massive recrystallization and exchange. Also, recall that Figures 3C and 5C illustrate that no correlation exists between CAS concentrations and Mn/Sr and $\delta^{18}O_{carbonate}$ values, thereby arguing against large-scale loss of sulfate during meteoric diagenesis. Therefore, we suggest that while some loss of sulfate may have occurred during diagenesis, it is unlikely that the low CAS concentrations in Death Valley carbonates can be entirely explained by this process alone. Within this context, we crudely estimate—based on $\delta^{34}S_{sulfate}$ variations and CAS concentrations recorded in Crystal Spring and Beck Spring carbonates—that late Mesoproterozoic–early Neoproterozoic oceanic sulfate concentrations were ~10% of modern values (e.g., ~3 mM). This estimate is consistent with calculations based on $\delta^{34}S_{CAS}$ variations recorded in mid Mesoproterozoic sediments from Arctic Canada (Kah et al., 2002).

Low Neoproterozoic Oceanic Sulfate Concentrations: Causes and Mass-Balance Implications

Finally, why were Neoproterozoic oceanic sulfate concentrations low? Traditionally, sulfate concentrations have been attributed to the oxidation state of Earth's ocean-atmosphere system. Prior to the evolution of atmospheric oxygen, sulfate concentrations should have been low because the primary source of seawater sulfate is river input, resulting in part from the oxidative weathering of sulfides on land. However, Walker and Brimblecombe (1985) suggested that even on the prebiotic

Earth under reducing conditions, photochemical oxidation of volcanic SO_2 (and H_2S) emissions may have resulted in an ocean with sulfate concentrations reaching ~3% of modern values (~1 mM). With the advent of oxygenic photosynthesis, sulfate concentrations should have increased as the oxidation state of Earth's ocean-atmosphere system evolved. As mentioned previously, Canfield and Teske (1996) proposed that an increase in the range of $\delta^{34}S_{pyrite}$ values during the Neoproterozoic resulted from a fundamental shift in the biogeochemical cycling of sulfur, facilitated by an increase in atmospheric oxygen concentrations to ~10% PAL. Therefore, our estimate of an early Neoproterozoic ocean with sulfate concentrations equal to ~10% of modern values appears to be consistent with their work. However, the exact relationship between atmospheric oxygen concentrations and oceanic sulfate levels is presently unknown, and other factors responsible for depressed oceanic sulfate concentrations deserve consideration.

Could Neoproterozoic oceanic sulfate concentrations have been ~10% of modern values while atmospheric oxygen levels were greater than 10% PAL—perhaps as high as 100% or 1× PAL? The Neoproterozoic carbon isotope record is dominated by enriched $\delta^{13}C_{carbonate}$ values, suggesting long periods of enhanced organic carbon burial (Knoll et al., 1986; Kaufman et al., 1997). This relationship might indicate that more organic carbon was available (Rothman et al., 2003) and that sulfate reduction rates in the Neoproterozoic ocean were much higher than at present. Therefore, assuming sufficient amounts of reactive iron were available (a big assumption), most of the sulfate entering the ocean would be sedimented as pyrite.

It is difficult to produce evidence for increased amounts of sulfur burial as pyrite because deep-water sections have received far less study then carbonate-dominated shallower ones. However, Condie et al. (2001) presented a compilation of black shale/total shale, as well as total black shale thicknesses through the Precambrian. They found the ratio of black shales to total shales, as well as the total thickness of black shales, to be higher in the Neoproterozoic relative to the Mesoproterozoic but lower compared to the Paleoproterozoic and late Archean. This, in combination with increased $\delta^{13}C_{carbonate}$ values, might indicate that Neoproterozoic sulfate consumption via BSR was higher relative to the Mesoproterozoic. However, enhanced organic carbon burial would have increased the potential for atmospheric oxygen enrichment (e.g., organic carbon is protected from oxidation), which then would have increased the potential for the weathering of sulfides on land and the riverine delivery of sulfate to the ocean. The exact relationship between atmospheric oxygen levels and its affect on sulfide weathering is unknown. For example, how does a 10% increase in atmospheric oxygen concentrations affect the amount of sulfate delivered to the oceans, and how much land mass was available for the weathering of sulfides in the Neoproterozoic?

Another possible explanation for low oceanic sulfate concentrations (and one that is extremely difficult to prove) is that more sulfate was pulled into and stored in oceanic crust as anhydrite. As discussed previously, it is believed that hydrothermal circulation of seawater sulfate at mid-ocean ridges has a minimal effect on the sulfur isotope composition of seawater sulfate (Alt, 1995). However, the recent discovery of gypsum pseudomorphed after anhydrite in the Macquarie Island ophiolite (ca. 10 Ma) indicates that this process may be more important than originally thought (Alt et al., 2003).

Although these questions remain to be answered, one that can be addressed theoretically is: what are the S isotope and mass-balance implications of a low sulfate ocean with background $\delta^{34}S_{sulfate}$ values similar to today (~20‰)? The simplest way to represent the sulfur isotope system is with a steady-state mass-balance equation arranged to yield $\delta^{34}S$ of seawater sulfate:

$$\delta^{34}S_{sulfate} = f_{py} * \Delta^{34}S + \delta^{34}S_{riv}$$

Here, f_{py} is the fraction of total sulfur burial occurring as pyrite, $\Delta^{34}S$ is the difference in isotope composition between $\delta^{34}S_{sulfate}$ and $\delta^{34}S_{pyrite}$, and $\delta^{34}S_{riv}$ is the sulfur isotope composition of the riverine input of sulfur to the oceans, which represents most of the sulfur delivered to the oceans. Modern values for $\delta^{34}S_{sulfate}$, f_{py}, $\Delta^{34}S$, and $\delta^{34}S_{riv}$ are ~20‰, 0.6, 35‰, and 6‰, respectively (Holser et al. 1988; Arthur 2000).

If Neoproterozoic oceanic sulfate concentrations were ~10% of modern values, then there is reason to believe that the fraction of total sulfur buried as pyrite was higher relative to today. An ocean with lower sulfate concentrations should produce fewer sulfate deposits (e.g., gypsum). The geologic record supports this contention (Strauss, 1993). Second, $\delta^{34}S_{riv}$ values may have been lower in the Neoproterozoic relative to today. If a larger fraction of sulfur was buried as pyrite in the Precambrian in general, we might expect $\delta^{34}S_{riv}$ to be more depleted than at present—perhaps 0‰. Third, based on limited $\delta^{34}S_{sulfate}$ data, Canfield and Teske (1996) showed that average $\Delta^{34}S_{sulfate-pyrite}$ values were lower than 35‰ (as they are today)—perhaps as low as 25‰. As discussed previously, lower $\Delta^{34}S$ values are typically associated with the activity of sulfate-reducing bacteria in the absence of intermediate sulfur species and sulfur disproportionation reactions associated with the oxidative portion of the sulfur cycle (Jørgensen, 1990; Canfield and Thamdrup, 1994). We suggest that a late Mesoproterozoic–early Neoproterozoic ocean with low sulfate concentrations and a S isotope composition similar to today (~20‰) might have the following values under steady-state conditions: f_{py} = 0.8, $\Delta^{34}S$ = 25‰, and $\delta^{34}S_{riv}$ = 0‰. These are just estimates (partly based on geologic evidence), but they provide a sense of the magnitude of change necessary to the inputs and outputs and isotopic composition of sulfur species involved in the sulfur cycle to compensate for an ocean with low sulfate concentrations and a $\delta^{34}S_{sulfate}$ value of ~20‰.

CONCLUSIONS

Our results indicate that the sulfur cycle behaved very differently in the early versus the late Neoproterozoic. More specifically, it appears that Neoproterozoic glacial events, whether global or not, forced high amplitude sulfur isotopic shifts and a substantial decrease in oceanic sulfate concentrations (in an ocean with already reduced sulfate concentrations). The positive $\delta^{34}S_{sulfate}$ values reflect either the enriched ^{34}S ocean resulting from BSR during glaciation or in the aftermath of glaciation as a result of BSR in an ocean with reduced sulfate concentrations *or* a combination of both.

While the $\delta^{34}S_{sulfate}$ variations expressed in late Mesoproterozoic and early Neoproterozoic carbonates are not as large as those in the later part of the era, the magnitude (~9‰) and rate of $\delta^{34}S$ change (<2 m.y.) appear to be consistent with an ocean with lower sulfate concentrations than today. Furthermore, based on $\delta^{34}S_{sulfate}$ patterns and CAS concentrations recorded in the Crystal Spring and Beck Spring formations, we estimate that late Mesoproterozoic to mid-Neoproterozoic oceanic sulfate concentrations were ~10% of modern values (e.g., ~3 mM).

ACKNOWLEDGMENTS

Financial support was provided by the NASA Astrobiology Institute Cooperative Agreement (NCC2-1057) to MTH and MAA. ARP was supported by the Carnegie Trust for the Universities of Scotland. The authors thank Sara Geleskie and Julie Johnson, who provided invaluable assistance in the laboratory, and Galen Halverson and Adam Maloof for helpful discussions. Reviews by Timothy Lyons, Jay Kaufman, and an anonymous reviewer significantly improved the clarity and content of this manuscript.

REFERENCES CITED

Alt, J.C., 1995, Sulfur isotopic profile through the oceanic crust: Sulfur mobility and seawater-crustal sulfur exchange during hydrothermal alteration: Geology, v. 23, p. 585–588, doi: 10.1130/0091-7613(1995)0232.3.CO;2.

Alt, J.C., Davidson, G.J., Teagle, D.A.H., and Karson, J.A., 2003, Isotopic composition of gypsum in the Macquarie Island ophiolite: Implications for the sulfur cycle and the subsurface biosphere in oceanic crust: Geology, v. 31, p. 549–552, doi: 10.1130/0091-7613(2003)0312.0.CO;2.

Arthur, M.A., 2000, Volcanic contributions to the carbon and sulfur geochemical cycles and global change, *in* Sigurdsson, H., Houghton, B.F., McNutt, S.R., Rymer, H., and Stix, J., eds., Encyclopedia of volcanoes: San Diego, Academic Press, p. 1045–1056.

Ault, W.U., and Kulp, J.L., 1959, Isotopic geochemistry of sulphur: Geochimica et Cosmochimica Acta, v. 16, p. 201–235, doi: 10.1016/0016-7037(59)90112-7.

Bischoff, J.L., and Dickson, F.W., 1975, Seawater-basalt interaction at 200°C and 500 bars: implications for origin of sea-floor heavy-metal deposits and regulation of seawater chemistry: Earth and Planetary Science Letters, v. 25, p. 385–397, doi: 10.1016/0012-821X(75)90257-5.

Brand, U., and Veizer, J., 1981, Chemical diagenesis of a multicomponent carbonate system -1) Trace elements: Journal of Sedimentary Petrology, v. 50, p. 1219–1236.

Burdett, J.W., Arthur, M.A., and Richardson, M., 1989, A Neogene seawater sulfur isotope age curve from calcareous pelagic microfossils: Earth and Planetary Science Letters, v. 94, p. 189–198, doi: 10.1016/0012-821X(89)90138-6.

Busenberg, E., and Plummer, L.N., 1985, Kinetic and thermodynamic factors controlling the distribution of SO_4 and Na in calcites and selected aragonites: Geochimica et Cosmochimica Acta, v. 49, p. 713–725, doi: 10.1016/0016-7037(85)90166-8.

Canfield, D.E., 1998, A new model for Proterozoic ocean chemistry: Nature, v. 396, p. 450–453, doi: 10.1038/24839.

Canfield, D.E., 2001, Isotope fractionation by natural populations of sulfate-reducing bacteria: Geochimica et Cosmochimica Acta, v. 65, p. 1117–1124, doi: 10.1016/S0016-7037(00)00584-6.

Canfield, D.E., and Teske, A., 1996, Late Proterozoic rise in atmospheric oxygen concentration inferred from phylogenetic and sulphur-isotope studies: Nature, v. 382, p. 127–132, doi: 10.1038/382127A0.

Canfield, D.E., and Thamdrup, B., 1994, The production of ^{34}S-depleted sulfide during bacterial disproportionation of elemental sulfur: Science, v. 266, p. 1973–1975.

Condie, K.C., Des Marais, D.J., and Abbott, D., 2001, Precambrian superplumes and supercontinents: a record in black shales, carbon isotopes, and paleoclimates?: Precambrian Research, v. 106, p. 239–260, doi: 10.1016/S0301-9268(00)00097-8.

Condon, D.J., Prave, A.R., and Benn, D.I., 2002, Neoproterozoic glacial-rainout intervals: Observations and implications: Geology, v. 30, p. 35–38, doi: 10.1130/0091-7613(2002)0302.0.CO;2.

Corsetti, F.A., and Kaufman, A.J., 2003, Stratigraphic investigations of carbon isotope anomalies and Neoproterozoic ice ages in Death Valley, California: GSA Bulletin, v. 115, p. 916–932.

Corsetti, F.A., and Kaufman, A.J., 2004, The relationship between the Neoproterozoic Noonday Dolomite and the Ibex Formation: New observations and their bearing on "snowball Earth," *in* Calzia, J.P., ed., Fifty years of Death Valley research: A volume in honor of Lauren A. Wright and Bennie Troxel: Earth-Science Reviews Special Issue, Volume 61 (in press).

Crowley, T.J., and Hyde, W.T., 2001, CO_2 levels required for deglaciation of a "Near-Snowball" Earth: Geophysical Research Letters, v. 28, p. 283–286, doi: 10.1029/2000GL011836.

Dehler, C.M., Elrick, M., Karlstrom, K.E., Smith, G.A., Crossey, L.J., and Timmons, J.M., 2001, Neoproterozoic Chuar Group (~800–742), Grand Canyon: a record of cyclic marine deposition during global cooling and supercontinent rifting: Sedimentary Geology, v. 141-142, p. 465–499, doi: 10.1016/S0037-0738(01)00087-2.

Derry, L.A., Kaufman, A.J., and Jacobsen, S.B., 1992, Sedimentary cycling and environmental change in the Late Proterozoic: evidence from stable and radiogenic isotopes: Geochimica et Cosmochimica Acta, v. 56, p. 1317–1329, doi: 10.1016/0016-7037(92)90064-P.

Detmers, J., Brüchert, V., Habicht, K.S., and Kuever, J., 2001, Diversity of sulfur isotope fractionations by sulfate-reducing prokaryotes: Applied and Environmental Microbiology, v. 67, p. 888–894, doi: 10.1128/AEM.67.2.888-894.2001.

Frank, T.D., and Lyons, T.W., 2000, The integrity of $\delta^{18}O$ records in Precambrian carbonates: A Mesoproterozoic case study, *in* Grotzinger, J.P., and James, N.P., eds., Carbonate Sedimentation and diagenesis in the evolving Precambrian world, Volume 67: Tulsa, Oklahoma, Society for Sedimentary Geology—SEPM, p. 315–326.

Gellatly, A.M., Lyons, T.W., and Kah, L.C., Carbonate-associated sulfate in Mesoproterozoic successions and modern carbonate systems: examining rapid shifts in paleoenvironmental conditions: Geological Society of America Abstracts with Programs, v. 33, no. 6, p. A-95.

Given, R.K., and Lohmann, K.C., 1985, Derivation of the original isotopic composition of Permian marine cements: Journal of Sedimentary Petrology, v. 55, p. 430–439.

Goldhaber, M.B., and Kaplan, I.R., 1974, The sulfur cycle, *in* Goldberg, E.D., ed., The sea, Volume 5: New York, New York, Wiley, p. 569–655.

Gorjan, P., Veevers, J.J., and Walter, M.R., 2000, Neoproterozoic sulfur-isotope variation in Australia and global implications: Precambrian Research, v. 100, p. 151–179, doi: 10.1016/S0301-9268(99)00073-X.

Grotzinger, J.P., and Kasting, J.F., 1993, New constraints on Precambrian ocean composition: Journal of Geology, v. 101, p. 235–243.

Habicht, K.S., and Canfield, D.E., 1997, Sulfur isotope fractionation during bacterial sulfate reduction in organic-rich sediments: Geochimica et Cosmochimica Acta, v. 61, p. 5351–5361, doi: 10.1016/S0016-7037(97)00311-6.

Habicht, K.S., Canfield, D.E., and Rethmeier, J., 1998, Sulfur isotope fractionation during bacterial reduction and disproportionation of thiosulfate and sulfite: Geochimica et Cosmochimica Acta, v. 62, p. 2585–2595, doi: 10.1016/S0016-7037(98)00167-7.

Habicht, K.S., Gade, M., Thamdrup, B., Berg, P., and Canfield, D.E., 2002, Calibration of sulfate levels in the Archean ocean: Science, v. 298, p. 2372–2374, doi: 10.1126/SCIENCE.1078265.

Halverson, G.P., Hoffman, P.F., Schrag, D.P., and Kaufman, A.J., 2002, A major perturbation of the carbon cycle before the Ghaub glaciation (Neoproterozoic) in Namibia: Prelude to snowball Earth?: Geochemistry Geophysics Geosystems, v. 3, doi: 10.1029/2001GC000244 (online publication).

Heaman, L.M., and Grotzinger, J.P., 1992, 1.08 diabase sills in the Pahrump Group, California: Implications for development of the Cordilleran miogeocline: Geology, v. 20, p. 637–640, doi: 10.1130/0091-7613(1992)0202.3.CO;2.

Hoffman, P.F., Kaufman, A.J., Halverson, G.P., and Schrag, D.P., 1998a, A Neoproterozoic snowball Earth: Science, v. 281, p. 1342–1346, doi: 10.1126/SCIENCE.281.5381.1342.

Hoffman, P.F., Kaufman, A.J., and Halverson, G.P., 1998b, Comings and goings of global glaciations in a Neoproterozoic tropical platform in Namibia: GSA Today, v. 8, no. 5, p. 1–9.

Hoffman, P.F., and Schrag, D.P., 2002, The snowball Earth hypothesis: testing the limits of global change: Terra Nova, v. 14, p. 129–155, doi: 10.1046/J.1365-3121.2002.00408.X.

Holser, W.T., 1977, Catastrophic chemical events in the history of the ocean: Nature, v. 267, p. 403–408.

Holser, W.T., Schidlowski, M., Mackenzie, F.T., and Maynard, J.B., 1988, Geochemical cycles of carbon and sulfur, *in* Gregor, C.B., Garrels, R.M., Mackenzie, F.T., and Maynard, J.B., eds., Chemical cycles in the evolution of the Earth: New York, Wiley, p. 105–173.

Hurtgen, M.T., Arthur, M.A., Suits, N.S., and Kaufman, A.J., 2002, The sulfur isotopic composition of Neoproterozoic seawater sulfate: Implications for a snowball Earth?: Earth and Planetary Science Letters, v. 203, p. 413–430, doi: 10.1016/S0012-821X(02)00804-X.

Hyde, W.T., Crowley, T.J., Baum, S.K., and Peltier, W.R., 2000, Neoproterozoic "snowball Earth" simulations with a coupled climate/ice-sheet model: Nature, v. 405, p. 425–429, doi: 10.1038/35013005.

Jørgensen, B.B., 1990, A thiosulfate shunt in the sulfur cycle of marine sediments: Science, v. 249, p. 152–154.

Kah, L.C., Lyons, T.W., and Chesley, J.T., 2001, Geochemistry of a 1.2 Ga carbonate-evaporite succession, northern Baffin and Bylot Islands: implications for Mesoproterozoic marine evolution: Precambrian Research, v. 111, p. 203–234, doi: 10.1016/S0301-9268(01)00161-9.

Kah, L.C., Lyons, T.W., and Frank T.D., 2002, Reservoir size, residence time, and the rate of isotopic change: examining the behavior of Mesoproterozoic marine sulfate: Geological Society of America Abstracts with Programs, v. 34, no. 6, p. 240.

Kampschulte, A., Bruckschen, P., and Strauss, H., 2001, The sulphur isotopic composition of trace sulphates in Carboniferous brachiopods: implications for coeval seawater, correlation with other geochemical cycles and isotope stratigraphy: Chemical Geology, v. 175, p. 149–173, doi: 10.1016/S0009-2541(00)00367-3.

Kaplan, I.R., and Rittenberg, S.C., 1964, Microbiological fractionation of sulphur isotopes: Journal of General Microbiology, v. 34, p. 71–91.

Karlstrom, K.E., Dehler, C.M., Sharp, Z.D., Geissman, J.W., Elrick, M.E., Timmons, J.M., Crossey, L.J., Bowring, S.A., Keefe, K., Knoll, A.H., Porter, S.M., Des Marais, D.J., and Weil, A.B., 2000, The Chuar Group of the Grand Canyon: Record of breakup of Rodinia, associated change in the global carbon cycle, and ecosystem expansion by 740 Ma: Geology, v. 28, p. 619–622, doi: 10.1130/0091-7613(2000)0282.3.CO;2.

Kaufman, A.J., Hayes, J.M., Knoll, A.H., and Germs, G.J.B., 1991, Isotopic compositions of carbonates and organic carbon from upper Proterozoic successions in Namibia: stratigraphic variation and the effects of diagenesis and metamorphism: Precambrian Research, v. 49, p. 301–327, doi: 10.1016/0301-9268(91)90039-D.

Kaufman, A.J., and Knoll, A.H., 1995, Neoproterozoic variations in the C-isotopic composition of seawater: stratigraphic and biogeochemical implications: Precambrian Research, v. 73, p. 27–49, doi: 10.1016/0301-9268(94)00070-8.

Kaufman, A.J., Knoll, A.H., and Narbonne, G.M., 1997, Isotopes, ice ages and terminal Proterozoic Earth history: Proceedings of the National Academy of Sciences of the United States of America, v. 94, p. 6600–6605, doi: 10.1073/PNAS.94.13.6600.

Kaufman, A.J., Varni, M.A., and Wing, B.A., 2002, Sulfur isotope constraints on a Snowball Earth: Geological Society of America Abstracts with Programs, v. 34, no. 6, p. 221.

Kennedy, M.J., 1996, Stratigraphy, sedimentology, and isotopic geochemistry of Australian Neoproterozoic postglacial cap dolostones: Deglaciation, $\delta^{13}C$ excursions, and carbonate precipitation: Journal of Sedimentary Research, v. 66, p. 1050–1064.

Kennedy, M.J., Runnegar, B., Prave, A.R., Hoffmann, K.-H., and Arthur, M.A., 1998, Two or four Neoproterozoic glaciations: Geology, v. 26, p. 1059–1063, doi: 10.1130/0091-7613(1998)0262.3.CO;2.

Kenny, R., and Knauth, L.P., 2001, Stable isotope variations in Neoproterozoic Beck Spring Dolomite and Mesoproterozoic Mescal Limestone paleokarst: Implications for life on land in the Precambrian: Geological Society of America Bulletin, v. 113, p. 650–658, doi: 10.1130/0016-7606(2001)1132.0.CO;2.

Kirschvink, J.L., 1992, Late-Proterozoic low-latitude global glaciation: The snowball earth, *in* Schopf, J.W., and Klein, C., eds., The Proterozoic biosphere: New York, Cambridge University Press, p. 51–52.

Knauth, L.P., and Epstein, S., 1976, Hydrogen and oxygen isotope ratios in nodular and bedded cherts: Geochimica et Cosmochimica Acta, v. 40, p. 1095–1108, doi: 10.1016/0016-7037(76)90051-X.

Knoll, A.H., and Canfield, D.E., 1998, Isotopic inferences on early ecosystems, *in* Manger, W.L., and Meeks, L.K., eds., Isotope paleobiology and paleoecology: The Paleontological Society Papers, v. 4, p. 212–243.

Knoll, A.H., Hayes, J.M., Kaufman, A.J., Swett, K., and Lambert, I.B., 1986, Secular variation in carbon isotope ratios from Upper Proterozoic successions of Svalbard and East Greenland: Nature, v. 321, p. 832–838.

Leather, J., Allen, P.A., Brasier, M.D., and Cozzi, A., 2002, Neoproterozoic snowball Earth under scrutiny: Evidence from the Fiq glaciation of Oman: Geology, v. 30, p. 891–894, doi: 10.1130/0091-7613(2002)0302.0.CO;2.

Lowenstein, T.K., Hardie, L.A., Timofeeff, M.N., and Demicco, R.V., 2003, Secular variation in seawater chemistry and the origin of calcium chloride basinal brines: Geology, v. 31, p. 857–860, doi: 10.1130/G19728R.1.

Lyons, T.W., Kah, L.C., and Gellatly, A.M., 2004, The Precambrian sulfur isotope record of evolving atmospheric oxygen, *in* Eriksson, P.G., et al., eds., The Precambrian Earth: Tempos and events: Developments in Precambrian geology: Amsterdam, Elsevier, p. 421–439.

Paytan, A., Kastner, M., Campbell, D., and Thiemens, M.H., 1998, Sulfur isotopic composition of Cenozoic seawater sulfate: Science, v. 282, p. 1459–1462, doi: 10.1126/SCIENCE.282.5393.1459.

Petsch, S.T., and Berner, R.A., 1998, Coupling the geochemical cycles of C, P, Fe, and S; the effect on atmospheric O_2 and the isotopic records of carbon and sulfur: American Journal of Science, v. 298, p. 246–262.

Poulsen, C., 2003, Absence of a runaway ice-albedo feedback in the Neoproterozoic: Geology, v. 31, p. 473–476, doi: 10.1130/0091-7613(2003)0312.0.CO;2.

Prave, A.R., 1994, The Pahrump Group: Alternative speculations on the Neoproterozoic geodynamic development of the Death Valley region: Geological Society of America Abstracts with Programs, v. 26, p. 82.

Prave, A.R., 1999, Two diamictites, two cap carbonates, two $\delta^{13}C$ excursions, two rifts: The Neoproterozoic Kingston Peak Formation, Death Valley, California: Geology, v. 27, p. 339–342, doi: 10.1130/0091-7613(1999)0272.3.CO;2.

Raab, M., and Spiro, B., 1991, Sulfur isotopic variations during seawater evaporation with fractional crystallization: Chemical Geology, v. 86, p. 323–333.

Roberts, M.T., 1982, Depositional environments and tectonic setting of the Crystal Spring Formation, Death Valley region, California, *in* Cooper, J.D., Troxel, B.W., and Wright, L.A., eds., Geology of selected areas in the San Bernardino Mountains, western Mojave Desert, and southern Great Basin, California: Geological Society of America Cordilleran Section Field Trip Guidebook and Volume: Shoshone, California, Death Valley Publishing Company, p. 143–154.

Ross, G.M., Bloch, J.D., and Krouse, H.R., 1995, Neoproterozoic strata of the southern Canadian Cordillera and the isotopic evolution of seawater sulfate: Precambrian Research, v. 73, p. 71–99, doi: 10.1016/0301-9268(94)00072-Y.

Rothman, D.H., Hayes, J.M., and Summons, R.E., 2003, Dynamics of the Neoproterozoic carbon cycle: Proceedings of the National Academy of Sciences of the United States of America, v. 100, p. 8124–8129, doi: 10.1073/PNAS.0832439100.

Seyfried, W.E., and Bischoff, J.L., 1979, Low temperature basalt alteration by seawater: An experimental study at 70 °C and 150 °C: Geochimica et Cosmochimica Acta, v. 43, p. 1937–1947, doi: 10.1016/0016-7037(79)90006-1.

Staudt, W.J., Oswald, E.J., and Schoonen, M.A.A., 1993, Determination of sodium, chloride and sulfate in dolomites: a new technique to constrain the composition of dolomitizing fluids: Chemical Geology, v. 107, p. 97–109, doi: 10.1016/0009-2541(93)90104-Q.

Staudt, W.J., and Schoonen, M.A.A., 1995, Sulfate incorporation into sedimentary carbonates, *in* Vairavamurthy, M.A., and Schoonen, M.A.A., eds., Geochemical transformations of sedimentary sulfur: Washington, D.C., American Chemical Society, p. 332–345.

Strauss, H., 1993, The sulfur isotopic record of Precambrian sulfates: new data and a critical evaluation of the existing record: Precambrian Research, v. 63, p. 225–246, doi:10.1016/0301-9268(93)90035-Z.

Taylor, B.E., Wheeler, M.C., and Nordstrom, D.K., 1984, Stable isotope geochemistry of acid mine drainage: experimental oxidation of pyrite: Geochimica et Cosmochimica Acta, v. 48, p. 2669–2678, doi: 10.1016/0016-7037(84)90315-6.

Ueda, A., and Krouse, H.R., 1986, Direct conversion of sulphide and sulphate minerals to SO_2 for isotope analyses: Geochemical Journal, v. 20, p. 209–212.

Varni, M.A., Kaufman, A.J., Misi, A., and Brito Neves, B.B., 2001, Anomalous $\delta^{34}S$ signatures in trace sulfate from a potential cap carbonate in the Neoproterozoic Bambui Group, Brazil: Geological Society of America Abstracts with Programs, v. 33, no. 6, p. A-96.

Veizer, J., Bruckschen, P., Pawellek, F., Diener, A., Podlaha, O.G., Carden, G.A.F., Jasper, T., Korte, C., Strauss, H., Azmy, K., and Ala, D., 1997, Oxygen isotope evolution of Phanerozoic seawater: Palaeogeography, Palaeoclimatology, Palaeoecology, v. 132, p. 159–172, doi: 10.1016/S0031-0182(97)00052-7.

Walker, J.C.G., and Brimblecombe, P., 1985, Iron and sulfur in the pre-biologic ocean: Precambrian Research, v. 28, p. 205–222, doi: 10.1016/0301-9268(85)90031-2.

Walker, J.C.G., and Lohmann, K.C., 1989, Why the oxygen isotopic composition of sea water changes with time: Geophysical Research Letters, v. 16, p. 323–326.

Wilkinson, B.H., Opdyke, B.N., and Algeo, T.J., 1991, Time partitioning in cratonic carbonate rocks: Geology, v. 19, p. 1093–1096, doi: 10.1130/0091-7613(1991)0192.3.CO;2.

Zempolich, W.G., Wilkinson, B.H., and Lohmann, K.C., 1988, Diagenesis of late Proterozoic carbonates: the Beck Spring Dolomite of eastern California: Journal of Sedimentary Petrology, v. 58, p. 656–672.

Manuscript Accepted by the Society March 28, 2004

Printed in the USA

Geological Society of America
Special Paper 379
2004

4 Ga of seawater evolution: Evidence from the sulfur isotopic composition of sulfate

Harald Strauss*
Geologisch-Paläontologisches Institut, Westfälische Wilhelms–Universität Münster, Corrensstrasse 24, 48149 Münster, Germany

ABSTRACT

Substantial changes in the global sulfur cycle are recorded in the sulfur isotopic composition of seawater sulfate. The Archean ocean was low in sulfate, with $\delta^{34}S$ values ~+4‰. Sulfate probably originated from the rainout of atmospheric sulfate aerosols, proposed on the basis of recorded mass-independent sulfur isotopic fractionation. Oxygenation of Earth's surface environments in the Paleoproterozoic changed the global sulfur cycle. The Proterozoic and Phanerozoic witnessed an increasing abundance of oceanic sulfate, resulting from oxidative weathering of sulfides on the continents. The sulfur isotopic composition has changed from an early Archean value at +4 to +32‰ at the Neoproterozoic-Cambrian transition. Temporal variations in the $\delta^{34}S$ of Phanerozoic oceanic sulfate between +11 and +32‰ indicate fluctuations in the fractional burial of reduced versus oxidized sulfur.

Keywords: sulfur isotopes, seawater, sulfate, Precambrian, Phanerozoic.

INTRODUCTION

The history of seawater sulfate can be reconstructed by studying its sulfur isotopic composition. Available proxy records are massive evaporitic calcium sulfates (e.g., Claypool et al., 1980; Strauss, 1993, 1997; Kampschulte et al., 1998) and barite (e.g., Cecile et al., 1983; Paytan et al., 1998), or trace quantities of sulfate in carbonates (e.g., Burdett et al., 1989; Kampschulte and Strauss, 2004) and phosphates (e.g., McArthur et al., 1986; Shields et al., 1999). Principle questions to be addressed include:

- origin and fate of seawater sulfate;
- temporal variations in the abundance of oceanic sulfate;
- implications for the oxidation state of the ocean-atmosphere system.

In the following sections, I will first introduce some principles of sulfur isotope geochemistry and evaluate the different proxy signals. This includes a brief compilation of our knowledge about the abundance of oceanic sulfate through time. A section on the sulfur isotopic composition of modern oceanic sulfate will be followed by a review of the Precambrian and Phanerozoic isotope records. In the final discussion, I will return to the three questions noted above.

SULFUR ISOTOPE SYSTEMATICS

A simplified view of the global sulfur cycle will serve as a base for introducing the principle reservoirs, processes, and associated isotope effects (Fig. 1). Oceanic sulfate represents the central reservoir. The average sulfur isotopic composition of modern marine dissolved sulfate lies at +21‰ (e.g., Rees et al., 1978; Longinelli, 1989). Riverine delivery of dissolved sulfate, derived from continental weathering of sulfides and sulfates, represents the principal input into this "reaction chamber." Its isotopic composition is reasonably well constrained for the modern world with a $\delta^{34}S$ value of +8‰ (Grinenko and Krouse, 1992). Additional input derives through the introduction of magmatic sulfur,

*hstrauss@uni-muenster.de

Strauss, H., 2004, 4 Ga of seawater evolution: Evidence from the sulfur isotopic composition of sulfate, *in* Amend, J.P., Edwards, K.J., and Lyons, T.W., eds., Sulfur biogeochemistry—Past and present: Boulder, Colorado, Geological Society of America Special Paper 379, p. 195–205. For permission to copy, contact editing@geosociety.org.

either via mid-ocean ridges or as volcanic sulfur. The former input function is rather poorly quantified. $\delta^{34}S_{H_2S}$ values between +3 and +13‰ for vent fluids (e.g., Shanks, 2001) might serve as a first approximation; however, this sulfate represents a mixture of recycled seawater sulfate, sulfur leached from oceanic crust, and genuine mantle sulfur. The input of volcanic sulfur is well quantified from present day measurements (Graf et al., 1998). Its isotopic composition varies between −10 and +10‰ (Shanley et al., 1998). It should be noted here that the input of magmatic sulfur has largely been neglected in modeling approaches (e.g., Kump, 1989; Berner, 2001; but see Hansen and Wallmann, 2003).

Removal of sulfate from the ocean occurs through two principal pathways: precipitation of dissolved seawater sulfate as marine evaporites including incorporation of sulfate into marine chemical precipitates (SO_4) and biological sulfur cycling and fixation as biogenic sulfur (BioS). In geological terms, these output functions are represented by marine evaporitic sulfates and by sedimentary pyrite and organic sulfur. The first process, the precipitation of evaporites, is not associated with any substantial isotope effect (e.g., Thode et al., 1961; Raab and Spiro, 1991). In contrast, microbial sulfur cycling and its subsequent fixation as sulfide (or as organic sulfur) results in a shift in the sulfur isotopic composition of variable magnitude but in general toward ^{34}S-depleted values for the resulting biogenic sulfur (e.g., Canfield, 2001). Isotopic fractionation between 2‰ and 42‰ has been measured for bacterial sulfate reduction under optimal growth conditions during experimental work (e.g., Detmers et al., 2001). Natural populations in modern marine sediments commonly display an isotopic fractionation of 18‰–45‰ for this process (e.g., Habicht and Canfield, 1997, 2001). On the other hand, isotopic fractionation appears to be greatly reduced under sulfate-limiting conditions, becoming minor at sulfate concentrations below 200 µmol (Habicht et al., 2002). Finally, iron sulfides in modern marine sediments reveal an overall isotopic fractionation of up to 70‰ (e.g., Habicht and Canfield, 2001). This is attributed to a combination of two principal processes: bacterial sulfate reduction and disproportionation of intermediate sulfur compounds (e.g., Habicht et al., 1998). Based on the overall sulfur isotopic fractionation between oceanic sulfate and sedimentary pyrite, measured in modern marine sediments, the importance of bacterial sulfate reduction has been quantified between 41% and 85%, with the remaining isotope effect being attributed to disproportionation reactions (Habicht and Canfield, 2001). Disproportionation reactions can be traced back in Earth history until 600–800 Ma (e.g., Canfield and Teske, 1996). Sulfur immobilized in sediments is returned to the oceanic reservoir via continental weathering and riverine delivery as dissolved sulfate.

A central objective for studying the sulfur isotopic composition of seawater sulfate through time is the quantification of temporal variations in the principle input or output functions. In a simple way, this can be addressed by an isotope mass balance:

$$\delta_{input} = f_{BioS}\delta_{BioS} + (1 - f_{BioS})\delta_{SO_4}$$

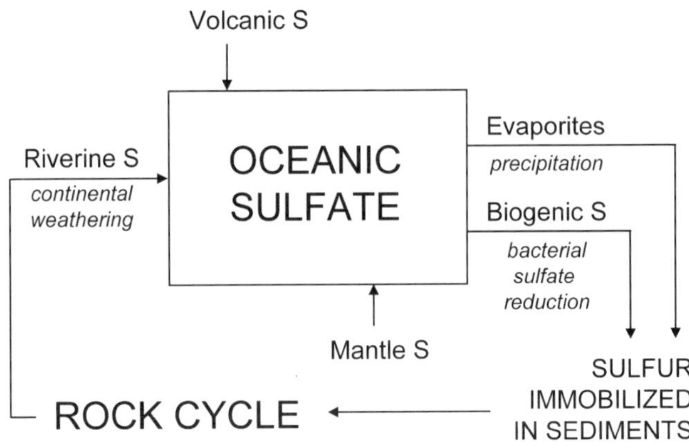

Figure 1. Schematic view of the global sulfur cycle.

Here, the two outputs and their isotopic compositions are set in relation to the average sulfur isotopic composition of crustal sulfur (δ_{input}). Its $\delta^{34}S$ value has been calculated at +2‰, and it is believed to have remained constant through time (e.g., Holser et al., 1988). The sulfur isotopic composition of seawater sulfate and biogenic sulfur (sedimentary pyrite, organically bound sulfur) for a given point in time allows solving the equation and determining the parameter f_{BioS}. This parameter is proportional to the size of the sulfur pool that is deposited in the sedimentary record primarily through microbially driven processes (such as bacterial sulfate reduction and/or disproportionation).

THE PROXY SIGNALS

In the modern ocean, both abundance and sulfur isotopic composition of dissolved oceanic sulfate are distributed with lateral and vertical homogeneity (e.g., Rees et al., 1978; Longinelli, 1989). As a consequence, a consistent isotope signal has been regarded as the prime criterion for the true, representative value for marine sulfate sulfur at any given period in time (Nielsen, 1989). Exceptions to this homogeneity are (semi)restricted basins or marginal seas, such as the Black Sea (e.g., Sweeney and Kaplan, 1980) or the Framvaren Fjord, Norway (e.g., Mandernack et al., 2003). Sulfate in these seas can be affected through contributions of riverine sulfate, expressed foremost in a $\delta^{34}S$ value that is lower than contemporaneous seawater sulfate. Alternatively, increasing $\delta^{34}S$ values reflect bacterial sulfate reduction in the water column.

Different proxy signals have been utilized to constrain the sulfur isotopic composition of ancient seawater sulfate. The traditional approach has been the analysis of marine evaporitic calcium sulfate. Based on empirical and experimental results (e.g., Holser and Kaplan, 1966), evaporitic calcium sulfate reflects the sulfur isotopic composition of the original brine, either without any isotopic difference or with a shift in $\delta^{34}S$ of 1‰–2‰ toward more positive values. Only during later stages of evaporation, such as in the potash facies, does this difference in $\delta^{34}S$ between parental brine and resulting sulfate mineral become larger (e.g., Holser and Kaplan, 1966, Raab and Spiro, 1991). Therefore, evaporitic calcium sulfate represents a suitable proxy signal for reconstructing the temporal evolution of $\delta^{34}S_{seawater\ sulfate}$ through time. A true drawback, however, is the irregular temporal distribution of evaporite deposits (e.g., Zharkov, 1984). Furthermore, evaporite deposits are frequently poorly dated. This has resulted in a poorly constrained and rather fragmentary sulfur isotope record, in particular for the Precambrian (e.g., Strauss, 1993; Lyons et al., 2004).

Few studies have considered barite. In order for barite to serve as proxy signal for seawater sulfate, several assumptions have to be made in respect to its origin. Barite can be found in modern marine sediments (Dehairs et al., 1980), precipitated as a primary mineral from seawater. This primary barite can be distinguished from diagenetic or hydrothermal barite through additional geochemical tracers (Paytan et al., 2002). Paytan et al. (1998) utilized marine barite to reconstruct the sulfur isotopic composition of Cenozoic seawater sulfate. Temporal variations in $\delta^{34}S$ are clearly discernible from their continuous isotope record with a time resolution of 1 Ma. Barite pseudomorphs after marine evaporitic calcium sulfates have been analyzed to constrain the sulfur isotopic composition of early Archean seawater sulfate (for a compilation, see Strauss, 1993; further data from Shen et al., 2001). These studies assume that pseudomorphic growth occurred without alteration of the sulfur isotope signal.

An increasing number of studies are exploiting the fact that sulfate represents a trace constituent in marine calcite, with concentrations ranging from a few tens of ppm in micritic carbonates to several thousand ppm in various biogenic carbonates (e.g., Volkov and Rozanov, 1983; Busenberg and Plummer, 1985; Staudt and Schoonen, 1995; Grossman et al., 1996). Sufficient evidence exists that this sulfate is present as structurally substituted sulfate (subsequently termed SSS) within the carbonate lattice (e.g., Takano, 1985; Staudt et al., 1994; Pingitore et al., 1995). This offers the potential for determining the sulfur isotopic composition of ambient seawater sulfate. The validity of this analytical approach has been demonstrated (e.g., Burdett et al., 1989; Kampschulte et al., 2001). $\delta^{34}S$ values of SSS in recent biogenic carbonates are equal to modern seawater sulfate. A comparison of $\delta^{34}S_{SSS}$ from biogenic carbonates, time-equivalent whole rock carbonates, and marine evaporitic sulfates of the same time interval has yielded similar results, suggesting that the sulfur isotopic composition of SSS faithfully reflects the sulfur isotopic composition of ambient seawater.

The sulfur isotopic composition of phosphate-bound sulfate was studied in order to elucidate phosphate genesis (e.g., Benmore et al., 1983; McArthur et al., 1986). The potential for studying seawater sulfate $\delta^{34}S$ exists, provided the signal has not been altered through diagenetic reactions. Bacterial sulfate reduction in particular can substantially change the sulfur isotopic composition of dissolved sulfate in pore waters during diagenesis. For example, Shields et al. (1999) noted a correlation between pyrite abundance and the sulfur isotopic composition of phosphate-bound sulfate in early Cambrian sediments from the Yangtze Platform, South China. Thus, a careful assessment of phosphate formation through petrographic and geochemical analyses should accompany a respective sulfur isotope study in order to validate this analytical approach.

THE ABUNDANCE OF OCEANIC SULFATE THROUGH TIME

The chemical composition of seawater has changed through time, including the abundance of oceanic sulfate (Fig. 2). Evidence stems, for example, from observations of primary carbonate (calcite versus aragonite seas: Sandberg, 1983) or evaporite mineralogy (KCl versus $MgSO_4$ seas: Hardie, 1996) or from the composition of fluid inclusions in halite (e.g., Lowenstein et al., 2001; Horita et al., 2002). Our knowledge about the temporal evolution of seawater sulfate abundance has improved substantially for the Phanerozoic but is still very limited for the Precambrian.

Secular changes in the concentration of sulfate in Phanerozoic seawater are evident from data for fluid inclusions in marine halite (e.g., Lowenstein et al., 2001; Horita et al., 2002). Oscillations in concentration between 8 and 28 mmol/kg H_2O over the past 600 Ma define three maxima: the modern ocean, the late Paleozoic–early Mesozoic transition, and the Neoproterozoic-Cambrian transition. In contrast, the Jurassic-Cretaceous time interval and much of the Paleozoic appear to be characterized by low seawater sulfate concentrations (e.g., Lowenstein et al. 2003, their Figure 1). Based on fluid inclusion data, the sulfate concentration in terminal Neoproterozoic seawater was ~23 mmol/kg (Horita et al., 2002).

Going back in time, the presence of thick, laterally extensive calcium sulfate evaporite deposits attests to a sufficiently high abundance of oceanic sulfate. Prominent examples include the Neoproterozoic Bitter Springs Formation, Central Australia (Stewart, 1974), or Mesoproterozoic marine evaporites in the Arctic Canadian Archipelago (Jackson and Ianelli, 1981).

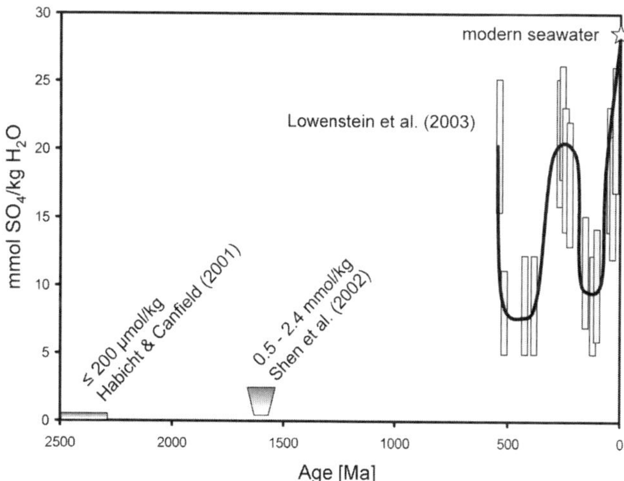

Figure 2. Variations in the concentration of oceanic sulfate through time.

No massive evaporites are known from Paleoproterozoic and Archean sedimentary successions.

Additional quantification of seawater sulfate abundances is largely indirect. It is generally based on the observed isotopic fractionation between Precambrian sulfate and pyrite, which is compared to data from comparable Phanerozoic or recent natural environments and/or experimental studies. Shen et al. (2002) suggested a range between 0.5 and 2.4 mmol/kg for late Paleoproterozoic oceanic sulfate. Habicht and Canfield (2001) proposed that Archean and early Paleoproterozoic seawater contained no more than 200 μmol/kg sulfate.

$\delta^{34}S$ IN MODERN OCEANIC SULFATE

The average sulfur isotopic composition of dissolved sulfate in the modern open ocean (Table 1) lies at 20.0 ± 0.25‰, based on a comprehensive review of available analytical data by Longinelli (1989). More recent measurements yielded slightly more positive $\delta^{34}S$ values at 20.8‰ (Burdett et al., 1989) or 20.9‰ (Kampschulte, 2002). Longinelli (1989, p. 227) stated that the average sulfur isotopic composition of dissolved oceanic sulfate is "...remarkably uniform both vertically and horizontally..." but is "limited to low and middle latitudes. At high latitudes (at least in the case of the Atlantic Ocean) isotopic fluctuations [are] considerably greater...." Sulfur isotope measurements by Kampschulte (2002) for two stations in the Arctic Ocean support this observation. However, an average $\delta^{34}S$ value of 21.1‰ ± 0.6‰ (n = 12) is in contrast to the notion that mean isotope values from higher latitudes are slightly lower than for the global ocean (Longinelli, 1989, p. 227).

A considerably larger spread in $\delta^{34}S_{seawater\ sulfate}$ with a slightly less negative mean isotopic composition has been recorded for dissolved sulfate from marginal seas (Table 1). A reasonable explanation is a greater influence on restricted basins of mixing with riverine sulfate. This observation is rather important for geological applications. It demands the unequivocal proof of the truly marine character for any studied ancient evaporite deposit. Proof could be provided, for example, through consistent $\delta^{34}S$ values for a variety of coeval deposits from different sedimentary basins. Independent evidence might be provided by trace element data (Kah et al., 2001) or the strontium isotopic composition of evaporites in comparison to those of contemporaneously deposited carbonates (e.g., Strauss, 1993; Kampschulte et al., 1998).

TABLE 1. $\delta^{34}S$ FOR DISSOLVED OCEANIC SULFATE

	$\delta^{34}S$ range (‰)	$\delta^{34}S$ avg (‰)	Std dev (‰)	n
World ocean				
Ault and Kulp (1959)		20.0	0.54	25
Thode et al. (1961)		20.1	0.38	16
Sasaki (1972)		20.0	0.21	20
Rees et al. (1978)		21.0	0.38	25
Longinelli (1989)		20.0	0.25	
Burdett et al. (1989)		20.8	0.36	
Kampschulte (2002)		20.9	0.50	25
Marginal seas				
Black Sea				
Vinogradov et al. (1962)	18.6–19.5	19.3		
Mekhtiyeva and Pankina (1969)	14.6–18.9			
Yeremenko and Pankina (1971)		19.3		
Sweeney and Kaplan (1980)	18.3–20.2	19.2		
Kampschulte (2002)		19.8		
Baltic Sea				
Hartmann and Nielsen (1969)		20.0		
Yeremenko and Pankina (1971)		18.3		
Kampschulte (2002)		20.3		
Gulf of Mexico				
Ault and Kulp (1959)	19.8–20.3	20.1		
Kampschulte (2002)		20.0		

THE $\delta^{34}S$ RECORD OF PRECAMBRIAN SEAWATER SULFATE

Recent comprehensive accounts of the sulfur isotope geochemistry of Precambrian sediments (e.g., Strauss, 2002; Lyons et al., 2004) provide a growing isotope record (Fig. 3). It is based on 678 sulfur isotope measurements for massive calcium and barium sulfates and trace quantities of sulfate in carbonates and phosphates. Throughout the Precambrian and into the Cambrian, the sulfur isotopic composition of seawater sulfate displays an increase in ^{34}S from early Archean values of ~+4.2‰ (±1.2‰, n = 80) to an average sulfur isotopic composition of +32.1‰ (±3.7‰, n = 134, not including trace sulfate data by Hurtgen et al., 2002) for the terminal Neoproterozoic and earliest Cambrian. The exact internal structure of this first order temporal trend, whether it is linear, episodic, or even cyclic, remains largely obscured because of the fragmentary nature of the record, particularly in the Archean and Paleoproterozoic. Higher order temporal variations could be resolved for some Mesoproterozoic and Neoproterozoic marine carbonate and evaporite successions (e.g., Kah et al., 2001; Strauss et al., 2001; A. Gellatly and T.W. Lyons, 2003, personal commun.; L.C. Kah, 2003, personal commun.). Extreme fluctuations in $\delta^{34}S$, including strongly ^{34}S enriched sulfur isotope values up to 51‰, have been recorded in trace sulfates from Neoproterozoic carbonates in Namibia (Hurtgen et al., 2002).

THE $\delta^{34}S$ RECORD OF PHANEROZOIC SEAWATER SULFATE

The sulfur isotopic composition of SSS in 331 whole rock and biogenic carbonates, such as brachiopods, belemnites, and foraminifera (Burdett et al., 1989; Strauss, 1999; Ohkouchi et al., 1999; Kampschulte et al., 2001; Kampschulte and Strauss, 2003), has resulted in a biostratigraphically constrained sulfur isotope record for Phanerozoic seawater sulfate (Fig. 4). It replaces the classical, yet much more fragmentary, sulfur isotope record that is based on 860 analyses of marine evaporites measured over the

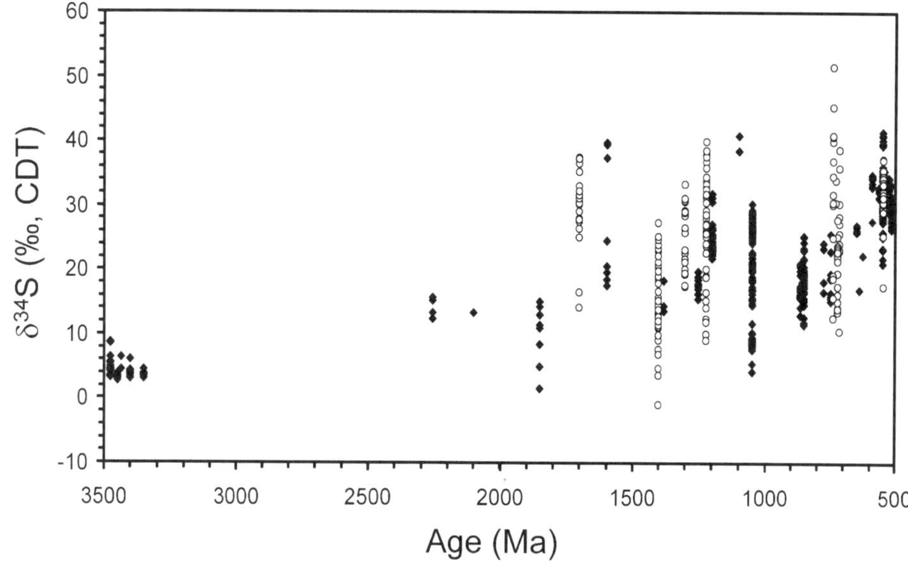

Figure 3. The sulfur isotopic composition of Precambrian seawater sulfate. Filled diamonds—massive sulfates; open circles—structurally substituted sulfate from carbonates and phosphates. Sources of data: Strauss, 1993; Shields et al., 1999; Kah et al., 2001; Shen et al., 2001; Strauss et al., 2001; Lyons et al., 2004; A. Gellatly and T.W. Lyons, 2003, personal commun. CDT—Cañon Diablo Troilite.

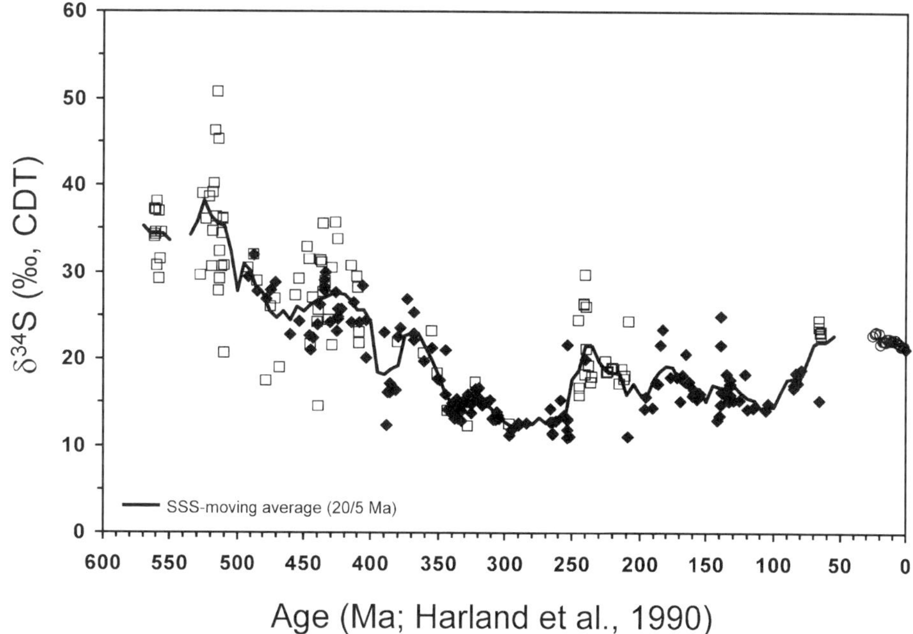

Figure 4. The sulfur isotopic composition of Phanerozoic seawater sulfate (data from Kampschulte and Strauss, 2004). Filled diamonds—biogenic carbonate, open squares—whole rock, open circles—biogenic carbonate (data from Burdett et al., 1989). CDT—Cañon Diablo Troilite.

past 50 yr (e.g., Claypool et al., 1980). Temporal resolution and continuity of the SSS sulfur isotope record is far superior to the traditional evaporite-based record. Clearly, the first-order trend displayed already by the evaporite-based data set, with a maximum in $\delta^{34}S$ around +32‰ in the early Cambrian followed by a decline toward a minimum at +12‰ in the Permo-Carboniferous and a subsequent rise in $\delta^{34}S$ toward a value of +21‰ for modern oceanic sulfate is mimicked by the SSS record. Additional previously undetected yet substantial variations in the sulfate sulfur isotopic composition, particularly in the Paleozoic, have been identified as a consequence of the much better temporal resolution.

EVOLUTION OF THE SULFUR ISOTOPIC COMPOSITION OF SEAWATER SULFATE

Tracing the temporal evolution of $\delta^{34}S_{seawater\ sulfate}$ through 4 Ga is strongly limited for the Archean and Paleoproterozoic. Our knowledge about the sulfur isotopic composition of seawater sulfate from the early part of Earth's history is solely based on data from early Archean barite occurrences from Australia, South Africa, and India (Strauss, 2003). They yield a consistent sulfur isotopic composition around +4‰. This is considerably lower than for any other time in Earth's history and rather close to the $\delta^{34}S$ value of magmatic sulfur (e.g., Clark and Fritz, 1997). Sulfate abundance in the global ocean was likely very low (<200 µM, Habicht et al., 2002), although contrasting views of a high-sulfate Archean ocean have been proposed (e.g., Ohmoto et al., 1993; Ohmoto, 1997). The interpretation of a low-sulfate ocean is based on the lack of a substantial sulfur isotopic fractionation between sulfate and sedimentary sulfide. Early Archean sedimentary sulfides display an average $\delta^{34}S$ value of 0.12 ± 2.87‰ (n = 351; e.g., Strauss, 2002), which is also quite similar to the value for magmatic sulfur. In conclusion, the early global sulfur cycle appears to reflect a largely magmatic signature, with sulfate deposition occurring as local accumulation as oxidized magmatic sulfur.

The discovery of mass-independent sulfur isotope fractionations (MIF) in Archean and early Paleoproterozoic sulfates and sulfides (nonzero values for $\Delta^{33}S = \delta^{33}S - 0.515 \times \delta^{34}S$; for further details, see Farquhar et al., 2000, and references therein) has renewed interest in the Archean sulfur cycle. MIF results from photochemical reactions, such as photodissociation of sulfur dioxide in the terrestrial atmosphere. This is in contrast to all known geological or biological processes on Earth. Modeling results indicate that such photochemical reactions occurred under largely reducing conditions (Pavlov and Kasting, 2002). Transfer to Earth's surface environments and—moreover—preservation of MIF in various reduced or oxidized sulfur-bearing compounds suggests low concentrations of oceanic sulfate. In fact, it is believed that the operational mode of the global sulfur cycle was largely based on photochemical processes in the atmosphere (Farquhar et al., 2000). Oceanic sulfate would have resulted from the rainout of atmospheric sulfate aerosols, carrying the signal of MIF into Earth's surface environment. Inorganic processes or even bacterial sulfate reduction would have formed sedimentary sulfides, preserving the signal of MIF. This scenario is very different from younger times, where oxidative continental weathering and riverine delivery regulates oceanic sulfate.

The deviation in $\delta^{34}S_{seawater\ sulfate}$ from a "magmatic" value close to 0‰ is thought to begin in the Paleoproterozoic (between 2.4 and 2.2 Ga). At the same time, mass-independent sulfur isotope fractionations disappear from the rock record (e.g., Farquhar et al., 2000; Mojszis et al., 2003; Bekker et al., 2004). Oxygenation of the ocean-atmosphere system (e.g., Holland, 1999, 2002) likely resulted in an increase in sulfate abundance in the global ocean. As a consequence, the importance of sulfate removal via bacterial sulfate reduction increased substantially (e.g., Canfield and Raiswell, 1999). This interpretation is supported by the appearance of large sulfur isotopic fractionations between sulfate and sulfide in early Paleoproterozoic times, including negative $\delta^{34}S_{pyrite}$ values (for recent compilations, see, e.g., Strauss, 2002; Lyons et al., 2004). Fractionation is of similar magnitude than in younger sediments. Coupled to global oxygenation, the global sulfur cycle changed from an early mode (>2.4–2.2 Ga) that was driven by magmatic or even photochemical processes to the present one that is coupled to the carbon cycle, buffered by oxygen, and largely driven by redox processes (e.g., Kump, 1989; Petsch and Berner, 1998; Berner, 2001).

Following a time interval of >1.5 Ga, currently without any relevant data, direct evidence for the sulfur isotopic composition of seawater sulfate becomes available again for the Mesoproterozoic and, in particular, the Neoproterozoic. These include sizeable new data sets for Mesoproterozoic marine evaporites (Kah et al., 2001) but more so for trace sulfate in carbonates (SSS, also termed CAS, or carbonate-associated sulfate) from late Paleoproterozoic to Mesoproterozoic sedimentary basins in northern Australia and the North American continent (Fig. 5 and Table 2;

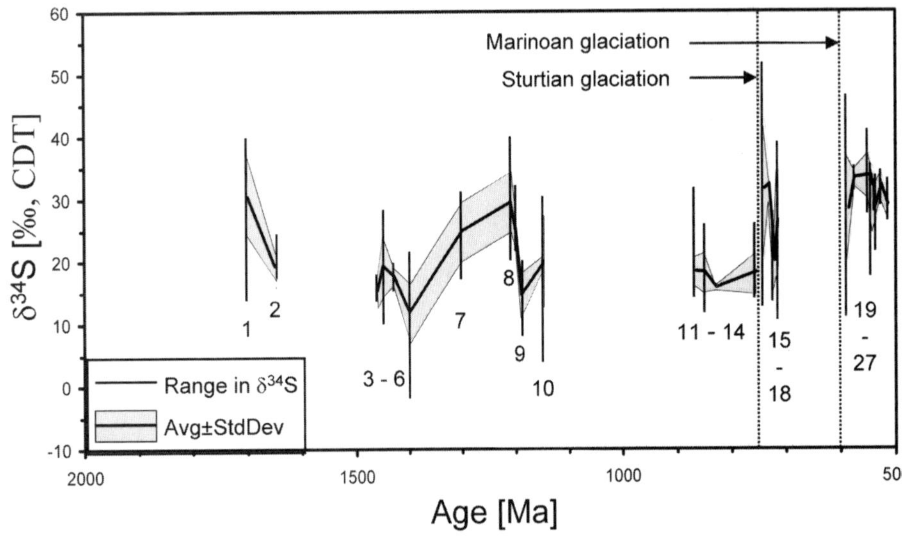

Figure 5. Temporal evolution of $\delta^{34}S_{sulfate}$ during Mesoproterozoic and Neoproterozoic times 1—Paradise Creek Formation, 2—McArthur Group, 3–6—Belt Supergroup, 7—Dismal Lakes Group, 8—Society Cliffs Formation, 9—Mescal Limestone, 10—Grenville metasediments, 11—Upper Roan Group, 12—Bitter Springs Formation, 13—Shaler Group, 14—Redstone River Formation, 15–18—Otavi Group, 19—Doushantuo Formation, 20—Siberian Platform (Precambrian), 2—Hanseran Formation, 22—Siberian Platform Precambrian + Lowest Cambrian), 23—Meishucunian, 24—Siberian Platform (Lower Cambrian), 25–26—Siberian Platform (upper Lower + lower Middle Cambrian; for source of data, see Table 2). CDT—Cañon Diablo Troilite.

TABLE 2. RANGES AND AVERAGE $\delta^{34}S$ VALUES FOR MESOPROTEROZOIC AND NEOPROTEROZOIC SULFATE

				$\delta^{34}S$ [‰, CDT]							
Unit no.	Stratigraphic unit	Age (Ma)	Material	Min	Max	Avg	Std dev	Avg – std dev	Avg + std dev	n	Ref. no.
27	Siberian Platform lower middle Cambrian	514	anh/gyp	26.4	33.3	28.9	1.4	27.5	30.3	36	15
26	Siberian Platform upper lower Cambrian	525	anh/gyp	29.0	34.4	32.0	2.0	30.0	34.0	13	16
25	Hormuz Formation	539	anh/gyp	21.0	33.0	28.0	3.2	24.8	31.2	25	18
24	Siberian Platform lower Cambrian	540	anh/gyp	30.7	35.5	32.7	1.6	31.1	34.3	8	16
23	Meishucunian Phosphorite	543	phosphorite	17.4	35.5	32.1	3.2	28.9	35.3	40	17
22	Siberian Platform Precambrian and lowest Cambrian	545	anh/gyp	33.0	35.0	33.7	0.9	32.8	34.6	4	16
21	Hanseran Formation	548	anh/gyp	27.5	41.4	33.7	3.6	30.1	37.3	48	15
20	Siberian Platform Precambrian	575	anh/gyp	31.2	34.8	33.4	1.2	32.2	34.6	17	14
19	Doushantuo Formation	585	CAS	10.8	46.4	28.3	8.2	20.1	36.5	37	13
18	Maieberg Formation	715	CAS	10.5	38.6	26.6	7.7	18.9	34.3	10	12
17	Ombaatjie Formation	721	CAS	13.4	27.8	20.1	4.6	15.5	24.7	21	12
16	Gruis Formation	730	CAS	30.5	33.9	32.2	2.4	29.8	34.6	3	12
15	Rasthof Formation	740	CAS	12.4	51.5	31.5	11.5	20.0	43.0	15	12
14	Redstone River Formation	755	anh/gyp	13.9	25.5	18.2	3.1	15.1	21.3	10	11
13	Shaler Group	830	anh/gyp			15.9		15.9	15.9	1	10
12	Bitter Springs Formation	850	anh/gyp	11.6	25.5	18.2	3.1	15.1	21.3	45	9
11	Upper Roan Group	865	anh/gyp	14.2	21.3	18.5	2.3	16.2	20.8	31	8
10	Grenville Series	1150	barite	4.2	30.2	20.1	7.2	12.9	27.3	103	7
10	Grenville Series, Unit 10	1150	barite	18.3	21.0	19.6	1.3	18.3	20.9	4	7
9	Mescal Limestone	1200	CAS	8.0	20.0	14.7	3.6	11.1	18.3	11	1
8	Society Cliffs Formation	1200	anh/gyp	21.8	31.8	24.5	2.6	21.9	27.1	33	6
8	Society Cliffs Formation	1200	CAS	20.1	39.8	29.5	4.7	24.8	34.2	44	5
7	Dismal Lakes Group	1300	CAS	17.3	31.1	24.9	4.7	20.2	29.6	21	5
6	Helena Formation	1400	CAS	–2.0	22.0	12.0	4.7	7.3	16.7	27	1
5	Spokane Formation	1430	barite	15.5	19.8	17.7	1.3	16.4	19.0	11	4
4	Newland Formation	1450	CAS	10.0	28.0	19.4	4.8	14.6	24.2	19	1
3	Newland Formation	1450	barite	13.6	18.3	15.4	2.5	12.9	17.9	3	3
2	McArthur River Group	1650	barite	17.5	24.5	19.4	1.9	17.5	21.3	13	2
1	Paradise Creek Formation	1700	CAS	14.0	40.0	30.4	6.1	24.3	36.5	23	1

Note: References: 1—Lyons et al., 2004; 2—Muir et al., 1985; 3—Strauss and Schieber, 1990; 4—Rye et al., 1983; 5—Kah et al., 2001; 6—L. Kah 2003, personal commun.; 7—Whelan et al. 1990; 8—Dechow and Jensen, 1965; Claypool et al., 1980; Strauss, 2001.; 9—Solomon et al., 1971, Claypool et al., 1980, Hayes et al., 1992; Strauss, 1993; Gorjan et al., 2000; 10—Claypool et al., 1980; 11—Ruelle, 1982; Hayes et al., 1992; Strauss, 1993; 12—Hurtgen et al., 2002; 13—Zhang et al., 2003; 14—Strauss, 2001; 15—Strauss et al., 2001; 16—Claypool et al., 1980; 17—Shields et al., 1999; 18—Houghton, 1980.
anh—anhydrite; CAS—carbonate-associated sulfate; gyp—gypsum; std dev—standard deviation.

data from Lyons et al., 2004; L. Kah, 2003, personal commun.) and the Neoproterozoic post-glacial intervals in Namibia (Hurtgen et al., 2002) and China (Zhang et al., 2003).

All of these studies involve continuous stratigraphic sampling. As a consequence, and for the first time in sulfur isotope geochemistry, secular variations in $\delta^{34}S$ have become apparent for many of the studied Proterozoic sedimentary basins, such as those recorded for the Society Cliffs Formation (Fig. 6). At present, observed variations need to be confirmed by data from coeval successions elsewhere in order to assure their global character. But, just as with carbon isotopes in the mid-1980s (e.g., Knoll et al., 1986; Magaritz et al., 1986, Lambert et al., 1987), the foundation for exciting discoveries has been established.

The magnitude of these oscillations over relatively short stratigraphic thicknesses is much bigger than in Phanerozoic times. Considering a steady-state scenario, observed short-term fluctuations in the sulfur isotopic composition of seawater sulfate suggest that the Proterozoic ocean contained less (if not much less) sulfate

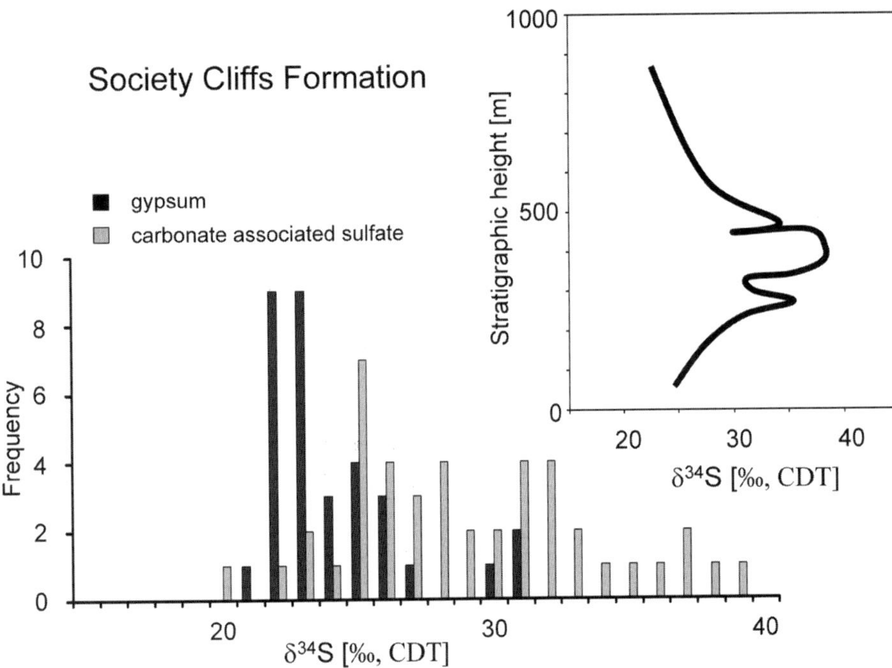

Figure 6. Secular variations in $\delta^{34}S$ for the Society Cliffs Formation. (Data from Kah et al., 2001 and L. Kah, 2003, personal commun.) CDT—Cañon Diablo Troilite.

than in Phanerozoic or modern times (e.g., Shen et al., 2002; Lyons et al., 2004; L. Kah, 2003, personal commun.). The average sulfate sulfur isotopic composition recorded in the Mesoproterozoic sediments fluctuates between +15‰ and +30‰. These fluctuations can be viewed as changes in the fractional burial of reduced (biogenic) sulfur (f_{BioS}). However, given a low-sulfate concentration for the Proterozoic ocean, and hence a limited sulfate reservoir, changes in f_{BioS} need not be very large. A paleoceanographic scenario that would favor high sulfate reduction is a stratified water body (e.g., Shen et al., 2002) in which the lower part turns anoxic due to high oxygen consumption. This would allow the process of sulfate reduction to occur also in the water column.

Extreme fluctuations in $\delta^{34}S_{sulfate}$, both in amplitude and duration (>15‰ in <1 m.y.), have been measured for the carbonates immediately overlying the Neoproterozoic glacial horizons in Namibia (e.g., Hurtgen et al., 2002). The principal mechanism for such changes involves a higher proportion of sulfate reduction and subsequent fractional burial of biogenic sulfur from a low-sulfate ocean; however, observed isotope excursions are difficult to reconcile under steady-state conditions. Whether a causal relationship to the proposed Snowball Earth scenario and its unusual paleoenvironmental (and geochemical) consequences is the principal factor for these variations in $\delta^{34}S$ is yet to be seen. What remains at present is the documentation of comparable oscillations in $\delta^{34}S$ from other coeval post-glacial successions in order to prove the global significance of these extreme fluctuations. Zhang et al. (2003) provide trace sulfate data from the Doushantuo carbonates overlying the Sinian glacial deposits of the Nantuo Formation, Yangtze Platform, South China. Some of their excursions amount to 20‰. Direct comparison of these Neoproterozoic post-glacial fluctuations in $\delta^{34}S_{sulfate}$, however, is currently somewhat difficult due to continuing discussion about the number and global correlation of glacial events (e.g., Kennedy et al., 1998).

The Phanerozoic record of $\delta^{34}S$ seawater sulfate (Fig. 4) displays secular variations on different time scales. A first order trend shows high values in the early Paleozoic, followed by a decrease to a minimum at +12‰ in the Permo-Carboniferous, and a subsequent rise in $\delta^{34}S$ to the modern value of +21‰. However, oscillations of higher order are clearly discernible. Compared to the Mesoproterozoic and Neoproterozoic, these secular variations are not as sizeable.

For the Phanerozoic, variations in $\delta^{34}S_{seawater\ sulfate}$ are readily interpreted as changes in the fractional burial of reduced sulfur (f_{BioS}) (e.g., Holser et al., 1988; Kump, 1989). Conceptually, this is based on the notion that the exogenic cycle is redox-buffered, limiting large fluctuations of atmospheric oxygen. Quantification via isotope mass balance (Fig. 7) identifies high fractional burial (2.5 × present day) for the early Paleozoic and very low values (0.5 × present day) for the late Paleozoic and early Mesozoic, followed by a rise to the modern value. Removal of sulfate via bacterial sulfate reduction must have been quite efficient in early Paleozoic times, while having been greatly reduced in the late Paleozoic. Additional proof is presented by high abundances of pyrite and high S/C ratios in early Paleozoic sedimentary rocks (Raiswell and Berner, 1986; Berner and Canfield, 1989). Fluctuations in f_{BioS} would also be consistent with fluctuations of the sulfate abundance in Phanerozoic seawater. As mentioned above, fluid inclusion data suggest high concentrations of oceanic sulfate at the Neoproterozoic-Cambrian transition, in the late Paleozoic, and for the modern ocean. A high sulfate ocean in the late

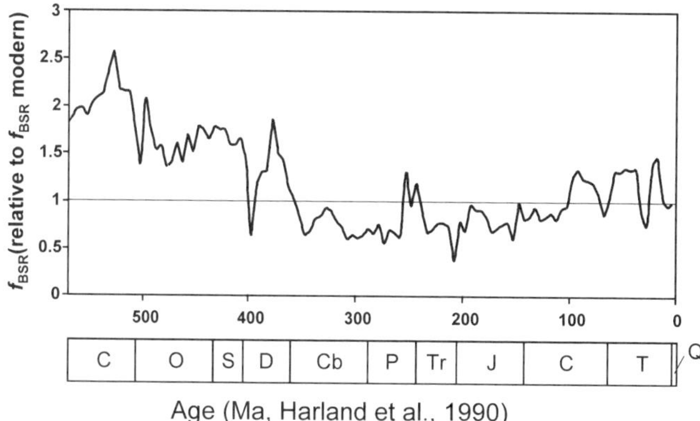

Figure 7. Variations in the fractional burial of reduced sulfur (f_{BioS}) during the Phanerozoic.

Paleozoic is also consistent with the notion of high atmospheric oxygen in the Permo-Carboniferous (e.g., Berner, 2001; Beerling et al., 2002), which would result in enhanced continental weathering and riverine sulfate delivery.

CONCLUSIONS

The sulfur isotopic composition of seawater sulfate records substantial changes in the global sulfur cycle through time. Most fundamental is a change in the principle operational mode.

Increasing evidence suggests that the Archean and early Paleoproterozoic global sulfur cycle was driven largely by inorganic processes in the widest sense. The Archean ocean was characterized by a low sulfate concentration. Initially released through magmatic processes, oceanic sulfate possibly originated from the rainout of atmospheric sulfate aerosols that had been affected by photochemical reactions. Local accumulations allowed the precipitation of marine evaporites.

Oxygenation of global surface environments during the Paleoproterozoic resulted in a concomitant increase in oceanic sulfate abundance. At the same time, the operational mode of the global sulfur cycle changed. Riverine delivery of dissolved sulfate, resulting from oxidative weathering on the continents, and sulfate removal via bacterial sulfate reduction have regulated the abundance of sulfate in the global ocean ever since. Strong secular variations in $\delta^{34}S$ of Proterozoic seawater sulfate suggest that oceanic sulfate abundance was still substantially lower than today. Short-term, high-amplitude excursions in $\delta^{34}S$, as evident from trace sulfate in carbonates, have been measured in particular for the Neoproterozoic post-glacial sediments.

Finally, variations in oceanic sulfate abundance and isotopic composition are also discernible for the Phanerozoic. Veizer et al. (1999) demonstrated that the sulfur isotopic composition of Phanerozoic seawater sulfate shows a strong positive correlation with the $^{87}Sr/^{86}Sr$ of respective coeval chemical sediments. This is interpreted to reflect the tectonic control of the exogenic cycle, most notably temporal changes in the importance of the hydrothermal versus riverine fluxes. In addition, a negative correlation between the isotopic compositions of sulfate sulfur and carbonate carbon attest to the biological redox control of the exogenic cycle.

The modern ocean is characterized by a homogeneous vertical and lateral distribution of sulfate abundance and isotopic composition.

ACKNOWLEDGMENTS

This contribution originates from a very stimulating session in honor of W.T. Holser at the 2001 GSA Annual Meeting in Boston. I am grateful to the organizers for inviting me to participate in this event. Research was supported by the Deutsche Forschungsgemeinschaft (DFG grants Str 281/7-1, Str 281/9-1, Str 281/11-1). Thanks go to Tim Lyons, Anne Gellatly, and Linda Kah for providing data from work in progress. Critical reviews by Linda Kah, Matt Hurtgen, and Tim Lyons greatly improved this contribution.

REFERENCES CITED

Ault, W.U., and Kulp, J.L., 1959, Isotopic geochemistry of sulphur: Geochimica et Cosmochimica Acta, v. 16, p. 201–235, doi: 10.1016/0016-7037(59)90112-7.

Beerling, D.J., Lake, J.A., Berner, R.A., Hickey, L.J., Taylor, D.W., and Royer, D.L., 2002, Carbon isotope evidence implying high O_2/CO_2 ratios in the Permo-Carboniferous atmosphere: Geochimica et Cosmochimica Acta, v. 66, p. 3757–3767, doi: 10.1016/S0016-7037(02)00901-8.

Bekker, A., Holland, H.D., Wang, P.-L., Rumble III, D., Stein, H.J., Hannah, J.L., Coetzee, L.L., and Beukes, N.J., 2004, Dating the rise of atmospheric oxygen: Nature, v. 427, p. 117–120.

Benmore, R.A., Coleman, M.L., and McArthur, J.M., 1983, Origin of sedimentary francolite from its sulphur and carbon isotope composition: Nature, v. 302, p. 516–518.

Berner, R.A., 2001, Modeling atmospheric O_2 over Phanerozoic time: Geochimica et Cosmochimica Acta, v. 65, p. 685–694, doi: 10.1016/S0016-7037(00)00572-X.

Berner, R.A., and Canfield, D.E., 1989, A new model for atmospheric oxygen over Phanerozoic time: American Journal of Science, v. 289, p. 333–361.

Burdett, J.W., Arthur, M.A., and Richardson, M., 1989, A Neogene seawater sulphate isotope age curve from calcareous pelagic microfossils: Earth and Planetary Science Letters, v. 94, p. 189–198.

Busenberg, E., and Plummer, L.N., 1985, Kinetic and thermodynamic factors controlling the distribution of SO_4^{2-} and Na^+ in calcites and selected aragonites: Geochimica et Cosmochimica Acta, v. 49, p. 713–725, doi: 10.1016/0016-7037(85)90166-8.

Canfield, D.E., 2001, Biogeochemistry of Sulfur Isotopes, *in* Valley, J.W., and Cole, D.R., eds., Stable Isotope Geochemistry: Reviews in Mineralogy and Geochemistry, v. 43, p. 607–636.

Canfield, D.E., and Raiswell, R., 1999, The evolution of the sulfur cycle: American Journal of Science, v. 299, p. 697–723.

Canfield, D.E., and Teske, A., 1996, Late Proterozoic rise in atmospheric oxygen concentration inferred from phylogenetic and sulphur-isotope studies: Nature, v. 382, p. 127–132, doi: 10.1038/382127A0.

Cecile, M.P., Shakur, M.A., and Krouse, H.R., 1983, The isotopic composition of western Canada barites and the possible derivation of oceanic sulphate $\delta^{34}S$ and $\delta^{18}O$ age curves: Canadian Journal of Earth Sciences, v. 20, p. 1528–1535.

Clark, I., and Fritz, P., 1997, Environmental Isotopes in Hydrogeology. Boca Raton, Louisiana, CRC Press LLC, 328 p.

Claypool, G.E., Holser, W.T., Kaplan, I.R., Sakai, H., and Zak, I., 1980, The age curve of sulfur and oxygen isotopes in marine sulfate and their mutual interpretation: Chemical Geology, v. 28, p. 199–260, doi: 10.1016/0009-2541(80)90047-9.

Dechow, E., and Jensen, M.L., 1965, Sulfur isotopes of some Central African sulfide deposits: Economic Geology and the Bulletin of the Society of Economic Geologists, v. 60, p. 894–941.

Dehairs, F., Chesselet, R., and Jedwab, J., 1980, Discrete suspended particles of barite and the barium cycle in the open ocean: Earth and Planetary Science Letters, v. 49, p. 528–550, doi: 10.1016/0012-821X(80)90094-1.

Detmers, J., Brüchert, V., Habicht, K.S., and Kuever, J., 2001, Diversity of sulfur isotope fractionations by sulfate-reducing prokaryotes: Applied and Environmental Microbiology, v. 67, p. 888–894, doi: 10.1128/AEM.67.2.888-894.2001.

Farquhar, J., Bao, H., and Thiemens, M., 2000, Atmospheric influence of Earth's earliest sulphur cycle: Science, v. 289, p. 756–758, doi: 10.1126/SCIENCE.289.5480.756.

Gorjan, P., Veevers, J.J., and Walter, M.R., 2000, Neoproterozoic sulfur-isotope variation in Australia and global implications: Precambrian Research, v. 100, p. 151–179, doi: 10.1016/S0301-9268(99)00073-X.

Graf, H.-F., Langmann, B., and Feichter, J., 1998, The contribution of Earth degassing to the atmospheric sulfur budget: Chemical Geology, v. 147, p. 131–145, doi: 10.1016/S0009-2541(97)00177-0.

Grinenko, V., and Krouse, H.R., 1992, Isotope data on the nature of riverine sulfates: Mitteilungen Geologisch-Paläontologisches Institut der Universität Hamburg, v. 72, p. 9–18.

Grossman, E.L., Mii, H.-S., Zhang, C., and Yancey, T.E., 1996, Chemical variation in Pennsylvanian brachiopod shells—diagenetic, taxonomic, microstructural, and seasonal effects: Journal of Sedimentary Research, v. 66, p. 1011–1022.

Habicht, K.S., and Canfield, D.E., 1997, Sulfur isotope fractionation during bacterial sulfate reduction in organic-rich sediments: Geochimica et Cosmochimica Acta, v. 61, p. 5351–5361, doi: 10.1016/S0016-7037(97)00311-6.

Habicht, K.S., and Canfield, D.E., 2001, Isotope fractionation by sulfate-reducing natural populations and the isotopic composition of sulfide in marine sediments: Geology, v. 29, p. 555–558, doi: 10.1130/0091-7613(2001)0292.0.CO;2.

Habicht, K.S., Canfield, D.E., and Rethmeier, J., 1998, Sulphur isotope fractionation during bacterial sulphate reduction and disproportionation of thiosulphate and sulfite: Geochimica et Cosmochimica Acta, v. 62, p. 2585–2595, doi: 10.1016/S0016-7037(98)00167-7.

Habicht, K.S., Gade, M., Thamdrup, B., Berg, P., and Canfield, D.E., 2002, A calibration of sulfate levels in the Archean Ocean: Science, v. 298, p. 2372–2374, doi: 10.1126/SCIENCE.1078265.

Hansen, K.W., and Wallmann, K., 2003, Cretaceous and Cenozoic evolution of seawater composition, atmospheric O_2 and CO_2: a model perspective: American Journal of Science, v. 303, p. 94–148.

Hardie, L.A., 1996, Secular variation in seawater chemistry: an explanation for the coupled variation in the mineralogies of marine limestones and potash evaporites over the past 600 my: Geology, v. 24, p. 279–283, doi: 10.1130/0091-7613(1996)0242.3.CO;2.

Harland, W.B., Armstrong, R.L., Cox, A.V., Craig, L.E., Smith, A.G., and Smith, D.G., 1990, A geological time scale 1989: Cambridge, N.Y., Cambridge University Press, 263 p.

Hartmann, M., and Nielsen, H., 1969, $\delta^{34}S$ Werte in rezenten Meeressedimenten und ihre Deutung am Beispiel einiger Sedimentprofile aus der westlichen Ostsee: Geologische Rundschau, v. 58, p. 621–655.

Hayes, J.M., Lambert, I.B., and Strauss, H., 1992, The sulfur-isotopic record, *in* Schopf, J.W. and Klein, C., eds., The Proterozoic Biosphere: A Multidisciplinary Study: Cambridge, Cambridge University Press, p. 129–131.

Holland, H.D., 1999, When did the Earth's atmosphere become oxic?: The Geochemical News, v. 100, p. 20–22.

Holland, H.D., 2002, Volcanic gases, black smokers, and the Great Oxidation Event: Geochimica et Cosmochimica Acta, v. 66, p. 3811–3826, doi: 10.1016/S0016-7037(02)00950-X.

Holser, W.T., and Kaplan, I.R., 1966, Isotope geochemistry of sedimentary sulfates: Chemical Geology, v. 1, p. 93–135, doi: 10.1016/0009-2541(66)90011-8.

Holser, W.T., Schidlowski, M., Mackenzie, F.T., and Maynard, J.B., 1988, Geochemical cycles of carbon and sulfur, *in* Gregor, C.B., Garrels, R.M., Mackenzie, F.T., and Maynard, J.B., eds., Chemical Cycles in the Evolution of the Earth: New York, Wiley, p. 105–173.

Horita, J., Zimmermann, H., and Holland, H.D., 2002, Chemical evolution of seawater during the Phanerozoic: implications from the record of marine evaporites: Geochimica et Cosmochimica Acta, v. 66, p. 3733–3756, doi: 10.1016/S0016-7037(01)00884-5.

Houghton, M.L., 1980, Geochemistry of the Proterozoic Hormuz Evaporites, southern Iran [M.Sc. Thesis]: Eugene, University of Oregon, 85 p.

Hurtgen, M.T., Arthur, M.A., Suits, N.S., and Kaufman, A.J., 2002, The sulfur isotopic composition of Neoproterozoic seawater sulfate: implications for a Snowball Earth?: Earth and Planetary Science Letters, v. 203, p. 413–429, doi: 10.1016/S0012-821X(02)00804-X.

Jackson, G.D., and Ianelli, T.R., 1981, Rift related cyclic sedimentation in the Neohelikian Borden Basin, N.W. Baffin Island: Geological Survey of Canada Paper 81-10, p. 269–302.

Kah, L.C., Lyons, T.W., and Chesley, J.T., 2001, Geochemistry of a 1.2 Ga carbonate-evaporite succession, northern Baffin and Bylot islands: Implications for Mesoproterozoic marine evolution: Precambrian Research, v. 111, p. 203–234, doi: 10.1016/S0301-9268(01)00161-9.

Kampschulte, A., 2002, Schwefelisotopenuntersuchungen an strukturell substituierten Sulfaten in marinen Karbonaten des Phanerozoikums—Implikationen für die geochemische Evolution des Meerwassers und die Korrelation verschiedener Stoffe, [Ph.D. dissertation]: Ruhr-Universität Bochom, 152 p. (in German).

Kampschulte, A., and Strauss, H., 2004, The sulphur isotopic evolution of Phanerozoic seawater based on the analysis of structurally substituted sulphate in carbonates: Chemical Geology, v. 204, p. 255–286.

Kampschulte, A., Buhl, D., and Strauss, H., 1998, The sulfur and strontium isotopic compositions of Permian evaporites from the Zechstein basin, northern Germany: Geologische Rundschau, v. 87, p. 192–199, doi: 10.1007/S005310050201.

Kampschulte, A., Bruckschen, P., and Strauss, H., 2001, The sulphur isotopic composition of trace sulphates in Carboniferous brachiopods: Implications for coeval seawater, correlation with other geochemical cycles and isotope stratigraphy: Chemical Geology, v. 175, p. 149–189, doi: 10.1016/S0009-2541(00)00367-3.

Kennedy, M.J., Runnegar, B., Prave, A.R., Hoffmann, K.-H., and Arthur, M.A., 1998, Two or four Neoproterozoic glaciations: Geology, v. 26, p. 1059–1063, doi: 10.1130/0091-7613(1998)0262.3.CO;2.

Knoll, A.H., Hayes, J.M., Kaufman, A.J., Swett, K., and Lambert, I.B., 1986, Secular variations in carbon isotope ratios from Upper Proterozoic successions of Svalbard and East Greenland: Nature, v. 321, p. 832–838.

Kump, L.R., 1989, Alternative modeling approaches to the geochemical cycles of carbon, sulfur, and strontium isotopes: American Journal of Science, v. 289, p. 390–410.

Lambert, I.B., Walter, M.R., Zang, W., Lu, S., and Ma, G., 1987, Paleoenvironment and carbon isotope stratigraphy of Upper Proterozoic carbonates of the Yangtze Platform: Nature, v. 325, p. 140–142, doi: 10.1038/325140A0.

Longinelli, A., 1989, Oxygen-18 and sulphur-34 in dissolved oceanic sulphate and phosphate, *in*: Fritz, P., and Fontes, J.Ch., eds., Handbook of Environmental Isotope Geochemistry—V. 3: Amsterdam, Elsevier, p. 219–255.

Lowenstein, T.K., Timofeev, M.N., Brennan, S.T., Hardie, L.A., and Demicco, R.V., 2001, Oscillations in Phanerozoic seawater chemistry: Evidence from fluid inclusions: Science, v. 294, p. 1086–1088, doi: 10.1126/SCIENCE.1064280.

Lowenstein, T.K., Hardie, L.A., Timofeev, M.N., and Demicco, R.V., 2003, Secular variation in seawater chemistry and the origin of calcium chloride basinal brines: Geology, v. 31, p. 857–860, doi: 10.1130/G19728R.1.

Lyons, T.W., Kah, L.C., and Gellatly, A.M., 2004, The Precambrian sulfur isotope record of evolving atmospheric oxygen, *in* Eriksson, P.G., et al., eds, The Precambrian Earth: Tempos and events: Developments in Precambrian geology, v. 12: Amsterdam, Elsevier, p. 421–439.

Magaritz, M., Holser, W.T., and Kirschvink, J.L., 1986, Carbon-isotope events across the Precambrian-Cambrian boundary on the Siberian Platform: Nature, v. 320, p. 258–259.

Mandernack, K.W., Krouse, H.R., and Skei, J.M., 2003, A stable sulfur and oxygen isotopic investigation of sulfur cycling in an anoxic marine basin, *Framvaren* Fjord, Norway: Chemical Geology, v. 195, p. 181–200, doi: 10.1016/S0009-2541(02)00394-7.

McArthur, J.M., Benmore, R.A., Coleman, M.L., Soldi, C., Yeh, H.-W., and O'Brien, G.W., 1986, Stable isotopic characterization of francolite formation: Earth and Planetary Science Letters, v. 77, p. 20–34, doi: 10.1016/0012-821X(86)90129-9.

Mekhtiyeva, V.I., and Pankina, R.G., 1969, Isotopic composition of sulfur in aquatic plants and dissolved sulfates. Geochemistry International, v. 5, p. 624–627.

Mojzsis, S.J., Coath, C.D., Greenwood, J.P., McKeegan, K.D., and Harrison, T.M., 2003, Mass-independent isotope effects in Archean (2.5 to 3.8 Ga) sedimentary sulfides determined by ion microprobe analysis: Geochimica et Cosmochimica Acta, v. 67, p. 1635–1658, doi: 10.1016/S0016-7037(03)00059-0.

Muir, M.D., Donnelly, T.H., Wilkins, R.W.T., and Armstrong, K.J., 1985, Stable isotope, petrological, and fluid inclusion studies of minor mineral deposits from the McArthur Basin: implications for the genesis of some sediment-hosted base metal mineralization from the Northern Territory: Australian Journal of Earth Sciences, v. 32, p. 239–260.

Nielsen, H., 1989, Local and global aspects of the sulphur isotope age curve of oceanic sulphate, *in* Brimblecombe, P., and Lein, A.Yu., eds., Evolution of the Global Biogeochemical Sulphur Cycle: New York, Wiley, p. 57–64.

Ohmoto, H., 1997, When did the Earth's atmosphere become oxic?: The Geochemical News, v. 93-Fall, p. 12–28.

Ohmoto, H., Kakegawa, T., and Lowe, D.R., 1993, 3.4-billion-year-old biogenic pyrites from Barberton, South Africa: Sulphur isotope evidence: Science, v. 267, p. 555–557.

Ohkouchi, N., Kawamura, K., Kajiwara, Y., Wada, E., Okada, M., Kanamatsu, T., and Taira, A., 1999, Sulfur isotope records around Livello Bonarelli (northern Apennines, Italy) black shale at the Cenomanian-Turonian boundary: Geology, v. 27, p. 535–538, doi: 10.1130/0091-7613(1999)0272.3.CO;2.

Pavlov, A.A., and Kasting, J.F., 2002, Mass-independent fractionation of sulphur isotopes in Archaean sediments: Strong evidence for an anoxic Archaean atmosphere: Astrobiology, v. 2, p. 27–41, doi: 10.1089/153110702753621321.

Paytan, A., Kastner, M., Campbell, D., and Thiemens, M.H., 1998, Sulfur isotopic composition of Cenozoic seawater sulfate: Science, v. 282, p. 1459–1462, doi: 10.1126/SCIENCE.282.5393.1459.

Paytan, A., Mearon, S., Cobb, K., and Kastner, M., 2002, Origin of marine barite deposits: Sr and S isotope characterization: Geology, v. 30, p. 747–750, doi: 10.1130/0091-7613(2002)0302.0.CO;2.

Petsch, S.T., and Berner, R.A., 1998, Coupling the geochemical cycles of C, P, Fe, and S: The effect of atmospheric O_2 and the isotopic records of carbon and sulfur: American Journal of Science, v. 298, p. 246–262.

Pingitore, N.E., Meitzner, G., and Lowe, K.M., 1995, Identification of sulfate in natural carbonates by X-ray absorption spectroscopy: Geochimica et Cosmochimica Acta, v. 59, p. 2477–2483, doi: 10.1016/0016-7037(95)00142-5.

Raab, M., and Spiro, B., 1991, Sulphur isotopic variations during seawater evaporation with fractional crystallization: Chemical Geology, v. 86, p. 323–333.

Raiswell, R., and Berner, R.A., 1986, Pyrite and organic matter in Phanerozoic normal marine shales: Geochimica et Cosmochimica Acta, v. 50, p. 1967–1976, doi: 10.1016/0016-7037(86)90252-8.

Rees, C.E., Jenkins, W.J., and Monster, J., 1978, The sulphur isotopic composition of ocean water sulphate: Geochimica et Cosmochimica Acta, v. 42, p. 377–381, doi: 10.1016/0016-7037(78)90268-5.

Ruelle, J.C.L., 1982, Depositional environments and genesis of stratiform copper deposits of the Redstone Copper Belt, Mackenzie Mountains, N.W.T, *in* Hutchinson, R.W., Spence, C.D., and Franklin, J.D., eds., Precambrian Sulphide Deposits: Geological Association of Canada Special Paper 25, p. 701–737.

Rye, R.O., Whelan, J.F., Harrison, J.E., and Hayes, T.S., 1983, The origin of copper-silver mineralization in the Ravalli Group as indicated by preliminary stable isotope studies: Montana Bureau of Mines Geology Special Publication, v. 90, p. 104–111.

Sandberg, P.A., 1983, An oscillatine trend in Phanerozoic non-skeletal carbonate mineralogy: Nature, v. 305, p. 19–22.

Sasaki, A., 1972, Variation in sulfur isotope composition of oceanic sulfate: 24th International Geological Congress, Sect. 10, p. 342–345.

Shanks, W.C., III, 2001, Stable isotopes in seafloor hydrothermal systems, *in* Valley, J.W., and Cole, D.R., eds., Stable Isotope Geochemistry: Reviews in Mineralogy and Geochemistry, v. 43, p. 607–636.

Shanley, J.B., Pendall, E., Kendall, C., Stevens, L.R., Michel, R.L., Phillips, P.J., Forester, R.M., Naftz, D.L., Liu, B., Stern, L., Wolfe, B.B., Chamberlain, C.P., Leavitt, S.W., Heaton, T.H.E., Mayer, B., Cecil, L.D., Lyons, W.B., Katz, B.G., Betancourt, J.L., McKnight, D.M., Blum, J.D., Edwards, T.W.D., House, H.R., Ito, E., Aravena, R.O., and Whelan, J.F., 1998, Isotopes as indicators of environmental change. *in* Kendall, C., and McDonell, J.J., eds., Isotopes Tracers in Catchment Hydrology: Amsterdam, Elsevier Science B.V., p. 761–816.

Shen, Y., Buick, R., and Canfield, D.E., 2001, Isotopic evidence for microbial sulphate reduction in the early Archaean era: Nature, v. 410, p. 77–81, doi: 10.1038/35065071.

Shen, Y., Canfield, D.E., and Knoll, A.H., 2002, Middle Proterozoic ocean chemistry: evidence from the McArthur Basin, northern Australia: American Journal of Science, v. 302, p. 81–109.

Shields, G.A., Strauss, H., Howe, S.S., and Siegmund, H., 1999, Sulphur isotope compositions of sedimentary phosphorites from the basal Cambrian of China: implications for Neoproterozoic–Cambrian biogeochemical cycling: Journal of the Geological Society [London], v. 156, p. 943–955.

Solomon, M., Rafter, T.A., and Dunham, K.C., 1971, Sulphur and oxygen isotope studies in the northern Pennines in relation to ore genesis: Transactions, Institute of Mining and Metallurgy, v. B, p. 259–275.

Staudt, W.J., and Schoonen, M.A.A., 1995, Sulfate incorporation into sedimentary carbonates, *in* Vairavamurthy, M.A., and Schoonen, M.A.A., eds., Geochemical Transformations of Sedimentary Sulfur: Washington, D.C., American Chemical Society, p. 332–345.

Staudt, W.J., Reeder, R.J., and Schoonen, M.A.A., 1994, Surface structural controls on compositional zoning of SO_4^{2-} and SeO_4^{2-} in synthetic single crystals: Geochimica et Cosmochimica Acta, v. 58, p. 2087–2098, doi: 10.1016/0016-7037(94)90287-9.

Stewart, A.J., 1974, Petrographic and geochemical study of the Ringwood evaporite deposit: Bureau of Mineral Resources Record, v. 1974, no. 145, p. 67.

Strauss, H., 1993, The sulphur isotopic record of Precambrian sulphates: New data and a critical evaluation of the existing record: Precambrian Research, v. 63, p. 225–246, doi: 10.1016/0301-9268(93)90035-Z.

Strauss, H., 1997, The isotopic composition of sedimentary sulphur through time: Paleogeography, Paleoclimatology, Paleoecology, v. 132, p. 97–118, doi: 10.1016/S0031-0182(97)00067-9.

Strauss, H., 1999, Geological evolution from isotope proxy signals—Sulphur: Chemical Geology, v. 161, p. 89–101, doi: 10.1016/S0009-2541(99)00082-0.

Strauss, H., 2001, 4 Ga of seawater evolution—evidence from sulfur isotopes: Geological Society of America Abstracts with Programs, v. 33, no. 6, p. A-94.

Strauss, H., 2002, The isotopic composition of Precambrian sulphides—Seawater chemistry and biological evolution: Special Publications of the International Association of Sedimentologists, v. 33, p. 67–105.

Strauss, H., 2003, Sulphur isotopes and the Early Archaean sulphur cycle: Precambrian Research, v. 126, p. 349–361, doi: 10.1016/S0301-9268(03)00104-9.

Strauss, H., and Schieber, J., 1990, A sulfur isotope study of pyrite genesis: The Mid-Proterozoic Newland Formation, Belt Supergroup, Montana: Geochimica et Cosmochimica Acta, v. 54, p. 197–204, doi: 10.1016/0016-7037(90)90207-2.

Strauss, H., Banerjee, D.M., and Kumar, V., 2001, The sulfur isotopic composition of Neoproterozoic to Early Cambrian seawater—Evidence from the cyclic Hanseran evaporites, NW India: Chemical Geology, v. 175, p. 17–28, doi: 10.1016/S0009-2541(00)00361-2.

Sweeney, R.E., and Kaplan, I.R., 1980, Stable isotope composition of dissolved sulfate and hydrogen sulfide in the Black Sea: Marine Chemistry, v. 9, p. 145–152, doi: 10.1016/0304-4203(80)90064-X.

Takano, B., 1985, Geochemical implications of sulfate in sedimentary carbonates: Chemical Geology, v. 49, p. 393–403, doi: 10.1016/0009-2541(85)90001-4.

Thode, H.G., Monster, J., and Dunford, H.B., 1961, Sulphur isotope geochemistry: Geochimica et Cosmochimica Acta, v. 25, p. 159–174, doi: 10.1016/0016-7037(61)90074-6.

Veizer, J., Ala, D., Azmy, K., Bruckschen, P., Buhl, D., Bruhn, F., Carden, G.A.F., Diener, A., Ebneth, S., Godderis, Y., Jasper, T., Korte, C., Pawellek, F., Podlaha, O.G., and Strauss, H., 1999, $^{87}Sr/^{86}Sr$, $\delta^{13}C$ and $\delta^{18}O$ evolution of Phanerozoic seawater: Chemical Geology, v. 161, p. 59–88, doi: 10.1016/S0009-2541(99)00081-9.

Vinogradov, A.P., Grinenko, V.A., and Ustinov, V., 1962, Isotopic composition of sulfur compounds in the Black Sea: Geochemistry International, v. 10, p. 973–997.

Volkov, I.I., and Rozanov, A.G., 1983, The sulphur cycle in oceans, *in* Ivanov, M.V., and Freeney J.R., eds., The Global Biogeochemical Sulphur Cycle: Chichester, Wiley & Sons, SCOPE 19, p. 357–448.

Whelan, J.F., Rye, R.O., deLorraine, W., and Ohmoto, H., 1990, Isotopic geochemistry of a Mid-Proterozoic evaporite basin: Balmat, New York: American Journal of Science, v. 290, p. 396–424.

Yeremenko, N.A., and Pankina, R.G., 1971, Variation of $\delta^{34}S$ in the sulfates of modern and ancient marine basins of the Soviet Union: Geochemistry International, v. 8, p. 45–54.

Zhang, T., Chu, X., Zhang, Q., Feng, L., and Huo, W., 2003, Variations of sulfur and carbon isotopes in seawater during the Doushantuo stage in late Neoproterozoic: Chinese Science Bulletin, v. 48, p. 1375–1380.

Zharkov, M.A., 1984, Paleozoic salt bearing formations of the world: Berlin, Springer-Verlag, 395 p.

Manuscript Accepted by the Society March 28, 2004

Printed in the USA